Springer Proceedings in Physics

Volume 147

For further volumes:
www.springer.com/series/361

Stéphane Sebban • Julien Gautier • David Ros •
Philippe Zeitoun
Editors

X-Ray Lasers 2012

Proceedings of the 13th International Conference on X-Ray Lasers, 11–15 June 2012, Paris, France

Springer

Editors
Stéphane Sebban
Laboratoire d'Optique Appliquée
ENSTA
Palaiseau, France

David Ros
LPGP
Université Paris-Sud
Orsay, France

Julien Gautier
Laboratoire d'Optique Appliquée
ENSTA
Palaiseau, France

Philippe Zeitoun
Laboratoire d'Optique Appliquée
ENSTA
Palaiseau, France

ISSN 0930-8989 ISSN 1867-4941 (electronic)
Springer Proceedings in Physics
ISBN 978-3-319-37609-7 ISBN 978-3-319-00696-3 (eBook)
DOI 10.1007/978-3-319-00696-3
Springer Cham Heidelberg New York Dordrecht London

Printed on acid-free paper

Springer is part of Springer Science+Business Media (www.springer.com)

Preface

These proceedings comprise invited and contributed papers presented at the 13th International Conference on X-Ray Lasers (ICXRL 2012) which was held 11–15 June 2012 in Paris, in the famous Quartier Latin, inside the historical Center of Cordeliers. This conference is part of a continuing series dedicated to recent developments and applications of x-ray lasers and other coherent x-ray sources with attention to supporting technologies and instrumentation.

New results in the generation of intense, coherent x-rays and progress towards practical devices and their applications are reported in these proceedings, including areas of research in plasma-based x-ray lasers, 4th generation accelerator-based sources and higher harmonic generation.

Recent achievements related to the increase of the repetition rate up to 100 Hz and shorter wavelength collisional plasma-based soft x-ray lasers down to about 7 nm are presented. Seeding the amplifying plasma with a femtosecond high-order harmonic of infrared laser was foreseen as the required breakthrough to break the picosecond frontier. Numerical simulations based on the Maxwell-Bloch model are presented in these proceedings, transposing the chirped pulse amplification technique to the x-ray domain in order to increase the time over which the femtosecond seed can be amplified.

These proceedings also include innovative applications of soft x-ray lasers based on techniques and diagnostics relevant to topical domains such as EUV lithography, inertial confinement fusion, or warm dense matter physics.

We would like to thank the International Advisory Board for helping to organize the scientific program and the Local Organizing Committee for operation of the meeting. Also we would like to thank the Conference Secretary and the supporting staff for running the conference smoothly and editing these conference proceedings.

Paris, France

Stéphane Sebban
Julien Gautier
David Ros
Philippe Zeitoun

Conference Organization

Conference Chair

Stéphane Sebban

Conference Co-Chairs

David Ros
Philippe Zeitoun

International Advisory Board

Attwood David (LBNL, USA)
Balmer J. (Univ. of Bern, Switzerland)
Bulanov S. (APRI, Japan)
Daido H. (JAEA, Japan)
Dunn J. (LLNL, USA)
Feldhauss J. (Desy, Germany)
Fill E. (MPQ, Germany)
Janulewicz K.A. (GIST, Korea)
Kato Y. (GPI, Japan)
Klisnick A. (ISMO, France)
Lewis C.L.S. (Queen's Univ., UK)
Nam C.H. (KAIST, Korea)
Nickles P.V. (MBI, Germany/GIST, Korea)
Pert G.J. (Univ. of York, UK)
Rocca J.J. (Colorado State Univ., USA)
Rus B. (PALS, Czech)
Sebban S. (LOA, France)
Suckewer S. (Princeton Univ., USA)
Tallents G.J. (Univ. of York, UK)
Vinogradov A. (LPI, Russia)
Xu Z.Z. (SIOM, China)
Zhang J. (STU, China)

Local Organizing Committee

Kevin Cassou (LPGP)
Brigitte Cros (LPGP)
Sylvie Jacquemot (LULI)
Sophie Kazamias (LPGP)
Annie Klisnick (ISMO)
Julien Gautier (LOA)
Olivier Guilbaud (LPGP)
Gilles Maynard (LPGP)
Moana Pittman (LUMAT)
Fabien Tissandier (LOA)

Sponsors

Institutions

Centre national de recherché Scientifique (CNRS)
Ecole Polytechnique
Ecole Nationale Supérieure des Techniques Avancées (ENSTA)
Triangle de la physique
Institut Lasers et Plasma
GDR APPLIX

Companies

Micro-Controle
Spectra-Physics
Plug in
BFI Optilas
Roper
Amplitude technologies
Optoprim
Imagine Optic
Andor technology
Source Lab
Quantel

Contents

1 Coherent Pulses from a Seeded Free-Electron Laser in the Extreme Ultraviolet 1
E. Allaria, D. Castronovo, G. De Ninno, S. Di Mitri, W. Fawley, E. Ferrari, L. Froehlich, L. Giannessi, B. Mahieu, G. Penco, C. Spezzani, and M. Trovo

2 Enhancement of Efficiency of XUV Generation in Atomic Gases Irradiated by Intense Laser Fields 7
A.V. Andreev, S.Y. Stremoukhov, and O.A. Shoutova

3 All-Optical Raman XFEL, Based on the Electron Emission in a Transverse High Intensity Optical Lattice 13
I.A. Andriyash, E. d'Humières, V.T. Tikhonchuk, and P. Balcou

4 Optical Transforms Related to Coherent Imaging of Inclined Objects 19
I.A. Artyukov, A.S. Busarov, N.L. Popov, and A.V. Vinogradov

5 Soft-X-Ray Lasing down to 6.85 nm in Ni-Like Samarium 29
J.E. Balmer, F. Staub, and F. Jia

6 Fourier Optics Study of Traveling-Wave Excitation at Short-Wavelength Plasma-Lasing 35
Davide Bleiner and Thomas Feurer

7 Coherent Short-Wavelength Plasma Radiation for Lab-scale Nano-inspection Tools 39
Davide Bleiner

8 Ion Temperature and Cross-relaxation Effects on OFI Soft-X-Ray Laser 45
A. Boudaa, G. Maynard, and B. Cros

9 Study of Particle Correlation Effects on Line Profiles of Ni-Like Collisional XUV Laser Amplifier 49
A. Calisti, S. Ferri, C. Mossé, B. Talin, A. Klisnick, L. Meng, D. Benredjem, and O. Guilbaud

10 The LUNEX5 Project in France 55
M.E. Couprie, C. Benabderrahmane, P. Betinelli, F. Bouvet, A. Buteau, L. Cassinari, J. Daillant, J.C. Denard, P. Eymard, B. Gagey, C. Herbeaux, M. Labat, A. Lestrade, A. Loulergue, P. Marchand, J.L. Marlats, C. Miron, P. Morin, A. Nadji, F. Polack, J.B. Pruvost, F. Ribeiro, J.P. Ricaud, P. Roy, T. Tanikawa, R. Roux, S. Bielawski, C. Evain, C. Szwaj, G. Lambert, A. Lifschitz, V. Malka, R. Lehe, A. Rousse, K. Ta Phuoc, C. Thaury, G. Devanz, M. Luong, B. Carré, G. LeBec, L. Farvacque, A. Dubois, and J. Lüning

11 High Harmonic Generation Driven by Two Quasi-collinear Pulses . 63
S. Daboussi, S. Kazamias, K. Cassou, O. Guilbaud, M. Pittman, O. Delmas, O. Neveu, B. Cros, G. Maynard, and D. Ros

12 Real-Time Observation of Laser Heated Metals with High Brightness Monochromatic X-Ray Techniques at Present and Their Future Prospects 69
H. Daido, T. Shobu, T. Yamada, S. Yamashita, K. Sugihara, A. Nishimura, and T. Muramatsu

13 Generation and Application of Coherent Radiation in the Water Window 77
L.V. Dao, C. Hall, H.L. Vu, K.B. Dinh, N. Gaffney, E. Balaur, P. Hannaford, and T.A. Smith

14 Time-Dependent Simulation of Carbon Illuminated by a High Intensity X-Ray Laser 83
Alberto G. de la Varga, Pedro Velarde, François de Gaufridy, Manuel Cotelo, David Portillo, and Philippe Zeitoun

15 Regenerative Laser Cavity Tuning for Efficient Soft-X-Ray Laser Operation 89
O. Delmas, K. Cassou, O. Guilbaud, S. Kazamias, S. Daboussi, M. Pittman, O. Neveu, J. Demailly, and D. Ros

16 X-Ray Laser Developments at PHELIX 97
B. Ecker, B. Aurand, D.C. Hochhaus, P. Neumayer, B. Zielbauer, K. Cassou, S. Daboussi, O. Guilbaud, S. Kazamias, T.T.T. Le, E. Oliva, L. Li, H. Zhao, Q. Jin, D. Ros, P. Zeitoun, and T. Kuehl

17 Parabolic Equation and Exact Transparent Boundary Conditions in X-Ray Optics—Application to Waveguides and Whispering Gallery Optics 105
R.M. Feshchenko and A.V. Popov

18 Harmonic Generation and Soft-X-Ray Laser with LASERIX: Source Development, Applications and Advanced Diagnosis 109
Olivier Guilbaud, Sophie Kazamias, Kevin Cassou, Moana Pittman, Sameh Daboussi, Olivier Delmas, Julien Demailly, Olivier Neveu, Chris Pouhe, Boris Vodungbo, Philippe Zeitoun, Lucy Wilson, Greg Tallents, Antonin Dusseix, Gabriel Richet, Aurelie Gense, Bich-Lien Nghiem, Brigitte Cros, Gilles Maynard, and David Ros

19 Observation of the Laser-Induced Surface Dynamics Using the Single-Shot Soft X-Ray Laser Probe 117
N. Hasegawa, Y. Ochi, T. Kawachi, M. Nishikino, M. Ishino, T. Imazono, T. Kaihori, T. Morita, A. Sasaki, K. Terakawa, Y. Minami, M. Baba, T. Tomita, M. Yamamoto, M. Yamagiwa, and T. Suemoto

20 Nano-meter Size Modification of Metal Surfaces Induced by Soft X-Ray Laser Single Pulse 121
Masahiko Ishino, Anatoly Faenov, Momoko Tanaka, Tatiana Pikuz, Satoshi Tamotsu, Noboru Hasegawa, Masaharu Nishikino, Sergei Starikov, Vladimir Stegailov, Genry Norman, Vladimir Fortov, Igor Skobelev, Takeshi Kaihori, and Tetsuya Kawachi

21 Speckle Statistics, Coherence and Polarization of a Collisional Soft X-Ray Laser 125
K.A. Janulewicz, C.M. Kim, P.V. Nickles, H. Stiel, M. Nishikino, N. Hasegawa, and T. Kawachi

22 Upscaling of X-Ray Laser Repetition Rate Using an OPCPA Architecture 137
Fei Jia, Felix Staub, and Jürg Balmer

23 Source Development of Novel Coherent X-Rays and Their Applications in JAEA 143
T. Kawachi, A. Sasaki, M. Nishikino, M. Ishino, N. Hasegawa, T. Imazono, Y. Ochi, M. Tanaka, A.Y. Faenov, T.A. Pikuz, A. Pirozhkov, T. Esirkepov, T. Nakamura, M. Kando, S.V. Bulanov, K. Kondo, K. Janulewicz, C.M. Kim, H. Stiel, G. Norman, T. Suemoto, T. Tomita, K. Namikawa, M. Yamagiwa, and Y. Kato

24 Characterization of Zn X-Ray Laser at PALS Centre, Its Applications in Dense Plasma Probing and Astrophysics 151
M. Kozlova, J. Nejdl, M. Krus, J. Prokupek, J. Dostal, B. Rus, A. Klisnick, L. Meng, F. Tissandier, C. Stehlé, R. Lefevre, U. Chaulagain, N. Champion, P. Barroso, F. Reix, P. Jagourel, J. Larour, F. Delmotte, C. Constancias, F. Suzuki-Vidal, and O. Acef

25 Development of High Reflective Multilayer Mirrors at "Water Window" Wavelengths in IPOE 161
Haochuan Li, Jingtao Zhu, and Zhanshan Wang

26 Time Resolved Holography Scheme Using a Table Top Soft X-Ray Laser . . . 165
E.B. Malm, N.C. Monserud, P.W. Wachulak, C. Brown, W. Chao, E. Anderson, H. Xu, C.P. Hains, G. Balakrishnan, C.S. Menoni, J.J. Rocca, and M.C. Marconi

27 Plasma Homogenization for Overcoming Refractive Losses in X-Ray Lasers . . . 171
Leili Masoudnia and Davide Bleiner

28 Temporal Coherence and Spectral Linewidth of Neon-Like XUV Lasers Pumped in the Quasi-steady State Regime . . . 175
L. Meng, A. Klisnick, M. Kozlova, K. Bohacek, M. Krus, J. Prokupek, L. Urbanski, M.C. Marconi, M. Berrill, J.J. Rocca, O. Guilbaud, F. Tissandier, S. Sebban, P. Zeitoun, A. Calisti, S. Ferri, C. Mossé, and B. Talin

29 Spectral Broadening of Ni-Like XUV Laser Lines . . . 181
L. Meng, A. Calisti, S. Ferri, C. Mossé, B. Talin, D. Benredjem, O. Guilbaud, and A. Klisnick

30 Visualization of Rapid Dynamic Interactions by Flash Soft X-Ray Microscopy . . . 185
C.S. Menoni, S. Carbajo, I.D. Howlett, W. Chao, E.H. Anderson, A.V. Vinogradov, I.A. Artioukov, K. Buchanan, M.C. Marconi, and J.J. Rocca

31 Using the X-FEL to Understand X-Ray Thomson Scattering for Partially Ionized Plasmas . . . 193
Joseph Nilsen, Walter R. Johnson, and K.T. Cheng

32 Development of Soft X-Ray Microscopy Using Fresnel Zone Plate for Observation of Laser-Induce Surface Dynamics . . . 199
M. Nishikino, M. Baba, T. Suemoto, N. Hasegawa, M. Ishino, T. Kaihori, T. Kawachi, and M. Yamagiwa

33 0.27 GW Soft X-Ray Pulse Using a Plasma-Based Amplification Chain . . . 203
E. Oliva, M. Fajardo, P. Velarde, D. Ros, S. Sebban, and P. Zeitoun

34 Harmonic Generation in Argon by Femtosecond Ti:Sapphire Laser . 209
Rabia Qindeel, Ricardo Elgul Samad, Anderson Zanardi de Freitas, Paulo Sergio Fabris de Matos, Edilson Lucena Falcão, and Nilson Dias Vieira Junior

35 Demonstration of a 100 Hz Repetition Rate Soft X-Ray Laser and Gain-Saturated Sub-10 nm Table-Top Lasers . . . 215
J.J. Rocca, B.A. Reagan, Y. Wang, D. Alessi, K. Wernsing, B.M. Luther, M.A. Curtis, M. Berrill, D. Martz, S. Wang, L. Yin, F. Furch, M. Woolston, D. Patel, V.N. Shlyaptsev, and C.S. Menoni

36 Optical Correction of X-Ray Laser Illumination for Short-Wavelength Microscopy 227
Mabel Ruiz-Lopez, Felix Staub, and Davide Bleiner

37 Repetitive XUV Discharge-Pumped Laser at 46.9 nm 231
Jiri Schmidt, Karel Kolacek, Oleksandr Frolov, Vaclav Prukner, and Jaroslav Straus

38 Line-Focus Generation for X-Ray Laser Pumping 235
F. Staub, D. Bleiner, F. Jia, and J.E. Balmer

39 Critical Components for XUV Probing of Laser Driven Shocks . . . 239
C. Stehlé, R. Lefèvre, U. Chaulagain, N. Champion, P. Barroso, F. Reix, P. Jagourel, J. Larour, E. Meltchakov, R. Mercier, F. Delmotte, M. Kozlova, J. Nejdl, M. Krus, J. Dostal, J. Prokupek, C. Constancias, F. Suzuki-Vidal, and O. Acef

40 High Density Optical-Field-Ionization Soft X-Ray Lasers 243
F. Tissandier, S. Sebban, M. Kozlova, J. Gautier, P. Zeitoun, A. Klisnick, and G. Maynard

41 Spectral Linewidth Measurement of a Ne-Like Ar Capillary Discharge Soft X-Ray Laser 257
L. Urbanski, M.C. Marconi, L.M. Meng, M. Berrill, O. Guilbaud, A. Klisnick, and J.J. Rocca

42 Defect Tolerant Talbot Nanopatterning 263
L. Urbanski, M.C. Marconi, A. Isoyan, A. Stein, C.S. Menoni, and J.J. Rocca

43 Imaging in Nanoscale Using Laser-Plasma Sources of Extreme Ultraviolet (EUV) 269
P.W. Wachulak, A. Bartnik, A. Baranowska-Korczyc, D. Pánek, P. Brůža, J. Kostecki, Ł. Węgrzyński, R. Jarocki, M. Szczurek, K. Fronc, D. Elbaum, and H. Fiedorowicz

44 Experiments to Diagnose Plasma with a Soft X-Ray Laser Double-Frequency Grating Interferometry 277
C. Wang, H.H. An, Z.H. Fang, J.R. Sun, W. Wang, W.D. Zheng, X.M. Qiao, and S.J. Wang

45 Heavy-Ion Spectroscopy with X-Ray Lasers at GSI 283
B. Zielbauer, B. Ecker, P. Neumayer, K. Cassou, S. Daboussi, O. Guilbaud, S. Kazamias, D. Ros, T. Kuehl, U. Eisenbarth, S. Goette, D. Winters, V. Bagnoud, and T. Stoehlker

Index 287

Contributors

O. Acef SYRTE, Observatoire de Paris, UPMC, CNRS, LNE, Paris, France

D. Alessi National Science Foundation ERC for Extreme Ultraviolet Science and Technology, Fort Collins, USA; Department of Electrical and Computer Engineering, Colorado State University, Fort Collins, CO, USA

E. Allaria Elettra-Sincrotrone Trieste, Basovizza, Trieste, Italy

H.H. An Shanghai Institute of Laser Plasma, Shanghai, China

E.H. Anderson NSF ERC for Extreme Ultraviolet Science and Technology, Department of Electrical & Computer Engineering, Colorado State University, Fort Collins, CO, USA; NSF ERC for Extreme Ultraviolet Science and Technology and Center for X-ray Optics, LBNL, Berkeley, CA, USA

A.V. Andreev Physics Department, M.V. Lomonosov Moscow State University, Moscow, Russia

I.A. Andriyash Univ. Bordeaux, CNRS, CEA, CELIA (Centre Lasers Intenses et Applications), Talence, France

I.A. Artyukov P.N. Lebedev Physical Institute, Moscow, Russia

B. Aurand Johannes-Gutenberg Universität Mainz, Mainz, Germany; ExtreMe Matter Institute EMMI, Darmstadt, Germany; GSI Helmholtzzentrum fuer Schwerionenforschung, Darmstadt, Germany

M. Baba Institute of Solid State Physics, The University of Tokyo, Kashiwa, Chiba, Japan

V. Bagnoud Helmholtz Institute Jena, Jena, Germany; GSI Helmholtz Center for Heavy-Ion Research, Darmstadt, Germany

G. Balakrishnan Department of Electrical Engineering, University of New Mexico and Center for High Technology Materials, Albuquerque, NM, USA

E. Balaur Department of Physics, La Trobe University, Melbourne, Australia; Australian Research Council Centre of Excellence for Coherent X-Ray Science, Melbourne, Australia

P. Balcou Univ. Bordeaux, CNRS, CEA, CELIA (Centre Lasers Intenses et Applications), Talence, France

J.E. Balmer Institute of Applied Physics, University of Bern, Bern, Switzerland

A. Baranowska-Korczyc Institute of Physics, Polish Academy of Sciences, Warsaw, Poland

P. Barroso GEPI, Observatoire de Paris, Univ. Paris-Diderot, CNRS, Paris, France

A. Bartnik Institute of Optoelectronics, Military University of Technology, Warsaw, Poland

C. Benabderrahmane Synchrotron SOLEIL, Saint-Aubin, France

D. Benredjem LAC, CNRS-Université Paris-Sud, Orsay, France

M. Berrill NSF-ERC for EUV Science and Technology, Colorado State University, Fort Collins, USA; National Science Foundation ERC for Extreme Ultraviolet Science and Technology, Fort Collins, USA; Department of Electrical and Computer Engineering, Colorado State University, Fort Collins, CO, USA; Oak Ridge National Laboratory, Oak Ridge, TN, USA

P. Betinelli Synchrotron SOLEIL, Saint-Aubin, France

S. Bielawski PhLAM/CERLA, Lille, France

D. Bleiner Institute of Applied Physics, University of Bern, Bern, Switzerland

K. Bohacek Institute of Physics of ASCR, Prague, Czech Republic

A. Boudaa Laboratoire de Physique et Chimie Quantique, Université M. Mammeri de Tizi-Ouzou, Hasnaoua, Algérie; Laboratoire de Physique des Gaz et des Plasmas, Univ. Paris-Sud, Orsay, France

F. Bouvet Synchrotron SOLEIL, Saint-Aubin, France

C. Brown Department of Electrical Engineering, Colorado State University and Engineering Research Center for Extreme Ultraviolet Science and Technology, Fort Collins, CO, USA

P. Brůža Czech Technical University, Prague, Czech Republic

K. Buchanan Physics Department, Colorado State University, Fort Collins, CO, USA

S.V. Bulanov Quantum Beam Science Directorate, JAEA, Kizugawa, Kyoto, Japan

A.S. Busarov P.N. Lebedev Physical Institute, Moscow, Russia

A. Buteau Synchrotron SOLEIL, Saint-Aubin, France

A. Calisti PIIM, CNRS-Aix-Marseille Université, Marseille, France

S. Carbajo NSF ERC for Extreme Ultraviolet Science and Technology, Department of Electrical & Computer Engineering, Colorado State University, Fort Collins, CO, USA

B. Carré CEA/DSM/IRAMIS/SPAM, Saclay, France

L. Cassinari Synchrotron SOLEIL, Saint-Aubin, France

K. Cassou CLUPS/LASERIX, Université Paris Sud XI, Orsay Cedex, France; LPGP, Univ. Paris-Sud, UMR 8578, CNRS, Orsay Cedex, France

D. Castronovo Elettra-Sincrotrone Trieste, Basovizza, Trieste, Italy

N. Champion Lerma, Observatoire de Paris, UPMC, CNRS, Paris, France

W. Chao NSF ERC for Extreme Ultraviolet Science and Technology, Department of Electrical & Computer Engineering, Colorado State University, Fort Collins, CO, USA; NSF ERC for Extreme Ultraviolet Science and Technology and Center for X-ray Optics, LBNL, Berkeley, CA, USA

U. Chaulagain Lerma, Observatoire de Paris, UPMC, CNRS, Paris, France

K.T. Cheng Lawrence Livermore National Laboratory, Livermore, CA, USA

C. Constancias CEA-LETI, Grenoble, France

Manuel Cotelo Instituto de Fusión Nuclear, Universidad Politécnica de Madrid, Madrid, Spain

M.E. Couprie Synchrotron SOLEIL, Saint-Aubin, France

B. Cros LPGP, Univ. Paris-Sud, UMR 8578, CNRS, Orsay Cedex, France

M.A. Curtis National Science Foundation ERC for Extreme Ultraviolet Science and Technology, Fort Collins, USA; Department of Electrical and Computer Engineering, Colorado State University, Fort Collins, CO, USA

E. d'Humières Univ. Bordeaux, CNRS, CEA, CELIA (Centre Lasers Intenses et Applications), Talence, France

S. Daboussi CLUPS/LASERIX, Université Paris Sud XI, Orsay Cedex, France; LPGP, Univ. Paris-Sud, UMR 8578, CNRS, Orsay Cedex, France

H. Daido Applied Laser Technology Institute Tsuruga Head Office, Japan Atomic Energy Agency, Tsuruga, Fukui, Japan; Quantum Beam Directorate, Japan Atomic Energy Agency, Kizugawa, Kyoto, Japan

J. Daillant Synchrotron SOLEIL, Saint-Aubin, France

L.V. Dao Centre for Atom Optics and Ultrafast Spectroscopy, Swinburne University of Technology, Melbourne, Australia; Australian Research Council Centre of Excellence for Coherent X-Ray Science, Melbourne, Australia

Anderson Zanardi de Freitas Centro de Lasers e Aplicações, IPEN, CNEN/SP, Sao Paulo, Brazil

François de Gaufridy Instituto de Fusión Nuclear, Universidad Politécnica de Madrid, Madrid, Spain; Institute of Physics, Czech Academy of Sciences, Prague, Czech Republic

Paulo Sergio Fabris de Matos Centro de Lasers e Aplicações, IPEN, CNEN/SP, Sao Paulo, Brazil

G. De Ninno Elettra-Sincrotrone Trieste, Basovizza, Trieste, Italy; Dept. of Physics, University of Nova Gorica, Nova Gorica, Slovenia

O. Delmas CLUPS/LASERIX, Université Paris Sud XI, Orsay Cedex, France; Amplitude Technologies, Lisses, France; LPGP, Univ. Paris-Sud, UMR 8578, CNRS, Orsay Cedex, France

F. Delmotte LCF, Institut d'Optique, CNRS, PXI, Palaiseau, France

J. Demailly CLUPS/LASERIX, Université Paris Sud XI, Orsay Cedex, France; Laboratoire de Physique des Gaz et Plasma, CNRS-Université Paris Sud, Orsay, France

J.C. Denard Synchrotron SOLEIL, Saint-Aubin, France

G. Devanz CEA/DSM/IRFU/SACM, Saclay, France

Alberto G. de la Varga Instituto de Fusión Nuclear, Universidad Politécnica de Madrid, Madrid, Spain

S. Di Mitri Elettra-Sincrotrone Trieste, Basovizza, Trieste, Italy

K.B. Dinh Centre for Atom Optics and Ultrafast Spectroscopy, Swinburne University of Technology, Melbourne, Australia; Australian Research Council Centre of Excellence for Coherent X-Ray Science, Melbourne, Australia

J. Dostal IoP/PALS Center, Prague, Czech Republic

A. Dubois LCPMR, Paris-VI, France

Antonin Dusseix Laboratoire de Physique des Gaz et Plasma, CNRS-Université Paris Sud, Orsay, France

B. Ecker Helmholtz Institute Jena, Jena, Germany; Johannes-Gutenberg Universität Mainz, Mainz, Germany

U. Eisenbarth GSI Helmholtz Center for Heavy-Ion Research, Darmstadt, Germany

D. Elbaum Institute of Physics, Polish Academy of Sciences, Warsaw, Poland

T. Esirkepov Quantum Beam Science Directorate, JAEA, Kizugawa, Kyoto, Japan

C. Evain PhLAM/CERLA, Lille, France

P. Eymard Synchrotron SOLEIL, Saint-Aubin, France

A.Y. Faenov Quantum Beam Science Directorate, JAEA, Kizugawa, Kyoto, Japan; Joint Institute for High Temperatures, Russian Academy of Science, Moscow, Russia

M. Fajardo Centro de Física dos Plasmas, Instituto Superior Técnico, Lisbon, Portugal

Edilson Lucena Falcão Physics Department, UFPE, Recife, Brazil

Z.H. Fang Shanghai Institute of Laser Plasma, Shanghai, China

L. Farvacque ESRF, Grenoble, France

W. Fawley Elettra-Sincrotrone Trieste, Basovizza, Trieste, Italy

E. Ferrari Elettra-Sincrotrone Trieste, Basovizza, Trieste, Italy; Dept. of Physics, University of Nova Gorica, Nova Gorica, Slovenia

S. Ferri PIIM, CNRS-Aix-Marseille Université, Marseille, France

R.M. Feshchenko P.N. Lebedev Physical Institute of RAS, Moscow, Russia

Thomas Feurer Institute of Applied Physics, University of Bern, Bern, Switzerland

H. Fiedorowicz Institute of Optoelectronics, Military University of Technology, Warsaw, Poland

Vladimir Fortov Joint Institute for High Temperatures, Russian Academy of Sciences, Moscow, Russia

L. Froehlich Elettra-Sincrotrone Trieste, Basovizza, Trieste, Italy

Oleksandr Frolov Pulse Plasma Systems Department, Institute of Plasma Physics AS CR, v.v.i., Prague, Czech Republic

K. Fronc Institute of Physics, Polish Academy of Sciences, Warsaw, Poland

F. Furch National Science Foundation ERC for Extreme Ultraviolet Science and Technology, Fort Collins, USA; Department of Physics, Colorado State University, Fort Collins, CO, USA

N. Gaffney Centre for Atom Optics and Ultrafast Spectroscopy, Swinburne University of Technology, Melbourne, Australia; Australian Research Council Centre of Excellence for Coherent X-Ray Science, Melbourne, Australia

B. Gagey Synchrotron SOLEIL, Saint-Aubin, France

J. Gautier Laboratoire d'Optique Appliquée, ENSTA, Palaiseau, France

Aurelie Gense Laboratoire de Physique des Gaz et Plasma, CNRS-Université Paris Sud, Orsay, France

L. Giannessi Elettra-Sincrotrone Trieste, Basovizza, Trieste, Italy; ENEA, Frascati, Italy

S. Goette GSI Helmholtz Center for Heavy-Ion Research, Darmstadt, Germany

O. Guilbaud CLUPS/LASERIX, Université Paris Sud XI, Orsay Cedex, France; LPGP, Univ. Paris-Sud, UMR 8578, CNRS, Orsay Cedex, France

C.P. Hains Department of Electrical Engineering, University of New Mexico and Center for High Technology Materials, Albuquerque, NM, USA

C. Hall Centre for Atom Optics and Ultrafast Spectroscopy, Swinburne University of Technology, Melbourne, Australia; Australian Research Council Centre of Excellence for Coherent X-Ray Science, Melbourne, Australia

P. Hannaford Centre for Atom Optics and Ultrafast Spectroscopy, Swinburne University of Technology, Melbourne, Australia; Australian Research Council Centre of Excellence for Coherent X-Ray Science, Melbourne, Australia

N. Hasegawa Quantum Beam Science Directorate, Japan Atomic Energy Agency, Kizugawa, Kyoto, Japan

C. Herbeaux Synchrotron SOLEIL, Saint-Aubin, France

D.C. Hochhaus ExtreMe Matter Institute EMMI, Darmstadt, Germany; Goethe-Universität Frankfurt, Frankfurt, Germany

I.D. Howlett NSF ERC for Extreme Ultraviolet Science and Technology, Department of Electrical & Computer Engineering, Colorado State University, Fort Collins, CO, USA

T. Imazono Quantum Beam Science Directorate, Japan Atomic Energy Agency, Kizugawa, Kyoto, Japan

M. Ishino Quantum Beam Science Directorate, Japan Atomic Energy Agency, Kizugawa, Kyoto, Japan

A. Isoyan Synopsys Inc., Hilsboro, OR, USA

P. Jagourel GEPI, Observatoire de Paris, Univ. Paris-Diderot, CNRS, Paris, France

K.A. Janulewicz Advanced Photonics Research Institute, Gwangju Institute of Science and Technology, Gwangju, Rep. of Korea

R. Jarocki Institute of Optoelectronics, Military University of Technology, Warsaw, Poland

F. Jia Institute of Applied Physics, University of Bern, Bern, Switzerland

Q. Jin Institute of Modern Physics, Chinese Academy of Sciences, Lanzhou, People's Republic of China

Walter R. Johnson University of Notre Dame, Notre Dame, IN, USA

T. Kaihori Quantum Beam Science Directorate, Japan Atomic Energy Agency, Kizugawa, Kyoto, Japan

M. Kando Quantum Beam Science Directorate, JAEA, Kizugawa, Kyoto, Japan

Y. Kato Quantum Beam Science Directorate, JAEA, Kizugawa, Kyoto, Japan

T. Kawachi Quantum Beam Science Directorate, Japan Atomic Energy Agency, Kizugawa, Kyoto, Japan

S. Kazamias CLUPS/LASERIX, Université Paris Sud XI, Orsay Cedex, France; LPGP, Univ. Paris-Sud, UMR 8578, CNRS, Orsay Cedex, France

C.M. Kim Advanced Photonics Research Institute, Gwangju Institute of Science and Technology, Gwangju, Rep. of Korea

A. Klisnick ISMO, CNRS-Université Paris-Sud, Orsay, France

Karel Kolacek Pulse Plasma Systems Department, Institute of Plasma Physics AS CR, v.v.i., Prague, Czech Republic

K. Kondo Quantum Beam Science Directorate, JAEA, Kizugawa, Kyoto, Japan

J. Kostecki Institute of Optoelectronics, Military University of Technology, Warsaw, Poland

M. Kozlova Institute of Physics of ASCR, Prague, Czech Republic; IoP/PALS Center, Prague, Czech Republic

M. Krus Institute of Physics of ASCR, Prague, Czech Republic; IoP/PALS Center, Prague, Czech Republic

T. Kuehl Helmholtz Institute Jena, Jena, Germany; GSI Helmholtz Center for Heavy-Ion Research, Darmstadt, Germany; ExtreMe Matter Institute EMMI, Darmstadt, Germany; Johannes-Gutenberg Universität Mainz, Mainz, Germany; GSI Helmholtzzentrum fuer Schwerionenforschung, Darmstadt, Germany

M. Labat Synchrotron SOLEIL, Saint-Aubin, France

G. Lambert Laboratoire d'Optique Appliquée, Palaiseau, France

J. Larour LPP, Ecole Polytechnique, UPMC, CNRS, Palaiseau, France

T.T.T. Le Laboratoire d'Optique Appliquée, Palaiseau cedex, France

G. LeBec ESRF, Grenoble, France

R. Lefevre Lerma, Observatoire de Paris, UPMC, CNRS, Paris, France

R. Lehe Laboratoire d'Optique Appliquée, Palaiseau, France

A. Lestrade Synchrotron SOLEIL, Saint-Aubin, France

Haochuan Li Institute of Precision Optical Engineering (IPOE), Physics Department, Tongji University, Shanghai, China

L. Li Laboratoire d'Optique Appliquée, Palaiseau cedex, France

A. Lifschitz Laboratoire d'Optique Appliquée, Palaiseau, France

A. Loulergue Synchrotron SOLEIL, Saint-Aubin, France

M. Luong CEA/DSM/IRFU/SACM, Saclay, France

B.M. Luther National Science Foundation ERC for Extreme Ultraviolet Science and Technology, Fort Collins, USA; Department of Electrical and Computer Engineering, Colorado State University, Fort Collins, CO, USA

J. Lüning LCPMR, Paris-VI, France

B. Mahieu Elettra-Sincrotrone Trieste, Basovizza, Trieste, Italy; Dept. of Physics, University of Nova Gorica, Nova Gorica, Slovenia

V. Malka Laboratoire d'Optique Appliquée, Palaiseau, France

E.B. Malm Department of Electrical Engineering, Colorado State University and Engineering Research Center for Extreme Ultraviolet Science and Technology, Fort Collins, CO, USA

P. Marchand Synchrotron SOLEIL, Saint-Aubin, France

M.C. Marconi NSF Engineering Research Center for Extreme Ultraviolet Science & Technology and Department of Electrical and Computer Engineering, Colorado State University, Fort Collins, CO, USA

J.L. Marlats Synchrotron SOLEIL, Saint-Aubin, France

D. Martz National Science Foundation ERC for Extreme Ultraviolet Science and Technology, Fort Collins, USA; Department of Electrical and Computer Engineering, Colorado State University, Fort Collins, CO, USA

Leili Masoudnia Institute of Applied Physics, University of Bern, Bern, Switzerland

G. Maynard LPGP, Univ. Paris-Sud, UMR 8578, CNRS, Orsay Cedex, France

E. Meltchakov LCF, Institut d'Optique, CNRS, PXI, Palaiseau, France

L.M. Meng ISMO, CNRS-Université Paris-Sud, Orsay, France

C.S. Menoni NSF Engineering Research Center for Extreme Ultraviolet Science & Technology and Department of Electrical and Computer Engineering, Colorado State University, Fort Collins, CO, USA; National Science Foundation ERC for Extreme Ultraviolet Science and Technology, Fort Collins, USA

R. Mercier LCF, Institut d'Optique, CNRS, PXI, Palaiseau, France

Y. Minami Institute of Solid State Physics, The University of Tokyo, Kashiwa, Chiba, Japan

C. Miron Synchrotron SOLEIL, Saint-Aubin, France

N.C. Monserud Department of Electrical Engineering, Colorado State University and Engineering Research Center for Extreme Ultraviolet Science and Technology, Fort Collins, CO, USA

P. Morin Synchrotron SOLEIL, Saint-Aubin, France

T. Morita Quantum Beam Science Directorate, Japan Atomic Energy Agency, Kizugawa, Kyoto, Japan

C. Mossé PIIM, CNRS-Aix-Marseille Université, Marseille, France

T. Muramatsu Applied Laser Technology Institute Tsuruga Head Office, Japan Atomic Energy Agency, Tsuruga, Fukui, Japan

A. Nadji Synchrotron SOLEIL, Saint-Aubin, France

T. Nakamura Quantum Beam Science Directorate, JAEA, Kizugawa, Kyoto, Japan

K. Namikawa Research Institute for Science & Technology, Tokyo University of Science, Noda, Chiba, Japan

J. Nejdl IoP/PALS Center, Prague, Czech Republic

P. Neumayer ExtreMe Matter Institute EMMI, Darmstadt, Germany

O. Neveu LPGP, Univ. Paris-Sud, UMR 8578, CNRS, Orsay Cedex, France; LASERIX, CLUPS, Univ. Paris-Sud, CNRS, Orsay Cedex, France

Bich-Lien Nghiem Laboratoire de Physique des Gaz et Plasma, CNRS-Université Paris Sud, Orsay, France

P.V. Nickles WCU Dept. of Nanobio Materials and Electronics, Gwangju Institute of Science and Technology, Gwangju, Rep. of Korea

Joseph Nilsen Lawrence Livermore National Laboratory, Livermore, CA, USA

M. Nishikino Quantum Beam Science Directorate, Japan Atomic Energy Agency, Kizugawa, Kyoto, Japan

A. Nishimura Applied Laser Technology Institute Tsuruga Head Office, Japan Atomic Energy Agency, Tsuruga, Fukui, Japan; Quantum Beam Directorate, Japan Atomic Energy Agency, Kizugawa, Kyoto, Japan

G. Norman Joint Institute for High Temperatures, Russian Academy of Sciences, Moscow, Russia; Moscow Institute of Physics and Technology, Moscow, Russia

Y. Ochi Quantum Beam Science Directorate, Japan Atomic Energy Agency, Kizugawa, Kyoto, Japan

E. Oliva Laboratoire d'Optique Appliquée, ENSTA ParisTech, École Polytechnique ParisTech, CNRS, UMR 7639, Palaiseau, France; Laboratoire d'Optique Appliquée, Palaiseau cedex, France

D. Patel National Science Foundation ERC for Extreme Ultraviolet Science and Technology, Fort Collins, USA; Department of Electrical and Computer Engineering, Colorado State University, Fort Collins, CO, USA

D. Pánek Czech Technical University, Prague, Czech Republic

G. Penco Elettra-Sincrotrone Trieste, Basovizza, Trieste, Italy

T.A. Pikuz Quantum Beam Science Directorate, JAEA, Kizugawa, Kyoto, Japan

Tatiana Pikuz Quantum Beam Science Directorate, Japan Atomic Energy Agency, Kyoto, Japan; Joint Institute for High Temperatures, Russian Academy of Sciences, Moscow, Russia

A. Pirozhkov Quantum Beam Science Directorate, JAEA, Kizugawa, Kyoto, Japan

M. Pittman LASERIX, CLUPS, Univ. Paris-Sud, CNRS, Orsay Cedex, France

F. Polack Synchrotron SOLEIL, Saint-Aubin, France

A.V. Popov Pushkov Institute of Terrestrial Magnetism, Ionosphere and Radiowave Propagation, Troitsk, Russia

N.L. Popov P.N. Lebedev Physical Institute, Moscow, Russia

David Portillo Instituto de Fusión Nuclear, Universidad Politécnica de Madrid, Madrid, Spain

Chris Pouhe Laboratoire de Physique des Gaz et Plasma, CNRS-Université Paris Sud, Orsay, France

J. Prokupek Institute of Physics of ASCR, Prague, Czech Republic; IoP/PALS Center, Prague, Czech Republic

Vaclav Prukner Pulse Plasma Systems Department, Institute of Plasma Physics AS CR, v.v.i., Prague, Czech Republic

J.B. Pruvost Synchrotron SOLEIL, Saint-Aubin, France

X.M. Qiao Institute of Applied Physics and Computational Mathematics, Beijing, China

Rabia Qindeel Centro de Lasers e Aplicações, IPEN, CNEN/SP, Sao Paulo, Brazil

B.A. Reagan National Science Foundation ERC for Extreme Ultraviolet Science and Technology, Fort Collins, USA; Department of Electrical and Computer Engineering, Colorado State University, Fort Collins, CO, USA

F. Reix GEPI, Observatoire de Paris, Univ. Paris-Diderot, CNRS, Paris, France

F. Ribeiro Synchrotron SOLEIL, Saint-Aubin, France

J.P. Ricaud Synchrotron SOLEIL, Saint-Aubin, France

Gabriel Richet Laboratoire de Physique des Gaz et Plasma, CNRS-Université Paris Sud, Orsay, France

J.J. Rocca NSF Engineering Research Center for Extreme Ultraviolet Science & Technology and Department of Electrical and Computer Engineering, Colorado State University, Fort Collins, CO, USA; National Science Foundation ERC for Extreme Ultraviolet Science and Technology, Fort Collins, USA; Department of Physics, Colorado State University, Fort Collins, CO, USA

D. Ros CLUPS/LASERIX, Université Paris Sud XI, Orsay Cedex, France; LPGP, Univ. Paris-Sud, UMR 8578, CNRS, Orsay Cedex, France

A. Rousse Laboratoire d'Optique Appliquée, Palaiseau, France

R. Roux Laboratoire de l'Accélérateur Linéaire, Orsay, France

P. Roy Synchrotron SOLEIL, Saint-Aubin, France

Mabel Ruiz-Lopez Institute of Applied Physics, University of Bern, Bern, Switzerland

B. Rus IoP/PALS Center, Prague, Czech Republic

Ricardo Elgul Samad Centro de Lasers e Aplicações, IPEN, CNEN/SP, Sao Paulo, Brazil

A. Sasaki Quantum Beam Science Directorate, Japan Atomic Energy Agency, Kizugawa, Kyoto, Japan

Jiri Schmidt Pulse Plasma Systems Department, Institute of Plasma Physics AS CR, v.v.i., Prague, Czech Republic

S. Sebban Laboratoire d'Optique Appliquée, ENSTA ParisTech, École Polytechnique ParisTech, CNRS, UMR 7639, Palaiseau, France

V.N. Shlyaptsev National Science Foundation ERC for Extreme Ultraviolet Science and Technology, Fort Collins, USA; Department of Electrical and Computer Engineering, Colorado State University, Fort Collins, CO, USA

T. Shobu Applied Laser Technology Institute Tsuruga Head Office, Japan Atomic Energy Agency, Tsuruga, Fukui, Japan; Quantum Beam Directorate, Japan Atomic Energy Agency, Kizugawa, Kyoto, Japan

O.A. Shoutova Physics Department, M.V. Lomonosov Moscow State University, Moscow, Russia

Igor Skobelev Joint Institute for High Temperatures, Russian Academy of Sciences, Moscow, Russia

T.A. Smith School of Chemistry, University of Melbourne, Melbourne, Australia; Australian Research Council Centre of Excellence for Coherent X-Ray Science, Melbourne, Australia

C. Spezzani Elettra-Sincrotrone Trieste, Basovizza, Trieste, Italy

Sergei Starikov Joint Institute for High Temperatures, Russian Academy of Sciences, Moscow, Russia; Moscow Institute of Physics and Technology, Moscow, Russia

F. Staub Institute of Applied Physics, University of Bern, Bern, Switzerland

Vladimir Stegailov Joint Institute for High Temperatures, Russian Academy of Sciences, Moscow, Russia; Moscow Institute of Physics and Technology, Moscow, Russia

C. Stehlé Lerma, Observatoire de Paris, UPMC, CNRS, Paris, France

A. Stein Center for Functional Nanomaterials, Brookhaven National Laboratory, Upton, NY, USA

H. Stiel Max-Born Institute for Nonlinear Optics and Short Pulse Spectroscopy, Berlin, Germany

T. Stoehlker Helmholtz Institute Jena, Jena, Germany; GSI Helmholtz Center for Heavy-Ion Research, Darmstadt, Germany

Jaroslav Straus Pulse Plasma Systems Department, Institute of Plasma Physics AS CR, v.v.i., Prague, Czech Republic

S.Y. Stremoukhov Physics Department, M.V. Lomonosov Moscow State University, Moscow, Russia

T. Suemoto Institute of Solid State Physics (ISSP), University of Tokyo, Kashiwa, Chiba, Japan

K. Sugihara Applied Laser Technology Institute Tsuruga Head Office, Japan Atomic Energy Agency, Tsuruga, Fukui, Japan

J.R. Sun Shanghai Institute of Laser Plasma, Shanghai, China

F. Suzuki-Vidal Imperial College, London, UK

M. Szczurek Institute of Optoelectronics, Military University of Technology, Warsaw, Poland

C. Szwaj PhLAM/CERLA, Lille, France

K. Ta Phuoc Laboratoire d'Optique Appliquée, Palaiseau, France

B. Talin PIIM, CNRS-Aix-Marseille Université, Marseille, France

Greg Tallents University of York, York, UK

Satoshi Tamotsu Division of Natural Science, Faculty, Nara Women's University, Nara, Japan

M. Tanaka Quantum Beam Science Directorate, Japan Atomic Energy Agency, Kyoto, Japan

T. Tanikawa Synchrotron SOLEIL, Saint-Aubin, France

K. Terakawa Institute of Solid State Physics, The University of Tokyo, Kashiwa, Chiba, Japan

C. Thaury Laboratoire d'Optique Appliquée, Palaiseau, France

V.T. Tikhonchuk Univ. Bordeaux, CNRS, CEA, CELIA (Centre Lasers Intenses et Applications), Talence, France

F. Tissandier LOA, Ecole Polytechnique, CNRS, Paris, France; Laboratoire d'Optique Appliquée, ENSTA, Palaiseau, France

T. Tomita Department of Ecosystem Engineering, The University of Tokushima, Tokushima, Japan; Faculty of Engineering, University of Tokushima, Tokushima, Japan

M. Trovo Elettra-Sincrotrone Trieste, Basovizza, Trieste, Italy

L. Urbanski NSF Engineering Research Center for Extreme Ultraviolet Science & Technology and Department of Electrical and Computer Engineering, Colorado State University, Fort Collins, CO, USA

P. Velarde Instituto de Fusión Nuclear, Universidad Politécnica de Madrid, Madrid, Spain

Nilson Dias Vieira Junior Centro de Lasers e Aplicações, IPEN, CNEN/SP, Sao Paulo, Brazil

A.V. Vinogradov P.N. Lebedev Physical Institute, Moscow, Russia

Boris Vodungbo Laboratoire d'Optique Appliquée, ENSTA ParisTech – CNRS – École Polytechnique, Palaiseau, France

H.L. Vu Centre for Atom Optics and Ultrafast Spectroscopy, Swinburne University of Technology, Melbourne, Australia; Australian Research Council Centre of Excellence for Coherent X-Ray Science, Melbourne, Australia

P.W. Wachulak Institute of Optoelectronics, Military University of Technology, Warsaw, Poland

C. Wang Shanghai Institute of Laser Plasma, Shanghai, China

S. Wang National Science Foundation ERC for Extreme Ultraviolet Science and Technology, Fort Collins, USA; Department of Electrical and Computer Engineering, Colorado State University, Fort Collins, CO, USA

S.J. Wang Shanghai Institute of Laser Plasma, Shanghai, China

W. Wang Shanghai Institute of Laser Plasma, Shanghai, China

Y. Wang National Science Foundation ERC for Extreme Ultraviolet Science and Technology, Fort Collins, USA; Department of Electrical and Computer Engineering, Colorado State University, Fort Collins, CO, USA

Zhanshan Wang Institute of Precision Optical Engineering (IPOE), Physics Department, Tongji University, Shanghai, China

K. Wernsing National Science Foundation ERC for Extreme Ultraviolet Science and Technology, Fort Collins, USA; Department of Electrical and Computer Engineering, Colorado State University, Fort Collins, CO, USA

Ł. Węgrzyński Institute of Optoelectronics, Military University of Technology, Warsaw, Poland

Lucy Wilson University of York, York, UK

D. Winters GSI Helmholtz Center for Heavy-Ion Research, Darmstadt, Germany

M. Woolston National Science Foundation ERC for Extreme Ultraviolet Science and Technology, Fort Collins, USA; Department of Electrical and Computer Engineering, Colorado State University, Fort Collins, CO, USA

H. Xu Department of Electrical Engineering, University of New Mexico and Center for High Technology Materials, Albuquerque, NM, USA

T. Yamada Applied Laser Technology Institute Tsuruga Head Office, Japan Atomic Energy Agency, Tsuruga, Fukui, Japan

M. Yamagiwa Quantum Beam Science Directorate, Japan Atomic Energy Agency, Kizugawa, Kyoto, Japan

M. Yamamoto Department of Ecosystem Engineering, The University of Tokushima, Tokushima, Japan

S. Yamashita Applied Laser Technology Institute Tsuruga Head Office, Japan Atomic Energy Agency, Tsuruga, Fukui, Japan

L. Yin National Science Foundation ERC for Extreme Ultraviolet Science and Technology, Fort Collins, USA; Department of Electrical and Computer Engineering, Colorado State University, Fort Collins, CO, USA

P. Zeitoun Laboratoire d'Optique Appliquée, ENSTA ParisTech, École Polytechnique ParisTech, CNRS, UMR 7639, Palaiseau, France

H. Zhao Institute of Modern Physics, Chinese Academy of Sciences, Lanzhou, People's Republic of China

W.D. Zheng Institute of Applied Physics and Computational Mathematics, Beijing, China

Jingtao Zhu Institute of Precision Optical Engineering (IPOE), Physics Department, Tongji University, Shanghai, China

B. Zielbauer GSI Helmholtz Center for Heavy-Ion Research, Darmstadt, Germany; Helmholtz Institute Jena, Jena, Germany; GSI Helmholtzzentrum fuer Schwerionenforschung, Darmstadt, Germany

Chapter 1
Coherent Pulses from a Seeded Free-Electron Laser in the Extreme Ultraviolet

E. Allaria, D. Castronovo, G. De Ninno, S. Di Mitri, W. Fawley, E. Ferrari, L. Froehlich, L. Giannessi, B. Mahieu, G. Penco, C. Spezzani, and M. Trovo

Abstract We report results of a free-electron laser operating with the high gain harmonic generation configuration in the wavelength rage from 60 to 20 nm. With the recent experiments done at the FERMI free-electron laser facility in Trieste we demonstrated the capability of the used method to produce highly coherent pulses allowing to improve the performances of FELs in the EUV spectral range overcoming some of the limitations that are typical of systems based on the self amplified spontaneous emission (SASE). Taking advantage of the seeding scheme adopted for FERMI high photon energy stability and a very high longitudinal and transverse coherence have been measured. The operation of the FERMI FEL facility will allow improving the use of the FEL light for those experiments that require a small spectral bandwidth or a precise control of the photon energy.

1 Introduction

FERMI@Elettra is a fourth generation light source under commissioning that has been built at the Elettra laboratory in Trieste. When in operation FERMI@Elettra will allow the Elettra user community to perform new kinds of experiments that could exploit the characteristic of such a new light source like short pulses and high peak intensity.

Free electron lasers have been recognized since the first proposal [1] and demonstration [2, 3] as possible sources that could be efficiently used for the generation of laser like pulses in the x-ray spectral range. Although it has been necessary to

E. Allaria (✉) · D. Castronovo · G. De Ninno · S. Di Mitri · W. Fawley · E. Ferrari · L. Froehlich · L. Giannessi · B. Mahieu · G. Penco · C. Spezzani · M. Trovo
Elettra-Sincrotrone Trieste, S.S. 14—km 163.5 in Area Science Park, 34012 Basovizza, Trieste, Italy
e-mail: enrico.allaria@elettra.trieste.it

G. De Ninno · E. Ferrari · B. Mahieu
Dept. of Physics, University of Nova Gorica, Nova Gorica, Slovenia

L. Giannessi
ENEA, Frascati, Italy

S. Sebban et al. (eds.), *X-Ray Lasers 2012*, Springer Proceedings in Physics 147,
DOI 10.1007/978-3-319-00696-3_1,

wait for the technology to be mature in order to support the requirements for x-rays FELs, the theory of high gain single pass FELs was developed [4, 5] showing how the self amplified spontaneous emission (SASE) FELs could be designed to operate in the x-ray spectral range. Since the first theoretical works, several projects have been done for demonstrating and use such a capability.

After first experiments in the infrared [6] and in the visible [7] the wavelength range has been rapidly extended from the VUV [8] up to the soft x-ray [9]. More recently the final goal to reach the hard x-ray spectral range has been achieved showing FEL operations at 1.5 and 1.2 Ångstrom [10, 11].

The recent progress on short wavelength FELs has been accompanied by the progress of other sources of laser-like pulses in the VUV and soft x-ray spectral range. Several methods have been proposed and studied for the generation of x-ray laser-like pulses that use the nonlinear response of materials to very high power laser pulses in the visible and near infrared [12–16].

In this work we will report about the recent results achieved at FERMI@Elettra with the first seeded FEL operated in the range between 65 and 20 nm.

2 The FERMI@Elettra Free Electron Laser

The FERMI FELs will produce photons in the ultraviolet and soft x-ray range, between 80 and 4 nm (Fig. 1).

Three experimental programs define the scientific case and have driven the design of the facility. The three FERMI experimental beamlines, Diffraction and Projection Imaging (DiProI), Elastic and Inelastic Scattering (EIS), Low Density Matter (LDM), require for their experiments high peak brightness, fully coherent, narrow and stable bandwidth photon pulses. Wavelength tunability and variable polarization, circular and linear, are also required [17, 18].

In order to cover the large spectral range FERMI has been designed with two FEL lines. FEL-1, based on a High Gain Harmonic Generation (HGHG) [19] single stage, is seeded by an external UV laser at about 260 nm [20]. FEL-1 produces photons in the wavelength range between 80 and 20 nm (15 eV to 62 eV). To reach 4 nm wavelength (310 eV), starting from an external seed laser in the UV range, a double stage HGHG cascade has been adopted for FEL-2.

Both FELs use the high energy and brightness electron bunches produced by a linear accelerator (Fig. 2). The accelerator is based on a normal conducting LINAC working at 3 GHz RF frequency and 50 Hz repetition rate. The high brightness electron beam is generated by a photocathode RF gun. Two stages of magnetic compression are used to get extremely short electron bunches (less than 1 ps) with high peak current.

A fourth harmonic RF structure (12 GHz, X-band) provides the longitudinal phase space linearization needed to optimize the compression process. At high compression factors, i.e. high peak bunch currents, micro-bunching instabilities are predicted; to cure them, a laser heater [21] is installed at 100 MeV to increase in a controlled way the incoherent energy spread of the electron beam.

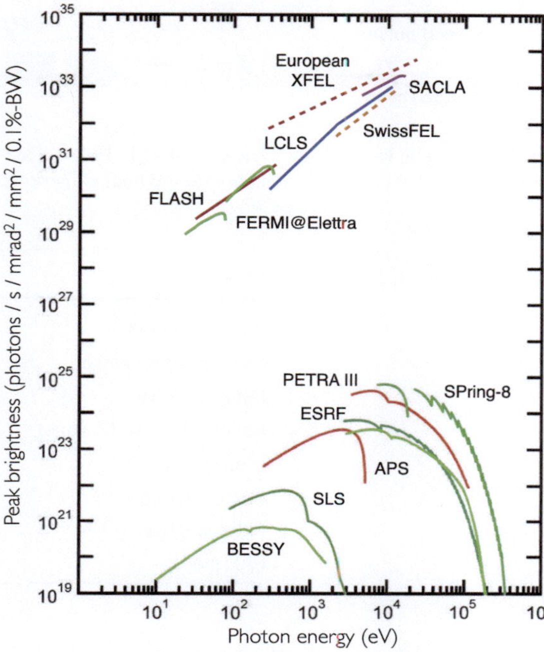

Fig. 1 Diagram showing the expected performance of FERMI in terms of peak brilliance as compared to other FEL and synchrotron radiation facilities (Image from: P. Schmüser, M. Dohlus, J. Rossbach, and C. Behrens, Free-Electron Lasers in the Ultraviolet and X-Ray Regime: Physical Principles, Experimental Results, Technological Realization, 2nd edition to be printed in Springer-Verlag, Berlin (2013))

Each stage of the two FEL lines is made up by the modulator, where the electron beam is seeded by the external laser, a dispersive bunching section, and the radiator, where the interaction between the electron beam and the produced coherent radiation produces the exponential growth of the FEL radiation. In Fig. 3 we show the schematic layout of the first FEL, FEL-1, that is based on a single stage harmonic generation.

For both FEL-1 and FEL-2 the final radiator is made up by six APPLE-II undulators, with magnetic periods of 55 and 35 mm respectively. APPLE-II undulators provide full control of the polarization of the FEL radiation and the variable gap allows changing the wavelength of the emitted radiation.

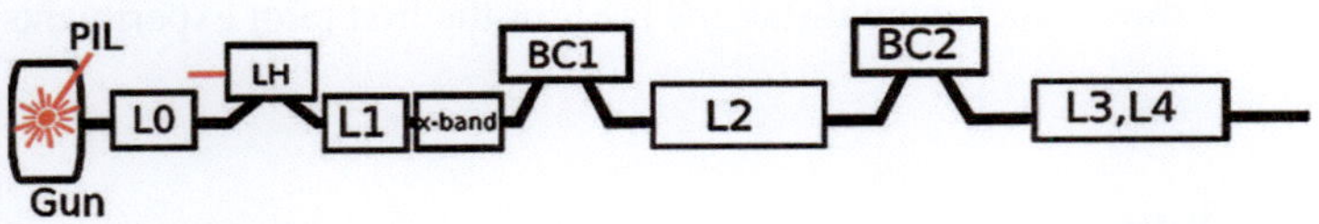

Fig. 2 Schematic layout of the FERMI linear accelerator. Electron bunches are produced in the *Gun* by 5 ps pulses from a Photo-Injector Laser (*PIL*). A first acceleration structure (*L0*) is used to accelerate the beam up to 100 MeV, where there is the laser heater (*LH*). Further acceleration is produced by *L1*, and an *x-band* cavity is used to linearize the phase space before the first bunch compressor (*BC1*). Further acceleration is produced by *L2* and *L3*, *L4* and if needed additional compression can be achieved using the second bunch compressor chicane (*BC2*)

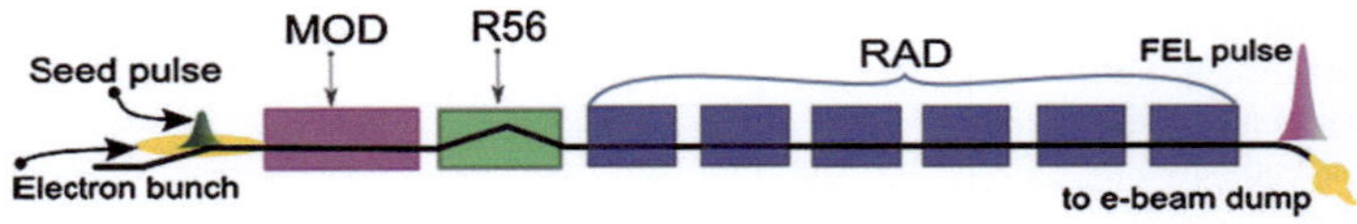

Fig. 3 Schematic layout of the FERMI first FEL, FEL-1. One modulator (*MOD*), a dispersive section (*R56*) and 6 APPLE-II undulators for the final radiator (*RAD*) define the first FEL of FERMI

Table 1 FEL parameters achieved so far with FERMI FEL-1

Parameter	Value achieved	Units
Wavelength range	65–20	nm
Photon energy range	19–62	eV
Bandwidth (rms)	22 (@32 nm)	meV
Relative bandwidth (rms)	5e–4 (@32 nm)	
Polarization	Hor/Circ	
Energy per pulse	20–50	μJ
Pulse length (FWHM)	∼100	fs (estimated)
FEL mode	TEM_{00}	

3 FEL Achieved Performance

During the first year of commissioning the FEL performance has been continuously improving. After few months from the first produced photons it has been possible to achieve the level of several tens of micro joules over the FEL-1 spectral range (Table 1). It has been also possible to demonstrate the high quality and coherence of FERMI FEL pulses [22]. This significantly improved what was generally available from other FEL sources working in a similar spectral range.

Taking advantage of the adopted scheme and of the variable gap undulators adopted for FERMI also the tunability of the FEL wavelength has been easily implemented and changes of wavelength over the whole spectral range have been done. Figure 4 reports a series of FEL spectra acquired during a shift where the FEL wavelength has been varied from about 20 nm up to 40 nm.

With the FEL parameters reported in Table 1, FERMI also started to provide photons to the three experimental stations [18] for the first pilot experiments [23].

4 Conclusions

The first FERMI FEL, FEL-1, has been operated since two years and recently reached fairly intense photon fluxes, producing tens of μJ, approaching the project design goals. With the already achieved photon flux the FEL-1 is showing the benefit of the seeding scheme adopted at FERMI demonstrating good single shot spectra

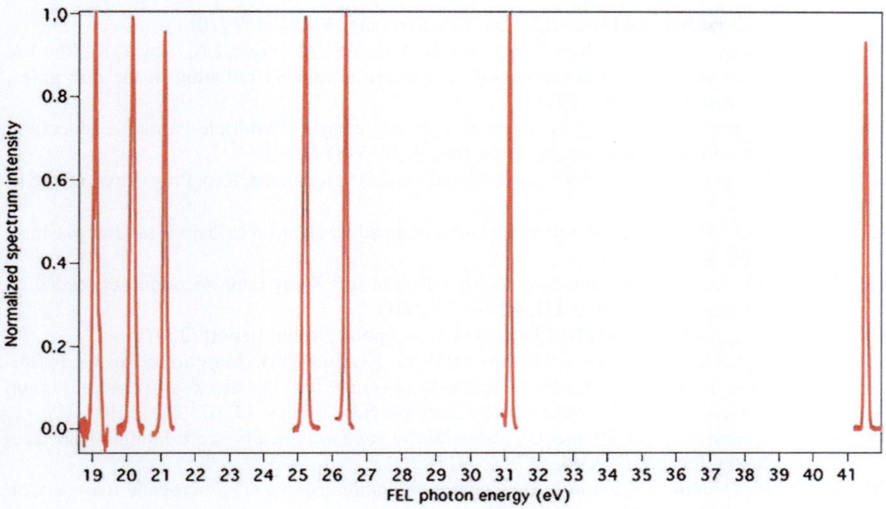

Fig. 4 Series of FEL spectra acquired during experiments that required the FEL tuning in the range 20–40 nm. Reported data are an average of 100 consecutive spectra showing that FERMI spectra are not only narrow bandwidth (tens of meV) but also very stable

with narrow bandwidth and excellent central wavelength stability. A very good degree of transverse coherence has been also observed. FEL tunability and variable polarization have been successfully set into operation.

References

1. Madey, J.M.J.: Stimulated emission of bremsstrahlung in a periodic magnetic field. J. Appl. Phys. **42**, 1906–1913 (1971)
2. Elias, L.R., Fairbank, W.M., Madey, J.M.J., Schwettman, H.A., Smith, T.I.: Observation of stimulated emission of radiation by relativistic electrons in a spatially transverse magnetic field. Phys. Rev. Lett. **36**, 717–720 (1976)
3. Deacon, D.A.G., et al.: First operation of a free electron laser. Phys. Rev. Lett. **38**, 892–894 (1977)
4. Colson, W.B.: Theory of a free electron laser. Phys. Lett. A **59**, 187–190 (1976)
5. Dattoli, G., Renieri, A., Torre, A.: Lectures on the Free Electron Laser Theory and Related Topics. World Scientific, Singapore (1993)
6. Hogan, M.J., et al.: Measurements of gain larger than 10^5 at 12 μm in a self amplified spontaneous emission free electron laser. Phys. Rev. Lett. **81**, 4867–4871 (1998)
7. Milton, S., et al.: Exponential gain and saturation of a self amplified spontaneous emission free electron laser. Science **292**, 2037–2041 (2001)
8. Ayvazyan, V., et al.: Generation of GW radiation pulses from VUV free-electron laser operating in the femtosecond regime. Phys. Rev. Lett. **88**, 104802 (2002)
9. Ackermann, W., et al.: Operation of a free-electron laser from the extreme ultraviolet to the water window. Nat. Photonics **1**, 336–342 (2007)
10. Emma, P., et al.: First lasing and operation of an angstrom-wavelength free-electron laser. Nat. Photonics **4**, 641–647 (2010)

11. Pile, D.: X-rays: first light from SACLA. Nat. Photonics **5**, 456–457 (2011)
12. McPherson, A., Gibson, G., Jara, H., Johann, U., Luk, T.S., McIntyre, I.A., Boyer, K., Rhodes, C.K.: Studies of multiphoton production of vacuum-ultraviolet radiation in the rare gases. J. Opt. Soc. Am. B **21**, 595 (1987)
13. Li, X.F., L'Huillier, A., Ferray, M., Lompré, L.A., Mainfray, G.: Multiple-harmonic generation in rare gases at high laser intensity. Phys. Rev. A **39**, 5751 (1989)
14. Daido, H.: Review of soft x-ray laser researches and developments. Rep. Prog. Phys. **65**, 1513 (2002)
15. Tallents, G.J.: The physics of soft x-ray lasers pumped by electron collisions in laser plasmas. J. Phys. D **36**, R259 (2003)
16. Zeitoun, Ph., et al.: A high-intensity highly coherent soft X-ray femtosecond laser seeded by a high harmonic beam. Nature **431**, 426–429 (2004)
17. Bocchetta, C.J., et al.: FERMI@Elettra FEL Conceptual Design Report (2007)
18. Allaria, E., Callegari, C., Cocco, D., Fawley, W.M., Kiskinova, M., Masciovecchio, C., Parmigiani, F.: The FERMI@Elettra free-electron-laser source for coherent x-ray physics: photon properties, beam transport system and applications. New J. Phys. **12**, 075002 (2010)
19. Yu, L.H.: Generation of intense UV radiation by subharmonically seeded single-pass free-electron lasers. Phys. Rev. A **44**, 5178 (1991)
20. Allaria, E., De Ninno, G.: Soft-X-ray coherent radiation using a single-cascade free-electron laser. Phys. Rev. Lett. **99**, 014801 (2007)
21. Huang, Z., Borland, M., Emma, P., Wu, J., Limborg, C., Stupakov, G., Welch, J.: Suppression of microbunching instability in the linac coherent light source. Phys. Rev. Spec. Top. Accel. Beams **7**, 074401 (2004)
22. Allaria, E., et al.: Highly coherent and stable pulses from the FERMI seeded free-electron laser in the extreme ultraviolet. Nat. Photonics **6**, 699–704 (2012)
23. Allaria, E., et al.: Tunability experiments at the FERMI@Elettra free-electron laser. New J. Phys. **14**, 113009 (2012)

Chapter 2
Enhancement of Efficiency of XUV Generation in Atomic Gases Irradiated by Intense Laser Fields

A.V. Andreev, S.Y. Stremoukhov, and O.A. Shoutova

Abstract We present the results of the theoretical study of the high-order harmonic generation (HHG) in atomic gases. It is shown that the photoemission spectra exhibit unusual behavior when the laser field strength approaches near-atomic values. In subatomic field strength the cut-off frequency increases linearly with laser pulse intensity. However, when the field strength approaches near-atomic region firstly cut-off frequency slows down and then saturates. To interpret such kind of photoemission spectrum behavior we have proposed the light-atom interaction theory based on the use of eigenfunctions of boundary value problem for "an atom in the external field" instead of the traditional basis of the "free atom" eigenfunctions.

1 Introduction

In spite of the twenty-year history, the effect of the HHG is still under a great interest of both experimentalists and theoreticians. From a practical point of view, the HHG is one of the effective mechanisms for producing a coherent emission in broad region of electromagnetic wave spectrum. The presence of plateau region in the harmonic amplitude distribution in extreme ultraviolet (XUV) region affords grounds for development of subfemtosecond pulse formation methods. As a result, the new frontiers are opened up in science by extending the nonlinear optics and time-resolved spectroscopy to the XUV region [1] and pushing ultrafast science to the attosecond domain, enabling XUV spectroscopy and imaging of molecular orbitals [2], surface dynamics [3], and electron motion. The HHG is the reliable route to produce attosecond light pulses [4, 5] and is therefore fundamental to attosecond science [6].

At present days, the efficiency of conversion to high-order harmonics is really too small to consider this emission as a real coherent XUV source for biology, plasma diagnostics, medicine, microscopy, photolithography, etc. Hence, the search

A.V. Andreev · S.Y. Stremoukhov (✉) · O.A. Shoutova
Physics Department, M.V. Lomonosov Moscow State University, Leninskie Gory, 1, build. 2, 119991, Moscow, Russia
e-mail: sustrem@gmail.com

S. Sebban et al. (eds.), *X-Ray Lasers 2012*, Springer Proceedings in Physics 147,
DOI 10.1007/978-3-319-00696-3_2,

for ways of increasing the cut-off frequency and the HHG efficiency in the XUV spectral range is still among the most topical problems of nonlinear optics.

The HHG effect was observed with the large number of periodic table elements having usually small and middle atomic numbers [7]. Up to now, there has been developed a number of different theoretical models to describe the dynamics of an atomic electron in a strong laser field. These models are based on different approximations, which are usually valid in the restricted area of laser pulse field strength. Here, we use the non-perturbative theory of light-atom interaction [8] which is equally applicable for both weak and strong laser fields.

2 Basic Statements of Theory

In the non-relativistic case, the theory of light-atom interaction is based on the time dependent Schrodinger equation TDSE:

$$i\hbar\frac{\partial\psi(\vec{r},t)}{\partial t} = \left[\frac{1}{2m}\left(\vec{p} - \frac{e}{c}\vec{A}(t)\right)^2 + U(r)\right]\psi(\vec{r},t), \tag{1}$$

where $U(\vec{r})$ is the potential energy of electron in the intra-atomic field, and $\vec{A}(t)$ is the vector potential of the external electromagnetic wave. The traditional approaches to the analysis of the TDSE are based on the expansion of the wave function in the series of "free atom" eigenfunctions.

The eigensolutions of the boundary value problem for Hamiltonian of the TDSE,

$$\left[\frac{1}{2m}\left(\vec{p} - \frac{e}{c}\vec{A}(t)\right)^2 + U(r)\right]\varphi_N(\vec{r},t) = E_N\varphi_N(\vec{r},t), \tag{2}$$

have been found in [8]. There are the following relationships between the eigenfunctions of the problem (2) and the "free atom" eigenfunctions $u_n(\vec{r})$:

$$\varphi_N(\vec{r},t) = u_n(\vec{r})\hat{V}^{-1}, \quad \hat{V} = \exp\left(-i\frac{q}{\hbar c}\vec{A}(t)\vec{r}\right). \tag{3}$$

The eigenvalues of the problem (2) coincide exactly with the "free atom" eigenvalues $E_N = E_n$.

The bases of the eigenfunctions of a "free atom" boundary value problem and the basis of $\varphi_n(\vec{r},t)$ are related one-to-one by the transformations [9, 10]. So, we can expand the wavefunction of the TDSE (1) into series of free atom eigenfunctions

$$\psi = \sum_{nlm} a_{nlm}(t)u_{nlm}(\vec{r})$$

and then use the relations of one-to-one transformations [9, 10] to calculate the matrix elements of operators. Executing some evident transformations we get the following set of differential equations for probability amplitudes of discrete and continuum spectrum states

$$i\hbar\frac{da_n}{dt} = \sum_{m,k} V_{nk}^{-1}E_kV_{km}a_m.$$

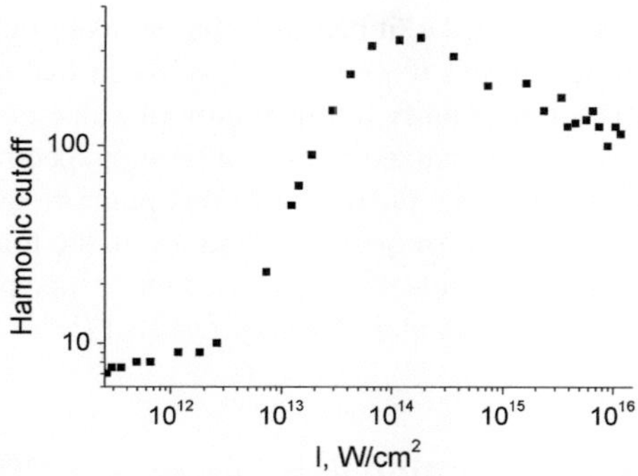

Fig. 1 Cut-off frequency as a function of the laser pulse intensity

It is well known, that the number of discrete states of any hydrogen-like atom is infinite and there is an innumerable number of continuum spectrum states. The developed approach provides the consistent mathematical procedure of a selection of atomic levels playing the main role in evolution of a state of the atom in an external laser field.

3 Numerical Research

In numerical simulations we assume that a single silver atom interacts with the pulse of Ti:Sapphire laser (the wavelength $\lambda = 800$ nm). The pulse duration is assumed as $\tau = 27$ fs. The amplitude of the laser field strength E_0 is varied in the wide range starting from a significantly subatomic to near-atomic values. As we have discussed in our previous work [9] if the laser field strength lies in the interval $I < 1.83 \cdot 10^{16}$ W/cm^2 then the approximately complete basis of states forms the following 284 levels of a silver atom: 5s (GS), 5p, 5d, 6p of the discrete spectrum and 280 states of continuum spectrum with orbital quantum numbers $l = 0$–3.

The spectrum of atomic response field in the far-field zone is proportional to the spectrum of the atomic current, which is given by

$$\vec{j}(t) = \frac{q}{m} \sum_{n,m,p,q} a_n^*(t) a_m(t) V_{np}^{-1}(t) \vec{p}_{pq} V_{qm}(t). \quad (4)$$

Our numerical calculations yield a series of harmonic emission spectra. In subatomic region the spectrum consists of the sequence of the odd harmonics of incident field carrier frequency. When the laser field strength approaches to near-atomic ones, the shape of the response spectrum is changed. There appear tendencies for growing of its width and formation of plateau. Further increase in the amplitude of the external field strength allows us to observe well-defined plateau with the definite cut-off frequency (CF). The results of computer calculations of the CF as a function of the laser pulse intensity is presented in Fig. 1. One can see that in the weak field range ($I < 2.9 \cdot 10^{13}$ W/cm^2) there is the quadratic growth of cut-off frequency with the field strength. However, at laser pulse intensity $I > 10^{14}$ W/cm^2 the CF is saturated, i.e. it ceases to be intensity dependent. The reason of such behavior is quite obvious if we take into account that the probability of electron ionization approaches to unity in this region of pulse intensity. It means that the atomic electron is mostly localized in the continuum spectrum states and it does not collide with its parent ion. So, the results of computer modeling show that the most probable reason of the cutoff frequency saturation is the total ionization of the irradiated atom. The behavior of cut-off frequency dependence presented in Fig. 1 is in a good agreement with experimental observed results [7].

4 Propagation and Dispersion Effects

In order to illustrate qualitatively the difference between the frequency-angular spectra of a single atom response and an ensemble of atoms we shall use further the simplest dispersion model.

In the case when the atoms of an ensemble are identical, the spectrum of the atomic current (4) is

$$\vec{J}(\vec{k},\omega) = \sum_{i=1}^{N} \vec{J}_i(\omega) \exp\left[i\frac{\omega}{c}\big(\vec{n}(\omega) - \vec{n}_0(\omega_0)\big)\vec{r}_i \right] = \vec{J}_0(\omega) f(\vec{k},\vec{k}_0),$$

where $\vec{J}_0(\omega)$ is the spectrum of a single atom current and the form factor $f(\vec{k},\vec{k}_0)$ is defined by

$$f(\vec{k},\vec{k}_0) = \int \rho_\omega(\vec{r}) \exp\left[i\frac{\omega}{c}\big(\vec{n}(\omega) - \vec{n}_0(\omega_0)\big)\vec{r} \right] dV. \tag{5}$$

Here, $\rho_\omega(\vec{r})$ is an effective density of responding atoms, which is the product of density of atoms and the ω/ω_0 power of the amplitude spatial distribution of driving laser pulse in irradiated volume.

According to the Lorentz classical theory the refractive index for electromagnetic wave propagating in ionized gaseous medium is given by

$$n(\omega) \approx 1 - \frac{1}{2}\left[\sum_\alpha \frac{\omega_{p\alpha}^2}{\omega^2 - \omega_{0\alpha}^2 + i\omega\gamma_\alpha} + \frac{\omega_{pe}^2}{\omega^2 + i\omega\gamma_e} \right],$$

where

$$\omega_{p\alpha(e)} = \sqrt{\frac{4\pi e^2 N_{\alpha(e)}}{mV}}$$

is the plasma frequency corresponding to neutral atoms ($\alpha = I$), ions ($\alpha = II$), and free electrons (e); N_α is the number of neutral atoms and ions in the irradiated volume V, and N_e is the number of free electrons.

In the case of the Gaussian incident beam the effective density of responding atoms can be approximated as follows

$$\rho_\omega(\vec{r}) = \frac{N_0}{V}\left[\exp\left(-\frac{z^2}{L^2} - \frac{\rho^2}{d^2} \right) \right]^{\omega/\omega_0},$$

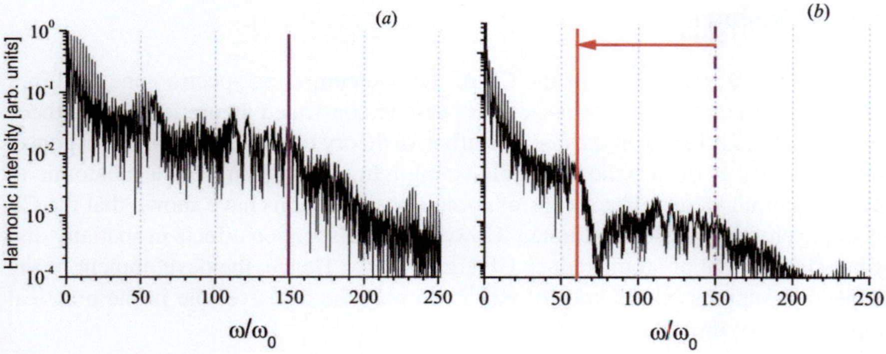

Fig. 2 The comparison of photoemission spectra generated in silver by a single atom (**a**) and an ensemble of atoms (**b**) at the laser pulse intensity $I = 2.37 \cdot 10^{15}$ W/cm^2

Fig. 3 The CF of an ensemble of atoms as a function of a laser pulse intensity

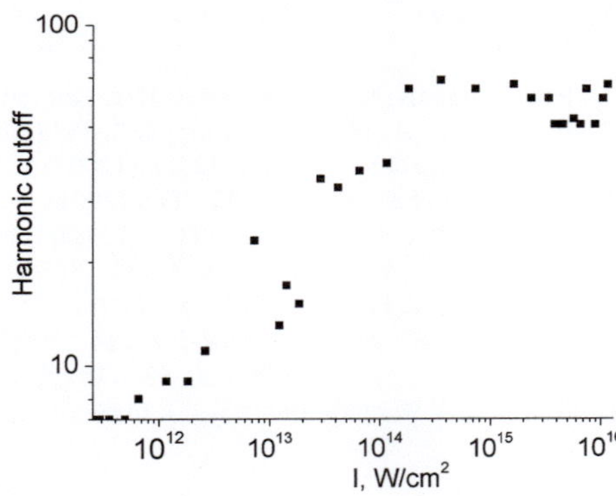

where d is the focal waist of a driving laser beam, and L depends both on the confocal parameter of driving laser beam and a spatial distribution of an atomic density in the laser plume produced by the pump laser pulse. We have assumed that the laser beam is propagated along the z axis.

By taking into account the typical experimental conditions we have calculated the function $|f(\vec{k}, \vec{k}_0)|^2$ defined by Eq. (5). Figure 2 shows in comparison the single atom response spectrum (Fig. 2a) and spatially distributed ensemble of atoms (Fig. 2b). One can see that the CF in atomic ensemble spectrum is significantly shifted in the long wave region in comparison with the CF in spectrum of single atom response. Figure 3 shows the integral dependency of the CF as function of laser pulse intensity which is calculated for spatially distributed ensemble of atoms. In these calculations we have taken into account that the ionization probability depends on the peak pulse intensity [9].

5 Conclusions

In this paper, the saturation of the CF of the photoemission spectra generated in a silver by a femtosecond laser pulses has been demonstrated theoretically. The theoretical analysis is based on the non-perturbative theory of nonlinear atomic response, which enables us to vary the laser field strength in wide region from a subatomic to an over-atomic values. The results of computer simulations have shown that the CF of single atom response is saturated. However, the dispersion effects in spatially distributed ensemble of atoms reduce CF significantly. Hence, the development of the dispersion-suppressed schemes of HHG can play the decisive role in the practical applications of this effect.

Acknowledgements This work was supported by the Russian Foundation for Basic Research, grant nos. 12-02-00969 and 12-02-31843.

References

1. Papadogiannis, N.A., et al.: Phys. Rev. Lett. **90**, 133902 (2003)
2. Itatani, J., et al.: Nature (London) **432**, 867 (2004)
3. Tobey, R.I., et al.: Opt. Lett. **32**, 286 (2007)
4. Paul, P.M., et al.: Science **292**, 1689 (2001)
5. Drescher, M., et al.: Science **291**, 1923 (2001)
6. Corkum, P.B., et al.: Nat. Phys. **3**, 381 (2007)
7. Ganeev, R.A.: Phys. Usp. **52**, 55 (2009) [UFN **179**, 65 (2009)]
8. Andreev, A.V.: Zh. Èksp. Teor. Fiz. **116**, 793 (1999) [JETP **89**, 421 (1999)]
9. Andreev, A.V., et al.: JETP Lett. **93**, 476 (2011)
10. Andreev, A.V., et al.: Eur. Phys. J. D **66** (2012)

Chapter 3
All-Optical Raman XFEL, Based on the Electron Emission in a Transverse High Intensity Optical Lattice

I.A. Andriyash, E. d'Humières, V.T. Tikhonchuk, and P. Balcou

Abstract We present a new scheme of X-ray free electron laser, based on the interaction between a moderately relativistic electron bunch, and a transverse high intensity optical lattice. Gain scaling laws in the small signal regime are derived analytically, and confirmed from Particle-In-Cell simulations. The nature of amplification and saturation are discussed. The resulting all-optical Raman X-ray laser opens perspectives for ultra-compact X-ray coherent light sources.

1 Introduction

Numerous applications of UV/X-ray radiation in science and industry demand compact sources of high-quality coherent light in very short wavelength ranges. A well known solution is represented by X-ray free electron lasers (XFEL), where electrons with relativistic energies scatter X-ray light in a magnetic undulator. Most conventional XFELs are based on the mechanism of stimulated inverse Compton scattering (SICS), and require long amplification lengths and highly mono-energetic, low divergent electron beams. Such restrictions result in large sizes and costs of such devices.

A more compact solution is possible using a laser wave as an "optical wiggler", and reaching a regime of parametric instability called stimulated Raman scattering (SRS). Resonant coupling of electron collective oscillations with the scattered and pump waves provides much higher gain than SICS and this scheme may be applied to the relativistic electron beams obtained via laser wake-field accelerators (LWFA). The challenge here is in the severe limitations on the quality of electron beam [1], which cannot be met in presently existing compact laser accelerating systems [2].

I.A. Andriyash (✉) · E. d'Humières · V.T. Tikhonchuk · P. Balcou
Univ. Bordeaux, CNRS, CEA, CELIA (Centre Lasers Intenses et Applications), 33400 Talence, France
e-mail: igor.andriyash@gmail.com

Present address:
I.A. Andriyash
CELIA, 351, Cours de la Liberation, 33405 Talence, France

S. Sebban et al. (eds.), *X-Ray Lasers 2012*, Springer Proceedings in Physics 147,
DOI 10.1007/978-3-319-00696-3_3,

A new promising scheme, recently proposed in [3], considers overlapping two transverse co-polarized laser pulses of equal frequency, thus obtaining a standing electromagnetic wave, called a *high intensity optical lattice*, with a wave-vector perpendicular to the beam axis [4]. The optical lattice provides a transverse confinement for electrons via the ponderomotive force and at the same time acts as a pump for the scattered radiation. Confinement of electrons prevents beams divergence along lattice wave-vector, and creates a new, controllable collective electron mode defined by the laser field. We investigate amplification in numerical simulations and describe it analytically using a hydrodynamic model of electron plasma interacting with laser and scattered fields.

2 Dynamics of Relativistic Electrons in a High Intensity Optical Lattice

We consider the geometry, where electrons are injected along z-axis into the optical lattice formed by two laser pulses, counter-propagating along x-axis. Laser radiation is considered of sub-relativistic intensity, $I \lesssim 10^{18}$ W/cm^2, so the dimensionless amplitude, $a_0 = 0.85 \cdot 10^{-9} \cdot \lambda_0 \sqrt{I}$ is small, where intensity I is in W/cm^2 and λ_0 is the laser wavelength in microns.

The dynamics of relativistic electrons in electromagnetic field is conveniently described in the reference system moving with an average electron velocity $v_b = c\beta_b \lesssim 1$. In this frame an observer "sees" laser beams as oblique to the beam axis with an angle, defined as $k_{0\perp}/k_{0\parallel} = 1/\gamma_b\beta_b \ll 1$, where $\gamma_b = (1-\beta_b^2)^{-1/2}$ is the beam Lorentz factor. Electron motion in this frame is non-relativistic, and the wavelength of the scattered electromagnetic wave is close to the one of the pump.

Injection of electrons into the lattice may be described as a relatively slow switch-on of the external electromagnetic field given by sum of the laser pulses

$$a_L = a_1 + a_2 = 2a(t)\sin(k_{0\perp}x)\cos(\omega_0 t + k_{0\parallel}z), \tag{1}$$

where the field amplitude grows during the time t_{inj} up to a_0. Electron with transverse velocity, $v_\perp \leq \sqrt{2}\,a_0 c$, may be trapped in the ponderomotive potential of the lattice and will oscillate along the closed phase trajectory $[x(t), p_x(t)]$. Particles with lower initial velocities, are trapped before the end of injection and the further growth of potential adiabatically compresses their phase trajectories thus increasing the velocity amplitude. If the time of injection is much bigger then the typical period of particle oscillation, the beam may be significantly compressed [4]. Electron oscillations in optical lattice potential $U_p(x) = m_e a_0^2 c^2 \sin^2 k_{0\perp}x$, follow elliptic phase trajectories. Approximating electron trajectories with harmonic law $x_e = x_m \cos(\Omega t)$, we may estimate the frequency as,

$$\Omega \simeq \Omega_0 (k_{0\perp}x_m)^{-1} \sin k_{0\perp}x_m, \tag{2}$$

where x_m is the oscillation amplitude and $\Omega_0 = \sqrt{2}\,a_0 k_{0\perp} c$ is a frequency of small-amplitude oscillations.

3 Collective Properties of Electrons, Trapped in a Lattice Potential

We study the collective electron dynamics in a fluid approach, i.e. electron motion follows the hydrodynamic equations, interacting with electromagnetic field via the ponderomotive force. In this simplified model, we neglect electrons electrostatic potential and longitudinal temperature, and consider the transverse temperature not to exceed the trapping limit. Interaction is described with the following set of equations:

$$\partial_t n_e + \nabla \cdot (n_e \mathbf{u}) = 0, \tag{3a}$$

$$\partial_t u_x + \nabla_x P_{xx}/m_e n_e + \nabla_x U_p = 0, \tag{3b}$$

$$\partial_t u_z + \nabla_z U_p = 0, \tag{3c}$$

$$\left((\partial_t - i\omega_0)^2 - c^2\nabla^2\right)\bar{a} = -\left(4\pi e^2/m_e\right)n_e\bar{a}, \tag{3d}$$

where $\mathbf{u}$ is an electron fluid velocity. Enveloped vector potential $\bar{a}$ is defined as $a = \Re[\bar{a}\exp(-i\omega_0 t + ik_0 z)]$ and it varies slowly in time creating the ponderomotive potential $U_p = m_e(c/2)^2|\bar{a}|^2$. Trapped beam follows adiabatic equation of state along the x-direction, so electron pressure reads $P_{xx} = m_e v_\perp^2 n^3/3n_0^2$, where $v_\perp = c\delta\beta_{\perp 0}$ is defined by the transverse velocity spread. Considering equilibrium between lattice potential and electron pressure, we may find the unperturbed transverse distribution of electron density

$$n_e = n_0\sqrt{1-\xi^2},$$

where $\xi = x/L$ is a coordinate normalized to the beam width $L = v_\perp/(\sqrt{2}a_0 k_{0\perp} c)$, and $n_0 = n_e(0)$ is a maximum of the electron density.

In the linear approach, we consider the collective electron modes as a first-order perturbation of the beam position Δ_x, defined as $\xi = (x + \Delta_x)/L$. Without a signal wave, there is no longitudinal force $\partial_z U_p$ in Eq. (3b) and from Eqs. (3a)–(3d) it follows:

$$\partial_t^2 \Delta_x = -\Omega^2 \Delta_x,$$

which describes the collective beam mode with a frequency $\omega_b = \Omega$.

The amplified small-amplitude signal wave a_s interacts with electrons via first-order perturbation of ponderomotive potential $U_p^{(1)} = m_e(c/2)^2\bar{a}_s\bar{a}_L^* + \text{c.c.}$, where c.c. stands for the complex conjugation. Assuming the plane signal wave we neglect diffraction terms in Eqs. (3a)–(3d) and linearize the system in Fourier domain to obtain the dispersion equation:

$$\left(\omega^2 - \Omega^2\right)\left((\omega-\omega_0)^2 - (k_z - k_{0\parallel})^2c^2 - \tilde{\omega}_p^2\right) = G\alpha\omega_0^4, \tag{4}$$

where $\alpha = (\omega_p a_0 k_z c k_{0\perp} L/\omega_0^2)^2$ is a coupling coefficient between electromagnetic and electron beam modes, and coefficient G is due to the coupling inhomogeneity. In case of a plane signal wave, the coupling is $\propto \xi^2\sqrt{1-\xi^2}$, and we may estimate $G \simeq 0.2$.

Amplification of the Stokes mode is described by the complex frequency in Eq. (4) with a real part $\omega \simeq \Omega$ and imaginary part Γ corresponding to an exponential growth rate. Typically, $\alpha \ll (\Omega/\omega_0)^3$, and the coupling of electromagnetic and electron modes is weak, and imaginary part of solution approximately reads:

$$\Gamma/\omega_0 = 0.5\sqrt{G\alpha\omega_0/\Omega}. \tag{5}$$

Such amplification may be saturated by the longitudinal trapping of electrons in the potential $|\bar{a}_s\bar{a}_L^*|$, resulting a full particle bunching $n_e \sim n_0 \exp(-i\Omega t) + \text{c.c.}$ From Eq. (3d), the maximal signal amplitude can be deduced as:

$$\langle a_s \rangle / a_0 \simeq (k_{0\perp} L)\omega_p^2/(\Omega\omega_0). \tag{6}$$

4 Numerical Simulations of SRS

Numerical investigation of SRS in the electron beam frame and verification of Eqs. (5), (6) was done with help of the reduced electromagnetic particle-in-cell 2D code EWOK. The code algorithm is based on a general particle-in-cell approach; it calculates the enveloped vector potential $\hat{a}$ from electromagnetic equation, reduced by neglecting second order terms $\partial_t^2\hat{a}$ and $\partial_z^2\hat{a}$, and Fourier transforming over x-coordinate $\partial_x^2\hat{a} \to -k_x^2\hat{a}_{k_x}$. The code is able to account for electrostatic field of electrons by solving the Poisson equation and it was verified that for the electron densities and energies, typical for LWFA beams, and short propagation distances, for which the space-charge effects on SRS development may be neglected.

For the numerical study we consider 4 kA current of electrons with average energies of 5 MeV; a normalized emittance of 1 mm·mrad; and a transverse size of the beam assumed around few micro-meters. The optical lattice is created by two laser beams at a wavelength of 1 μm, intensity $1.6 \cdot 10^{16}$ W/cm^2 and linear amplitude ramp of 20 μm. A plane seed wave with amplitude $a_s = 10^{-6}$ is injected into the simulation domain at the left boundary and it circulates in the system due to the periodic boundary conditions. The scattered field, measured at the right boundary, is presented in Fig. 1a; it exhibits an exponential growth with a rate $\Gamma/\omega_0 = 1.8 \cdot 10^{-3}$. Amplification saturates after 600 μm at a signal amplitude $a_s = 1.4 \cdot 10^{-4}$, when modulation of electron density results in a full bunching (see Fig. 1b).

The spectral-spatial characteristics of electron-scattered light interaction may be studied by measuring electron density and scattered wave along the x-axis of simulation domain and projecting temporal oscillations to the Fourier domain. Electron collective modes are mainly localized at the frequency $\Omega = \sqrt{2}a_0 k_{0\perp} c = 0.016\omega_0$ and their transverse profile is in a good qualitative agreement with the one predicted in analytic study $\propto \xi^2\sqrt{1-\xi^2}$ (see Fig. 1c). The second harmonic of electron oscillations 2Ω can be also observed on the beam axis, but does not contribute to the amplification. In Fig. 1d, one may see the amplified scattered light at Stokes frequency $\omega_s = \omega_0 - \Omega$, with a quasi-plane profile, slightly modulated due to the electron density distribution.

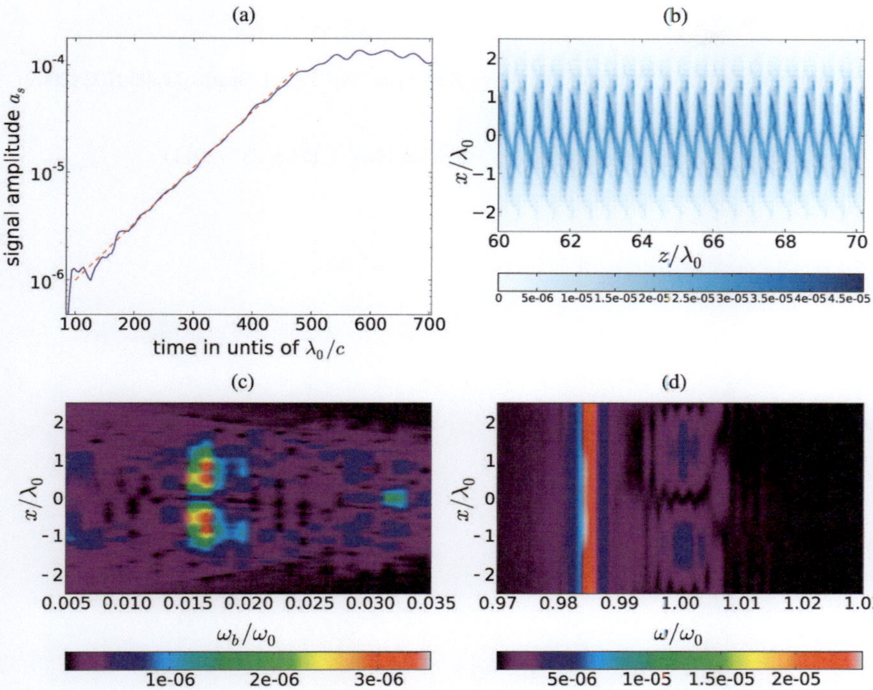

Fig. 1 Evolution of the signal dimensionless amplitude at the right boundary at the beam axis (**a**), fitted exponential growth with rate $\Gamma/\omega_0 = 1.9 \cdot 10^{-3}$ (*dashed line*); electron density distribution at saturation $600\lambda_0/c$ (**b**); spectral-spatial distribution of electron beam (**c**) and electromagnetic (**d**) modes

Growth rate and saturation level, provided by Eqs. (5), (6), read $\Gamma/\omega_0 = 1.7 \cdot 10^{-3}$ and $a_s = 1.4 \cdot 10^{-4}$ and are in perfect agreement with the values observed in the simulations. These estimates may be used for more realistic parameters—laser-accelerated electron beam of 20 pC, with average electron energy 40 MeV interacts with an optical lattice, created by laser light with intensity of $2.5 \cdot 10^{17}$ W/cm^2. For a case of cold electrons amplification proceeds on a millimeter length and results into the generation of a beam of 17 keV photons, with a total energy of the order of 1 μJ.

To conclude, we have demonstrated theoretically a new Raman process of amplification of coherent X-rays in the interaction between a relativistic electron bunch and a high intensity optical lattice. Presented estimates indicate the possibility for experimental demonstration of the process in a compact setup. The Raman XFEL should further studied as a way to create a ultra-compact coherent X-ray sources, up to the hard X-ray range.

References

1. Sprangle, P., Hafizi, B., Peñano, J.R.: Phys. Rev. Spec. Top. Accel. Beams **12**, 050702 (2009)
2. Lundh, O., et al.: Nat. Phys. **7**, 219 (2011)
3. Balcou, P.: Eur. Phys. J. D **59**, 525 (2010)
4. Andriyash, I.A., Balcou, P., Tikhonchuk, V.T.: Eur. Phys. J. D **65**, 533 (2011)

Chapter 4
Optical Transforms Related to Coherent Imaging of Inclined Objects

I.A. Artyukov, A.S. Busarov, N.L. Popov, and A.V. Vinogradov

Abstract The theory of coherent reflection imaging of inclined objects at grazing angles of illumination is developed on the basis of the parabolic wave equation. A new optical transformation connecting image and object field distributions is presented. The codes for simulation the properties of practical optical systems are created. Simulation results for parameters close to an X-ray laser experiment are given.

1 Introduction

The development of laboratory X-ray lasers [1, 2] and accelerator based 3rd and 4th generation light sources [3, 4] makes coherent X-ray beams available to many researches and various applications. As for imaging and microscopy the coherence of novel X-ray sources offers several advantages and opportunities: phase retrieval technique and lensless imaging with wavelength resolution; quantitative, diffraction and interferometric microscopy; reduction of radiation loading on the investigated objects and environment and possibility to capture a diffraction pattern or an image with subpicosecond exposure prior to its degradation etc. The theoretical basis for coherent X-ray imaging, including Fresnel propagation theory and phase retrieval is well developed and routinely used for transparent objects [5, 6]. However many objects are opaque and cannot be studied in the transmission mode. Therefore there is a quest for coherent imaging methods in a reflection mode which should be suitable for grazing incidence illumination of objects as far as we speak about soft and hard X-rays [7–9]. In these papers the efforts have been made to modify the Kirchhoff diffraction integral for application to lensless imaging of obliquely illuminated reflective objects. Another approach based on the parabolic wave equation (PWE)

$$2ik\frac{\partial u}{\partial z} + \frac{\partial^2 u}{\partial x^2} = 0 \tag{1}$$

for slowly varying wave field $u(x, z) = e^{-ikz}E(x, z)$ is given in [10–13]. In the present paper we extend this approach to imaging of inclined objects with an ideal

I.A. Artyukov (✉) · A.S. Busarov · N.L. Popov · A.V. Vinogradov
P.N. Lebedev Physical Institute, 53 Leninsky Prospekt, 119991 Moscow, Russia
e-mail: iart@sci.lebedev.ru

S. Sebban et al. (eds.), *X-Ray Lasers 2012*, Springer Proceedings in Physics 147,
DOI 10.1007/978-3-319-00696-3_4,

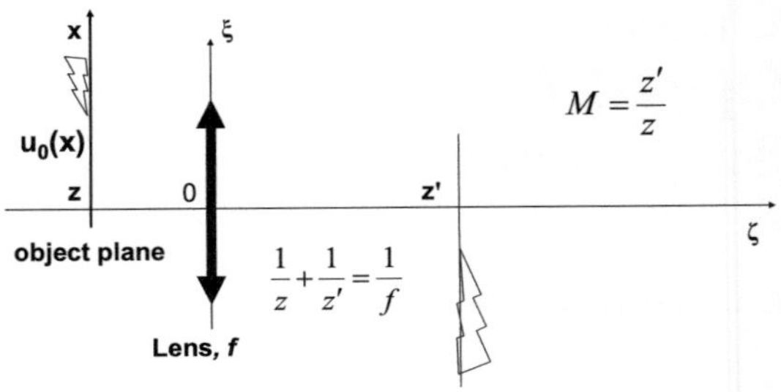

Fig. 1 Coordinate system for paraxial optics Fresnel transform (2)

lens. As a result new optical transformations are obtained. They extend the general relations of Fresnel propagation theory [5, 14] to the case of objects tilted to the axis of an optical system.

2 Imaging Transform in Fresnel Optics

The Fresnel propagation theory gives the following formula for the optical field in the space (ξ, ζ) behind the lens [14] (see Fig. 1) in terms of the field distribution $u_0(x)$ of the object located at distance z from the lens:

$$u(\xi,\zeta) = C(\xi,\zeta)\int_{-\infty}^{\infty} dx \cdot u_0(x) \cdot \exp\left\{\frac{ik}{2}\left[\frac{x^2}{z} - \frac{(\frac{\xi}{\zeta}+\frac{x}{z})^2}{\frac{1}{\zeta}-\frac{1}{z'}}\right]\right\}, \tag{2}$$

where z' being optically conjugated with z, so as $\frac{1}{z}+\frac{1}{z'}=\frac{1}{f}$, f is the focal length of an ideal lens represented as usual by a phase factor $T(x) = \exp\{-\frac{ikx^2}{2f}\}$ and

$$C(\xi,\zeta) = \sqrt{\frac{k}{2\pi i z\zeta}}\frac{\exp\{\frac{ik\xi^2}{2\zeta}\}}{\sqrt{\frac{1}{\zeta}-\frac{1}{z'}}}. \tag{3}$$

Formulas (2), (3) can be obtained by successive application of Fresnel propagation integral to the object field and then to the field distribution on the surface of the lens. Important that the field $u_0(x)$ in (2) is taken in the plane normal to the beam k vector which coincides with the optical axis throughout this paper. For $\zeta \to z'$ formula (2) converges to

$$u(\xi) = \frac{\exp\{\frac{ik\xi^2}{2z'}(\frac{1}{M}+1)\}}{i\sqrt{M}}u_0\left(-\frac{\xi}{M}\right), \quad M = \frac{z'}{z}, \tag{4}$$

i.e. represents (as it should be) the magnified and inversed image of the object $u_0(s)$ [14].

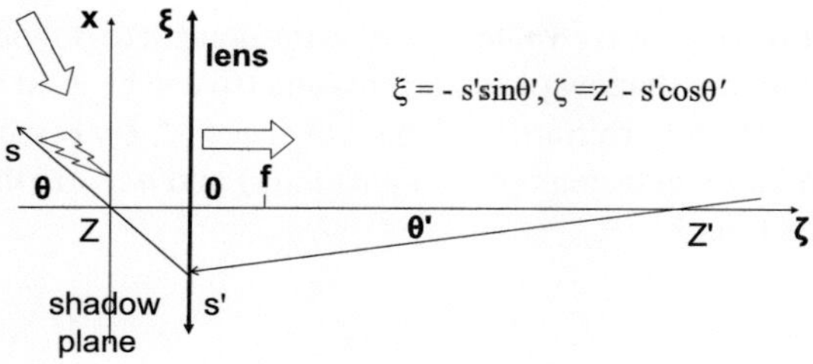

Fig. 2 Coordinate system for paraxial wave propagation and imaging in the case of inclined object (Sects. 2–4)

3 Free Space Propagation from a Tilted Object

Suppose now that an object is inclined to the optical axis at an angle θ (see Fig. 2). Then as was shown in [12, 13] the field in the vertical ("shadow") plane located at the right edge of the object is given by

$$u_0(x) = \sqrt{\frac{k}{2\pi i \cos\theta}}\, x \int_0^\infty \frac{u_0(s)}{s^{\frac{3}{2}}} ds \cdot \exp\left\{\frac{ik(x - s\cdot\sin\theta)^2}{2s\cdot\cos\theta}\right\}, \quad x > 0; \qquad (5)$$
$$u_0(x) = 0, \quad x < 0,$$

s here is the coordinate in the tilted object plane. Formula (5) is the exact solution to PWE (1) for $x > 0$ corresponding to object field $u_0(s)$ and a reasonable approximation to the diffracted field of a restricted ($u_0(s) = 0$ for $s < 0$) object for $x < 0$.[1] The result (5) extends the Fresnel propagation integral to a tilted object, giving the shadow field distribution corresponding to the tilted object $u_0(s)$.

4 The Wave Field of a Tilted Object Behind the Lens

To find the wave field of a tilted object transmitted through the lens one has to install (5) into (2), integrate over x in the shadow plane and obtain:

$$u(\xi,\zeta) = \frac{1}{2\pi}\sqrt{\frac{l(\zeta)}{z\zeta\cos\theta}}\exp\left\{\frac{ik\xi^2}{2\zeta}\left[1 - \frac{l(\zeta)}{\zeta}\right]\right\}$$
$$\times \int_0^\infty \frac{ds\cdot u_0(s)}{s^{\frac{3}{2}}(A + i\gamma)}\exp\left\{\frac{ik}{2}\frac{s\cdot\sin^2\theta}{\cos\theta}\right\}F(\eta)\exp(\eta^2). \qquad (6)$$

$$\frac{1}{l(\zeta)} = \frac{1}{\zeta} - \frac{1}{z'}, \quad A(s) = \frac{1}{z} - \frac{l(\zeta)}{z^2} + \frac{1}{s\cdot\cos\theta} = \frac{z\cdot z' - \zeta(z+z')}{z^2(z'-\zeta)} + \frac{1}{s\cdot\cos\theta},$$

$$\eta(s) = i\sqrt{\frac{ik}{2A(s)}}B, \quad B = \frac{\xi\cdot l(\zeta)}{z\zeta} + \mathrm{tg}\,\theta, \qquad (7)$$

$$F(\eta) = \exp(-\eta^2) - \eta\sqrt{\pi}\left[1 - \Phi(\eta)\right], \quad \Phi(\eta) = \frac{2}{\sqrt{\pi}}\int_0^\eta \exp(-t^2)\cdot dt,$$

[1]The details concerning application of PWE in X-ray optics and other fields can be found in [5, 15, 16].

a small positive value $\gamma \to 0$ is introduced to provide the convergence of the integral over the shadow plane. Expression (6) can be used to find the radiation field entering a detector arbitrarily oriented in space (ξ, ζ). Further in this paper basing on (6) the relation between optical fields $u_0(s)$ and $u(s')$ in the object and detector planes will be found.

5 The Image Field Distribution

The evident question is: what is the optimal location of the detector plane providing a perfect image of a tilted object $u_0(s)$? The intuitive answer is: the best similarity between image and object is expected when the detector plane is optically conjugated with the object plane. This plane is exhibited in Fig. 1 by a ray traveling along the tilted object and refracted in the lens. The following relation is valid if object and detector planes are optically conjugated:

$$\xi = (\zeta - z')\,\mathrm{tg}\,\theta', \quad \text{where } \mathrm{tg}\,\theta' = \frac{z}{z'}\,\mathrm{tg}\,\theta, \text{ and}$$
$$\xi = -s' \cdot \sin\theta', \ \zeta = z' - s' \cdot \cos\theta'. \tag{8}$$

Taking into account (8) and (7) we observe that in the detector plane:

$$B = 0, \quad \eta = 0 \quad \text{and} \quad A(s) = \frac{1}{\cos\theta}\left(\frac{1}{s} - \frac{1}{s^*(s')}\right), \tag{9}$$

where

$$s^*(s') = \frac{zs' \cdot \sin\theta'}{z' \cdot \sin\theta - s' \cdot \sin(\theta + \theta')} \tag{10}$$

is the counter image of s'.

With (8)–(10) formula (6) for the field of the image is considerably simplified and results in

$$u(s') = -\frac{s^*}{2\pi}\sqrt{\frac{M\cos\theta}{s'\cos\theta'}} \cdot \exp\left\{-\frac{iks'\sin^2\theta'}{2\cos\theta'}\right\}$$
$$\times \int_0^\infty \frac{ds}{\sqrt{s}}\frac{u_0(s)}{s - s^* - i\gamma}\exp\left\{\frac{iks \cdot \sin^2\theta}{2\cos\theta}\right\}. \tag{11}$$

The values θ' and s^* here are determined by the geometrical parameters of the system according to (8) and (10).

Formula (11) gives the optical field $u(s')$ at the detector plane in terms of the field distribution $u_0(s)$ at the object. It is the direct extension of Fresnel paraxial formula (4) to the case of a tilted object. The qualitative difference between (4) and (11) is that for tilted object in general case the proportionality (similarity) between the image ($|u(s')|^2$) and object ($|u_0(s)|^2$) does not exist. Of course for $\theta \to \pi/2$ formula

(11) transforms into (4). To be convinced, note that for $\theta \to \pi/2$ the exponential factor in the integrand in (11) is large and the following relation holds:

$$\lim_{t\to\infty} \frac{e^{itx}}{x - i\gamma} = 2i\pi\delta(x). \tag{12}$$

Then formula (11) is reduced to

$$u(s') = \frac{1}{i}\sqrt{\frac{Ms^*\cos\theta}{s'\cos\theta'}} \cdot \exp\left\{\frac{iks'^2 \operatorname{tg}\theta \sin(\theta+\theta')}{2M[z'\sin\theta - s'\sin(\theta+\theta')]}\right\} u_0(s^*). \tag{13}$$

The common features of (4) and (13) are: (a) the object intensity distribution $|u_0(s)|^2$ is reproduced with some scaling in the detector plane and (b) the phase factor depends only on the optical scheme geometry and does not depend on the structure of the object $u_0(s)$. These features do not appear for general formula (11). The dimensionless parameter responsible for transition from (11) to (13) is $\Phi = \frac{2\lambda\cos\theta}{a\sin^2\theta}$ (a is the object feature size), introduced earlier in [11]. So for $\Phi \ll 1$, that means normally illuminated object, $\theta \approx \pi/2$, and wavelength $\lambda = 2\pi/k$ small as compared to the feature size a, we arrive from general wave optics expression (11) to formula (13) representing the wave field which is characteristic for paraxial geometrical optics. It is easy to check that for $\theta \approx \pi/2$ formula (13) transforms exactly into (4). In the opposite case $\Phi \gg 1$, small feature size a and low tilt angle $\theta \to 0$, the image field (11) takes the form

$$u(s') = -\frac{s^*}{2\pi}\sqrt{\frac{M\cos\theta}{s'\cos\theta'}} \cdot \exp\left\{-\frac{iks'\sin^2\theta'}{2\cos\theta'}\right\}$$
$$\times \left[i\pi\frac{u_0(s^*)}{\sqrt{s^*}} + \int_0^\infty \frac{u_0(s)\,ds}{\sqrt{s}} P\frac{1}{s-s^*}\right], \tag{14}$$

consisting of two parts. The first, local additive, reproduces the shape of the object field $u_0(s)$ and the second, integral additive, introduces the blurring of the ideal image. So it clearly demonstrates that tilting of an object not only distorts (formula (13)) but also blurs the image.

6 Simulation Results

To illustrate the obtained results the calculations have been made for an idealized optical scheme containing a zone plate with the focal length $f = 0.1$ cm and other parameters close to those achieved for laboratory X-ray lasers: wavelength $\lambda = 10$ nm, $z = z' = 0.2$ cm (one to one imaging). Numerical aperture was taken as large as $NA = 0.5$ in order to separate and investigate the influence of the object tilting on the image quality. The calculated images of a table of figures in the planes optically conjugated to the tilted object are shown in Fig. 3 for grazing angles $\theta = 0.1$ and $\theta = 0.01$. The sizes of the original table are 460 μ × 325 μ. The size of a number is proportional to its value. The large inlets are zoomed versions of the corresponding boxes in the left upper corners of the pictures. It can be observed that for $\theta = 0.1$

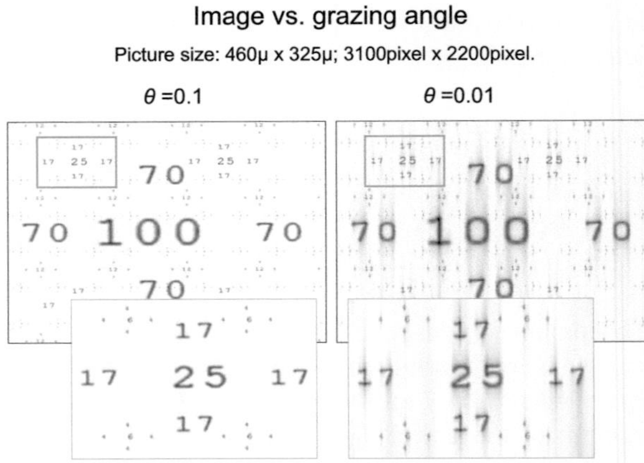

Fig. 3 Coordinate calculated images of a table inclined at grazing angles 0.1 and 0.01 rad. The wavelength $\lambda = 10$ nm, the parameters of the optical system are: $f = 0.1$ cm, $NA = 0.5$, $z = z' = 0.2$ cm (see Fig. 2 and Sect. 5) in the case of inclined objects (Sects. 2–4)

the image of a tilted table in optically conjugated plane is rather perfect, whereas for $\theta = 0.01$ some blurring appears. Similar effect was observed in calculated images of the text pages. The results of such calculations can be used for simulation and evaluation of spatial resolution and field of view of practical optical systems.

7 Discussion

Consider the conditions for the applicability of our approach. The use of PWE implies the following approximation to the plane wave solution of the wave equation [15, 16]:

$$E(x, z) = e^{iqx + i\sqrt{k^2 - q^2}z} \approx e^{ikz}u(x, z), \quad u(x, z) = \exp\left\{iqx - i\frac{q^2}{2k}z\right\}, \quad q \ll k. \tag{15}$$

This wave corresponds to the space harmonic $u_0(s) = e^{i\nu s}$ in the tilted object plane where $\nu = 2\pi/a$ can be found by taking $x = s\sin\theta$ and $z = -s\cos\theta$ in (15):

$$\nu = q\sin\theta + \frac{q^2}{2k}\cos\theta, \quad \text{or} \quad q = \frac{\nu}{\sin\theta}\Psi(\Phi), \tag{16}$$

where

$$\Psi(\Phi) = \frac{2}{\Phi}(\sqrt{1+\Phi} - 1) = \begin{cases} 1, & \Phi \ll 1 \\ \frac{2}{\sqrt{\Phi}}, & \Phi \gg 1 \end{cases} \tag{17}$$

and $\Phi = \frac{2\nu\cos\theta}{k\sin^2\theta}$ is the same parameter mentioned in the end of the previous section and introduced in [11]. Formulas (15)–(17) can be used to find the divergence of the beam diffracted by a tilted object:

$$\Delta\theta = \frac{q}{k} = (\sqrt{1+\Phi} - 1)\,\mathrm{tg}\,\theta = \begin{cases} \frac{\lambda}{a\sin\theta}, & \Phi \ll 1, \\ \sqrt{\Phi}\,\mathrm{tg}\,\theta, & \Phi \gg 1, \end{cases} \tag{18}$$

where $\lambda = 2\pi/k$ is the wavelength and $a = 2\pi/\nu$—the feature size in the object plane. For $\theta = \pi/2$ formula (18) gives the result of the Fresnel propagation theory: $\Delta\theta = \lambda/a$. For $\Phi \ll 1$ it gives the intuitively evident estimate for the case of a tilted object: $\Delta\theta = \lambda a \sin\theta$. And for $\Phi \gg 1$ it leads to a new result valid for objects tilted at very low angles.

The evident condition for using PWE is low beam divergence: $\Delta\theta = q/k \ll 1$ (see (15)). To investigate it in more details introduce a small value $\alpha \ll 1$ to write the applicability condition as:

$$\frac{q}{k} < \alpha. \tag{19}$$

Taking q from (16) one obtains from here the following requirement for the validity of PWE approach:

$$\frac{a}{\lambda} > \frac{2}{\alpha^2 \cos\theta + 2\alpha \sin\theta}. \tag{20}$$

Combining (20) with the condition $\Phi \gg 1$ gives the range of grazing angles and feature sizes where the diffraction patterns of a tilted object are expected to considerably differ from those obtained at coherent normal incidence illumination:

$$\frac{2\cos\theta}{\sin^2\theta} > \frac{a}{\lambda} > \frac{2}{\alpha^2 \cos\theta + 2\alpha \sin\theta}. \tag{21}$$

Inequality (20) as well as right-hand side of (21) are important as they express the applicability condition of PWE for an arbitrary observation angle θ which we consider to be specular relative to the angle of object illumination. As it is seen from (21) it changes from $a > \lambda/\alpha$ for $\theta = \pi/2$—Fresnel propagation theory, to $a > 2\lambda/\alpha^2$—the imaging at very low angles $\theta \to 0$. Finally apply (15)–(17) to investigate the spatial resolution for paraxial (in the frames of validity of PWE) imaging of inclined objects. For this purpose $\Delta\theta = q/k$ in (18) can be considered as the numerical aperture *NA* of an optical system used for imaging. Then one obtains the spatial resolution δ in the form:

$$\delta \approx \frac{2\lambda}{(NA)^2 \cos\theta + 2NA \sin\theta}, \tag{22}$$

which for $\theta = \pi/2$ reduces to the well-known criteria $\delta \approx \lambda/NA$. It is worth noting that estimations of this section are valid to the order of magnitude especially taking into account that unlike the Fresnel theory according to (15), (16) the diffraction diagram for a tilted object is asymmetrical and shifted away from the object surface even for a symmetrical feature. Also formula (22) implies that the apertures of the diffracted beam (corresponding to the feature size a) and that of the optical system are matched and therefore all the necessary information is collected at the detector. However the similarity between an object and image does not necessarily occur. Even in the ideal case their field distributions are related by the general transform (11), which provides reproduction of the object only for $\Phi \ll 1$.

8 Conclusions

The developed theory and obtained simulation results reveal the following features of coherent reflection imaging at grazing angles:

1. The coherent image of an inclined object appears in an optically conjugated surface.
2. The image quality depends on the value of $\Phi = 2\lambda \cos\theta/(a \sin^2\theta)$, a—is feature size, θ—grazing angle.
3. Calculations showed that image-to-object similarity keeps up to very small grazing angles.
4. The method can be extended to practical optical schemes. Gratings or crystals have to be added for tilting of the wavefront.

Acknowledgements The authors are grateful to A.V. Kraiskii, S.I. Mikhailov and I.G. Zubarev for fruitful discussions. The work was supported in part by RFBR Project # 10-02-00991-a.

References

1. Suckewer, S., Jaeglé, P.: X-Ray laser: past, present, and future. Laser Phys. Lett. **6**(6), 411–436 (2009)
2. Berrill, M., Alessi, S., Wang, Y., Domingue, S., Martz, D., Luther, B., Liu, Y., Rocca, J.: Improved beam characteristics of solid-target soft x-ray laser amplifiers by injection-seeding with high harmonics. Opt. Lett. **35**, 2317 (2010)
3. Schmuser, P., Dohlus, M., Rossbach, J.: Ultraviolet and Soft X-Ray Free-Electron Lasers. Springer Tracts in Modern Physics, vol. 229. Springer, Berlin (2008)
4. Kim, K.J., Shvyd'ko, Yu.V., Lindberg, R.R.: An X-ray free-electron laser oscillator for record high spectral purity, brightness, and stability. Synchrotron Radiat. News **25**(1), 25–31 (2012)
5. Paganin, D.M.: Coherent X-ray Optic. Clarendon Press, Oxford (2005)
6. Nugent, K.: Coherent methods in the X-ray sciences. Adv. Phys. **59**(1), 1–99 (2010)
7. Fenter, F., Park, C., Kohli, V., Zhang, Z.: Image contrast in X-ray reflection interface microscopy: comparison of data with model calculations and simulations. J. Synchrotron Radiat. **15**, 558–571 (2008)
8. Marathe, S., Kim, S.S., Kim, S.N., Kim, C., Kang, H.C., Nickles, P.V., Noh, D.Y.: Coherent diffraction surface imaging in reflection geometry. Opt. Express **18**, 7253–7262 (2010)
9. Roy, S., Parks, D., Seu, K.A., Su, R., Turner, J.J., Chao, W., Anderson, E.H., Cabrini, S., Kevan, S.D.: Lensless X-ray imaging in reflection geometry. Nat. Photonics **5**, 243–245 (2011)
10. Artyukov, I.A., Popov, A.V., Vinogradov, A.V.: Wave field transformation at coherent imaging of a flat reflection mask. In: Proc. SPIE, art. id 745114 (2009)
11. Artyukov, I.A., Mitrofanov, A.N., Popov, A.V., Vinogradov, A.V.: Theory and computation towards coherent reflection imaging of tilted objects. In: X-ray Lasers 2010: Proceedings of the 12th International Conference on X-ray Lasers, pp. 329–340. Springer, Berlin (2010)
12. Artyukov, I.A., Mikhailov, S.I., Popov, N.L., Vinogradov, A.V.: Theoretical study of coherent reflection imaging at grazing angles. In: Proc. SPIE, vol. 8140 (2011)
13. Artyukov, I.A., Vinogradov, A.V., Popov, N.L., Seleznev, V.N.: Simulation of grazing-incidence coherent imaging. Quantum Electron. **42**(2), 140–142 (2012)
14. Papoulis, A.: Systems and Transforms with Applications in Optics. McGraw-Hill, New York (1968)

15. Vinogradov, A.V., Popov, A.V., Kopylov, Yu.V., Kurokhtin, A.N.: Numerical Simulation of X-Ray Diffractive Optics. A&B Publishing House, Moscow (1999)
16. Mireille, L.: Parabolic Equation Methods for Electromagnetic Wave Propagation. Institution of Electrical Engineers, Michel Faraday House, London (2000)

Chapter 5
Soft-X-Ray Lasing down to 6.85 nm in Ni-Like Samarium

J.E. Balmer, F. Staub, and F. Jia

Abstract We report on recent progress achieved in X-ray laser research at the Institute of Applied Physics of the University of Bern. Using the 10-TW Nd:glass CPA (chirped-pulse amplification) laser system and the grazing-incidence pumping (GRIP) scheme, intense soft-X-ray lasing has been obtained on the 4d $\rightarrow$ 4p, $J = 0$–1 line of samarium (Sm, $Z = 62$) at 7.36 nm, with weak lasing observed at 6.85 nm. This was achieved with main pulse energies of 10–12 J and a pulse duration of 1.5 ps. Crucial to these results was the introduction of a second, relatively intense (22.7 %) prepulse $\sim$30 ps before the main pulse, in addition to the 13.4 % prepulse incident on target several nanoseconds before the main pulse.

1 Introduction

The recent introduction of the grazing-incidence pumping (GRIP) scheme has greatly improved the efficiency of soft-X-ray lasers operating in the 10–20 nm region. Given the dramatic reduction in pump energy required—substantially less than 1 J, for example, in the case of the 18.9-nm laser of Ni-like molybdenum—the repetition rates of these lasers have increased from single shot to the 10-Hz regime [1–3]. With pump energies required for saturated lasing as low as 2–3 J, the 11.9- and 11.4-nm lasers in Ni-like Sn and Sb demonstrated in single-shot experiments are potential candidates for operation at few-Hz repetition rates [4, 5], as is the 10.9-nm laser in Ni-like Te pumped with a total energy on target of less than 4 J at 1-Hz repetition rate [6]. When attempting to move towards shorter wavelengths, one is confronted with the strong wavelength scaling of pump energy and optimum density for gain. Taking the 4d $\rightarrow$ 4p lasing lines as given, shorter wavelength goes with higher atomic number, Z. As the atomic number gets larger, both the optimum density and electron temperature increase. The strong wavelength scaling of the electron temperature leads to a rapid increase in pump energy required. Simulations [7] predict that saturated sub-10-nm lasing should be feasible with 5–10 J

J.E. Balmer (✉) · F. Staub · F. Jia
Institute of Applied Physics, Sidlerstrasse 5, 3012 Bern, Switzerland
e-mail: balmer@iap.unibe.ch

S. Sebban et al. (eds.), *X-Ray Lasers 2012*, Springer Proceedings in Physics 147,
DOI 10.1007/978-3-319-00696-3_5,

of pump energy, and ~50 J would be required in the GRIP scheme to generate an X-ray laser in the water window (2.5–4.4 nm). The optimum density has been predicted to increase from $\sim 2 \times 10^{20}$ cm^{-3} for Ni-like silver ($Z = 47$, $\lambda = 13.9$ nm) to $\sim 10^{21}$ cm^{-3} for Ni-like samarium ($Z = 62$, $\lambda = 7.36$ nm) [8]. The latter equals the critical density for Nd:glass lasers, with the consequence that the optimum density for gain becomes higher than the critical density for Nd:glass (and Ti:Sapphire) driver lasers when moving towards still shorter wavelengths. The maximum is however relatively broad, so that reasonable gain is expected at densities within a factor of ~2× of the optimum. With regard to the GRIP scheme, the optimum angle, given by $n_e = n_c \sin^2\theta$, where n_c is the critical density for the pump wavelength, approaches 90°, i.e., normal incidence, as the optimum density approaches n_c. Obviously, in this case, the main advantages of grazing-incidence pumping such as enhanced absorption due to increased path length in the plasma are lost and one is left with the standard transient-collisional-excitation (TCE) scheme.

In this work, we report on X-ray lasing close to saturation on the 4d → 4p, $J = 0$–1 line of Sm at a wavelength of 7.36 nm, using the GRIP scheme at an angle of 50°. This was achieved with a main pumping pulse energy of 11 J in a 1.5-ps pulse, applying a triple-pulse scheme in which a first (few-percent) prepulse, preceding the main pulse by several nanoseconds, is followed by a second, relatively intense (22.7 %) prepulse 26 ps before the main pulse. As in most of our previous work, the prepulses are collinear with the main pulse, compressed to 1.5 ps, and irradiate the target at the GRIP angle.

2 Experimental Setup

The experiments were conducted using the existing 1054-nm Nd:glass CPA laser system (BeAGLE facility) that is capable of providing a maximum output energy of ~20 J in a 1.5-ps duration pulse at a rate of 1 shot every 25 minutes. The maximum energy of the system is limited by the damage threshold of the 190 × 350 mm^2 size, 1740-lines/mm compressor gratings, given as 250 mJ/cm^2 by the manufacturer. Prepulses are generated as in our previous experiments by introducing beamsplitters with reflectivities of 0.5, 2.8, 4.5, 8, or 16 % into the double-passed beam path of the final amplifier (see Fig. 1). This scheme has the advantage that the prepulses are propagating along the same path as the main pulse, thus minimizing the variations in pulse overlap at the line focus. The drawback is that this scheme generates a series of post-pulses that, although not influencing the X-ray laser output itself, represent some loss in overall efficiency. If we denote by E_0 the pulse energy entering the prepulse generator equipped with two beamsplitters having reflectivities R_1 and R_2, respectively, the relevant pulse sequence at the line focus consists of three pulses having energies of $E_1 = R_1E_0$, $E_2 = R_2(T_1)^2E_0$, and $E_M = (T_1T_2)^2E_0$, where E_M denotes the energy of the main pulse and $T_i = 1 - R_i$. As an example, the two beamsplitters having reflectivities of 8 % and 16 % used in the experiments described below, generate prepulses of 13.4 % and 22.7 % of the main pulse energy, respectively.

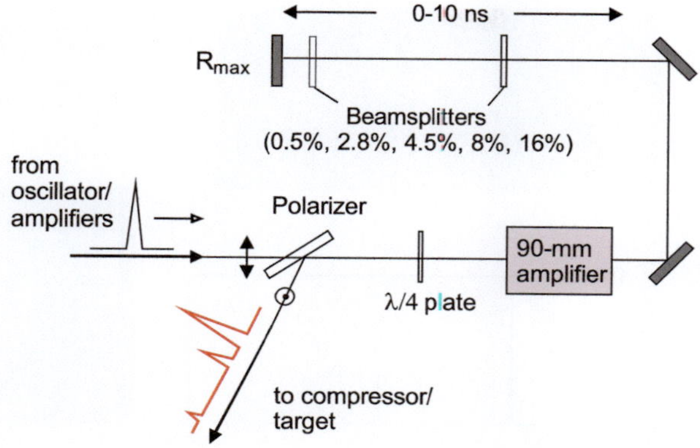

Fig. 1 Schematic of the prepulse scheme used in the present experiments

The beam containing the main pulse as well as the prepulses is expanded to 130 mm in the last spatial filter telescope, compressed to 1.5 ps, and focused by a combination of an f = 80 cm doublet and an f = 30.4 cm (12″) spherical mirror to produce a 12-mm long and 50-μm wide line focus (FWHM, measured at low power) at a GRIP angle of 50° (see Fig. 2). This configuration results in a traveling-wave velocity of $v_{exc} = 1.41c$ along the target, and thus a temporal mismatch of $\Delta t \approx 12$ ps between the pump pulse and the X-ray laser pulse at the output end of the target. A slight tilt (0.09°) and shift (11 mm) of the second compressor grating is used to reduce the traveling-wave velocity to $v_{exc} = c$ [9].

The main diagnostics of the soft X-ray emission from the plasma is an on-axis, time-integrating XUV spectrometer that consists of a 1200-lines/mm, aberration-corrected Hitachi grating (radius of curvature: 5649 mm), working at a grazing-incidence angle of 3°. The grating disperses the incident radiation onto a 40-mm diameter P20 phosphor screen, which is imaged to a cooled CCD camera having a pixel size of 23 × 23 μm^2. The wavelength coverage of the spectrometer is

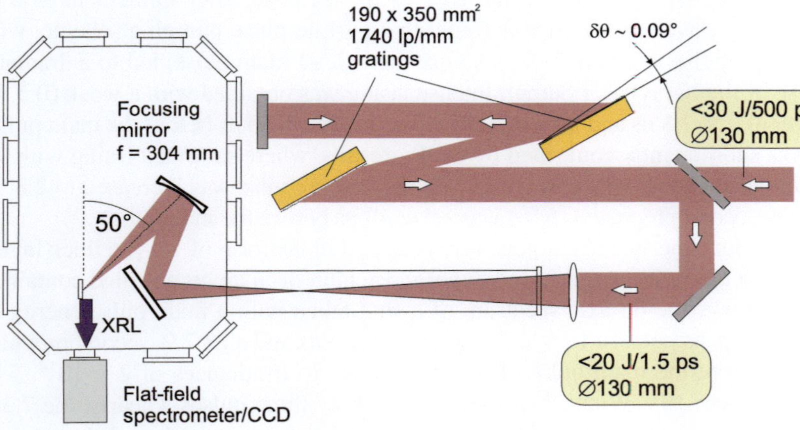

Fig. 2 Schematic of the focusing scheme used in the present experiments

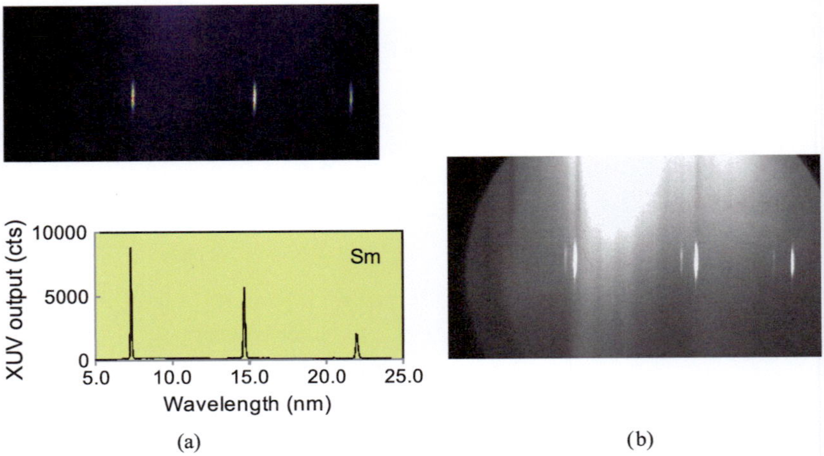

Fig. 3 (**a**) On-axis spectrum showing the 7.36-nm, 4d → 4p, J = 0–1 lasing line of Ni-like Sm in first, second, and third order for a pump energy of 12.6 J in a 1.5-ps pulse, (**b**) overexposed spectrum showing the short-wavelength 4d(3/2, 3/2) → 4p(3/2, 1/2), J = 0–1 lasing line at 6.85 nm

~5–25 nm with a spectral resolution of ~0.2 nm. The relatively poor resolution is a consequence of the slitless operation of the spectrometer, which results in a measured linewidth being dominated by source broadening.

3 Results and Discussion

The experiments on Ni-like samarium lasing were guided by our previous results demonstrating saturated lasing in Ni-like Sn and Ba at wavelengths of 11.9 and 9.2 nm, respectively [4, 10]. In the case of the Sn laser, after implementation of the traveling-wave scheme, it was found that a triple-pulse pumping scheme, with a second prepulse less than 100 ps before the ~2.5-J main pulse, led to a dramatic increase of the X-ray laser output. Intense lasing was obtained with a weak (0.5 %) first prepulse at 2.8 ns and a second, 16 % prepulse at ~50 ps before the main pulse. This was subsequently confirmed by the Ba results, where saturated lasing with 9 J of main pulse energy was observed when the first prepulse was increased to 2.8 %, while the second prepulse (16 %) was set at 43 ps before the main pulse.

For the Sm experiments, targets were prepared in the form of ~1-μm thick layers coated onto glass slides and stored either in vacuum or in an argon-filled container. Figure 3 shows the on-axis spectrum of a shot taken with a main pulse energy of 12.6 J in a 1.5-ps pulse, a 13.4 % first prepulse at 3 ns and a 22.7 % second prepulse at 26 ps before the main pulse. This corresponds to irradiances of 2×10^{14}, 3×10^{14}, and 1.4×10^{15} W/cm^{-2}, respectively, for the three pulses. Clearly, the 7.36-nm lasing line is seen to dominate the spectrum in first, second, and third order. According to the previous calibration of our diagnostics, the output energy of this

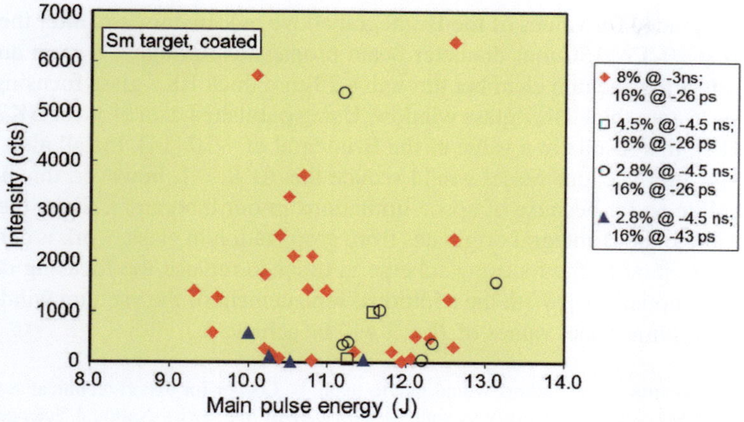

Fig. 4 7.36-nm line intensity of Ni-like Sm vs. main pulse pumping energy, showing large output variation

shot is 3–4 μJ. Comparison with the Ni-like Sn and Ba data, taking into account the scaling of the saturation irradiance as a function of atomic number [11] implies that this is very close to or even exceeding saturation.

Starting from the prepulse configuration of the 9.2-nm Ba laser discussed above, a series of shots was taken to optimize the 7.36-nm Sm laser output for a range of the amplitudes and delays of the two prepulses. Due to the low repetition rate of the driver laser system, only a limited scan in this multi-dimensional parameter space (delay and amplitude of the two prepulses, energy of the main pulse) was practically possible. Figure 4 shows the measured 7.36-nm laser intensity as a function of the main pulse energy for various time delays of the main pulse relative to the peaks of the first and second prepulse. It is seen that lasing is achieved for main pulse energies as low as ~9 J, but that very large output variations occur even under nominally identical pumping conditions.

Output variations may be caused, for example, by spurious prepulses generated in the amplifier chain, well-known in the case of regenerative amplifiers such as the one used in our system. Indeed, the prepulse monitor, a fast photodiode (risetime 0.7 ns) coupled to a 1-GHz oscilloscope, reveals a prepulse of 5×10^{-6} of the main pulse energy at ~6 ns before the arrival of the main pulse. However, since this pulse is at a very low level and observed on all the shots, it may be excluded as the main cause for the output variations. More likely, on the other hand, are main pulse irradiance variations in the line focus caused by excessive values of the B-integral [12] of the beam exiting the compressor. The B-integral is a measure of the nonlinear phase shift accumulated during propagation, defined as

$$B = \frac{2\pi}{\lambda} \int \gamma I(z)\, dz$$

where λ denotes the wavelength, γ the nonlinear index of refraction in SI units and I the irradiance. As has been shown in [13], the quality of the (line) focus becomes severely degraded for values of the B-integral above ∼4. In our case, after the compressor, the 10-TW/130-mm diameter beam propagates through ∼5 m in ambient air and enters the vacuum chamber through a 28-mm thick BK7 glass focusing doublet and a 25-mm thick BK7 glass window. Using published data of γ for BK7 glass and air [12], this results in a value of the B-integral of ∼10 [14]. Installation of the compressor in a vacuum vessel would reduce this to $B < 1$, however, this has not been possible so far because of space limitations in our laboratory. Since the main contribution to the B-integral originates from propagation in glass, work is currently under way to modify the focusing scheme in order to replace the focusing doublet by reflective optics [14]. With the additional replacement of the vacuum window by a somewhat thinner one, values of $B < 3$ will be achieved.

Acknowledgements The authors would like to thank B. Locher for expert technical assistance and target preparation. This work was supported in part by the Swiss National Science Foundation. F. Staub was supported by the Swiss National Science Foundation under D. Bleiner's Förderungsprofessur (grant PP00P2-133564/1).

References

1. Keenan, R., et al.: High-repetition-rate grazing-incidence pumped X-ray laser operating at 18.9 nm. Phys. Rev. Lett. **94**, 103901 (2005)
2. Larotonda, M.A., et al.: Characteristics of a saturated 18.9-nm tabletop laser operating at 5-Hz repetition rate. IEEE J. Sel. Top. Quantum Electron. **10**, 1363 (2004)
3. Rocca, J.J., et al.: Saturated 13.2-nm high-repetition-rate laser in nickel-like cadmium. Opt. Lett. **30**, 2581 (2005)
4. Grünig, M., et al.: Saturated x-ray lasing in Ni-like Sn at 11.9 nm using the GRIP scheme. Opt. Commun. **282**, 267 (2009)
5. Imesch, C., et al.: Gain-saturated Ni-like antimony laser at 11.4 nm in grazing-incidence pumping geometry. Opt. Commun. **283**, 66 (2010)
6. Alessi, D., et al.: Gain-saturated 10.9-nm tabletop laser operating at 1 Hz repetition rate. Opt. Lett. **35**, 414 (2010)
7. Shlyaptsev, V.N., et al.: Numerical study of transient and capillary x-ray lasers and their applications. In: Proc. 9th Int. Conf. X-Ray Lasers. IOP Conf. Ser., vol. 186, p. 325 (2005)
8. Pert, G.J.: Optimizing the performance of nickel-like collisionally pumped x-ray lasers. Phys. Rev. A **73**, 033809 (2006)
9. Chanteloup, J.-C., et al.: Pulse-front control of 15-TW pulses with a tilted compressor and application to the subpicosecond traveling-wave pumping of a soft-x-ray laser. J. Opt. Soc. Am. B **17**, 151 (2000)
10. Staub, F., et al.: Soft-x-ray lasing in nickel-like barium at 9.2 nm using the grazing-incidence scheme. Opt. Commun. **285**, 2118 (2012)
11. Sasaki, A., et al.: The gain distribution of the transient collisional excited X-ray lasers. J. Quant. Spectrosc. Radiat. Transf. **71**, 665 (2001)
12. Brown, D.C.: High-Peak-Power Nd: Glass Laser Systems. Springer, Berlin (1981)
13. Seznec, S., et al.: Towards the 10^{19}–10^{20} W/cm^{-2} regime with amplified chirped pulses: basic limitations and solutions. Opt. Commun. **87**, 331 (1992)
14. Staub, F., et al.: Line-focus generation for X-ray laser pumping. In: Sebban, S., et al. (eds.) X-Ray Lasers 2012. Springer Proceedings in Physics, vol. 147. Springer, Cham (2013). Chapter 38 in this book

Chapter 6
Fourier Optics Study of Traveling-Wave Excitation at Short-Wavelength Plasma-Lasing

Davide Bleiner and Thomas Feurer

Abstract Traveling-wave excitation close to the speed of light implies small-angle target-irradiation. Yet, short-wavelength lasing needs large irradiation angles. Pulse-front back-tilt is considered to overcome such trade-off. Pulse-front tilt by means of compressor misalignment was found effective only if coupled with a strong front-end imaging/focusing component.

1 Introduction

Laser-induced plasma-columns serve as gain media for short-wavelength coherent emission. In traveling-wave excitation (TWE) one matches the propagation of the amplified spontaneous emission (ASE) along the plasma column to the speed of light c. For a given irradiation angle γ the nominal TWE speed is obtained as $v_{TWE} = c/\cos\gamma$. However, for *short-wavelength* amplification the pump pulse must be deposited deeper into the plasma, where higher electron density is found. For this purpose target irradiation must be at large angle, as the optical penetration scales as $n_c \sin^2\gamma$, where n_c is the critical density. The latter requirement of *large* angles poses a trade-off with TWE close to $1c$ that requires *small* angle irradiation.

In chirped-pulse amplification (CPA) a stretched-pulse is amplified and then re-compressed in a grating pair (compressor) [1–4]. The two compressor gratings must be perfectly parallel to achieve optimum recompression over the entire beam diameter. In this work the grating pair is intentionally misaligned, in order to tune a controlled amount of pulse-front *back*-tilt. Thus, for large irradiation angles one could in principle still have the TWE speed close to c as the *pulse-front* gives the effective TWE speed. Aim of this work was to investigate quantitatively such technique. A one-dimensional Fourier optics code was written for computing the parametric dependence between the compressor misalignment and the pulse characteristics. Experimental data on the pulse structure and TWE speed were available and used for comparison.

D. Bleiner (✉) · T. Feurer
Institute of Applied Physics, University of Bern, Sidlerstrasse 5, Bern, 3012, Switzerland
e-mail: bleiner@iap.unibe.ch

S. Sebban et al. (eds.), *X-Ray Lasers 2012*, Springer Proceedings in Physics 147,
DOI 10.1007/978-3-319-00696-3_6,

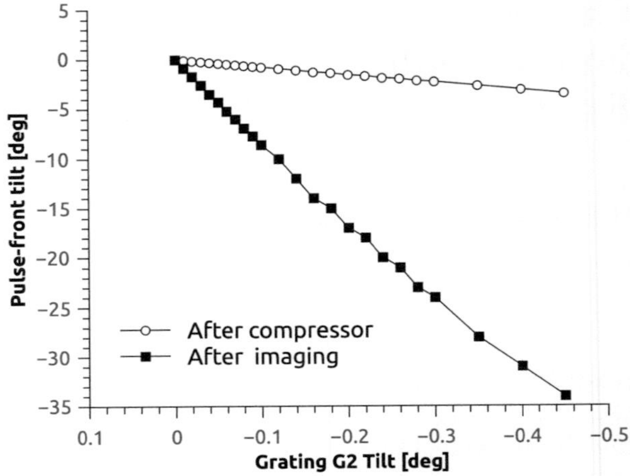

Fig. 1 Pulse-front back-tilt right after the compressor and at the target ("after imaging") as a function of G2 tilt. The magnification is essential to achieve TWE compensation

2 Fourier Optics Model

Fourier optics considers [5], within the paraxial approximation, a set of superposed plane waves in order to obtain the natural modes of the propagation medium itself, i.e. the E field characteristics over space and time. The free-space propagation (FSP) transfer function is expressed as follows, with propagation in the z-axis and radial isotropy over the transverse dimension r:

$$E_{FSP}(k_r, \Omega) = E_o(k_r, \Omega) \exp\left[-ikz - \frac{iz}{2k} k_r^2\right] \tag{1}$$

The grating-dispersion (GD) transfer function is expressed as follows:

$$E_{GD}(k_r, \Omega) = \frac{1}{b} E_o\left(\frac{k_r + \gamma\Omega}{b}, \Omega\right) \tag{2}$$

where the constant $b = \cos\alpha / \cos\beta_o$ is the beam size change upon diffraction, with the angle of incidence α and the diffraction angle β_o at the center frequency ω_o. The coupling between wavevector and frequency is determined by $\gamma = 2\pi M / (\omega_o G \cos\beta_o)$, with the diffraction order M and the grating period G. The relative frequency $\Omega = \omega - \omega_o$ is used throughout. Repeated application of Eqs. (1) and (2) computes the E_{out} at setup's exit. The second order phase term (chirp) of a grating compressor is obtained by the following expression:

$$\phi_2 = -\frac{i\Delta}{k} \gamma_1^2 \Omega^2 = \frac{\lambda_o}{2\pi c^2} \left(\frac{\lambda_o}{G}\right)^2 \frac{L}{\cos^3\beta_o} \Omega^2 \tag{3}$$

with the orthogonal grating separation L. Tilting the second grating by ε causes all subsequent angles of incidence to change accordingly. Finally, the target is some distance away from the last grating and the transfer function of an imaging/focusing component with the focal length f is as follows:

$$E_{img}(r, \Omega) = E_o(x, \Omega) \exp\left[\frac{ik}{2f} r^2\right] \tag{4}$$

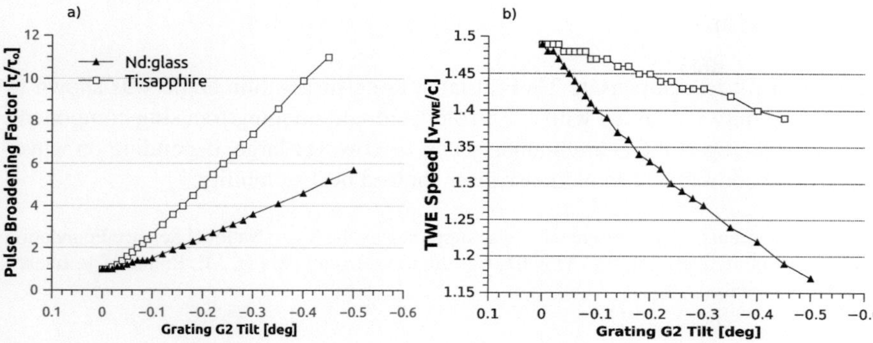

Fig. 2 Comparison of pro and contra of the discussed technique: **a**) pulse broadening at the beam boundary with respect to the best compression as a function of G2 tilt; **b**) Traveling-wave excitation (TWE) speed in units of speed of light as a function of G2 tilt

The computations were performed for the Nd:glass laser wavelength of 1054 $\pm$ 2 nm. For comparison results have been generated for a Ti:sapphire laser. The Gaussian beam radius was 66 mm and focused using a $f = 609.6$ mm mirror under an angle of 48°. The compressor had gratings with $G = 1740$ lpmm, and the slant distance Δ was optimized by the code for a pulse stretch per bandwidth of 296 ps/nm. The roof mirror has a half-distance $d = 1360$ mm. The angle of incidence of the stretched pulse on the first grating was 60.8°.

3 Traveling-Wave Compensation with Misaligned Compressor

Figure 1 indicates that pulse back-tilt due to compressor misalignment is much smaller than the few tens of degrees required for TWE compensation, and mostly is enhanced by the imaging stage proportionally to the magnification. In fact, for given beam width a shorter line focus means a steeper back-tilt of the pulse-front. The misalignment of the compressor causes concomitant degradation of the pulse compression. Figure 2.a shows that the grating tilt can degrade the pulse temporal width. Clearly, short wavelength and small bandwidth mitigate the effect.

Finally, the combined effect on the TWE speed is shown, as a function of compressor misalignment, on Fig. 2.b. The nominal TWE speed at 48° is 1.5c but for half-degree grating tilt, and considering a 5-fold demagnification, a reduction of TWE speed down to 1.1c is computed for the 1054 nm wavelength. Experimental validation of the TWE speed was done using a streak camera connected to the target by means of optical fibers to the two extremes of the line focus. The data confirmed a TWE of 1.1c. Experimental measurements of the pump pulse duration, using an autocorrelator showed a pulse duration of 2.7 $\pm$ 0.8 ps, whereas the theoretical Fourier-limited pulse duration is computed as 0.82 ps. This indicates a broadening of a factor of 3 by means of misalignment.

4 Conclusions

Pulse-front tilt to compensate TWE at large target-irradiation angle was shown to be effective only if coupled with a >5-fold front-end imaging/focusing component. Pulse broadening at the beam boundary can be however large, depending on wavelength and bandwidth. The Nd:glass laser showed best suitability.

Acknowledgements The present work was supported by the Swiss National Science Foundation under the grant number PP00P2-133564/1. Contributions from F. Staub, J.E. Balmer, Ch. Imesch are acknowledged.

References

1. Fiorini, C., Sauteret, C., Rouyer, C., Blanchot, N., Seznec, S., Migus, A.: IEEE J. Quantum Electron. **30**, 1662 (1994)
2. Pretzler, G., Kasper, A., Witte, K.J.: Appl. Phys. B **70**, 1 (2000)
3. Chanteloup, J.C., Salmon, E., Sauteret, C., Migus, A., Zeitoun, Ph., Klisnick, A., Carillon, A., Hubert, S., Ros, D., Nickles, P., Kalachnikov, M.: J. Opt. Soc. Am. B **17**, 151 (2000)
4. Akturk, S., Gu, X., Zeek, E., Trebino, R.: Opt. Express **12**, 4399 (2004)
5. Diels, J.C., Rudolph, W.: Ultrashort Laser Pulse Phenomena. Elsevier, Amsterdam (2006)

Chapter 7
Coherent Short-Wavelength Plasma Radiation for Lab-scale Nano-inspection Tools

Davide Bleiner

Abstract Coherent short-wavelength plasma radiation demonstrated a number of advantages for nano-inspection. Contending technologies are here discussed and strategies for X-ray laser optimizations are highlighted, thus bridging the gap between proof-of-principle or enabling tools.

1 Introduction

The continuing progress towards even smaller devices put forward a challenge to the nano-fabrication and nano-inspection "tools" that have to be made available before the next generation of nano-devices comes into realization. This is the premise for short-wavelength light sources given their potential for high-resolution. A number of applications are thus depending on such *enabling tools*. For instance, *magnetization dynamics* bears enormous technological significance for a number of new devices. We need however more resolution to understand basic interactions, such as spin-torque phenomena or the laser-assisted spin-manipulation. We need also nano-fabrication tools to develop new magnetic materials of improved purity. Also, *coherent diffraction* by nano-structures, such as protein nano-crystals and 2D crystals, bears the interest of resolving new structures and provide insights on molecular functions. Such "*lensless imaging*" has the interest of overcoming limitations brought by X-ray optics. As an additional example, *spectroscopy of correlated electron materials* shows its relevance to investigate metal-insulator transitions, high-temperature superconductivity, giant magneto-resistance, etc, if tools with the required meV resolution are accessible. Increasing evidence of the importance of nanoscale heterogeneities points at nano-scale spatial resolution. A similar consideration applies to time-resolution.

To address such needs, several countries invested hundreds of millions for the construction of X-ray Free-electron Lasers (xFEL). The FLASH (Germany) for instance, offering radiation in the extreme UV and soft X-ray domains, began operation in 2005. The LCLS system (USA) pushed the limit into the hard X-ray in

D. Bleiner (✉)
Institute of Applied Physics, University of Bern, Sidlerstrasse 5, Bern, 3012, Switzerland
e-mail: bleiner@iap.unibe.ch

S. Sebban et al. (eds.), *X-Ray Lasers 2012*, Springer Proceedings in Physics 147,
DOI 10.1007/978-3-319-00696-3_7,

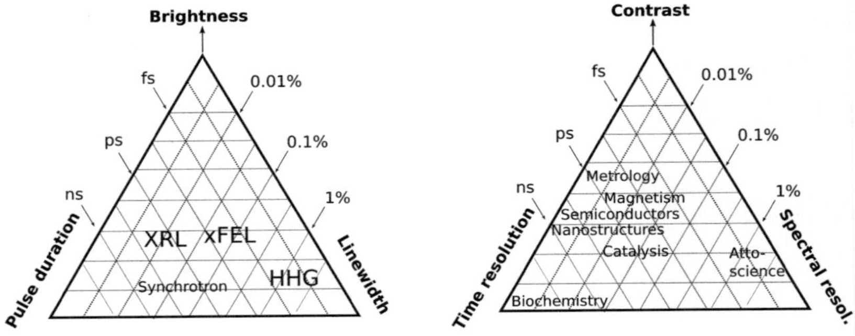

Fig. 1 Capabilities of the main short-wavelength sources and specification for several scientific cases

2009. The SACLA system in Japan, also in the hard X-ray, initiated in 2011. Other xFEL projects are making their way in Switzerland, Italy, Germany, and South Korea. Figure 1 compares a number of short-wavelength sources with the requirements for some nano-science cases. Plasma-based X-ray lasers (XRL) stretch over quite a few of the mentioned applications. This background offers a promising perspective for *laboratory-scale* nano-inspection tools.

2 XRL Bridging the Gap Between Proof-of-Principle and Enabling Tools

Proof-of-principle research tools are: (i) science-oriented, (ii) system-upgradable, (iii) of strategic interest. On the other hand, *enabling* tools are: (i) practice-oriented, (ii) platform-accessible, in terms of footprint and cost, and (iii) of productivity-interest. For instance, actinic metrology tools for EUV lithography belong to the latter case, whereas the xFEL to the former. The XRL has some intrinsic advantages, such as the low cost and footprint, or the spectral purity, to make it useful for *lab-scale* proof-of-principle studies as well as for *actinic* nano-inspection. Figure 2 summarizes the development for lab-scale XRL systems, showing how development matched with computational predictions. Table 1 quantifies productivity and cost, based on bibliometric data for the xFEL and the XRL communities. The publication output for the XRL was as low as 37 % of the xFEL community output. The latter advanced at a pace of a paper every 7.9 days, versus the 21.5 days to see a new XRL paper. Citations are 23 % for XRL versus xFEL. Nevertheless, the cost associated with xFEL instrumentation is remarkably higher, namely 300 times higher than that of a single XRL. Hence, one XRL paper costed approx. 100 kUSD versus the 2,300 kUSD for a xFEL one.

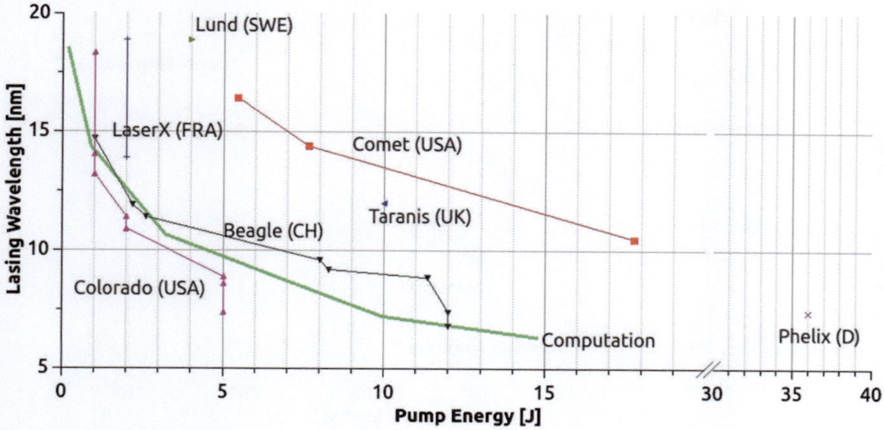

Fig. 2 Major laboratory-scale X-ray lasers systems based on coherent plasma emission. The computational prediction agrees with experimental achievements: a 10-fold pump increase for a 2-fold wavelength reduction

Table 1 Bibliometric analysis for xFEL and XRL communities in the last 5 year, realized on May 9th, 2012, using the *Web of Knowledge* interface accessing all data bases

	Communities		
	xFEL	XRL	xFEL/XRL
Publications	230	85	2.7x
Publication rate [p.a.]	46	17	2.7x
Days to paper	7.9	21.5	0.4x
Citations	1957	443	4.4x
Citation/Publications	8.5	5.2	1.6x
Construction Cost [M USD]	539	15	35.9x
Cost of paper [M USD]	2.3	0.1	23x
Cost of citation [k USD]	275	14	19.6x

3 Perspectives for XRL Development

Figure 3 compares the main short-wavelength sources in average output power and optical étendue and shows that one can identify specific domains of implementations. Besides, the potential for the XRL to fill the existing gap with contending technologies is shown. The XRL could reduce the gap with accelerator-based facilities on three steps: (i) reduction of the optical étendue, (ii) increase of the repetition rate, (iii) increase of conversion efficiency. The reduction of the étendue implies optimization of the source size as well as the beam divergence. Several groups have shown improvements of the collimation by means of seeding or line-focusing [3]. Ruiz et al. [4] have shown illumination improvements by either relay-imaging or telecentric correction. The increase of repetition rate from single shot operation to

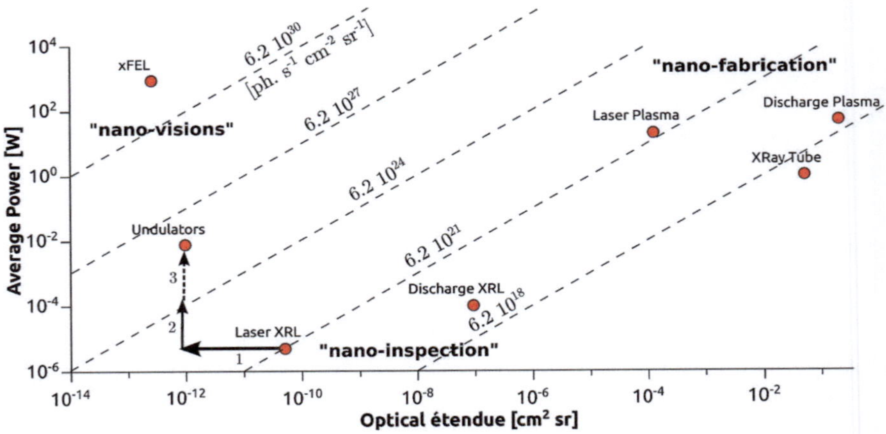

Fig. 3 Average source power versus optical étendue for a number of short-wavelength sources, and brightness in photons at 100 eV (*dashed lines*). The potential for the XRL to fill the existing gap with accelerator sources is indicated as: (1) reduction of source size and X-ray collimation, (2) increase of repetition rate, (3) optimization of plasma geometry and waveguiding

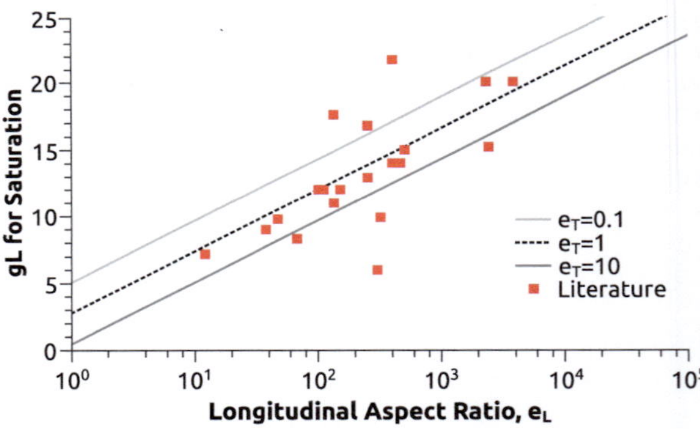

Fig. 4 Gain-length product at saturation as a function of longitudinal plasma aspect ratio (e_L), as predicted with own scaling law (see text). Literature data are plotted for benchmarking

10 Hz [5] or even 100 Hz [1, 6] is subject of ongoing research. Finally, the enhancement of the conversion efficiency by means of seeding and optimization of active volume geometry is to be mentioned [2].

The gain-length product for saturation is a function of active-volume geometry, i.e. longitudinal (e_L) and transverse (e_T) aspect ratio, and we proposed the following scaling-law: $G_{sat} = 2.7 + 2\,\mathrm{Ln}[e_L/\sqrt{e_T}\,]$ [7]. This expression shows that for an ordinary line focus with a plasma column of $e_L = 300$ and $e_T \approx 1$, then $G_{sat} \approx 14$, in agreement with experimental observations (Fig. 4). In terms of conversion efficiency, the problem is addressed considering both the efficient deposition of the pump pulse and the efficient extraction of the short-wavelength signal. Masoudnia et al. [8] showed that the pump delivery angle can alter the active volume geometry and thus the saturation gain-length product. The propagation of the short-wavelength signal can be described with a complex wave-vector, $k = \beta + i\alpha$, obtained plugging the expression for the plasma dielectric constant into the dispersion for a transverse wave. Fill [9] has shown that $\alpha = \frac{-g}{2}$ and $\beta = \frac{2\pi}{\lambda}\sqrt{\varepsilon}$, where g is the gain, λ is the wavelength, and ε is the optical permittivity. Thus k is steered by an electron density gradient term β and a gain-guiding term α. Fill has also given an expression to determine the relative importance of α versus β, namely α is dominating for $\frac{\lambda n_e}{g} < 8.75 \cdot 10^{11}$. For $\lambda = 12$ nm and $n_e \approx 10^{20}$ one obtains that gain-guiding is dominating only when the gain is above 137 cm^{-1}. Both forms of refraction however point at the identical solution: *plasma homogenization*. The use of pre-pulses indeed helped to relax plasma gradients prior to the delivery of the main pulse. The use of hollow cylinder targets is being investigated in our group to confine the plasma. Weber et al. [10] has reported the performance for transverse irradiation of micro-cavities, through a side opening. Masoudnia et al. [8] have shown computational predictions for the case of longitudinal irradiation at GRIP condition. Plasma reflection of incoming beam would not cause pump loss, since trapped in the cavity. The solution is interesting also for high repetition rate XRL, because the hollow target can be rotated such that ablation and re-deposition can lead to target self-healing.

4 Conclusions

Scientific cases proposed for the xFELs are partly suitable for the XRL. In comparison to the former, the latter has shown a productivity of 37 %, but for a modest cost of a few percent of the xFEL. Besides such investment factors, the unmatched linewidth of $<10^{-4}$, the pulse duration of a few ps, and the high single-shot brightness are interesting technical factors. At the moment, XRL are limited by the optical étendue and the repetition rate. Strategies for reducing the gap have been discussed. In terms of average power, the capillary discharge XRL shows a degree of maturity for nano-science applications [11]. The limitation of not being scalable to shorter wavelengths is irrelevant in imaging and spectroscopy, since the most of applications require sources with photon energy of 10–25 eV (or 120–50 nm in wavelength).

Acknowledgements The present work was supported by the Swiss National Science Foundation under the grant number PP00P2-133564/1. Contributions from F. Staub, J.E. Balmer, L. Masoudnia, M. Ruiz-Lopez, Th. Feurer are acknowledged.

References

1. Curtis, A.H., Reagan, B.A., Wernsing, K.A., Furch, F.J., Luther, B.M., Rocca, J.J.: Opt. Lett. **36**, 2164 (2011)
2. Oliva, E., Zeitoun, Ph., Sebban, S., Fajardo, M., Velarde, P., Cassou, K., Ros, D.: Opt. Lett. **34**, 2640 (2009)
3. Bleiner, D., Balmer, J.E., Staub, F.: Appl. Opt. **50**, 6689 (2011)

4. Ruiz-Lopez, M., Staub, F., Bleiner, D.: Optical correction of X-ray laser illumination for short-wavelength microscopy. In: Sebban, S., et al. (eds.) X-Ray Lasers 2012. Springer Proceedings in Physics, vol. 147. Springer, Cham (2013). Chapter 36 in this book
5. Guilbaud, O., et al.: Opt. Lett. **35**, 1326 (2010)
6. Tuemmler, J., Jung, R., Stiel, H., Nickles, P.V., Sandner, W.: Opt. Lett. **34**, 1378 (2009)
7. Bleiner, D.: Saturation gain-length product during short-wavelength plasma lasing. Appl. Phys. Lett. **101**, 081105 (2012)
8. Masoudnia, L., Bleiner, D.: Hohlraum target for overcoming refractive losses in plasma x-ray lasers. Laser Phys. (IOP) **23**, 056003 (2013)
9. Fill, E.E.: Opt. Commun. **67**, 441 (1998)
10. Weber, R., Cunningham, P.F., Balmer, J.E.: Appl. Phys. Lett. **53**, 2596 (1988)
11. Rocca, J.J.: Rev. Sci. Instrum. **70**, 3799 (1999)

Chapter 8
Ion Temperature and Cross-relaxation Effects on OFI Soft-X-Ray Laser

A. Boudaa, G. Maynard, and B. Cros

Abstract We present a theoretical investigation on the influence of ion temperature with cross-relaxation (CR) effects upon a collisionally pumped, optical-field-ionized (OFI) soft-X-ray laser (SXRL). Our results indicate that ion heating can induce a significant modification of the SXRL energy. However, whereas the CR rate is large in an OFI plasma, its contribution on the SXRL energy is found to be small.

1 Introduction

Soft-X-Ray lasers (SXRL) amplified in a plasma generated through OFI (Optical Field Ionization) have interesting properties in terms of application. In particular saturated amplification of spontaneous emission (ASE) with high gain has been experimentally demonstrated by focussing a high intensity infra-red (IR) beam in a xenon [1] or a krypton [2] gas target, at wavelength of respectively 41.8 nm and 32.8 nm. Moreover it has been shown that guiding [3] and seeding techniques [4] can lead to SXRL pulses having μJ energies together with high spatial and spectral coherence. The maximum of the SXRL energy has been obtained at relatively low pressure, around 20 mbar without guiding [1, 2] and 5 mbar with guiding [3]. At these pressures collisional broadening is small leading to a narrow line profile and to a high gain. Therefore the Doppler broadening can be dominating already at an ion temperature of few eV. It has been shown in [5] that such temperatures are reached in xenon or krypton plasmas generated by OFI, through the relaxation of strong collective interaction. In Sect. 2 we present an evaluation of the temperature effect on the ASE energy for a krypton target using typical values for the

A. Boudaa
Laboratoire de Physique et Chimie Quantique, Université M. Mammeri de Tizi-Ouzou, Hasnaoua 15000, Algérie

A. Boudaa · G. Maynard (✉) · B. Cros
Laboratoire de Physique des Gaz et des Plasmas, Univ. Paris-Sud, 91405 Orsay, France
e-mail: gilles.maynard@u-psud.fr

S. Sebban et al. (eds.), *X-Ray Lasers 2012*, Springer Proceedings in Physics 147,
DOI 10.1007/978-3-319-00696-3_8,

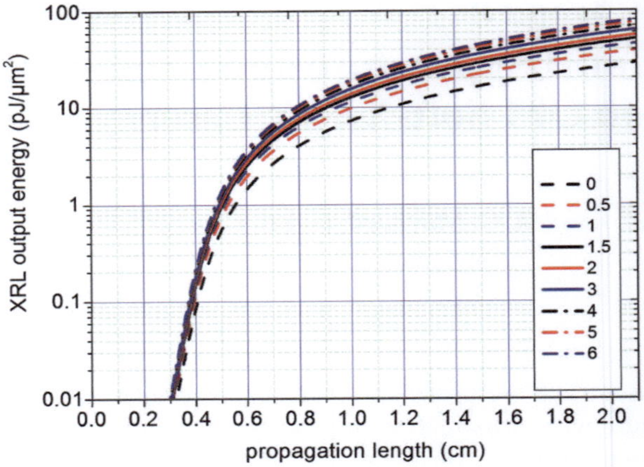

Fig. 1 ASE energy versus plasma length for the 32.8 nm line of Kr^{8+} at 5 mbar. Each curve corresponds to a specific value of the ratio T_i/T_{eq}, indicated in the *legend*

experimental parameters. The calculations have been done by taking into account an inhomogeneous broadening in the gain.

In [5], it was shown that in the OFI plasma, the ions are strongly correlated. It means that the rate of elastic ion-ion collisions is quite high. These elastic collisions induce cross-relaxation (CR) effects that redistribute the excited ions over the Maxwellian velocity distribution. This relaxation has to be introduced in the Maxwell-Bloch equations. In our calculation, it has been done using the "strong-collision" model which assumes that the redistribution is uniform in velocity. In Sect. 3, we analyse how the CR can affect the results of Sect. 2.

2 Temperature Effect on ASE Without Cross-relaxation

To analyse the temperature effect on the ASE energy, we have introduced the inhomogeneous broadening in the COFIXE numerical code described in [3]. Here we present the results of a simplified calculation in 1D geometry, for a Kr gas with a pressure of 5 mbar irradiated by an IR beam with a maximum intensity of 9×10^{16} W cm^{-2} and 35 fs of duration. At such intensity, the Kr atoms get an ionization state of 8+, collisional excitation in the plasma then leads to ASE of the 32.8 nm line of the $3d^9 4d\,^1S_0$-$3d^9 4p\,^1P_1$ transition. From [5] we deduce that the temperature T_i of the Kr^{8+} ions reaches an equilibrium value T_{eq} of 3.3 eV. In Fig. 1, are reported our results for the SXRL energy versus the plasma length, at several values of the ratio T_i/T_{eq}.

We can observe on Fig. 1 a significant influence of T_i on the ASE energy as soon as T_i becomes close to T_{eq}. It indicates that the ion temperature should indeed be included when considering SXRL in an OFI plasma.

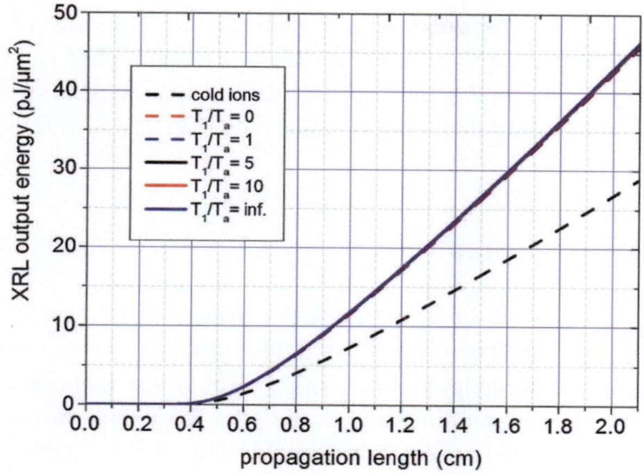

Fig. 2a Same as Fig. 1, taking into account the CR effect and with $T_i/T_{eq} = 1$. Each curve corresponds to a specific value of the ratio T_1/T_a, given in the *legend*

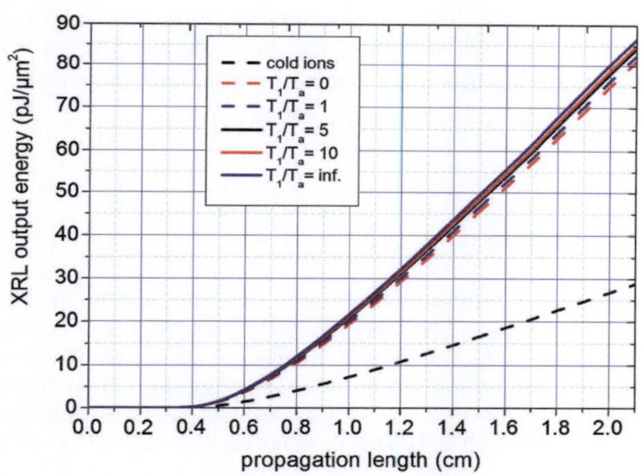

Fig. 2b Same as Fig. 2a, with $T_i/T_{eq} = 6$

3 Cross-relaxation Effects

CR of the ion population has been introduced in the COFIXE code through a relaxation term toward the Maxwellian distribution. The importance of the CR depends on the characteristic relaxation time noted T_a, which has to be compared to the characteristic gain recovery time noted T_1. For $T_1/T_a \ll 1$, CR is negligible, we recover the standard inhomogeneous broadening. In the opposite limit $T_1/T_a \gg 1$, the CR is dominating, the velocity distribution of the amplifying ions remains Maxwellian. In that case the frequency profile of the gain is unchanged, as in the homogeneous case, with the difference that the profile is Maxwellian instead of being Lorentzian. A precise determination of T_1/T_a is a rather complex task (see [6]), but its value should be large, considering the strong coupling of the ions in an OFI plasma. As a first step, we have analysed the influence of the CR on the output ASE energy for several values of T_1/T_a between 0 and infinity. The results are reported on Fig. 2a for $T_i = T_{eq}$ and on Fig. 2b at a higher ion temperature $T_i = 6T_{eq}$.

We can observe on Fig. 2a, that, for $T_i/T_{eq} = 1$, the CR has a negligible effect on the ASE energy. In fact it is only at a much higher value of T_i, that the CR yields a significant modification of the ASE energy. As seen on Fig. 2b, even at $T_i/T_{eq} = 6$, the difference in energy considering either $T_1/T_a = 0$ or $T_1/T_a = \infty$ remains small.

4 Conclusion

The influence of strong ion-ion correlation (IIC) on ASE output energy in an OFI plasma have been investigated. Our results show that the main influence of IIC is to increase the Doppler broadening through the increase of the ion temperature up to few eV. The modification of the gain frequency profile through cross relaxation is a second order effect, which yields a significant contribution only for much higher temperatures. However cross relaxation can have a clearer signature when considering the amplification of a seeded signal. Moreover, as shown in [6], strong IIC should also be considered for the evaluation of collisional broadening. Therefore more investigations are needed in order to determine the full influence of strong IIC on SXRL amplification in a dense plasma.

References

1. Sebban, S., Haroutunian, R., et al.: Saturated amplification of a collisionally pumped optical-field-ionization soft X-ray laser at 41.8 nm. Phys. Rev. Lett. **86**, 3004–3007 (2001)
2. Sebban, S., Mocek, T., et al.: Demonstration of a Ni-like Kr optical-field-ionization collisional soft X-ray laser at 32.8 nm. Phys. Rev. Lett. **89**, 253901 (2002)
3. Cros, B., Mocek, T., et al.: Characterization of the collisionally pumped optical-field-ionized soft-x-ray laser at 41.8 nm driven in capillary tubes. Phys. Rev. A **73**, 033801 (2006)
4. Goddet, J.P., Sebban, S., et al.: Aberration-free laser beam in the soft x-ray range. Opt. Lett. **34**, 2438–2440 (2009)
5. Maynard, G., Lambert, F., et al.: Determination of the ion temperature in a plasma created by optical field ionization. Contrib. Plasma Phys. **47**, 352–359 (2007)
6. Calisti, A., et al.: Line profiles of Ni-like collisional XUV laser amplifiers: particle correlation effects. High Energy Density Phys. **9**, 516–522 (2013)

Chapter 9
Study of Particle Correlation Effects on Line Profiles of Ni-Like Collisional XUV Laser Amplifier

A. Calisti, S. Ferri, C. Mossé, B. Talin, A. Klisnick, L. Meng, D. Benredjem, and O. Guilbaud

Abstract We discuss the effects of particle correlations on the spectral broadening due to the radiator motion (Doppler broadening) for a Ni-like XUV laser line pumped in two different regimes (transient and quasi-steady state regimes) of collisional excitation. In a medium with gain, radiative transport effects modify the observed profile and these modifications depend on the homogeneous or inhomogeneous nature of the intrinsic profile (before amplification). The intrinsic line profile is usually described by a Voigt profile, which is the convolution of a Lorentzian profile due to the different homogeneous broadenings (electron collision-induced transitions and spontaneous emission) and an inhomogeneous Gaussian profile due to Doppler shifts (due to the radiator motion in the free-particle limit). In this paper, it is shown that accounting for the correlations between particles modifies noticeably the radiator-motion broadened profiles, whatever the densities and temperatures values.

1 Introduction

A better characterization of the spectral properties of existing plasma-based XUV lasers is necessary in order to improve the understanding of some characteristics of these lasers. In particular, the spectral width of the lasing line is an important parameter since it controls the temporal coherence of the source, as well as the shortest duration (Fourier-transform limit) that can be reached for a fully coherent pulse.

A. Calisti (✉) · S. Ferri · C. Mossé · B. Talin
PIIM, CNRS-Aix-Marseille Université, Marseille, France
e-mail: annette.calisti@univ-provence.fr

A. Klisnick · L. Meng
ISMO, CNRS-Université Paris-Sud, Orsay, France

D. Benredjem
LAC, CNRS-Université Paris-Sud, Orsay, France

O. Guilbaud
LPGP, CNRS-Université Paris-Sud, Orsay, France

S. Sebban et al. (eds.), *X-Ray Lasers 2012*, Springer Proceedings in Physics 147,
DOI 10.1007/978-3-319-00696-3_9,

Moreover, the relative contribution of homogeneous versus inhomogeneous broadening controls the existence of re-broadening of the laser linewidth when amplified in the saturation regime. XUV lasers pumped by collisional excitation of Ni-like ions are generated in plasmas with electronic densities and ionic and electronic temperatures very dependent on the parameters of the laser pulse used to induce plasma heating and collisional pumping ($N_e \sim 5 \times 10^{19}$ to 8×10^{20} cm^{-3}, $T_e \sim 500$ to 1500 eV and $T_i \sim 20$ to 500 eV). For such conditions, plasmas are strongly coupled (plasma coupling parameter, $\Gamma \gg 1$) and the ionic interactions have to be taken into account.

In laser-produced plasmas, the main causes of spectral line broadenings are the spontaneous emission, the electronic collisions, the ionic Stark effect and the Doppler frequency shift. It is usual to consider that the spontaneous emission and electronic collisions contribute to the homogeneous broadening and ionic Stark effect and Doppler broadening contribute to inhomogeneous broadening as they are due to local inhomogeneities. Due to the variety of plasma conditions, the relative contributions of the different broadenings are very distinct.

In this article, the PPP line shape code [1] has been used to perform a detailed analysis of the various broadening mechanisms of the spectral profiles of the $4d - 4p$ ($J = 0$–1) Ni-like Ag lasing line at $\lambda = 13.9$ nm for conditions relevant for both transient and quasi-steady state (QSS) regimes of collisional excitation. A study on the accuracy of the free-particle Doppler approximation, versus densities and temperatures has been done by using classical molecular dynamics (MD) simulations.

2 Spectral Line Shape Modeling

Including the contribution of ionic emitter motion, the line shape function is given by:

$$I(\omega) = \Re e \frac{1}{\pi} \int_0^\infty e^{i\omega t} \langle e^{i\vec{k}.\vec{r}(t)} \vec{d}(t).\vec{d}(0) \rangle \, dt \qquad (1)$$

with $\vec{d}$ the radiator dipole operator and $k = 2\pi/\lambda$.

Ignoring correlations between the ion translation $\vec{r}(t)$ and the dipole moment $\vec{d}(t)$:

$$I(\omega) = \Re e \frac{1}{\pi} \int_0^\infty e^{i\omega t} S_s(k,t) C(t) \, dt. \qquad (2)$$

$S_s(k,t) = \langle e^{i\vec{k}.\vec{r}(t)} \rangle$ is the self-structure factor and $C(t) = \langle \vec{d}(t).\vec{d}(0) \rangle$ the dipole autocorrelation function. In this work, the PPP line shape code [1] is used to calculate $C(t)$. The PPP code is a multi-electron radiator line broadening code developed to calculate theoretical spectral line profiles for a general emitter in a plasma, using data for atomic energy levels and radial matrix elements generated by atomic structure programs [2]. The line profile calculations are done in the framework of the standard theory (quasi-static ion and electron impact approximations) or if necessary including the effects of ionic perturber dynamics by using the Frequency Fluctuation Model [3, 4]. The self-structure factor is well known in the free-particle limit (Doppler effect) resulting from the hypothesis that each radiating ion moves at constant velocity $\vec{r}(t) = \vec{v}t$ with a Maxwellian distribution of velocities, and is given by:

$$S_s(k,t) = e^{-k^2t^2/2\beta m} \tag{3}$$

with $\beta = 1/k_B T$ and m the ion mass. Classical molecular dynamics simulation technique (MD) is a straightforward way to compute $S_s(k,t)$ taking into account interactions between ions. The plasma model consists of classical point ions interacting together through a coulombic potential screened by electrons and localized in a cubic box of side L with periodic boundary conditions. Newton's equations of particle motion are integrated by using a velocity-Verlet algorithm using a time-step consistent with energy conservation. Integrating the Newton's equation gives access to the positions and velocities, $\{\vec{r}^N(t), \vec{v}^N(t)\}$, of the N ions as a function of time and thus to the associated static and dynamic statistical properties such as structure factors, velocity correlation functions, diffusion coefficients, ion-ion collision rates, etc.

3 Results and Discussion

Spectral line shape calculations have been performed for Ni-like Ag laser $4d - 4p$ ($J = 0$–1) line at 13.9 nm for density and temperature ranges chosen to be consistent with population inversion induced by collisional excitation. Two series of results corresponding respectively to transient ($T_i = 20$ eV, $T_e = 200$ eV) and QSS ($T_i = T_e = 200$ eV) pumping regimes, are presented for different electronic densities. Figures 1 and 2 show the full widths at half maximum (FWHM) corresponding to the different mechanisms of broadening taken separately and together as a function of the electronic density for the transient and QSS pumping regimes respectively. It has been checked that the Stark effect associated with ion microfield has a negligible contribution to the line profile. Electronic collisions and spontaneous emission give rise to the lifetime broadening (blue chain plus crosses) which is homogeneous and increases linearly with N_e at a given T_e. The motion of an emitting ion yields a Doppler shift of the X-ray laser line. If we assume that each radiating ion moves at constant velocity $\vec{r}(t) = \vec{v}t$ with a Maxwellian distribution of velocities, the line profile is inhomogeneous and shows a Gaussian shape with a width (dashed black line) given by a simple analytical formula depending only on T_i. Accounting for both of the previous effects, the resulting spectral profile will be the convolution of the homogeneous and inhomogeneous profile, leading to the so-called Voigt profile with a width which is a complex combination of both homogeneous and inhomogeneous linewidths (dotted red line plus crosses). Similar results for the spectral profile of the $4d - 4p$ ($J = 0$–1) lasing line in Ni-like Mo have been obtained and are discussed in the framework of amplification in [5].

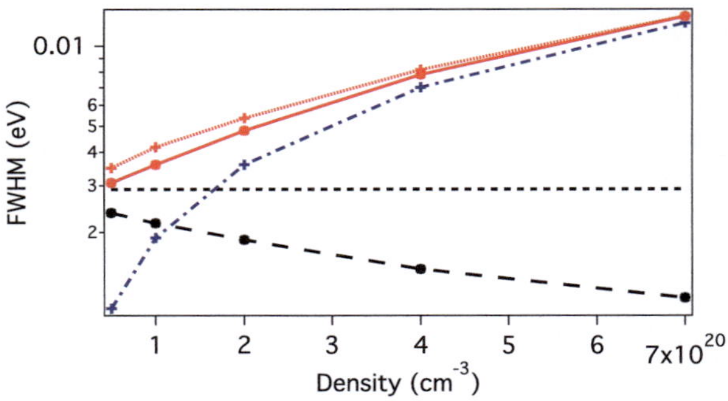

Fig. 1 Full width at half maximum versus density of the $4d - 4p$ laser line in the transient XUV laser case. Broadening due to: radiator motion in the free particle limit, i.e. Doppler broadening (*dashed black line*), accounting for particle interactions (*broken black line plus stars*), lifetime broadening (*blue chain plus crosses*) and total broadening (radiator motion + finite lifetime effects) in the free particle limit (*dotted red line plus crosses*) and accounting for interactions (*full red line plus stars*)

Fig. 2 Same as Fig. 1 in the QSS case

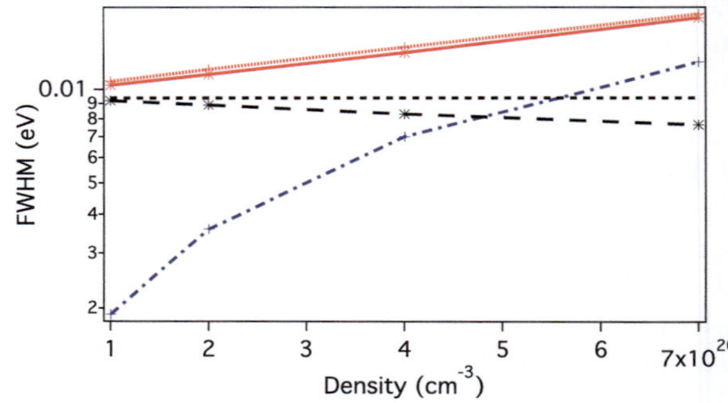

In order to account for velocity changing effects, the broadening due to the translational motion of the emitter (broken black line plus stars) has been obtained by using the self-structure factors computed by MD simulations. It can be seen that taking into account interactions between ion emitter and other ions of the plasma gives rise to a narrowing of the profile whatever the considered densities and this narrowing increases with the density. The line profile is no longer Gaussian, so the overall profile is no longer a Voigt profile. Moreover if during their effective lifetimes, the radiating and absorbing ions sample many velocities, not just one as it is supposed in the Doppler free particle limit, the effect of this velocity redistribution will be to homogenize the Doppler component of the intrinsic line profile. Figure 3 shows the evolution in time of the frequency distribution function of a population of radiating emitters chosen to have an initial velocity such as $\vec{k} \cdot \vec{v}(0) = 0$. Four different timescales of evolution have been considered, the largest one corresponding to the radiative lifetime (inverse of the lifetime broadening). It can be seen that the velocity redistribution is nearly complete during the radiative lifetime. Similar results have been obtained in the transient pumping regime case.

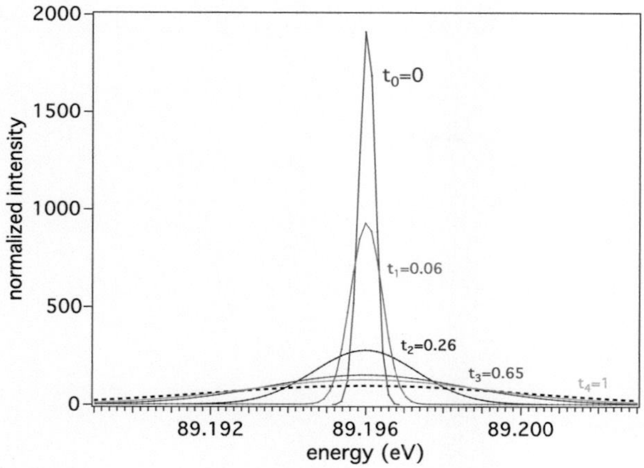

Fig. 3 Frequency distribution functions for conditions of the QSS pumping regime at $N_e = 10^{20}$ cm^{-3}. Times are in units of radiative lifetimes. The *black broken line* corresponds to the Doppler free particle limit

4 Conclusion

This paper concerns a study of the effects of particle correlations on the line broadening due to the radiator motion in the particular case of XUV laser lines. Two different pumping regimes have been considered, transient pumping regime for which ionic temperature is relatively low, so the plasma coupling parameter is large, and quasi steady state regime for which the ionic temperature is higher and the plasma coupling parameter of the order of 1. It has been shown that the effect of correlations cannot be neglected in evaluating Doppler effect for the two cases of interest. For all the considered densities and temperatures, by taking into account correlations between particles, the radiator-motion broadened profiles are modified in being homogenized by velocity redistribution in all the cases and narrowed essentially in the transient regime.

References

1. Calisti, A., et al.: Phys. Rev. A **42**, 5433 (1990)
2. Grant, I.P., et al.: Comput. Phys. Commun. **21**, 207 (1980)
3. Talin, B., et al.: Phys. Rev. A **51**, 1918 (1995)
4. Calisti, A., et al.: Phys. Rev. E **81**, 016406 (2010)
5. Meng, L., et al.: In: X-Ray Lasers 2012. Springer Proceedings in Physics, vol. 147. Springer, Cham (2013). Chapter 28 in this book

Chapter 10
The LUNEX5 Project in France

M.E. Couprie, C. Benabderrahmane, P. Betinelli, F. Bouvet, A. Buteau, L. Cassinari, J. Daillant, J.C. Denard, P. Eymard, B. Gagey, C. Herbeaux, M. Labat, A. Lestrade, A. Loulergue, P. Marchand, J.L. Marlats, C. Miron, P. Morin, A. Nadji, F. Polack, J.B. Pruvost, F. Ribeiro, J.P. Ricaud, P. Roy, T. Tanikawa, R. Roux, S. Bielawski, C. Evain, C. Szwaj, G. Lambert, A. Lifschitz, V. Malka, R. Lehe, A. Rousse, K. Ta Phuoc, C. Thaury, G. Devanz, M. Luong, B. Carré, G. LeBec, L. Farvacque, A. Dubois, and J. Lüning

Abstract LUNEX5 (free electron Laser Using a New accelerator for the Exploitation of X-ray radiation of 5th generation) aims at investigating the production of short, intense, and coherent pulses in the soft x-ray region. The single Free Electron Laser (FEL) line fed by either a superconducting Linear Accelerator of 400 MeV or a 0.4–1 GeV Laser Wake Field Accelerator (LWFA) enables seeding (High order Harmonic in Gas and Echo Enabled Harmonic Generation) for the advanced fourth generation source with the conventional accelerator, and single spike operation

M.E. Couprie (✉) · C. Benabderrahmane · P. Betinelli · F. Bouvet · A. Buteau · L. Cassinari · J. Daillant · J.C. Denard · P. Eymard · B. Gagey · C. Herbeaux · M. Labat · A. Lestrade · A. Loulergue · P. Marchand · J.L. Marlats · C. Miron · P. Morin · A. Nadji · F. Polack · J.B. Pruvost · F. Ribeiro · J.P. Ricaud · P. Roy · T. Tanikawa
Synchrotron SOLEIL, Saint-Aubin, France
e-mail: couprie@synchrotron-soleil.fr

R. Roux
Laboratoire de l'Accélérateur Linéaire, Orsay, France

S. Bielawski · C. Evain · C. Szwaj
PhLAM/CERLA, Lille, France

G. Lambert · A. Lifschitz · V. Malka · R. Lehe · A. Rousse · K. Ta Phuoc · C. Thaury
Laboratoire d'Optique Appliquée, Palaiseau, France

G. Devanz · M. Luong
CEA/DSM/IRFU/SACM, Saclay, France

B. Carré
CEA/DSM/IRAMIS/SPAM, Saclay, France

G. LeBec · L. Farvacque
ESRF, Grenoble, France

A. Dubois · J. Lüning
LCPMR, Paris-VI, France

S. Sebban et al. (eds.), *X-Ray Lasers 2012*, Springer Proceedings in Physics 147,
DOI 10.1007/978-3-319-00696-3_10,

for the fifth generation source with the LWFA. Two pilot user experiments for time-resolved studies of isolated species and solid state matter will take benefit of LUNEX5 FEL.

1 Introduction

Accelerator based X-ray light sources rely on the emission of synchrotron radiation by relativistic electrons (of normalized energy γ) in bending magnets and in periodic permanent magnetic structures, such as undulators. The so-called first generation took advantage of the parasitic synchrotron radiation emitted in the storage rings initially built for high energy physics. The second generation was developed on dedicated storage ring accelerators. The third generation arose with reduced emittances (product of the beam size by the divergence) and high number of installed undulators leading to a brightness increase and a partial transverse coherence. Indeed, a planar undulator creating a sinusoidal vertical field (amplitude B_o, period λ_o), emits on axis at the resonant wavelength λ and its odd harmonics on axis, as a result from the interferences with the electron bunches through the successive periods: $\lambda = \lambda_o(1 + K^2/2)/2\gamma^2$ with $K = 0.94\lambda_o(\text{cm})B_o(\text{T})$. Fourth Generation Light Sources (4G) enable longitudinal coherence by setting in phase the emitting electrons thanks to the FEL process. A light wave of wavelength λ (spontaneous emission progressing along the undulator or stored in an optical cavity, or external seed) interacts with the electron bunch in the undulator, inducing an energy modulation of the electrons; which is gradually transformed into density modulation at the wavelength λ and leads to a coherent radiation emission. The first FEL experience was carried out in 1977 [1] on MARK-III at Stanford, USA using an oscillator configuration in the infra-red. Nowadays, two FEL on linear accelerator, LCLS [2] (Stanford, USA) and SACLA [3] (Harima, Japan), are operating in the 1 Å region with 100–10 fs pulse and GW peak power. In the Self Amplified Spontaneous Emission configuration the spontaneous emission progress along the undulator and is amplified towards saturation [4]. The uncorrelated trains of radiation, which result from the interaction of electrons progressing jointly with the previously emitted spontaneous radiation, lead to spiky longitudinal and temporal distributions, apart from single spike operation for low charge regime [5]. Seeding either with amplified spontaneous emission from first stage undulators sent through a monochromator (so-called self seeding [6]) or with an external laser or a short wavelength coherent light source, such as High order Harmonics generated in gas, allows the saturation length to be shortened, the jitter to be reduced, and the longitudinal coherence to be improved [7–9]. The Echo Enabled Harmonic Generation (EEHG) scheme with a double electron–laser interaction can push the spectral range towards shorter wavelengths when operating on a high order harmonic of the seed wavelength [10–12]. The so-called Fifth Generation (5G) replacing the conventional linac by a Laser Wakefield Accelerator (LWFA), which provides GV/m of acceleration with very short bunches [13–15] is under progress with first observation of the spontaneous undulator radiation [16, 17] and FEL projects [18, 19].

2 LUNEX5 Constituting Elements

LUNEX5 (see Fig. 1) consists in a 400 MeV conventional superconducting accelerator and a 0.4–1 GeV LWFA accelerator, both feeding a common FEL line. It aims at investigating the production of short, intense, and coherent pulses in the soft x-ray region on advanced 4GLS using a Conventional Linear Accelerator (CLA) with the latest seeding schemes and ultra-short pulses of a few fs as well as on 5GLS first for qualifying LWFA with a FEL application, and for providing shorter FEL pulses.

The CLA includes a superconducting L band linac with a Desy-Zeuthen type photo-gun, a 1.3 GHz cryomodule for accelerating up to 200 MeV, a third harmonic cavity, a laser heater, a magnetic chicane, a diagnostic section, a second 1.3 GHz for accelerating module up to 400 MeV and the space for a possible additional third cryomodule. The cryomodules will be designed for future CW operation, the first nominal case of operation being at 50 Hz. 1.3 GHz 20 kW CW solid state amplifiers will be developed at SOLEIL. Lasers from CILEX (Centre Interdisciplinaire de Lumière Extrême) with APOLLON 10 PW laser and its "proximity centers" such as the 60 TW laser of LOA will be used for the LWFA before the dedicated one will be available. Transport to FEL line is achieved with variable permanent magnet quadrupoles. The FEL line comprises transport magnetic elements, dump dipole and in-vacuum undulators with periods of 30 (modulators) or 15 mm (radiators), with a cryogenic option [15]. At 3 mm gap, a peak field of 1.5 T (1.7 T) will be produced at room temperature with NdFeB magnet (at 77 K with PrFeB magnets) for a 15 mm period. The seeding laser system with a regenerative amplifier will be used for producing the High order Harmonics in Gas (HHG), generated in a gas cell for the Echo scheme (after tripling) and for the pump probe experiments of the pilot users. The FEL light will hit a first toroidal mirror (to be interchanged with a second one for the echo source point) for focusing on the exit slit; it will then pass through a double rotation monochromator operating in the Petersen mode to select and stabilize the photon energy and through two focusing mirror systems to be dispatched on the two experimental set-ups. Diagnostics will consist in standard bunch charge monitors, Optical Transition Radiation and fluorescent screens for beam transverse dimensions, standard stripline Beam Position Monitors (BPM) for the linac and cavity BPMs [16] for the undulator sections, a beam arrival time monitor, bunch length monitors and a RF deflecting cavity.

3 LUNEX5 Performances

Figure 2 shows the expected performances of LUNEX5. The spectrum covers the 4–40 nm range with the first, third and fifth harmonics, a fundamental peak power between 10 and 100 MW, corresponding to more than 10^{11} photons/pulse and 10^{27} peak brightness and harmonic peak power from 1 MW down to a few hundreds W. Each wavelength can be obtained with different configurations (amplifier, cascade with a High Order Harmonic seed, echo with 266 nm lasers, ...). The FEL saturates

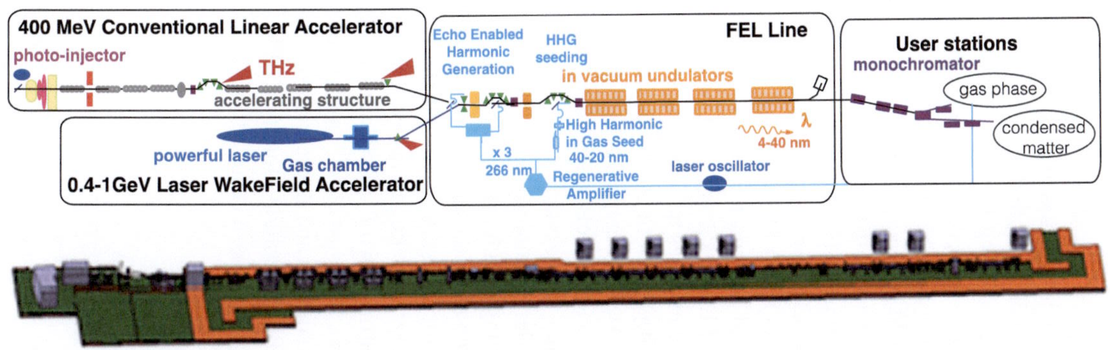

Fig. 1 LUNEX5 scheme and CATIA sketch. From the *left*: electron gun, superconducting accelerating sections (yellow) with the compression chicane in between, LWFA chamber, two echo chicanes and modulator-, radiator-undulators, monochromator, pilot user experimental stations, shielding walls in orange

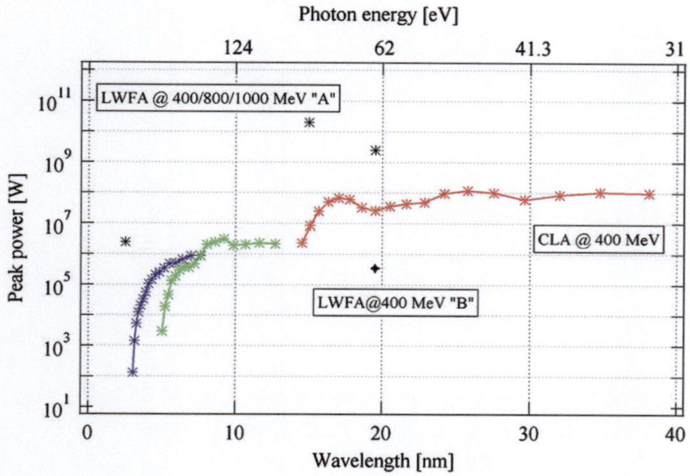

Fig. 2 FEL peak power versus wavelength. Calculations with an 400 MeV electron beam, with a 1.5π mm mrad emittance, a peak current of 400 A, a slice energy spread of 0.02 % (CLA) and A: 0.1, B: 0.5 % (LWFA), a bunch duration of 1 ps (CLA) and A: 2, B: 20 fs-RMS (LWFA)

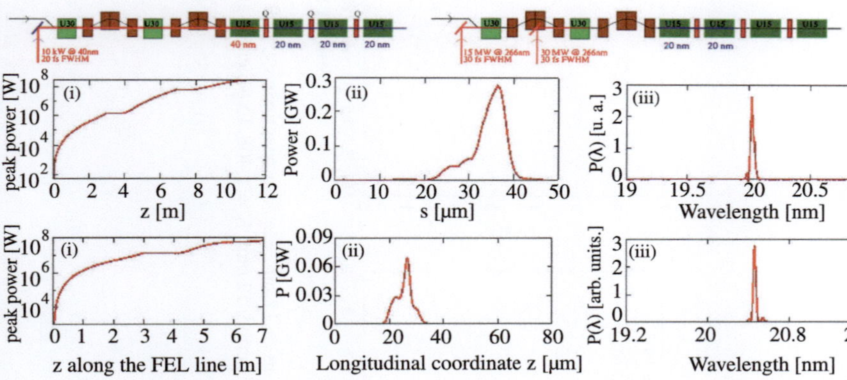

Fig. 3 GENESIS 4G+ FEL simulations at 20 nm. *Top*: cascade and echo schemes. Electron beam energy: 400 MeV, 0.01 % energy spread, 1.5π mm mrad emittance, 400 A peak current, 1 ps pulse duration. Seed: (*middle*: cascade): 20 fs FWHM with 10 kW peak power at 40 nm (*bottom*: echo) 30 fs FWHM, 15 MW and 30 MW at 266 nm. Exponential growth (*left*), FEL temporal (*center*) and spectral profile (*right*)

earlier in the echo case than in the cascade one (7 wrt 11 m), with slightly lower power (65 MW wrt 0.27 GW), longer pulses (24 wrt 17 fs) at the Fourier limit as shown in Fig. 3. Advanced 4GLS FEL can be tuned by a combined variation of the seed wavelength and the undulator gaps. 5G FEL performances critically depend on the electron beam quality and on the optimization of the transport line (see Fig. 4).

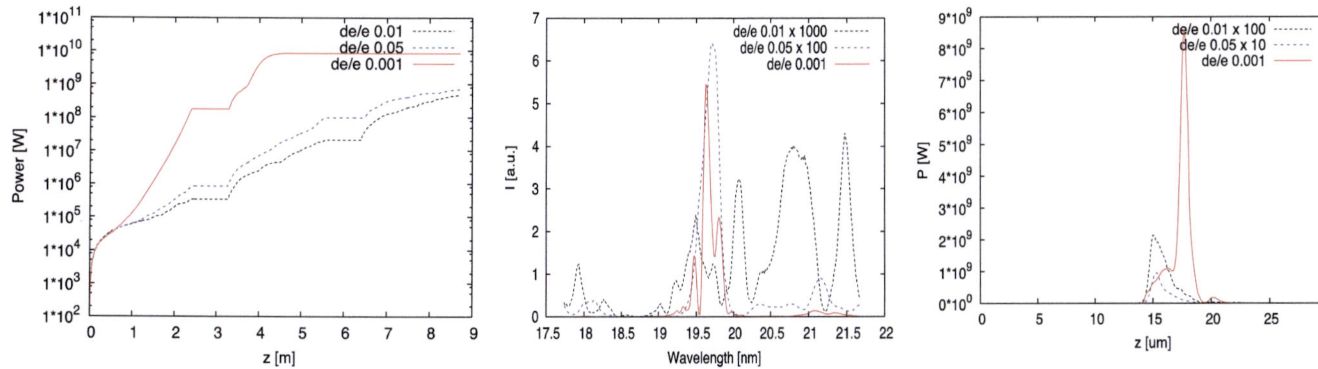

Fig. 4 SASE GENESIS 5G FEL simulations at 17 nm. Electron beam energy: 400 MeV, 0.1/0.5/1 % energy spread, 1π mm mrad emittance, 10 kA peak current, 2 fs RMS pulse duration. Undulator period: 12 mm; $K = 1.408$. Exponential growth (*left*), FEL temporal profile (*center*), spectral line (*right*)

4 LUNEX5 Experimental Stations

Pilot user experiments on isolated species and condensed matter will validate the LUNEX5 demonstrator radiation from a user point of view. So far, no user experiments have been carried out neither with HHG seeded FEL nor with Echo Enabled Harmonics. LUNEX5 has two end stations for time resolved pump-probe studies of isolated species (TR-AMO) and for condensed matter imaging exploiting the coherence. The TR-AMO end station will consist in a high resolution velocity map imaging (VMI) spectrometer allowing for spectroscopy of cold atoms/molecules, clusters or nanoparticles, issued from a multi-purpose source, combined with the full momenta characterization of both electrons and ions using a COLTRIMS type of spectrometer based on time-of-flight and particle 2D position detection (covariance mapping will be used for data analysis). Lensless imaging technique also relies on very stable setup and accurate algorithms to extract real space images. In first experiments magnetization dynamics will be triggered through intense fs laser irradiation and then probed by resonant x-ray magnetic scattering.

5 Conclusion

LUNEX5 is aiming for ultra short FEL pulses quest, production and use on a coupled CLA-LWFA based test facility for complementary use and test of new ideas, such as the recently proposed scheme [20] with three electron photon interactions which could enable to reach the 1 nm region. It will pave the way towards the next generation of light source (4GLS+, 5GLS).

References

1. Deacon, D.A.G., et al.: First operation of a free electron laser. Phys. Rev. Lett. **38**, 892–894 (1977)
2. Emma, P., et al.: First lasing and operation of an Å wavelength free-electron laser. Nat. Photonics **4**, 641 (2010)
3. Tanaka, H., Yabashi, M., et al.: A compact X-ray free-electron laser emitting in the sub-ångström region. Nat. Photonics **6**, 540–544 (2012)
4. Bonifacio, R., Pellegrini, C., Narducci, L.M.: Collective instabilities and high gain regime in a free electron laser. Opt. Commun. **50**, 373–378 (1984)
5. Din, Y., et al.: Measurements of ultralow emittance and ultrashort electron beams in the linac coherent light source. Phys. Rev. Lett. **102**, 254801 (2009)
6. Emma, P.: Hard X-ray self seeding at LCLS. In: Proceed. IPAC 12, New Orleans, USA, May 2012 (2012)
7. Yu, L.H., et al.: High-gain harmonic-generation free-electron laser. Science **289**, 932 (2000)
8. Lambert, G., et al.: Injection of harmonics generated in gas in a free electron laser providing intense and coherent extreme-UV light. Nat. Phys. **4**, 296–300 (2008)
9. Togashi, T., et al.: Extreme ultraviolet free electron laser seeded with high-order harmonic of Ti:sapphire laser. Opt. Express **1**, 317–324 (2011)

10. Stupakov, G.: Using the beam-echo effect for generation of short-wavelength radiation. Phys. Rev. Lett. **102**, 074801 (2009)
11. Xiang, D., et al.: Demonstration of the echo-enable harmonic generation technique for short-wavelength seeded free electron lasers. Phys. Rev. Lett. **105**, 114801 (2010)
12. Zhao, Z.T., et al.: Progress in SDUV-FEL and development of X-ray FELs in Shanghai. In: Proceeding 32nd Internat. FEL Conference, Malmö, Sweden, August 25–27, 2010, pp. 15–19 (2010)
13. Lundh, O., et al.: Few femtosecond, few kiloampere electron bunch produced by a laser–plasma accelerator. Nat. Phys. **7**, 219–222 (2011)
14. Nakajima, K.: Towards a table-top free-electron laser. Nat. Phys. **4**, 92 (2008)
15. Leemans, W., Esarey, E.: Laser-driven plasma-wave electron accelerators. Phys. Today **62**(3), 44 (2009)
16. Schlenvoigt, H.-P., et al.: A compact synchrotron radiation source driven by a laser-plasma wakefield accelerator. Nat. Phys. **4**, 130–133 (2008)
17. Fuchs, M., et al.: Laser-driven soft-X-ray undulator source. Nat. Phys. **5**, 826 (2009)
18. Fawley, W.M., et al.: A design for an XUV FEL driven by the laser-plasma accelerator at the LBNL LOASIS facility. In: Proceed. FEL06, Berlin, Germany, pp. 455–458 (2006)
19. Anania, M.P., et al.: The ALPHA-X beam line: towards a compact FEL. In: Proceedings IPAC10, Kyoto, Japan, p. 2263 (2010). http://www.mpq.mpg.de/APS/gruener.php
20. Xiang, D., Stupakov, G.: Triple modulator–chicane scheme for seeding sub-nanometer x-ray free-electron lasers. New J. Phys. **13**, 093028 (2011)

Chapter 11
High Harmonic Generation Driven by Two Quasi-collinear Pulses

S. Daboussi, S. Kazamias, K. Cassou, O. Guilbaud, M. Pittman, O. Delmas, O. Neveu, B. Cros, G. Maynard, and D. Ros

Abstract We have investigated the high harmonic generation driven by two quasi-collinear IR pulses separated by picosecond delay. Experiments were performed at the LASERIX facility using 50 fs pulses at 800 nm with energy up to 25 mJ. We have studied the generation of two high harmonic pulses from the same medium. In particular, we show how the ratio in energy of each pulse can be used to balance the perturbation induced by the first pulse in the generating gas.

1 Introduction

High order harmonic generation (HHG) from the interaction of an intense infrared (IR) laser and a gas target provides a coherent table-top radiation source in the extreme ultraviolet (XUV) range. Such a source can have a large scientific impact, particularly concerning applications. One of the most significant among them is the seeding of various types of soft x-ray laser (SXRL) amplifiers with high harmonic pulses from plasma-based x-rays [1, 2] to free electron lasers [3]. Traditionally, these amplifiers work in the amplification of spontaneous emission regime (ASE) which presents some limitations in terms of beam optical properties [4, 5]. A straightforward idea explored in several laboratories around the world is to use harmonic radiation to seed the SXRL thus combining the high output energy of SXRL sources with the spatial and temporal coherence of the harmonic source [6, 7]. In this work we concentrated on the possibility to generate from the same generating medium a double temporal structure with variable delay in the picosecond range in order to study more deeply the plasma-based SXRL seeding temporal aspects like polarization [8] and temporal gain construction [9]. For this type of experiment, the relevant temporal range of interest goes from few fs up to 10 ps.

S. Daboussi (✉) · S. Kazamias · K. Cassou · O. Guilbaud · O. Delmas · O. Neveu · B. Cros · G. Maynard · D. Ros
LPGP, Univ. Paris-Sud, UMR 8578, CNRS, bat 210, campus d'Orsay, 91405 Orsay Cedex, France
e-mail: sameh.daboussi@u-psud.fr

S. Daboussi · S. Kazamias · K. Cassou · O. Guilbaud · M. Pittman · O. Delmas · O. Neveu · D. Ros
LASERIX, CLUPS, Univ. Paris-Sud, CNRS, campus d'Orsay, 91405 Orsay Cedex, France

S. Sebban et al. (eds.), *X-Ray Lasers 2012*, Springer Proceedings in Physics 147,
DOI 10.1007/978-3-319-00696-3_11,

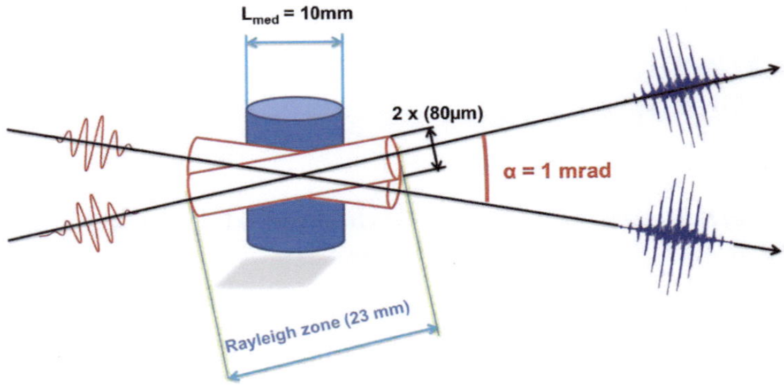

Fig. 1 Quasi-collinear process

2 Description of the Experiment

The pumping of two HHG beams by two close following infrared short pulses in the picosecond range is a non-trivial issue due to the ionization problem which is deleter for phase-matching [10]. A key issue for the present work is to investigate the possibility to generate two XUV pulses from the same medium and to study its temporal response. For this purpose it is required to characterize each pulse independently, in terms of total energy and also spectrum and beam characteristics. Extreme UV streak cameras could reach such temporal resolution but the complex working conditions they offer, together with their small dynamic range make them hard to use. We made the choice to use a slightly quasi-collinear double pulse excitation configuration. For this, two infrared beams generated in a Mach-Zehnder type interferometer with variable delay are focused, in the gas cell, with a small angle (1 mrad), so that they behave in the gas cell such as if they were collinear (see Fig. 1). The experimental investigation was carried out at LASERIX, a high-power laser facility fully dedicated to the development and applications of SXRL beams in the range of 7 to 30 nm [11]. The IR source used in our experiment is based on a low energy sampling from the amplified Ti:Sapphire laser system that provides 50 fs duration pulses centered at a wavelength of 800 nm. Due to pulse air compression, the maximum energy dedicated to HHG is limited to 25 mJ at a repetition rate of 10 Hz.

The schematic of our setup for quasi-collinear HHG is shown on Fig. 2. The incoming laser pulse is set into a Mach-Zehnder type system with an adjustable length arm in order to introduce a controllable delay (-100 ps, $+100$ ps) between the reflected and the transmitted pulses. The temporal resolution of the system was 6 fs due to motorized translation stage accuracy. A polarizer associated with a zero order half wave plate was installed in the path of the reflected beam in order to adjust its energy. Variable apertures were placed on both the reflected and the transmitted beams to adjust intensities and focusing geometries [12]. The two parallel beams are recombined with a 50/50 beam splitter with a small lateral separation and then

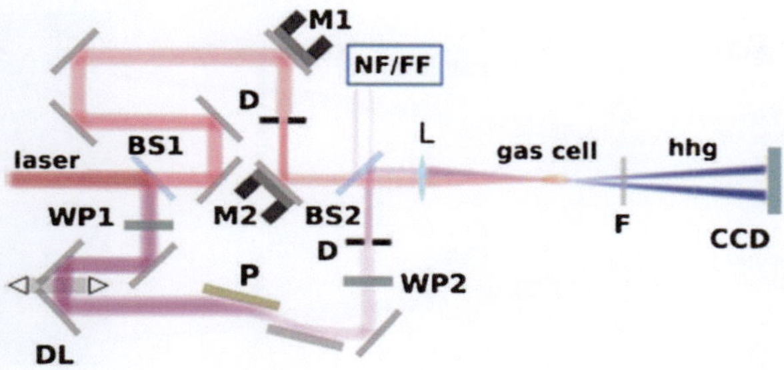

Fig. 2 Experimental setup for quasi-collinear HHG: *BS*, beamsplitter 50/50; *WP*, half wave plates; *P*, Brewster Polarizer; *NF*, real time near field control; *FF*, real time far field control. The polarization of the reflected pulse is controlled using the wave plate *WP2*. Reflected energy sent on the cell is adjusted with $WP1 + P$

focused into the gas cell with 1100 mm lens with a small angle (1 mrad). The footprints of the two corresponding XUV beams were recorded 3 m further from the source on a backlighted XUV CCD camera. The scattered laser light and high harmonic orders higher than 47 were blocked by an aluminum filter (0.2 μm width). After energy optimization in argon for HHG signal separately for each pulse, only a small portion of the available energy (10 mJ) was used in our configuration. The estimated energy per harmonic beam is 70 pJ using the CCD and filter calibration and referring to previous XUV NIST calibrated photodiode data obtained in similar conditions. In those conditions a precise laser alignment is crucial to achieve precise superposition of the beams in the focal region. A real time control of near and far fields showed that the two focal spots were superposed throughout the cell when varying the delay.

3 Temporal Response of the Generating Medium

3.1 Experimental Investigation

After optimization of each of the two individual HHG beams (Fig. 3(a, b)), the influence of the delay was analyzed for equivalent generation conditions on both arms (15 mbar of argon, IR intensity near 1.5×10^{14} W/cm^2). Figure 4 shows the delay scan from 500 fs up to 100 ps of the normalized integrated signal for each beam as recorded with the CCD camera. Each of the two curves, reported in Fig. 4 yields the ratio between the measurement of the total integrated XUV signal generated by the first (resp. the second) IR beam in the double (Fig. 3(c)) and single pulse configurations (Fig. 3(a, b)). Each measurement was averaged over five laser shots and the experiment was repeated over several days. The represented error bars are attributed to HHG signal fluctuations due to shot to shot variations of laser pointing and mechanical instabilities of the delay line. As expected, the first pulse has a ratio which

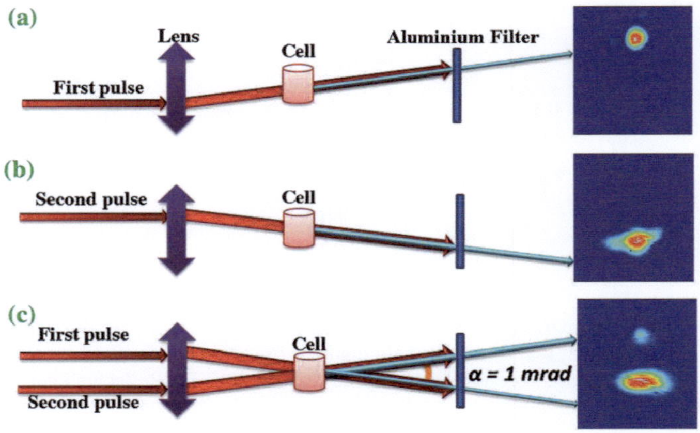

Fig. 3 (**a**, **b**) HHG generation with the first and the second beam respectively when the other driving beam is blocked. (**c**) HHG generation with the two IR pulses from the same

is closed to one, showing that it is not affected by the presence of the second one. The ratio slightly higher than one is not significant since it represents the ratio over two integrated signals measured with an interval of few tens of minutes. On the other hand, we can observe in Fig. 4 that the second pulse has a ratio value close to 0.5 for any kind of delay in the range studied. This indicates that HHG phase-matching for the second pulse is strongly perturbed by the ionization of the gas target generated by the first one. The phase matching condition has been optimized for one beam, that is in our case for the first HHG pulse. Therefore, due to ionization, the phase matching condition is no more satisfied for the second HHG pulse leading to a strong reduction of its energy. Moreover, as this reduction remains constant, one can deduce that the ionization rate does not seem to depend on the delay, even for a time as long as 100 ps.

3.2 Theoretical Estimations for Plasma Recombination Characteristic Time

In order to confirm this experimental observation, theoretical estimations were made for the evolution of the argon plasma at a pressure of 25 mbar with an optical field ionization rate around 5 % which represents the gas medium state after interaction with the first pulse. The ions and atoms are supposed to be at room temperature and the electron average temperature inferred from OFI calculations is close to 0.5 eV. In that case, neither hydrodynamic expansion nor ionization and excitation due to collisions have time to play a significant role. Therefore we have only to consider recombination processes. The direct recombination to the fundamental state having a time characteristic close to the millisecond, in fact only the recombination to highly excited states can have a significant rate. These states having a long life time, the

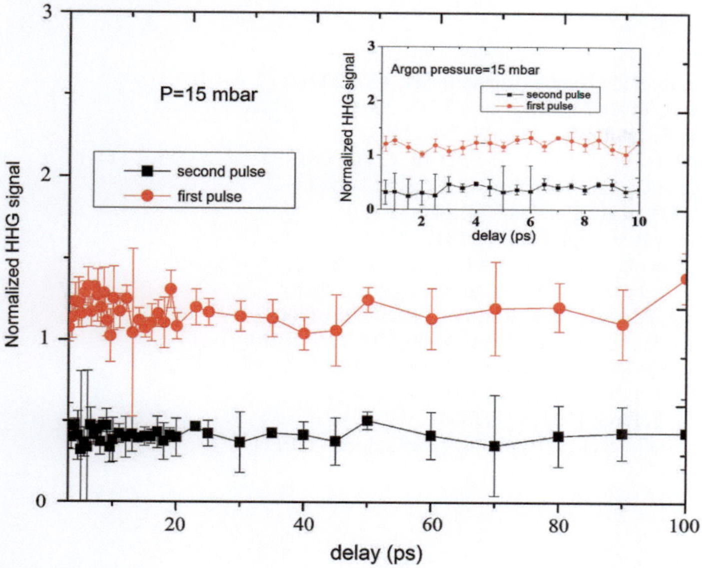

Fig. 4 Measured HHG signal generated by the first pulse (*circle*) and by the second pulse (*square*) as a function of the delay in 15 mbar of argon normalized with the respective single HHG beam. These measures correspond to encircled energy recorded on the CCD

radiative decay has a negligible contribution during 100 ps. However for HHG generation, the highly excited states act as ionized states because they are ionized by the IR beam at very low intensity well before the maximum of the harmonic generation. From our theoretical estimations [13–15] we have deduced that, in the considered condition, the plasma recombination time is of the order of 2 nanoseconds. This confirms the experimental observations that the plasma ionization remains constant within a delay range of few hundreds of picoseconds.

4 Conclusion

In conclusion, a full temporal study in the picosecond range of two XUV pulses generated from the same medium was done thanks to the quasi-collinear scheme. This double pulse can be used in the XUV-XUV pump-probe experiment or the SXRL seeding experiment. One advantage of our system is that, depending on the applications, it can be used even in quasi or fully collinear configuration. We theoretically demonstrated in this work that the recombination time of the plasma generated after interaction between the IR laser and the gas is longer than 2 ns, this was experimentally verified up to 100 ps. This result shows that IR driving laser with a repetition rate larger than 0.5 GHz are not recommended for efficient high harmonic generation in gas cell.

References

1. Zeitoun, P., et al.: Nature (London) **431**, 426 (2004)
2. Wanget, Y., et al.: Nat. Photonics **2**, 94 (2008)
3. Lambert, G., et al.: Nat. Phys. **4**, 296 (2008)
4. Guilbaud, O., et al.: Eur. Phys. J. D **40**, 125 (2006)
5. Guilbaud, O., et al.: Europhys. Lett. **74**, 823 (2006)
6. Goddet, J.P., et al.: Opt. Lett. **34**, 2438 (2009)
7. Berrill, M., et al.: Opt. Lett. **35**, 2317 (2010)
8. Oliva, E., et al.: Phys. Rev. A **84**, 13811 (2011)
9. Mocek, T., et al.: Phys. Rev. Lett. **95**, 173902 (2005)
10. Kazamias, S., et al.: Phys. Rev. Lett. **90**, 193901 (2003)
11. Zielbauer, B., et al.: Appl. Phys. B **100**, 731–736 (2010)
12. Kazamias, S., et al.: Eur. Phys. J. D **21**, 353 (2002)
13. Cros, B., et al.: Phys. Rev. A **73**, 033801 (2006)
14. Vlcek, V.: J. Phys. D **22**, 623 (1989)
15. Kapper, M.G., et al.: J. Appl. Phys. **109**, 113308 (2011)

Chapter 12
Real-Time Observation of Laser Heated Metals with High Brightness Monochromatic X-Ray Techniques at Present and Their Future Prospects

H. Daido, T. Shobu, T. Yamada, S. Yamashita, K. Sugihara, A. Nishimura, and T. Muramatsu

Abstract We present the x-ray techniques for characterizing laser heated metals for welding and cutting techniques. At present, with an undulator (70 keV) as well as bending magnet (30 keV) sources at SPring-8 as a probe source, CW 300 W Ytterbium fiber laser irradiates an Aluminum slab as a sample. Simultaneously the x-ray beam probes the sample for real time observation of a molten pool. We observe the convection indicated by the motion of tungsten based particles as a tracer in the molten pool. During the cooling phase, the molten metal is solidified with residual stresses which are affected by the heating and convection processes. In this experiment the time and space resolution are ∼milli-second and several tens of μm, respectively. On the other hand, microscopic short transient phenomena also play a significant role for the quality of a solidified material. For this purpose, we need high energy short pulse x-ray sources. We try to discuss on the capability and limitation of present x-ray sources and the prospect of an ultra high brightness x-ray source as a complementary source for full characterization of the laser heated and cooling processes of metals.

1 Introduction

Precise characterization of laser heated metal is useful and important for high quality welding and thick metal cutting relevant to maintenance of nuclear plants and decommissioning. In this paper, we present x-ray techniques for real time observation of melting and solidification processes and their interpretation with use of a computer simulation technique.

H. Daido (✉) · T. Shobu · T. Yamada · S. Yamashita · K. Sugihara · A. Nishimura · T. Muramatsu
Applied Laser Technology Institute Tsuruga Head Office, Japan Atomic Energy Agency, Tsuruga, Fukui 914-8585, Japan
e-mail: daido.hiroyuki@jaea.go.jp

H. Daido · T. Shobu · A. Nishimura
Quantum Beam Directorate, Japan Atomic Energy Agency, Kizugawa, Kyoto 619-0215, Japan

S. Sebban et al. (eds.), *X-Ray Lasers 2012*, Springer Proceedings in Physics 147,
DOI 10.1007/978-3-319-00696-3_12,

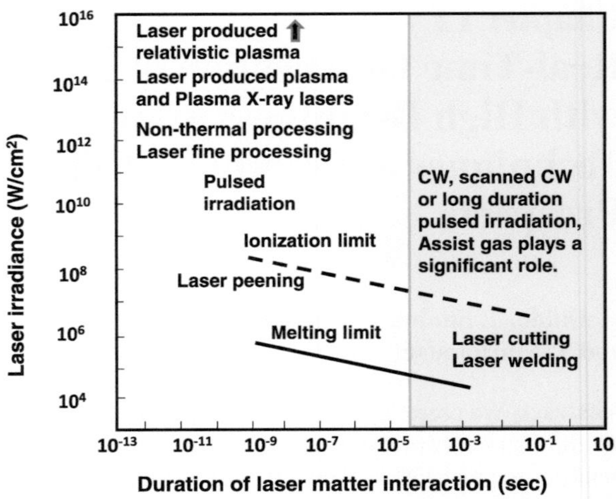

Fig. 1 Wide ranges of laser parameters are used for nuclear engineering such as laser peening, welding and cutting. In the pulsed irradiation regime, ultra-high intensity laser applications (beyond upper limit of the laser irradiance) and laser plasma based x-ray lasers are placed

Figure 1 shows wide range of laser matter interaction studies [1] which contribute to the science and industries. Above the intensity of 10^{18} W/cm^2 which is beyond the figure, one can see the area of the ultra-high intensity laser matter interaction where the relativistic or nonlinear effects play a significant role [2]. We also find the intensity regime of plasma based soft x-ray lasers [3, 4]. In the pulsed irradiation regime, for the nuclear engineering, the laser peening [5, 6] is one of the most attractive techniques to make artificial stress distribution which strengthens the plant components. The non-thermal processing is useful for removal of materials without excessive melting [7, 8]. The ultra-short laser pulse makes nanometer structures on the surface of many materials [9]. The dashed and solid lines respectively represent ionization and melting limits. On the other hand, CW (continuous Wave), scanned CW or long duration pulsed irradiation is also quite useful for the nuclear plant engineering, such as laser welding and cutting. In this region an assist gas plays a significant role for melting and solidification.

The laser cutting of thick metallic materials are necessary for dismantling nuclear power plant and other nuclear facilities because a present fiber laser is powerful enough while it can enter into the narrow gap. The laser can melt and cut the thick metal without reactive force. Successful cutting is realized if the laser hits a fresh metal surface continuously with a help of assist gas flow which blow off the molten and vaporized metals. High quality welding is also necessary for the nuclear engineering. The detailed characterization of the laser welding phenomena is useful for it.

2 Experimental Setup

Dynamics of a molten pool surface taken by a high speed camera contribute to characterize a molten pool. However, visible light can provide only the surface of the

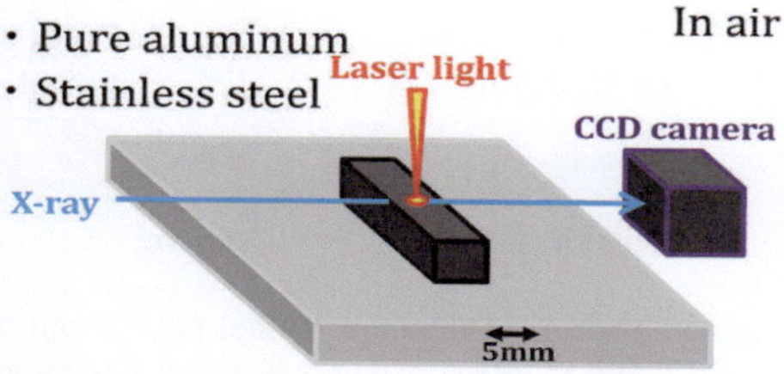

Fig. 2 Experimental setup of the real time observation of a CW 300 W laser heated Aluminum or stainless steel molten pool probed by an intense 30 keV or 70 keV x-ray beam, respectively with a band width of $<10^{-4}$ (the band width/the central energy). The laser illuminates for 5∼10 seconds. The sample width is ∼1∼5 mm

molten pool. If we need to see the interior of the molten pool, we need an intense high energy x-ray beam for imaging. In addition to this, if we would like to see a clear boundary between the solid and liquid phases we need an intense monochromatic x-ray source. For example, mass density difference between the solid and liquid iron is typically ∼10 % and we need ∼milli-second temporal resolution with long time continuous operation as well as a several tens of μm spatial resolution. An intense monochromatic x-ray beam delivered from SPring-8 whose significant feature is its ultra high brightness is suitable for this purpose. For real time observation of a laser heated Aluminum and iron based materials with > a few mm thickness to see hydrodynamic phenomena, we need an intense 30 keV and 70 keV monochromatic x-ray beam, respectively. Figure 2 shows an example of the experimental setup.

3 Scenario of the Physical Processes and Experimental Result

Figure 3 show the expected physical processes of laser heated metals. A laser light is absorbed. This process includes fast and complicated processes such as free electron contribution to the absorption. Successive laser heating causes a phase change to make a molten pool. In this case, we call the heat conduction type melting. If the laser intensity increases, the regime moves into the key hole type [10]. In the pool, the convection is driven by the buoyancy for Aluminum or by the surface tension for stainless steel [11]. When the laser heating stops, the molten pool is cooled down and solidified. Then we find the residual stress distribution which is expected to be influenced by the processes happened previously.

The series of experiments have been performed with use of the setup shown in Fig. 2 [12]. Figure 4 shows an x-ray snap shot of the laser heated molten pool. The bending magnet based source coupled with a pair of crystal spectrometers delivers 30 keV x-ray with an intensity of 10^8 photons/second/mm^2 and the band width of 10^{-4}. The temporal resolution is ∼10 msec which depends upon a frame rate of a camera. The dynamic behavior of laser heated molten pool is visible.

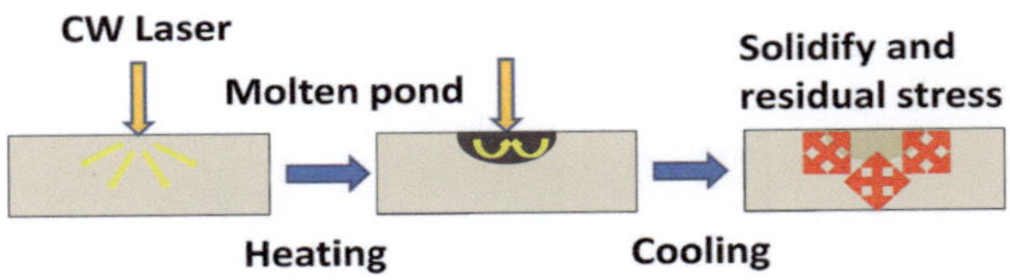

Fig. 3 The scenario of the laser heated Aluminum molten pool. From the *left* to the *right*, the laser heated pool whose solid liquid interface expands in radial direction with convection inside the pool. The laser stops. The molten pool begins to be solidified. The area of the pool decreases rapidly by cooling. The solidification is completed and the residual stress appears around the laser heated region

Fig. 4 An intense 30 keV x-ray beam makes it possible to make a real time movie of laser heated molten pool. The laser illuminates an Aluminum sample from the top and its power is 300 W. The thickness of the sample is 1 mm and the laser spot diameter is 1 mm. The solid liquid interface is clearly visible mainly owing to the monochromaticity of the beam. The *white horizontal line* represents 7.4 mm

4 Comparison Between the Experimental and the Computational Results

We have developed relevant multi-dimensional hydrodynamic simulation code SPLICE [13] based on the incompressible fluid model including solid, liquid and gaseous states. As we have described each processes briefly in Fig. 3. The details of the solidification processes should be given by other aspect of the physical models [14]. We have observed almost whole processes as an x-ray movie including heating, melting and a formation of molten pool, convection in a pool, solidification with residual stress formation.

Firstly, the dynamical behavior of the laser heated and solidified iron surface is described for the comparison between the experimental and computational results. The upper panel of Fig. 5 shows the precise observation of the surfaces using a high speed video camera in the visible spectral range having a frame rate of 12500 frames/second. The corresponding result obtained by the 3-dimensional hydrodynamic code is shown in the lower panel of Fig. 5. In this material, the temperature gradient causes unbalanced surface tension and the flow is driven by this force. If we switch off the surface tension artificially in the code, just a flat surface profile is visible. The computed result shows similar surface profile as observed by the high speed camera.

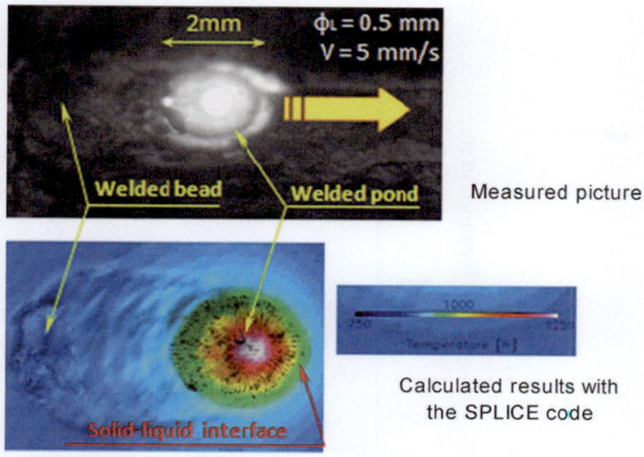

Fig. 5 The *upper panel* represents the snap shot of the 300 W, CW laser illuminated iron surface recoded by a high speed video camera in the visible spectral range. The laser illuminates from a reader to the paper. The scanning speed and its direction are 5 mm/second and the direction of the arrow, respectively. The spot size is 0.5 mm. The *lower panel* represents the result of the computer simulation. The welded beads are visible in both the panels

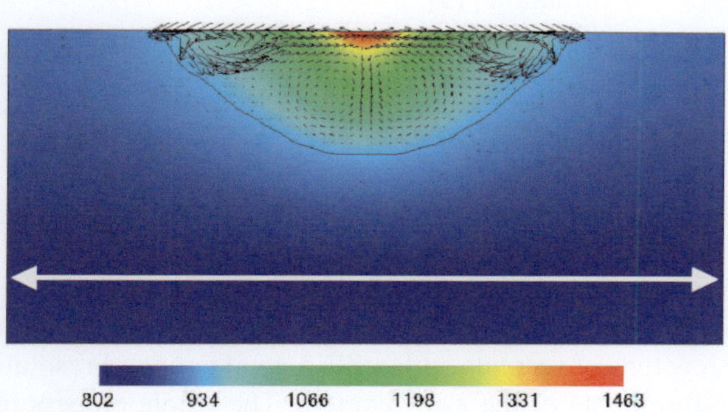

Fig. 6 The result of the SPLICE code which simulates the experiment shown in Fig. 4. The laser illuminates from the top to the central part of Aluminum sample where the high temperature region can be seen. The molten pool with clear liquid to solid interface is visible. The *white horizontal line* represents 10 mm. The scale of this figure is the same as that of Fig. 4. The size of the molten pool reasonably coincides with the experimental results. The *arrows* in the molten pool represent the directions of the convection which also coincide with the experimental results

Secondly, cross-sectional view of the molten pool is depicted. The size of the molten pool in the experiment is compared with the simulation result as shown in Fig. 6. The directions of the convection coincide with each other. The convective flow in the molten pool is clearly visible in this x-ray movie such as ~20 mm/second. The simulation result shows that the speed of the tracers is ~40 mm/second which is faster than the experimental one. After the complete solidification, the residual stress has been measured in the experiment. In this technique, if the stress is created at specific place inside the sample, lattice structure changes

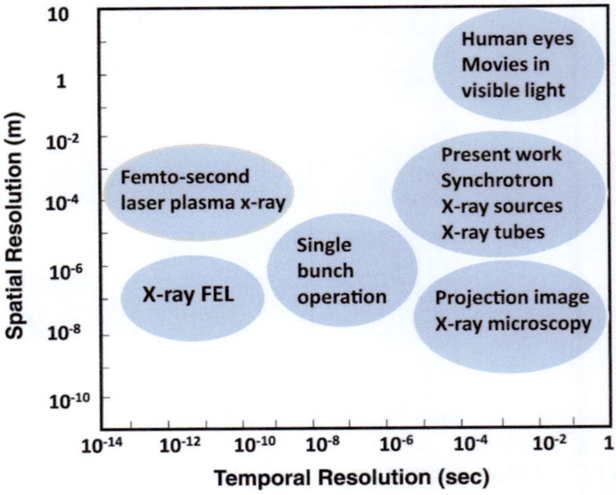

Fig. 7 Spatial resolution vs. temporal resolution for various x-ray sources. Another important aspects of imaging include a field of view and duration of successive imaging (x-ray movie)

and the diffraction angle is broadened. The spectral broadening is interpreted into the residual stress. Subsequent scan inside the sample makes it possible to create a 3-dimensional stress distribution [15].

5 Outlook of the Ultra-intense X-Ray Techniques

Figure 7 represents the map of x-ray techniques in respect to the temporal and spatial resolutions including imaging [16] and diffraction.

High speed x-ray camera coupled with an undulator provides wide field of view (a few cm) with several tens of μm resolution and long time continuous operation of more than a few tens of seconds with milli-second temporal resolution which depends upon the frame rate of the x-ray camera. The whole process including laser irradiation and cooling can be identified quite successfully as described previously. However, if one would like to see precisely the specific parts, for example as shown in Fig. 4, the melting front is not resolved sufficiently. We need higher temporal and spatial resolutions such as $<$ a few μ-second and $<$ a few μm. During the solidification phase, we also need much higher resolutions for a real time observation of crystallization where a typical dendrite size is $\sim$10 μm. Such precise imaging may not need relatively wide field of view. A single bunch operation of a Synchrotron source is one of the possible candidates. The parameters for imaging are complementary.

The creation of residual stresses have routinely measured with the scanning x-ray technique [12] which needs a long time operation and it might be observed with time resolved x-ray diffraction made by an ultra-intense coherent x-ray. An x-ray Free Electron Laser [17] might provide such a beam if it can deliver pulses with higher photon energy. Compact ultra short pulse laser driven x-ray sources with a pump-probe technique have also contributed to the real time observation of phase transitions with ultra fast x-ray diffraction techniques [18]. Such laser driven sources might be also applicable to the imaging with a single shot based operation. Full characterization of melting and solidification of metals is in progress with the technique described here. In addition, these techniques are applicable to the non-metallic materials.

6 Summary

The real time imaging of a laser irradiated metallic molten pool using a Synchrotron facility has been realized. An undulator x-ray source is an almost ideal source for characterizing the macroscopic hydrodynamic behavior of the process. The result is compared with the multi-dimensional fluid simulation. Real time imaging of solidification process has also been realized. We also have made off-line measurement of residual stress distribution. Based on the present achievement, extension of the ultra high brightness x-ray techniques to characterizing the finer and faster components is discussed.

References

1. Von Allmen, M., Blatter, A.: Laser-Beam Interactions with Materials, 2nd edn. Springer, Berlin (1995)
2. Daido, H., Nishiuchi, M., Pirozhkov, A.S.: Review of laser-driven ion sources and their applications. Rep. Prog. Phys. **75**, 056401 (2012) (71 pages)
3. Daido, H.: Review of soft x-ray laser researches and developments. Rep. Prog. Phys. **65**, 1513–1576 (2002)
4. Suckewer, S., Jaegle, P.: X-ray lasers: past, present, future. Laser Phys. Lett. **6**, 411–436 (2009)
5. Sakino, Y., Sano, Y., Kim, Y.-C.: Application of laser peening without coating on steel welded joints. Int. J. Struct. Integr. **2**, 332–344 (2011)
6. Sano, Y., et al.: Residual stress improvement in metal surface by underwater laser irradiation. Nucl. Instrum. Methods Phys. Res. B **121**, 432–436 (1997)
7. Nishimura, A., et al.: Ablation of work hardening layers against stress corrosion cracking of stainless steel by repetitive femtosecond laser pulses. SPIE Proc. **5562**, 673–677 (2004)
8. Nishimura, A., et al.: Development of inspection and repair technology for the micro cracks on heat exchange tubes. J. Jpn. Laser Process. Soc. **17**, 207–212 (2010)
9. Azuma, H., et al.: Single-shot creation of nanometer-sized silicon tadpoles by ultrahigh-intensity laser. Jpn. J. Appl. Phys. **43**, L1172–L1174 (2004)
10. Fujinaga, S., et al.: Direct observation of keyhole behaviour during pulse modulated high-power Nd:YAG laser irradiation. J. Phys. D: Appl. Phys. **33**, 492–497 (2000)
11. Fujii, H., Sogabe, N., Nogi, K.: Convection in weld pool under microgravity and terrestrial conditions. Mater. Sci. Forum **512**, 301–304 (2006)
12. Yamada, T., et al.: Phenomenological evaluation of laser-irradiated welding processes with a combined use of higher-accuracy experiments and computational science methodologies (3) In-situ observations of welded pool using an intense x-ray beam. In: Proc. the 12th International Symposium on Laser Precision Microfabrication (2011). Paper number # 11-055 Tu3-O-7, Att0091

13. Yamashita, S., et al.: Phenomenological evaluation of laser-irradiated welding processes with a combined use of higher-accuracy experiments and computational science methodologies (5) Numerical simulation of the welding processes with a multi-dimensional multi-physics analysis code SPLICE. In: Proc. International Conference on Nuclear Engineering, May 16–19, 2011 (2011). Paper number ICONE19-43939
14. Kou, S., Le, Y.: The effect of quenching on the solidification structure and transformation behaviour of stainless steel welds. Metall. Trans. A **13A**, 1141–1152 (1982)
15. Muramatsu, T., et al.: Study on fundamental processes of laser welded metals observed with intense x-ray beams. In: Lee, J., Nam, C.H., Janulewicz, K.A. (eds.) X-Ray Lasers 2010. Springer Proceedings in Physics, vol. 136, pp. 245–255 (2011)
16. Zewail, A.H., Thomas, J.M.: 4D Electron Microscopy. Imperial College Press, London (2009), Chap. 1
17. Tanaka, H.: Next-generation light source gets boost from powerful new analysis technique. RIKEN home page, http://www.riken.go.jp/engn/r-world/research/results/2010/100922/image/100922.pdf (September 2010)
18. Rousse, A., Rischel, C., Gauthier, J.-C.: Colloquium: femtosecond x-ray crystallography. Rev. Mod. Phys. **73**, 17–31 (2001)

Chapter 13
Generation and Application of Coherent Radiation in the Water Window

L.V. Dao, C. Hall, H.L. Vu, K.B. Dinh, N. Gaffney, E. Balaur, P. Hannaford, and T.A. Smith

Abstract Coherent extreme ultraviolet radiation around the water window region (~4.4 nm) is generated by using an infrared driving pulses at 1400 nm, with energy 2.5 mJ and duration 40 fs in a semi-infinitive helium gas cell. The beam profile and spatial coherence characteristics of the source in this region are shown to be excellent, enabling this source to be used for coherent diffractive imaging.

1 Introduction

Femtosecond laser-driven high-harmonic generation (HHG) has become an increasingly important source for many applications in time-resolved studies of ultrafast dynamics and in coherent diffractive imaging [1–3]. For biological applications the generation of radiation in the "water window" region (4.4–2.3 nm, or 280–540 eV) is an attractive feature because in this wavelength range the radiation is not absorbed by water and nitrogen while absorption of carbon is strong. Intense ultrashort pulses in the water-window wavelength region would allow the capture of images of live cells in a real environment, preserving structural information that can be lost in the sample's preparation process for other imaging methods. This soft x-ray radiation can be used to perform various experiments with high spatial and temporal resolution. Thus, light sources in this wavelength

L.V. Dao (✉) · C. Hall · H.L. Vu · K.B. Dinh · N. Gaffney · P. Hannaford
Centre for Atom Optics and Ultrafast Spectroscopy, Swinburne University of Technology, Melbourne, Australia
e-mail: dvlap@swin.edu.au

E. Balaur
Department of Physics, La Trobe University, Melbourne, Australia

T.A. Smith
School of Chemistry, University of Melbourne, Melbourne, Australia

L.V. Dao · C. Hall · H.L. Vu · K.B. Dinh · N. Gaffney · E. Balaur · P. Hannaford · T.A. Smith
Australian Research Council Centre of Excellence for Coherent X-Ray Science, Melbourne, Australia

S. Sebban et al. (eds.), *X-Ray Lasers 2012*, Springer Proceedings in Physics 147,
DOI 10.1007/978-3-319-00696-3_13,

region have been targeted by many researchers including life scientists and physicists.

In this paper we report the use of a bismuth triborate (BiB_3O_6, or BIBO) nonlinear crystal to construct a high power optical parameter amplifier pumped by 8 mJ pulses from a multi-pass Ti:sapphire amplifier at 1 kHz repetition rate. High energy (2.5 mJ) IR pulses at 1400 nm and with short duration ($<$40 fs) are used for phase-matched generation of radiation of high spatial coherence around the water window region in a semi-infinitive cell filled with helium gas. The coherence of the source is studied by measurements of the beam profile and the interference pattern from a Young's double slit. The broadband source is tested for applicability for coherent diffractive imaging.

2 Experiments and Results

The output of a 1 kHz multi-stage, multi-pass, chirped-pulse amplifier system that produces 8 mJ pulses of 30 fs duration and centered at 810 nm is used to pump the OPA system. A collinear OPA scheme is used in our high-energy IR laser system in order to efficiently amplify the OPA signals. One-third of the output of the multistage amplifier (2.5 mJ) pumps a commercial OPA (Quantronix, Palitra), which is designed with a two-stage configuration seeded by a white light continuum. The remaining 5.5 mJ is used to pump the power amplifier stage. In the high power OPA stage we employed a 2-mm thick BIBO crystal with an aperture of $15 \times 15\ mm^2$, which was placed about 50 cm away from the commercial OPA. The power amplifier OPA stage is seeded by the amplified signal in the commercial OPA whose output is carefully adjusted for a collimated signal beam. The maximum conversion efficiency for the power amplifier stage was about 60 %, producing a total (signal and idler) output of $\sim$4 mJ. An IR pulse duration of 40 fs was measured through sum-frequency mixing with a 30 fs fundamental pulse.

The OPA output pulses at 1400 nm are focused by a 300 mm focal length fused silica lens into a 250 mm-long gas cell with a 3 mm-thick fused silica window at the entrance and a 200 μm pinhole at the exit [4]. We use helium gas for HHG in the water window region because of its high ionization energy and because theoretical calculations [5] indicate that it is possible to obtain phase matching for the generation of radiation in the water window region in helium gas using 1400 nm radiation. A high gas pressure (up to 5 atm) is used for generation of radiation around 4.3 nm. A 300 nm-thick aluminum filter with high transmission in the wavelength range $<$5 nm ($\sim$10 % at 4 nm), and a 300 nm-thick silver filter with transmission in the wavelength range 4–12 nm are used to remove the residual fundamental beam. The high harmonic beam passes through a 0.5 mm-wide, 20 mm-high entrance slit of a grazing incidence spectrometer before being dispersed by a 1200 lines/mm diffraction grating and detected by a back-illuminated XUV charge-coupled device (CCD) camera. The far-field profile of the harmonic beam is detected directly by

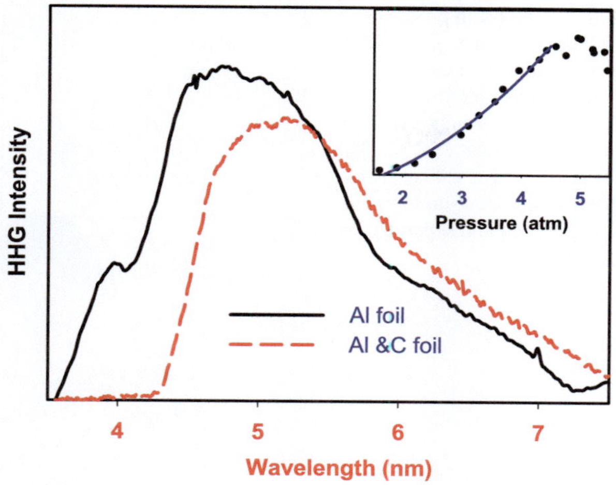

Fig. 1 XUV HHG spectra from He at 5 atm. The C-edge absorption can be seen (*dashed line*) when a 150 nm thick C-foil is added. The *inset* shows the harmonic yield versus the gas pressure: the *dots* show the experimental data and the *line* represents the best fit with P^2 where P is the gas pressure

a 13.3 × 13.3 mm CCD chip of pixel size 13 × 13 μm (PIXIS-XO-1024B, Roper Scientific). For studies of the spatial coherence of the high harmonic beam in more detail a Young double slit (YDS) comprising two parallel slits 4 μm wide, 100 μm high and slit separation of 20 μm is used. The YDS is located 1000 mm from source. The interference fringes are detected by a CCD camera located 800 mm from the slit.

Figure 1 shows the XUV spectrum emitted from helium at 5 atm He pressure with an exposure time of 20 s for each spectrum. The C-edge absorption can be seen (dashed line) when a 150 nm thick C-foil is added. Since the position of the focus of the driving laser beam is close to the exit pinhole of the gas cell the HHG radiation is produced mostly in regions in the gas cell before the focal point along the axis of the fundamental beam. The bandwidth and beam profile of the HHG radiation depend on the gas pressure, the focus position and the diameter of the fundamental beam. The beam profile is remarkably improved for pressures >3 atm. The intensity of the IR pulses ($<5 \times 10^{14}$ W/cm^2) is adjusted for low ionization of the gas ($\eta < 1$ %) to generate high harmonic radiation from the neutral atoms. The inset in Fig. 1 shows the total photon number in the spectral range 4–5 nm for different gas pressures where a 300 nm Al foil has been used. The closed circles show the experimental data and the solid line shows a fit based on P^2. For low pressures ($P < 4$ atm), the HHG signal increases quadratically with pressure as is expected for phase-matched HHG emission. For high pressures (>5 atm) the phase mismatch is increased because of the high absorption restricting further HHG yield increases.

Figure 2 shows the spatial profile of the HHG beam at a distance of 1800 mm from the source. The beam diameter was measured to be 0.8 mm (FWHM). The almost perfect Gaussian profile of the HHG beam suggests there is no density disturbance due to ionization in the interaction region. The observed dependence of the HHG output on pressure and the observed spatial profile indicate that macroscopic phase matching is satisfied along the propagation axis in our geometrical config-

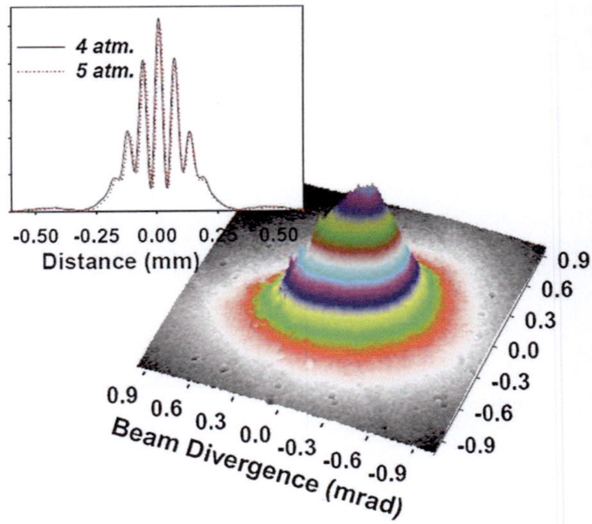

Fig. 2 The beam profile of radiation at a distance of 1800 mm from the source. The *inset* displays interference fringes of the Young's double slit measurement (YDS). The YDS device consists of two parallel slits of 4 μm in width, 100 μm high and separated by 20 μm

uration. The photon flux around the maximum of the beam profile, measured by the CCD, in the wavelength range 4–5 nm is about 7×10^7–10^8 photons/(cm^2 s) when the transmission of the Al foil is assumed to be ∼10 %. To study the spatial coherence of the source in more detail, a YDS is used. The inset in Fig. 2 shows a clear interference pattern resulting from the YDS, which consists of two parallel slits 4 μm wide, 100 μm high and 20 μm separation. The beam diameter at the slit position is ∼0.5 mm (FWHM) which is much larger than the slit separation. When the gas pressure is high (>3.5 atm) the contrast at the center of the interference pattern is nearly independent of the gas pressure. We note that the YDS was illuminated from a broad wavelength beam (an Al foil was used to filter out the fundamental beam) and therefore the contrast of the fringes is not very high. The interference pattern is dependent on the gas pressure indicating a variation of the harmonic spectrum with pressure. The Gaussian beam profile and high contrast of the YDS interference fringes confirm that the HHG radiation is highly coherent.

The diffraction imaging from a pinholes sample illuminated by a broadband source is shown in Fig. 3. Close inspection reveals that the diffraction pattern is blurred because of the broad band nature of the source (4–5 nm). In the present reconstruction procedure we require that the bandwidth of the source is less than 10 % of the center wavelength and the coherence degree should be >0.9. New reconstruction procedures need to be developed and also the bandwidth of the source needs to be reduced with a quasi-phase matching configuration. These improvements are under development.

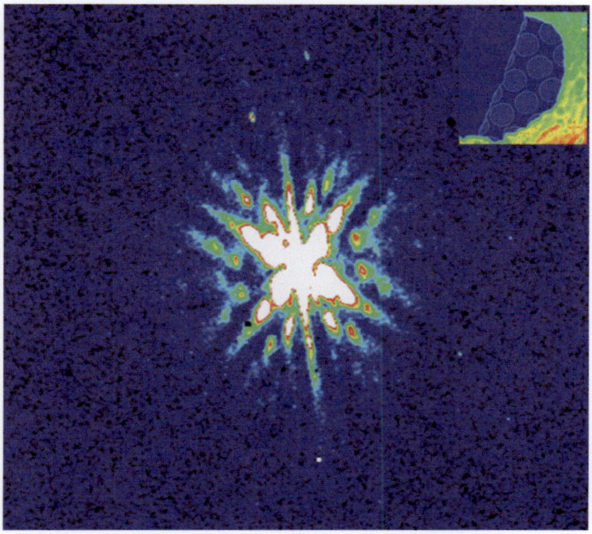

Fig. 3 Diffraction pattern from the pinholes sample illuminated by the source at 4–5 nm. A scanning electron microscope (SEM) image of the sample is shown in the *inset*

References

1. Gagnon, E., et al.: Soft X-ray-driven femtosecond molecular dynamics. Science **317**, 1374–1378 (2007)
2. Cavalieri, A.L., et al.: Attosecond spectroscopy in condensed matter. Nature **449**, 1029 (2007)
3. Chen, B., et al.: Multiple wavelength diffractive imaging. Phys. Rev. A **79**, 023809 (2009)
4. Dao, L.V., et al.: Generation of high flux, highly coherent extreme ultraviolet radiation in a gas cell. J. Appl. Phys. **104**, 023105 (2008)
5. Chen, M.-C., et al.: Bright, coherent, ultrafast soft x-ray harmonics spanning the water window from a tabletop light source. Phys. Rev. Lett. **105**, 173901 (2010)

Chapter 14
Time-Dependent Simulation of Carbon Illuminated by a High Intensity X-Ray Laser

Alberto G. de la Varga, Pedro Velarde, François de Gaufridy, Manuel Cotelo, David Portillo, and Philippe Zeitoun

Abstract We simulate a biological cell composed of solid density carbon illuminated by a high intensity X-ray laser with a time-dependent model. This first version is a simple model that neglects inverse bremsstrahlung absorption by free electrons, electron conduction or hydrodynamic effects. Atomic data needed for the simulations can be generated with the flexible atomic code (FAC) or the screened hydrogenic model (SHM).

1 Introduction

Taking advantage of the recently developed collisional-radiative module in the code bigbart [1] we have implemented the module `muffinman` to study time-dependent radiation transport in carbon as an approximation to simulate the response of biological tissue.

We expect inverse bremsstrahlung absorption by free electrons, electron conduction or hydrodynamic effects to be negligible. At first hand these assumptions seem reasonable. Due to the non-equilibrium nature of the problem free electrons are born with high energies which, in addition to the high energy photons from the laser pulse, produces low resonant free electron absorption that comes mainly from low energy electrons. The short time scale employed leads in principle to low hydrodynamic and electron conduction effects. The influence of this effects will be studied in future versions in order to check our assumptions.

A.G. de la Varga (✉) · P. Velarde · F. de Gaufridy · M. Cotelo · D. Portillo
Instituto de Fusión Nuclear, Universidad Politécnica de Madrid, José Gutiérrez Abascal 2, 28006 Madrid, Spain
e-mail: agardelavarga@gmail.com

F. de Gaufridy
Institute of Physics, Czech Academy of Sciences, Prague, Czech Republic

P. Zeitoun
Laboratoire d'Optique Appliquée, UMR CNRS/ENSTA ParisTech/Ecole Polytechnique ParisTech, Chemin de la hunière, 91671 Palaiseau, France

S. Sebban et al. (eds.), *X-Ray Lasers 2012*, Springer Proceedings in Physics 147,
DOI 10.1007/978-3-319-00696-3_14,

There is no treatment of escaped or trapped electrons [2], all electrons are considered trapped and contribute to the average free electron energy and collision rates. For the sake of simplicity we assume that free electrons have time for thermalization at a certain calculated thermalization temperature.

Energy exchange involving ions, free electrons and the radiation field is computed in each timestep to check the consistency of the model and to obtain the free electron temperature in the next timestep.

2 Description of the Time-Dependent Radiation Transport Module

The module `muffinman` implements an explicit model that computes the variation in configuration populations and the energy exchange between timesteps. Configurations are connected through all possible collisional and radiative processes. Processes and their inverse considered in `muffinman` are spontaneous emission and resonant photoabsorption (multipole transitions E1, E2 and M1), collisional excitation and collisional de-excitation, radiative recombination and photoionization, collisional ionization and three body recombination, and autoionization and resonant capture.

Spontaneous emission, collisional excitation, radiative recombination, collisional ionization and autoionization rates may be computed with the relativistic atomic code FAC [3] or SHM [4, 5]. Collisional de-excitation, three body recombination and resonant capture are calculated by detailed balance [6]. For transitions due to interaction with the radiative field, resonant photoabsorption and photoionization, formulas from [7] are used.

Once we have computed the transition rates, the system of equations that yields the change in population of configuration α is written as

$$\frac{dn_\alpha}{dt} = -n_\alpha \sum_{\beta \neq \alpha}^{N} W_{\alpha\beta} + \sum_{\beta \neq \alpha}^{N} n_\beta W_{\beta\alpha} \quad (\mathrm{s}^{-1}) \tag{1}$$

where n_α is the population of configuration α normalized to unity and $W_{\alpha\beta}$ refers to the transition rate from configuration α to configuration β. The system is solved with an explicit method. Also energy exchange is calculated for each process that produces energy transfer between ions, free electrons and the laser pulse. The variation of energy of each component c due to process γ between configurations α and β with an energy exchange of $\Delta e^{\gamma}_{\alpha\beta}$ in a timestep Δt is

$$E_c^{n+1} = E_c^n + \sum_{\gamma} \sum_{\beta} \sum_{\alpha} \Delta e^{\gamma}_{\alpha\beta} W^{\gamma}_{\alpha\beta} n_\alpha \Delta t \quad (\mathrm{eV}) \tag{2}$$

$$\Delta E_c = E_c^{n+1} - E_c^n \tag{3}$$

where E_c can be the energy of the ions E_{bound}, free electrons E_{free} and laser pulse E_{pulse}. Consistency in energy transfer implies in each timestep

$$\Delta E_{bound} + \Delta E_{free} + \Delta E_{pulse} = 0 \tag{4}$$

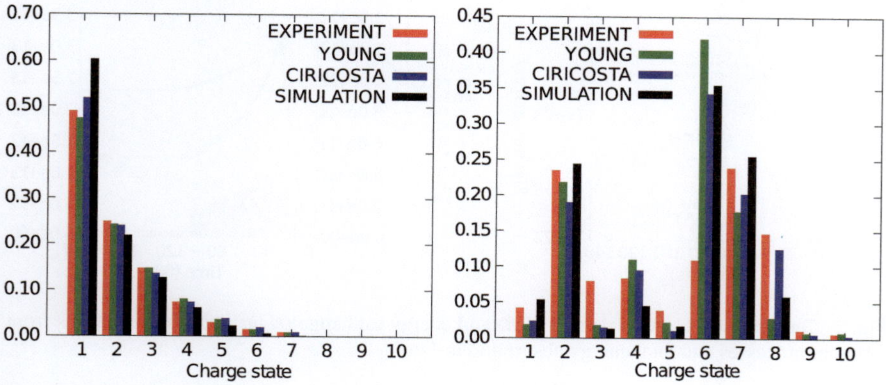

Fig. 1 Charge-state yields in the 800 eV (*left*) and 1050 eV (*right*) cases

From the free electron density n_e and the free electron energy density ρ_E we compute the average free electron energy

$$\langle E_{av} \rangle = \frac{\rho_E}{n_e}.$$

Equivalent thermalization temperature is obtained by

$$\frac{3}{2} k T_e = \langle E_{av} \rangle.$$

3 Experimental Benchmark

In order to test the accuracy of the developed module we compared experimental neon charge-state yields measured at LCLS with theoretical calculations [8, 9]. The neon gas (10^{19}) was irradiated at intensities up to 10^{18} W/cm^2 and photon energies of 800 eV (below the K-edge), 1050 eV (just above the K-edge) and 2000 eV (far above the K-edge).

Here we consider the 800 and 1050 eV cases for comparison with a 0D model. As in the theoretical model in [8] we consider photoionization and autoionization as predominant over collisional processes. This is approximately true at this densities, although collisional ionization may compete with photoionization in the 800 eV case [9].

In Fig. 1 (left) theoretical models correctly reproduce experimental charge-state yields. This is due to the simple absorption mechanism with dominance of valence shell photoionization, with no K-shell vacancies and therefore no autoionization. In Fig. 1 (right) our model performs in the range of the other two theoretical models but none is close to the experimental measurement. Production of Ne^{2+}, Ne^{4+} and Ne^{6+} is mainly due to autoionization. Inner shell photoionization is the main mechanism for Ne^{1+}, Ne^{3+} and Ne^{5+} production, introducing other mechanisms due to

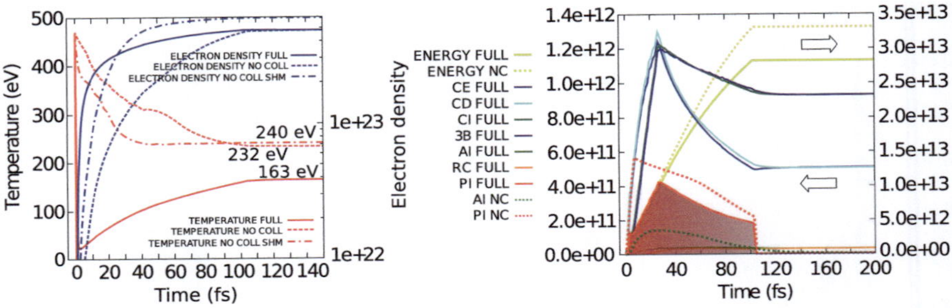

Fig. 2 Temperature, electron density (*left*) and energy exchange (*right*) with free electrons from different processes and total energy absorption in J/m^3

Table 1 Temperatures (eV) after 10^{17} W/cm^2 and 1 keV photon energy irradiation

FAC	FAC NC	SHM NC	Whittaker et al.
163	232	240	230

inner-shell vacancies not considered in neither of the theoretical models which may explain the discrepancies.

4 High-Intensity X-Ray Simulations

The aim of the developed module is to perform 1D simulations of solid carbon illuminated by ultraintense and ultrashort x-ray pulses as a first approximation to the irradiation of biological tissue.

First we will compare 0D simulations of solid carbon with a SHM model [10]. In their article Whittaker et al. investigated temperatures following from irradiation of 100 fs flat-top pulses at different intensities and photon energies, assuming that after irradiation and before the expansion of the sample, approximately one picosecond, the sample reaches full thermalization. Only photoionization and autoionization were considered in their model.

In Fig. 2 and Table 1 it can be seen that FAC (FULL) and SHM with no collisions (NC) agree quite well with results by Whittaker et al., but when collisional processes are included the temperature reached is lower. This happens because electron density increases rapidly and so collisional ionization competes with photoionization, reducing the photon absorption and hence the energy absorbed by the sample.

In the second selected case at 2 keV and 10^{18} W/cm^2, see Table 2, we included calculations with a modified SHM model (SHM MOD) as in Whittaker et al., where the Kramer's photoionization was scaled as the inverse square and not the inverse cube. This greatly increases photoionization cross sections far from the ionization potential of a transition, resulting in a great enhancement of the photon absorption leading to high temperatures. Our modified model agrees with their results and FAC and SHM models predict much lower temperatures after irradiation.

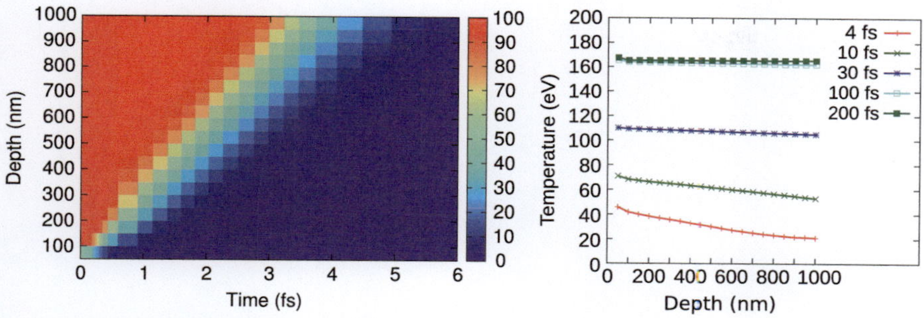

Fig. 3 Absorption (%) and temperature profiles in 1 μm of solid carbon

Table 2 Temperatures (eV) after 10^{18} W/cm^2 and 2 keV photon energy irradiation

FAC	SHM	SHM MOD	SHM MOD NC	Whittaker et al.
192	90	538	584	560

Finally we performed a simulation of a 1 μm carbon layer at 2×10^{16} W/cm^2 and wavelength $\lambda = 13$ nm, consistent with $I\lambda^2$ values from FEL experiments [8, 11]. The absorption through the layer, see Fig. 3, decreases to almost zero after less than 10 fs. Temperature profile in the sample needs a few tens of fs to reach uniformity.

Acknowledgements The authors would like to thank the financial support provided by the Spanish Ministerio de Educación y Cultura program no. ENE2009-09837 and also to the European Commission project SFINX-LASERLAB-II.

References

1. de la Varga, A.G., et al.: High Energy Density Phys. **7**, 163 (2011)
2. Hau-Riege, S.P., et al.: Phys. Rev. E **69**, 051906 (2004)
3. Gu, M.F.: Can. J. Phys. **86**, 675 (2008)
4. More, R.M.: J. Quant. Spectrosc. Radiat. Transf. **27**, 345 (1982)
5. Faussurier, G., et al.: J. Quant. Spectrosc. Radiat. Transf. **58**, 233 (1997)
6. Salzmann, D.: Atomic Physics in Hot Plasmas. Oxford University Press, New York (1998)
7. Rose, S.J.: High Energy Density Phys. **5**, 23 (2009)
8. Young, L., et al.: Nature **466**, 56 (2010)
9. Ciricosta, O., et al.: High Energy Density Phys. **7**, 111 (2011)
10. Whittaker, D.S., et al.: Phys. Plasmas **18**, 013105 (2011)
11. Ackermann, W., et al.: Nat. Photonics **1**, 336 (2007)

Chapter 15
Regenerative Laser Cavity Tuning for Efficient Soft-X-Ray Laser Operation

O. Delmas, K. Cassou, O. Guilbaud, S. Kazamias, S. Daboussi, M. Pittman, O. Neveu, J. Demailly, and D. Ros

Abstract The influence of temporal pump laser profile on the Soft X-ray laser (SXRL) efficiency is investigated in double grazing incidence pumping (DGRIP) transient collisional scheme. We have demonstrated a simple method that allows to produce and control one or several small prepulses and the ASE level of the pump laser beam by modifying the regenerative amplifier parameters. By changing the temporal structure of the pump beam we observed both an increase of the SXRL energy and an improvement of the target lifetime. Measurements made for different configurations are presented and discussed.

1 Introduction

The recent progress made on high intensity lasers and laser-pumped XUV sources allowed the construction of the LASERIX soft-X-ray laser (SXRL) facility at the Paris Sud University. SXRL pulses considered here are generated in a plasma in the transient collisional excitation scheme. For this, the plasma is created by a long laser pulse (100 ps to 1 ns) focused on solid target, which in turn is excited by a short pulse (in the ps range) creating then a population inversion through electrons/ions collisions in the soft-X-ray domain. It has recently been demonstrated on LASERIX that low intensity precursor pulses and a low amount of ASE in a nanosecond range are favourable for XUV signal level and target lifetime [1]. In the present work, we study the effect of the forefront temporal contrast on the SXRL generation by adding

O. Delmas (✉) · K. Cassou · O. Guilbaud · S. Kazamias · S. Daboussi · M. Pittman · O. Neveu · J. Demailly · D. Ros
CLUPS/LASERIX, Université Paris Sud XI, 91405 Orsay Cedex, France
e-mail: olivier.delmas@u-psud.fr

O. Delmas · K. Cassou · O. Guilbaud · S. Kazamias · S. Daboussi · O. Neveu · J. Demailly · D. Ros
LPGP, Bât 210, Université Paris Sud XI, 91405 Orsay Cedex, France

O. Delmas
Amplitude Technologies, 2-4 rue du Bois Chaland, 91090 Lisses, France

S. Sebban et al. (eds.), *X-Ray Lasers 2012*, Springer Proceedings in Physics 147,
DOI 10.1007/978-3-319-00696-3_15,

precursor pulses by simply tuning the output synchronization of the regenerative amplifier (RGA) of the laser chain.

2 Experimental Set-up

2.1 Laserix Facility Overview

The LASERIX facility provides two XUV beamlines. The first is based on High Harmonic Generation in rare gas (Ar, Ne...) while the second is based on transient collisional SXRL with Ni or Ne like targets (Mo, Ag, Ti). The SXRL is pumped by a 10 Hz–100 TW class Ti:Sa laser with a double pulse structure. The pulse to pulse energy stability of SXRL has been improved by using a "Double Grazing Incidence Pumping" (DGRIP) configuration, where the long pulse is created before amplification in order to produce two pulses within the same laser beam [2] which is more stable and convenient than the classical GRIP configuration [3] which requires the use of two different beams (Fig. 1). A disadvantage of this method is a loss of energy for the long pulse during the compression process which can be estimated to be of about 15 % of the total energy for a typical energy ratio of 70/30 % in favour of the main pulse. Another disadvantage of the DGRIP configuration for LASERIX, which can propose multibeam configurations mainly for pump-probe experiments, lies in the double pulse structure which degrades the temporal resolution for femtosecond IR or XUV based HHG probe.

2.2 Temporal Pulse Shaping

The setup used to produce the two following pulses for the DGRIP configuration is illustrated on Fig. 2. After passing through the Offnër stretcher, the beam is split into two parts with an energy ratio control at the input of a Mach-Zehnder interferometer like setup. In one arm, corresponding to the first pulse, the beam passes through a Martinez stretcher in order to increase its spectral phase dispersion [4]. In the other arm, corresponding to that of the main pulse, the beam is sent into a motorized delay line in order to control the temporal delay with respect to the first. The two beams are then recombined in one with a beam splitter which is sent into the RGA. The cavity configuration of the RGA induces a transverse mode filtering which allows a perfect spatial recombination between the two pulses. After three stages of multipass power amplification, the two pulses are injected in the compressor so as to finally obtain a long pulse having a duration fixed by the Martinez stretcher followed by a short pulse only compressed in the 0.5 to 10 ps range.

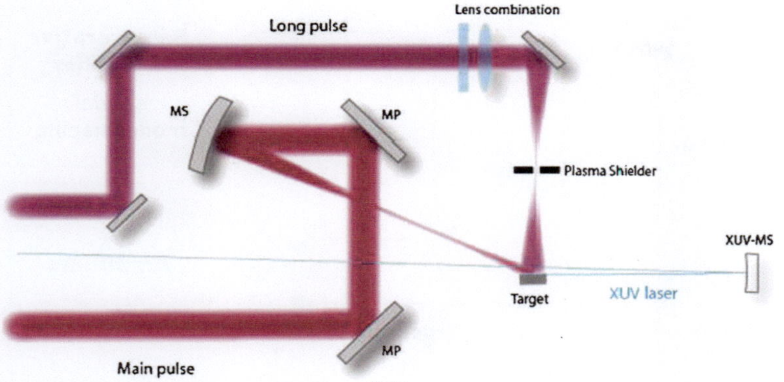

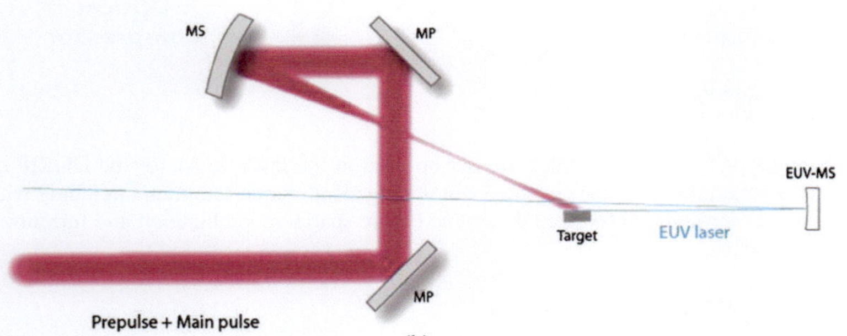

Fig. 1 Experimental set-up; (**a**) In the GRIP scheme two different beams are focused on the target, the nanosecond pulse (*purple*) arrives normally to the target surface for plasma generation while the picosecond pulse (*red*) arrives later at a grazing incidence to pump the plasma [3]; (**b**) In the DGRIP scheme, the two pulses arrive in the same beam at a grazing incidence on the target [2]

2.3 Regenerative Amplifier Tuning

It is generally admitted that a front nanosecond pedestal or even a prepulse in the temporal profile can change dramatically the SXRL performance [5–7]. In the present approach, we generate prepulses in the RGA of the Ti:Sa front-end laser (Fig. 3) by a fine adjustment of the trigger delay of the output Pockels cell (PK_{OUT}). In the same way, we add one or more round trips after the maximum pulse extraction from the cavity, in order to favour front ASE pedestal, by a coarse delay adjustment of PK_{OUT}. This way, since the synchronization of all the pump lasers remains unchanged with respect to the initial maximum extraction from RGA, the following power amplifiers will increase an ASE signal in that temporal region. By carefully adjusting first the coarse delay for ASE generation then fine delay for prepulses we could evidence strong improvement of the SXRL emission and a non negligible extension of the target lifetime.

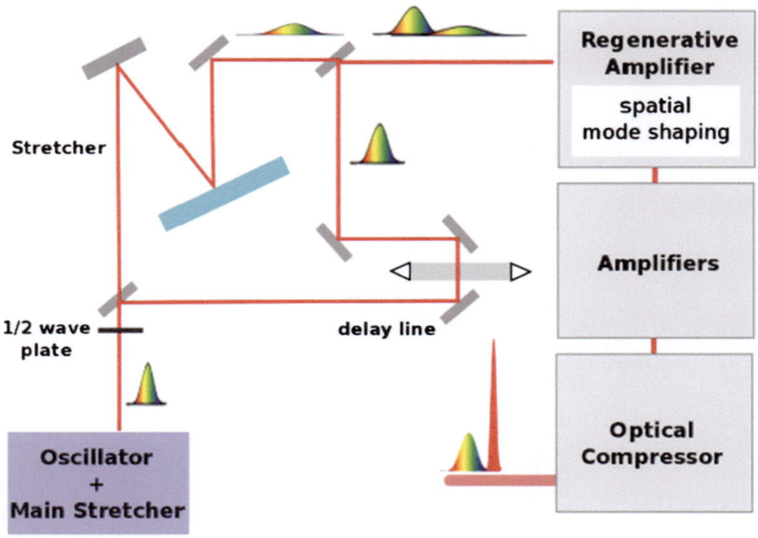

Fig. 2 Scheme of the Mach Zehnder like set up used in the laser driver for the DGRIP configuration. After separation, one part of the beam undergoes a second temporal stretching while the other part passes through a motorized delay line before spatial recombination and injection in the amplification chain

3 SXRL Optimization Experiments

3.1 Lifetime Improvement

In a first approach, we are interested in the improvement of the target lifetime with respect to ASE level by modifying the number of round trips added in the RGA. Three different cases are considered to observe the role played by the ASE level (Fig. 4). In all cases, the single shot SXRL emission is recorded at 10 Hz with a XUV 16 bits camera, and integrated energy is calculated. With no round trips added in the cavity (blue plot), the average integrated XUV energy is around 3000 shots. This configuration shows an unstable energy regime, then a progressive decrease of the signal after 200 shots. When we delay by 2 round trips the pulse extraction (violet plot), the SXRL energy is two times greater. Even if the emission decreases faster, reflecting a higher consumption velocity (Fig. 4(b)), SXRL is still effective over 500 shots. Thus, we are not only able to obtain higher SXRL energy but we can also use the same target position for a greater number of efficient shots. On the contrary, with 3 supplementary round trips (red curve), the target lifetime is distinctly worse than the two previous cases. The SXRL energy drops back to the initial level but rapidly falls after only 150 shots.

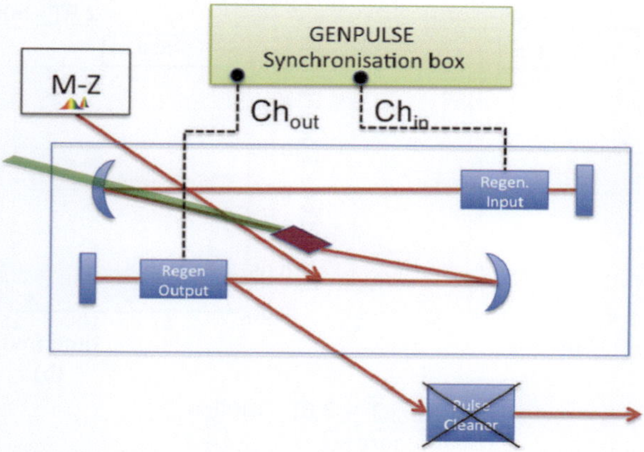

Fig. 3 Regenerative amplifier scheme. After passing through the Mach-Zehnder setup to get the double pulse structure the beam is injected in the RGA. The coarse and the fine adjustments of the pulse extraction delay allow to modify respectively ASE level and small prepulse intensity in the temporal laser profile. For this experiment the pulse cleaner located at the output of the RGA for high contrast pulse generation has to be bypassed in order not to trunk the complex shaped temporal profile

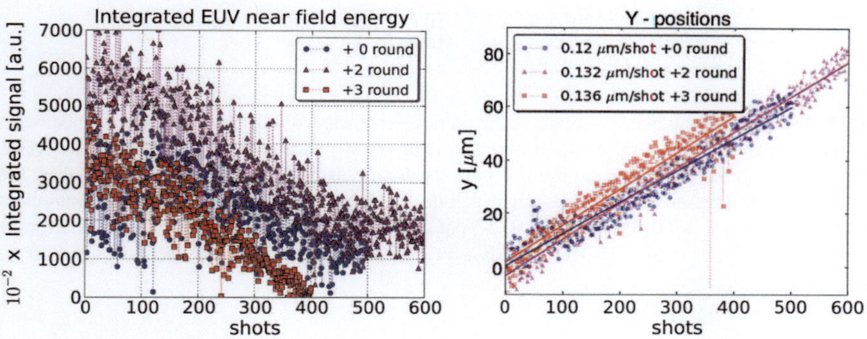

(a) (b)

Fig. 4 (**a**) Integrated EUV near field energy with different number of round trips (RT) added in the RGA from nominal pulse extraction. With 2 RT (*violet curve*), SXRL energy and target lifetime are improved compared to the nominal RGA configuration (*blue curve*). With 3 more RT, maximum energy is not higher and the lifetime behaviour is really poor showing no SXRL emission after 400 shots. (**b**) Evolution of the position of the near-field SXRL emission for different number of added RT. The target consumption increases with the number of RT, while the slope of the penetration into the target increases from the *blue curve* to the *red*

3.2 Short Prepulse Effect

The short prepulse comes from a leak from the RGA due to a small amount of polarization induced ellipticity of the laser pulse when passing through the PK during the high voltage slope. The energy level of the short prepulse can be coarsely controlled

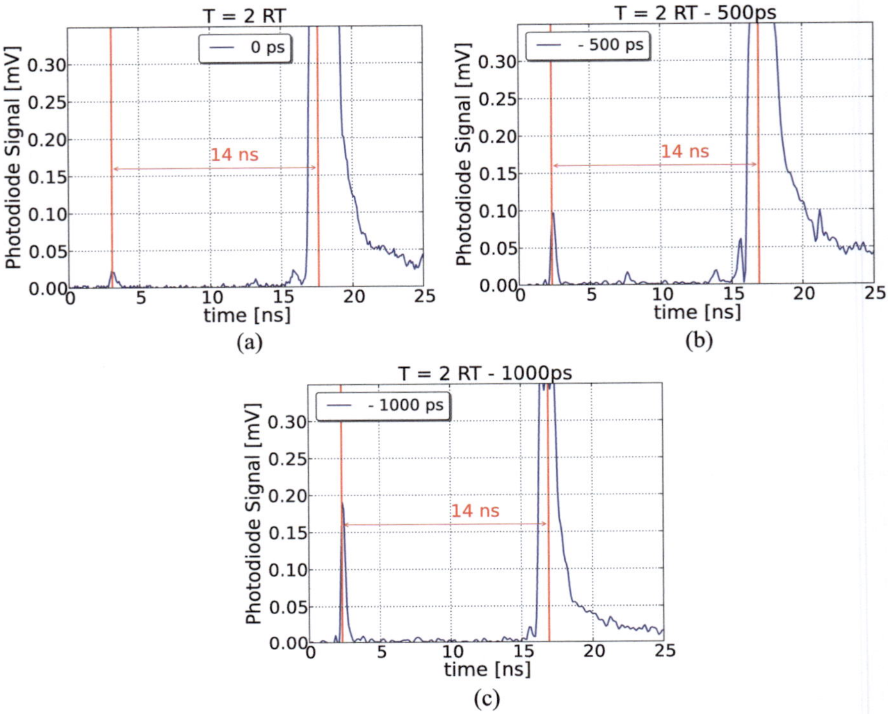

Fig. 5 Fast photodiode measurements of the prepulse intensity for different fine delay adjustments of PK_{OUT} from the temporal position corresponding to two additional passes (RT) in the RGA. The main saturated pulse is composed of both the long plasma pulse of 250 ps followed by the pump pulse compressed down to 3 ps which can not be temporally resolved with this measurement system. (**a**) $Ch_{out} = 2$ RT, low prepulse (0.5 % of main pulse). (**b**) $Ch_{out} = 2$ RT − 500 ps, average prepulse intensity >20 mJ (2 % MP). (**c**) $Ch_{out} = 2$ RT − 1000 ps, High Prepulse intensity >40 mJ (4 % MP)

by a fine adjustment of PK_{OUT} timing with a precision limited by the intrinsic jitter of the PK trigger. The prepulse exiting the RGA will then be amplified in the following succession of multipass amplifiers.

The prepulses are observed with an ultrafast photodiode (30 ps rising time) associated with a 2.4 GHz oscilloscope. Measurements have been done in three different cases by tuning prepulse intensity by delaying the timing by 500 ps steps. Figure 5 shows that a negligible prepulse of 0.5 % of the main pulse is located 14 ns before the main pulse. This configuration corresponds to the optimal coarse delay for SXRL generation located 2 round trips after the maximum peak of RGA amplification as depicted in the previous section. The round trip duration being of 14 ns, it evidences that prepulse takes place at the round trip preceding the main extraction. We can also observe that the shift of PK triggering by 500 ps increases four times the prepulse energy (b), which doubles once more when adding 500 ps once again (c).

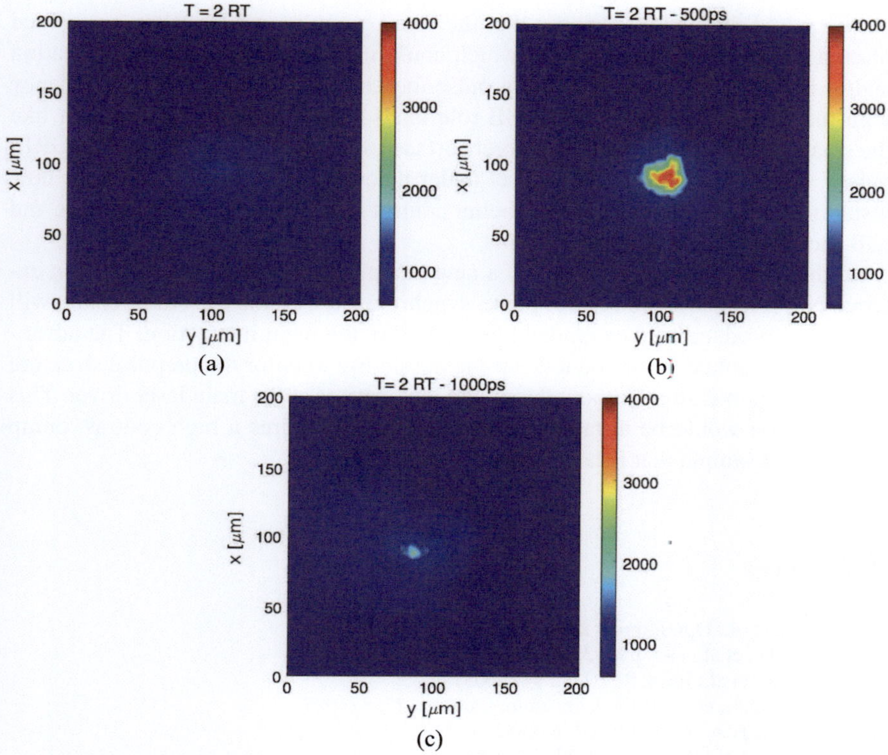

Fig. 6 Integrated SXRL near field profile measured with a 16 bits XUV CCD camera for different fine delays corresponding to the prepulse intensities of Fig. 5 in the configuration corresponding to 2 supplementary round trips in the RGA. (**a**) Very low prepulse intensity (0.5 % MP). (**b**) Average prepulse intensity (2 % MP). (**c**) High prepulse intensity (4 % MP)

Figure 6 shows near field SXRL measurements made for the best ASE configuration described previously (RGA tuned with 2 supplementary round trips with respect to nominal extraction) for the three fine delays of 0.500 ps and 1 ns corresponding to the prepulse levels of Fig. 5. This figure evidences a strong energy enhancement with the presence of an average prepulse (Fig. 5(b)) by comparison with the cases of no (Fig. 5(a)) or intense (Fig. 5(c)) prepulse.

4 Conclusion and Future Work

This empirical approach confirms that SXRL efficiency strongly depends on the pump laser temporal structures in the ten's of nanosecond range. In this paper, we saw that RGA cavity tuning allows control on both ASE level and a picosecond prepulse intensity. By modifying these parameters, we observed an improvement of the SXRL efficiency and a better target lifetime. Interpretations are still in discussion, but we could reasonably suppose that the small prepulse creates a low preionized plasma in front of the solid target which could increase the main plasma creation and/or its characteristics by the long pulse then increasing the coupling efficiency with the short pump pulse. The ASE role is also discussed, but it could act, like the small prepulse, to prepare the target surface for a better interaction with DGRIP pulses. All these suppositions are yet under theoretical investigation. We are now installing a new system allowing to better control stability of the small prepulse and will soon reproduce these experiments.

In the same frame, we will study a new pumping configuration in which an external Nd:YAG nanosecond laser pulse, synchronized with the CPA laser chain, will be used to produce cold expanded plasma before the main interaction. The advantage of this method would be to keep the possibility to tailor the prepulse structure for SXRL optimization without degrading the contrast of the main laser driver. This configuration would be attractive if an experiment requires a high contrast pump irradiating a sample that is later probed by the SXRL.

References

1. Habib, J., et al.: Opt. Express **20**(9), 10128–10137 (2012)
2. Zimmer, D., et al.: Opt. Lett. **35**(4), 450–452 (2010)
3. Keanan, R., et al.: Phys. Rev. Lett. **94** (2005)
4. Martinez, O.E., et al.: IEEE J. Quantum Electron. **23**, 8 (1987)
5. Janulewicz, K.A., et al.: Phys. Rev. E **82**, 5 (2010)
6. Kim, H.T., et al.: Phys. Rev. A **77**, 2 (2008)
7. Janulewicz, K.A., et al.: Phys. Rev. A **68**, 5 (2003)

Chapter 16
X-Ray Laser Developments at PHELIX

B. Ecker, B. Aurand, D.C. Hochhaus, P. Neumayer, B. Zielbauer, K. Cassou, S. Daboussi, O. Guilbaud, S. Kazamias, T.T.T. Le, E. Oliva, L. Li, H. Zhao, Q. Jin, D. Ros, P. Zeitoun, and T. Kuehl

Abstract We present results of two experimental campaigns dedicated to improve the performance of our Mo XRL source. A first beamtime was dedicated to study the XRL output when the target is pumped with three collinear pulses. The results show that the first pulse can be used to improve the spatial profile of the XRL beam. The second project focused on an injection-seeding experiment utilizing a double-stage XRL setup. Here we were able to characterize the amplifier medium by a temporally resolved measurement of the size of the gain zone. The amplified XRL beam exhibited an increased beam quality of almost two orders of magnitude in brilliance.

B. Ecker · B. Zielbauer
Helmholtz Institute Jena, Helmholtzweg 4, 07743, Jena, Germany

B. Ecker · B. Aurand · T. Kuehl
Johannes-Gutenberg Universität Mainz, Saarstr. 21, 55122, Mainz, Germany

B. Aurand · D.C. Hochhaus · P. Neumayer
ExtreMe Matter Institute EMMI, Planckstr. 1, 64291, Darmstadt, Germany

B. Aurand · B. Zielbauer (✉) · T. Kuehl
GSI Helmholtzzentrum fuer Schwerionenforschung, Planckstr. 1, 64291, Darmstadt, Germany
e-mail: b.zielbauer@gsi.de

K. Cassou · S. Daboussi · O. Guilbaud · S. Kazamias · D. Ros
LASERIX-CLUPS, Université Paris-Sud 11, 91400, Orsay, France

T.T.T. Le · E. Oliva · L. Li · P. Zeitoun
Laboratoire d'Optique Appliquée, 91761 Palaiseau cedex, France

D.C. Hochhaus
Goethe-Universität Frankfurt, 60438 Frankfurt, Germany

H. Zhao · Q. Jin
Institute of Modern Physics, Chinese Academy of Sciences, 73000 Lanzhou, People's Republic of China

S. Sebban et al. (eds.), *X-Ray Lasers 2012*, Springer Proceedings in Physics 147,
DOI 10.1007/978-3-319-00696-3_16,

1 Motivation

The main motivation for X-ray laser (XRL) research at GSI is to perform spectroscopy experiments on highly-charged heavy-ions stored in the experimental storage ring (ESR) of the GSI accelerator facility. The first experiment of this kind will aim at measuring the $2s_{1/2} - 2p_{1/2}$ transition in Li-like ions. For ions of an atomic number between 50 (Sn) and 92 (U), this transition energy lies between 100 eV and 300 eV [1], which corresponds to wavelengths between 12 nm and 4 nm. Setting up the experiment in a way, that the XRL is counter propagating to the ion bunch, one can exploit the relativistic Doppler effect. This way, the XRL wavelength as seen by the ion bunch will be blue-shifted according to $\lambda' = \lambda \cdot \sqrt{\frac{1-\beta}{1+\beta}}$. Here λ' and λ denote the Doppler-shifted and original wavelength of the XRL, respectively. The velocity β of the ion bunch stored in the ESR can be as high as 0.7. For a Mo XRL of 18.9 nm, changing the ion velocity between 0.4 and 0.7 will correspond to Doppler-shifted wavelengths between 12 nm and 8 nm. This will be sufficient to excite the $2s_{1/2} - 2p_{1/2}$ transition in ions of atomic numbers up to 67 (Li-like Ho). In addition, adjusting the velocity of the ion bunch allows for tuning the otherwise fixed XRL wavelength to resonance with the ionic transition. The precise measurement of this transition energy in high-Z ions will provide a valuable benchmark for QED theories, which aim at modeling the complex electromagnetic interactions within such few-electron systems.

In order to support the spectroscopy experiments with the most suitable XRL source, a number of preparations and developments are required. The XRL pulse energy for this experiment has been estimated to at least 1 μJ. In addition, the XRL beam quality is of concern, since the XRL pulse needs to propagate over approximately ten meters, ensuring good spatial overlap with the ion beam at the same time. Due to radiation safety measures, access to the ESR area during the experiment will not be possible without a significant loss of time. Thus, the complete XRL setup needs to be remote-controlled and support hours of stable XRL operation.

To address these issues, we pursued two different approaches. In a first experimental campaign, we performed multi-parameter scans in order to optimize our conventional Mo XRL source, especially in terms of output energy. Within this context, we examined the influence of a third pumping pulse, which is applied to the target before the conventional double-pulse. This technique has already been studied both theoretically [2] and experimentally. The third pulse has proven to have a significant impact on the minimum pumping energy required for XRL operation [3, 4], on target lifetime [5] and on the size of the gain zone [6].

Our second approach aimed at operating two Mo XRL targets in a seed-amplifier configuration. It has been demonstrated in former experiments that such a setup allows for a significant increase in beam quality [7–9]. In our opinion, this approach is an interesting alternative to high-order harmonic (HH) seeding, where the pulse duration of the seed (usually few tens of fs) is significantly smaller compared to the lifetime of the gain within the amplifier medium (few ps). Several groups already succeeded in demonstrating the amplification of a HH seed in various XRL media [10–14]. However, the spectral and especially the temporal mismatch of the HH seed

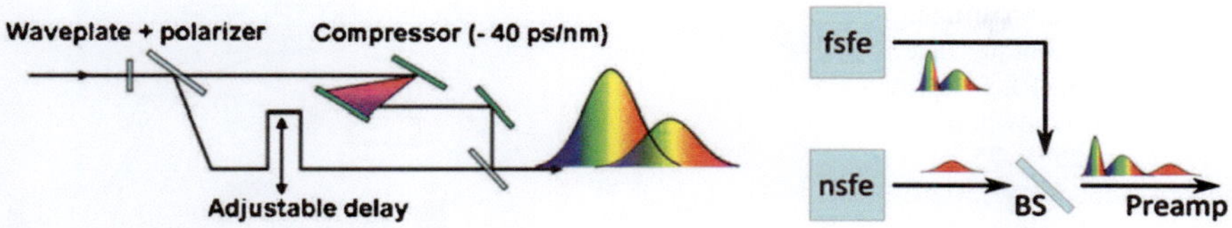

Fig. 1 *Left side*: Sketch of the Mach-Zehnder interferometer creating the double-pulse. *Right side*: The combination of both frontends using a beam splitter (BS) leads to three collinear pulses

might ultimately cause constrains in efficiently extracting the energy stored in the amplifying XRL medium [15]. In this paper, we present the results of both beam-times dedicated to the optimization and characterization of our Mo XRL source.

2 The PHELIX Preamplifier System

The PHELIX laser system provides two independent frontends delivering pulses of either nanosecond duration (ns-frontend) or femtosecond duration (fs-frontend). The fs-frontend is based on chirped pulse amplification (CPA). The original ~100 fs pulses delivered by the master oscillator are stretched to several nanoseconds before being injected in two consecutive regenerative amplifiers. The fs-frontend delivers pulses of up to 20 mJ at a repetition rate of 10 Hz. Between the stretcher and the first amplifier, a Mach-Zehnder interferometer allows for the creation of a double-pulse structure (left side of Fig. 1). The energy ratio of the two pulses can be varied continuously by use of a wave plate and a polarizer and the timing between the two pulses can be adjusted between 0 ns and 2 ns by a delay stage. One of the two arms of the interferometer provides a pulse compressor, which enables setting up different pulse durations after the final compression stage of the CPA chain.

Due to an upgrade of the timing system of the PHELIX laser within the last year, it is now possible to combine this double-pulse with a third pulse created by the ns-frontend. This delivers ~20 mJ pulses at a repetition rate of 0.5 Hz, while the pulse duration can be varied between 1.5 ns and more than 15 ns. This allows for setting up a series of three collinear pulses with variable energy ratios and delays (right side of Fig. 1).

The pulse series is injected into the preamplifier section, providing a total energy of up to 5 J at a repetition rate of one shot in two minutes. Within the preamplifier, a serrated aperture is used to define the size as well as the geometrical shape of the beam. Besides circular beams of various diameters, one can choose an aperture that provides two vertically separated, square beams. This double-beam option, originally intended to support pump-probe experiments, has been used for the double-stage XRL experiment. The experimental results presented in this paper have been performed in the X-ray laboratory of the PHELIX facility. It comprises the final pulse compressor as well as the target chamber. More information on the complete PHELIX system, including the main amplifier and the Petawatt target area, can be found in [16].

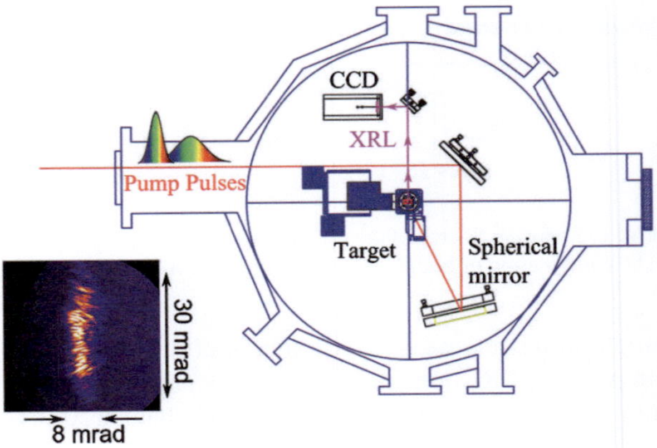

Fig. 2 XRL setup utilizing the DGRIP technique. The *inset* shows a XRL footprint obtained with the optimized parameters: 25 %–75 % energy ratio; delay 250 ps; main pulse duration 3 ps

3 DGRIP Mo XRL Performance

In order to optimize the output of our conventional double-pulse grazing incidence (DGRIP, [17, 18]) Mo XRL source, we performed scans of the most important pump parameters. The simple experimental configuration is shown in Fig. 2. The collinear double-pulse is focused on the target by a spherical mirror. The XRL output is detected by a footprint camera positioned 40 cm away from the target. A flat 45° XUV mirror was used both to align the XRL output on the CCD and to suppress the background of the plasma emission. In addition, Al filters of various thicknesses have been employed.

Two delay scans have been performed for energy ratios of 25 %–75 % and 50 %–50 % (prepulse–main pulse), varying the delay between 150 ps and 500 ps in steps of 50 ps or even 25 ps. For the optimum energy ratio of 25 %–75 % and 250 ps delay, we performed a main pulse duration scan between 2 ps and 5 ps, finding the highest XRL pulse energy at 3 ps. Five consecutive shots have been recorded for each set of parameters at a repetition rate of one shot every two minutes.

The target was irradiated by a total pump energy of 600 mJ, which was focused to a line of 6 mm length and 70 μm width under a GRIP angle of 25°. The pulse durations of 200 ps and 3 ps and the energy ratio of 25 %–75 %, lead to intensities of $3 \times 10^{11}\ \mathrm{W/cm^2}$ and $6 \times 10^{13}\ \mathrm{W/cm^2}$ for the prepulse and main pulse, respectively. The obtained pulse energies of the XRL output were in the order of several hundreds of nanojoules, corresponding to conversion efficiencies slightly below 10^{-6}. This pulse energy is not sufficient to support the ESR spectroscopy experiment. However, a gain length measurement series showed that the XRL has not been saturated. Setting up the experiment with a line focus length larger than 6 mm may thus allow for achieving higher XRL pulse energies, especially considering that the PHELIX preamplifier was not yet operated at its maximum energy output. Another concern is the non-ideal beam quality of the XRL source. Its divergence in the direction away from the target added up to 8 mrad. In the other direction, the XRL exhibited a significantly larger divergence of 30 mrad. This aspect ratio does not allow for an efficient spatial overlap with the ion beam. This issue has been addressed by applying the third pumping pulse.

4 Triple-Pulse Pumping

As described in Sect. 2, the additional pulse of the ns-frontend is combined with the double-pulse before the preamplifier section, to form a series of three collinear pulses. Thus, no change of the setup as depicted in Fig. 2 was necessary for switching from double-pulse pumping (DGRIP) to triple-pulse pumping (TGRIP). Since the two frontends are temporally synchronized, the delay between the double-pulse and the third pulse could easily be varied by use of a Stanford delay generator. In the following, this delay is defined as the time between the maximum of the early third pulse and the maximum of the main pulse.

Starting with the optimized DGRIP parameters concerning the double-pulse, we performed several scans with different durations (1.5 ns, 3 ns and 5 ns) and energies (250 mJ and 500 mJ) of the third pulse. For each energy and pulse duration setting, we varied the delay of the third pulse from 7.5 ns to 0.5 ns in steps of 1 ns. Afterwards, the delay between the double-pulse was varied from the original 250 ps to 500 ps and 700 ps, performing another set of third-pulse delay scans. Again, five consecutive shots for each parameter setting have been measured to increase the quality of the data.

In terms of XRL pulse energy, the optimum conditions were found with the original double-pulse optimized in DGRIP and a 1.5 ns long third pulse of 500 mJ, which resulted in an intensity of $1 \times 10^{11}\ \mathrm{W/cm^2}$. The optimum delay of this pulse was confined to a time window between 2 ns and 3 ns. For larger delays (corresponding to an earlier arrival of the third pulse), the XRL output was significantly suppressed. With a delay of 2 ns or 3 ns, the third pulse affected the shape of the XRL output, leading to a symmetric XRL footprint with a divergence of 15 mrad in both directions. The energy of the TGRIP XRL was in the same order of magnitude than the original XRL pulses pumped with two pulses. However, the change in the aspect ratio of the XRL pulse is of interest for the ESR spectroscopy experiments, as it allows for a good spatial overlap between the XRL pulse and the ion beam. For delays smaller than 2 ns we observed the transition back to the original footprint patterns known from the DGRIP scheme.

5 The Double-Stage Mo XRL Experiment

The double-stage XRL experiment has been carried out using the Butterfly configuration [19]. As depicted in Fig. 3(a), it is based on two XRL media that are pumped in the DGRIP scheme, providing both XRL targets with TW excitation. Additionally, DGRIP allows for a very compact and simple injection-seeding setup relying only on two beams and one target. The seed XRL is injected into the amplifying

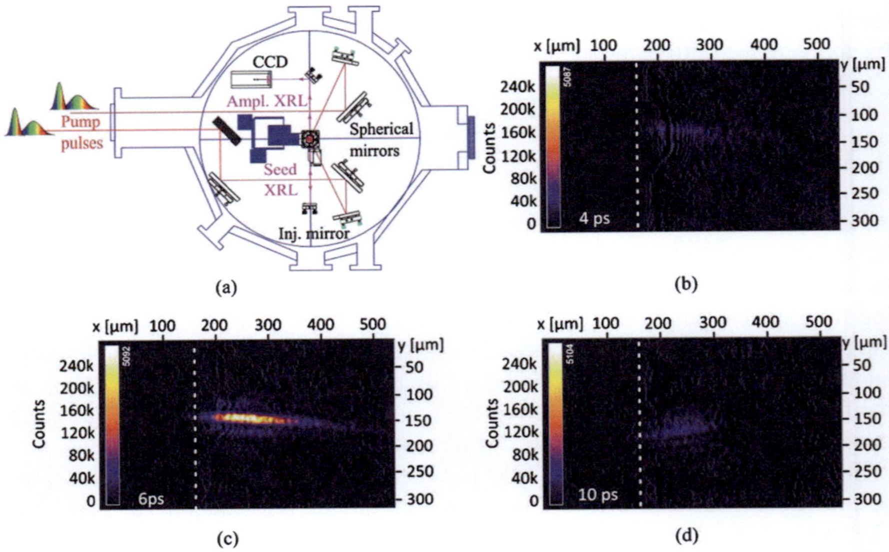

Fig. 3 (**a**): Sketch of the Butterfly setup. (**b–d**): Signal of the amplified seed for different delays between the seed and amplifier heating pulse. The *white dashed lines* indicate the target surface

medium by a spherical XUV mirror. A 45° XUV mirror is then used to align the amplified seed to the footprint camera.

The beams are created using the PHELIX double-beam option described in Sect. 2. Each of the beams is carrying an identical double-pulse structure and is focused onto the same Mo slab target under a GRIP angle of 25°, vertically separated by a few millimeters. The size of the line focus providing the seed (amplifier) was 5 mm × 90 µm (2.5 mm × 220 µm). The pump energy was 300 mJ for each beam, which resulted in an intensity of 8×10^{10} W/cm^2 (prepulse) and 2×10^{13} W/cm^2 (main pulse) for the seed XRL. The amplifier target was irradiated with 7×10^{10} W/cm^2 (prepulse) and 1×10^{13} W/cm^2 (main pulse). The injection mirror had a focal length of 120 mm and was positioned 275 mm away from the target. The resulting spatial coupling of the seed XRL into the amplifier medium was 2 %.

The timing between the two IR beams has been adjusted by a fast rise-time photo diode and a 12 GHz oscilloscope. This method allowed for synchronizing the main pulses of each IR beam within an accuracy of ±15 ps, which is much larger than the lifetime of the gain and the XRL pulse duration. The absolute timing between the two beams is therefore not defined. Only the relative delay, which is set by a motorized delay stage positioned outside the target chamber and provides sub-picosecond resolution, can be interpreted. In order to achieve a simultaneous arrival of the seed XRL and the IR main pulse at the amplifier target, the delay of the two IR main pulses has been set to 1835 ps, corresponding to 2 × 275 mm/c. This will be referred to as a delay of 0 ps. However, due to the large uncertainty of the synchronization method, this absolute value of the delay is not significant. The optimum timing had to be found by scanning the delay of the IR main pulses within the 30 ps window. In the following, smaller delays correspond to an earlier arrival time of the seed at the amplifier and vice versa.

Figure 3(b)–(d) depict three exemplary results of a timing scan. The complete measurement series comprises a time window of 12 ps, scanned with 1 ps steps. The data shows the temporal development of the amplifier medium, in terms of size, distance to the target and magnitude of the gain. The scale in Fig. 3(b)–(d) corresponds to the exit plane of the target and has been calibrated using a 50 μm wire attached to the target. In addition to the determination of the lifetime of the gain of the Ni-like Mo XRL to 4 ps FWHM, these results could provide valuable input for hydro simulations. The data obtained at 6 ps (Fig. 3(c)) shows the maximum size of the gain zone, extending more than 200 μm in the direction away from the target. In the perpendicular direction, the size of the gain medium does not vary for different delays and amounts to 25 μm, which is significantly smaller than the width of the line focus (220 μm). This could indicate that the intensity of the IR pulses was not large enough to create a population inversion in the complete amplifier plasma.

The energy of the amplified pulse added up to 90 nJ, which is comparable to the 140 nJ of the seed pulse. The divergence of the amplified beam was 4 mrad × 0.5 mrad, which is significantly improved compared to the 8 mrad × 30 mrad of the single XRL-target setup. Hence, the amplification process helped to increase the brilliance of the source by a factor of 80.

6 Conclusions and Outlook

Performing detailed multi-parameter scans, we optimized our single-target Mo XRL source pumped in DGRIP. Adding a third pulse of nanosecond duration before the original double-pulse did not lead to an increase in photon numbers. However, it was possible to improve the quality of the XRL beam with respect to its aspect ratio. This is of interest for the ESR spectroscopy experiments, where a good spatial overlap of the XRL and the ion beam is required.

In addition, we carried out an injection-seeding experiment using the Butterfly configuration. The data allowed for following the temporal development of the gain in terms of its geometrical size, lifetime and magnitude. The amplified XRL exhibited a brilliance 80 times larger compared to the single-target geometry. Further experiments of this kind will aim at determining saturation effects, increasing the spatial coupling efficiency and measuring the spatial coherence of the amplified XRL beam.

References

1. Kim, Y.-K., et al.: Phys. Rev. A **44**, 148 (1991)
2. Janulewicz, K.A., Kim, C.M.: Phys. Rev. E **82**, 056405 (2010)

3. Balmer, J., et al.: In: Sebban, S., et al. (eds.) X-Ray Lasers 2012. Springer Proceedings in Physics, vol. 147. Springer, Cham (2013). Chapter 5 in this book
4. Ursescu, D., et al.: Ag X-ray laser pumped with one long and two short pulses
5. Habib, J., et al.: Opt. Express **20**, 10128–10137 (2012)
6. Cassou, K., et al.: Opt. Lett. **32**, 139–141 (2007)
7. Sebban, S., et al.: Phys. Rev. A **61** (2000)
8. Nishikino, M., et al.: Phys. Rev. A **68**, 1–4 (2003)
9. Tanaka, M., et al.: Opt. Lett. **28**, 1680–1682 (2003)
10. Ditmire, T., et al.: Phys. Rev. A **51**, 6 (1995)
11. Zeitoun, Ph., et al.: Nature **431** (2004)
12. Hasegawa, N., et al.: Phys. Rev. A **76** (2007)
13. Mocek, T., et al.: Phys. Rev. Lett. **95**, 19–22 (2005)
14. Wang, Y., et al.: Nat. Photonics **2**, 94–98 (2008)
15. Zeitoun, P., et al.: Time-dependent Bloch-Maxwell modelling of high harmonic seeding in gas and solid amplifier
16. Bagnoud, V., et al.: Appl. Phys. B **100**, 1 (2010)
17. Zimmer, D., et al.: Opt. Express **13**, 2403 (2008)
18. Ros, D., et al.: Nucl. Instrum. Methods Phys. Res., Sect. A **653**(1), 76–79 (2011)
19. Ecker, B., et al.: In: Proc. SPIE, vol. 8140 (2011). doi:10.1117/12.897277

Chapter 17
Parabolic Equation and Exact Transparent Boundary Conditions in X-Ray Optics—Application to Waveguides and Whispering Gallery Optics

R.M. Feshchenko and A.V. Popov

Abstract In this paper we report a novel transparent boundary (TBC) condition for the 2D parabolic wave equation in which the dielectric constant is linearly dependent on the transversal coordinate. The proposed condition does not involve special functions and can be a basis for an efficient finite difference approximation.

1 Introduction

Leontovitch and Fock introduced the parabolic wave equation (PWE, also known as the Fresnel equation) as applied to radio wave propagation more then half a century ago [1]. It was widely used in oceanic acoustics for modeling sound wave propagation [2, 3]. Further applications have been found in optics, including X-ray and EUV spectral band [4] where it is used to describe the propagation of weakly divergent light beams in inhomogeneous media. However any numerical solution of PWE, for instance by finite difference approximation, requires appropriate truncation conditions as realistic computational domains must be necessarily finite, while the original wave field is sought in the whole space.

In this paper we will be only concerned with the linear 2D parabolic wave equation having the following form:

$$2ik\frac{\partial u}{\partial z} + \frac{\partial^2 u}{\partial x^2} + k^2\alpha(x,z)u = 0, \quad \alpha(x,z) = \alpha_0(x,z) + axg(z), \tag{1}$$

where x and z are, respectively, longitudinal and transversal coordinates, a is a parameter and $g(z)$ is a real function, $k = 2\pi/\lambda$ is the wave number and $\alpha_0(x,z)$ is a finite arbitrary function. Equation (1) describes, for instance, the propagation of light in a weakly bent X-ray waveguide or optical fiber in a curvilinear system

R.M. Feshchenko (✉)
P.N. Lebedev Physical Institute of RAS, Moscow, Russia
e-mail: rusl@sci.lebedev.ru

A.V. Popov
Pushkov Institute of Terrestrial Magnetism, Ionosphere and Radiowave Propagation, Troitsk, Russia

S. Sebban et al. (eds.), *X-Ray Lasers 2012*, Springer Proceedings in Physics 147,
DOI 10.1007/978-3-319-00696-3_17,

of coordinates. Function $\alpha(x, z)$ in this case is a sum of a true dielectric constant $\alpha_0(x, z)$, which, for simplicity, is supposed to vanish at the boundary of the computational domain and beyond it, and an additional linear term resulting from bending and proportional to the curvature $g(z)$.

There are two main types of boundary conditions for parabolic equation (1): transparent boundary conditions (TBCs) [5, 6], and perfectly matched layer (PML) conditions [7]. The TBCs are generally non-local Neumann-to-Dirichlet or Dirichlet-to-Neumann mappings relating the wave field boundary values with its first derivative by x, whereas PMLs are local.

In the case of standard 2D parabolic wave equation in free space ($\alpha = 0$), the TBC now is known for more than 20 years as Baskakov-Popov-Papadakis (BPP) condition [3, 5–8]. A TBC for Eq. (1) with $g(z) = 1$ has been known for a long time as well [6]. However, involving integration of a ratio of Airy functions to obtain the kernel, it is rather complicated and difficult for numerical implementation [9].

In this paper we contrive to obtain a much simpler TBC for the 2D parabolic equation (1) with a linear potential, which does not involve special functions and additional integration, having a kernel explicitly expressed via elementary functions. We also try to preserve as much generality as possible by considering the case when the curvature $g(z)$ is an arbitrary function.

2 Transparent Boundary Condition

Let us consider the parabolic wave equation (1) where function $\alpha_0 = 0$. It can be transformed by introducing a new variable $\varphi(x, z)$:

$$u(x,z) = \varphi(x,z)\exp\left[i\frac{kax}{2}G(z) - i\frac{ka^2}{8}\int_0^z G^2(\xi)\,d\xi\right], \quad G(z) = \int_0^z g(\xi)\,d\xi. \tag{2}$$

By substituting (2) into (1) one obtains an equivalent equation for the function φ:

$$2ik\frac{\partial\varphi}{\partial z} + \frac{\partial^2\varphi}{\partial x^2} + ikaG(z)\frac{\partial\varphi}{\partial x} = 0. \tag{3}$$

The advantage of this equation is independence of its coefficients of x, which allows us to obtain a transparent boundary condition by applying Laplace transform by this variable, which yields

$$\frac{\partial F}{\partial z} = \left[\frac{iw^2}{2k} - \frac{aw}{2}G(z)\right]F - \left[\frac{iw}{2k} - \frac{a}{2}G(z)\right]\varphi_0(z) - \frac{i}{2k}\varphi_0'(z), \tag{4}$$

where $\varphi_0(z) = \varphi(0, z)$, $\varphi_0'(z) = \varphi_x'(0, z)$, w is Laplace variable and $F = F(w, z)$ is the Laplace transform of φ.

We shall assume that the initial wave field $u(x, 0) = 0$ for $x > 0$—the sources of the wave field are located where $x < 0$. This means that $F(w, 0) = 0$ as well. Now the unique solution of (4) can be written as

$$F = -\int_0^z \exp\left[\frac{iw^2}{2k}(z-\xi) - \frac{aw}{2}\int_\xi^z G(\mu)\,d\mu\right] \times \left\{\left(\frac{iw}{2k} - \frac{a}{2}G(\xi)\right)\varphi_0(\xi) + \frac{i}{2k}\varphi_0'(\xi)\right\} d\xi. \tag{5}$$

Applying the inverse Laplace transform to (5) and substituting in (2) we write down the following two equivalent TBCs:

$$u(x_0,z) = -\frac{1}{\pi\sigma}\int_0^z \exp\big[i\Phi(z,\xi)\big]\frac{u_x'(x_0,\xi)}{\sqrt{z-\xi}}\,d\xi + \frac{iak}{2\pi\sigma}\int_0^z \exp\big[i\Phi(z,\xi)\big]\frac{u(x_0,\xi)(\bar{G}-G(\xi))}{\sqrt{z-\xi}}\,d\xi, \tag{6}$$

$$\begin{aligned} u_x'(x_0,z) &+ \frac{\sigma a}{4}\int_0^z \exp\big[i\Phi(z,\xi)\big]\frac{u_x'(x_0,\xi)(\bar{G}-G(z))}{\sqrt{z-\xi}}\,d\xi \\ &= -\sigma\frac{\partial}{\partial z}\int_0^z \exp\big[i\Phi(z,\xi)\big]\frac{u(x_0,\xi)}{\sqrt{z-\xi}}\,d\xi \\ &+ \frac{i\sigma ka^2}{8}\int_0^z \exp\big[i\Phi(z,\xi)\big]\frac{u(x_0,\xi)G(z,\xi)(\bar{G}-G(z))}{\sqrt{z-\xi}}\,d\xi \\ &+ \frac{i\sigma kax_0}{2}g(z)\int_0^z \exp\big[i\Phi(z,\xi)\big]\frac{u(x_0,\xi)}{\sqrt{z-\xi}}\,d\xi, \end{aligned} \tag{7}$$

where

$$\Phi(z,\xi) = \frac{kax_0}{2}G(z,\xi) - \frac{ka^2}{8}(z-\xi)\big[\langle G^2\rangle - \bar{G}^2\big], \quad G(z,\xi) = \int_\xi^z g(\tau)\,d\tau, \tag{8}$$

$$\bar{G} = \frac{1}{z-\xi}\int_\xi^z G(\tau)\,d\tau, \quad \langle G^2\rangle = \frac{1}{z-\xi}\int_\xi^z G^2(\tau)\,d\tau, \quad \sigma = \sqrt{\frac{2k}{\pi i}}. \tag{9}$$

In a waveguide with a constant curvature $g(z) = 1$ we have: $G(z) = z$, $\bar{G} = (z+\xi)/2$, $\langle G^2\rangle = (z^2 + z\xi + \xi^2)/3$ and $\langle G^2\rangle - \bar{G}^2 = (z-\xi)^2/12$. After substituting these expressions into (6) and (7) we obtain boundary conditions having the form of convolutions. In the particular case $a = 0$, they are reduced to the ordinary BPP conditions (see [5]).

3 Conclusion

A novel TBC for the parabolic wave equation with a linear dependence of the dielectric susceptibility on the transversal coordinate x and an arbitrary dependence on the longitudinal coordinate z is obtained. It is different from the previously known [6] transparent condition for the constant linear potential. The latter has a kernel that is a complex integral of the logarithmic derivative of an Airy function that can not be expressed through elementary functions. In contrast, the kernels of all integrals in the TBC proposed here are elementary. Even more important is the applicability of this new TBC to the case of variable curvature of the artificial boundary, including possible change of sign. That fact makes it very promising for the simulation of realistic bent X-ray waveguides and optical fibers.

Acknowledgements This work was supported in part by the basic research program of Russian Academy of Sciences "Femtosecond optics and new optical materials" (subprogram "New optical materials"), and by the Russian Foundation for Basic Research (RFBR) grant No. 10-02-00991-a.

References

1. Fock, V.A.: Electromagnetic Diffraction and Propagation Problems. Pergamon, Tarrytown (1965)
2. Tappert, F.D.: The parabolic approximation method. In: Keller, J.B., Papadakis, J.S. (eds.) Wave Propagation and Underwater Acoustics. Lecture Notes in Physics, vol. 70, pp. 224–287 (1977)
3. Papadakis, J.S.: Exact nonreflecting boundary conditions for parabolic type approximations in underwater acoustics. J. Comput. Acoust. **2**, 83–98 (1994)
4. Attwood, D.: Soft X-rays and Extreme Ultraviolet Radiation. Cambridge University Press, Cambridge (1999)
5. Baskakov, V.A., Popov, A.V.: Implementation of transparent boundaries for numerical solution of the Schrodinger equation. Wave Motion **14**, 123–128 (1991)
6. Popov, A.V.: Accurate modelling of transparent boundaries in quasi-optics. Radio Sci. **31**, 1781–1790 (1996)
7. Berenger, J.: A perfectly matched layer for the absorption of electromagnetic waves. J. Comput. Phys. **114**, 185–200 (1994)
8. Yevick, D., Friese, T., Schmidt, F.: A comparison of transparent boundary conditions for the Fresnel equation. J. Comput. Phys. **168**, 433–444 (2001)
9. Ehrhardt, M., Mickens, R.E.: Solutions to the discrete Airy equation: application to parabolic equation calculations. J. Comput. Phys. **72**, 183–206 (2004)

Chapter 18
Harmonic Generation and Soft-X-Ray Laser with LASERIX: Source Development, Applications and Advanced Diagnosis

Olivier Guilbaud, Sophie Kazamias, Kevin Cassou, Moana Pittman, Sameh Daboussi, Olivier Delmas, Julien Demailly, Olivier Neveu, Chris Pouhe, Boris Vodungbo, Philippe Zeitoun, Lucy Wilson, Greg Tallents, Antonin Dusseix, Gabriel Richet, Aurelie Gense, Bich-Lien Nghiem, Brigitte Cros, Gilles Maynard, and David Ros

Abstract The LASERIX facility has restarted operation in the middle of the year 2010. Since then, important work has been achieved on the XUV sources performance, reliability and stability, which lead to the construction of two beamlines. The first one is based on transient collisional soft X-ray laser. In the second one, the high order harmonic generation process is used to convert infrared laser into coherent XUV radiations. Characteristics and some recent works on both of them will be presented. Besides, a complete setup for pump-probe experiments has been developed and successfully tested. Some examples involving this device and dealing with plasma opacity, radiobiology and nanomagnetism will be presented. We will conclude this paper by a discussion on how application experiments can be turned into valuable diagnostics for XUV sources.

O. Guilbaud (✉) · S. Kazamias · K. Cassou · S. Daboussi · O. Delmas · J. Demailly · O. Neveu · C. Pouhe · A. Dusseix · G. Richet · A. Gense · B.-L. Nghiem · B. Cros · G. Maynard · D. Ros
Laboratoire de Physique des Gaz et Plasma, CNRS-Université Paris Sud, Orsay, France
e-mail: olivier.guilbaud@u-psud.fr

M. Pittman · D. Ros
Centre Laser de l'Université Paris-Sud, Université Paris-Sud, Orsay, France

O. Delmas
Amplitude Technologies, Evry, France

B. Vodungbo · P. Zeitoun
Laboratoire d'Optique Appliquée, ENSTA ParisTech – CNRS – École Polytechnique, Palaiseau, France

L. Wilson · G. Tallents
University of York, York, UK

S. Sebban et al. (eds.), *X-Ray Lasers 2012*, Springer Proceedings in Physics 147,
DOI 10.1007/978-3-319-00696-3_18,

1 Introduction

As demonstrated by the papers presented during this conference, laser-based soft X-ray sources have now reached a sufficient level of maturity to be considered as valuable and available scientific tools for a large user community. XUV sources are sophisticated systems encompassing both the laser driver, the conditioning optics, the conversion medium, and specific XUV optics. In the Laserix facility two laser-based XUV sources are now in operation: a soft X-ray laser beamline and a high order harmonics beamline. We will briefly present these two devices, not only including some fundamental aspect of the wavelength conversion but also the technical developments on the laser driver and optical systems to achieve higher control on these sources. One particular feature of laser-based sources is their natural synchronization with the infrared laser. They are therefore highly interesting for pump-probe experiments and we will therefore give an overview of activities on this subject with Laserix. Due to their spectral range, specific tools are required to characterised the radiation properties of these sources. Among them, the temporal characterisation is one of the most challenging. As a prospect we will discuss the opportunity of using some ultrafast phenomena as a tool for soft X-ray laser temporal profile reconstruction.

2 Laserix Present Status

All the laser-based XUV sources of the facility are driven by the same high power laser chain. This system is a commercial Titanium-Sapphire laser delivering 800 nm pulses at 10 Hz repetition rate and that can be recompressed to duration as short as 35 fs (confirmed from SPIDER measurements). In a recent laser upgrade, KDP doubling crystals of the Nd-YAG pumping lasers have been replaced by LBO crystals leading to an increase of doubling efficiency and of the output energy before compression from 2 J to 3 J. This change leads also to a significant reduction of warming time and to a global improvement of the laser beam profile. Addition of remote monitoring and control of the laser beam at critical points of the laser chain enables a faster laser preparation. A schematic of the beamline is presented in Fig. 1. The laser output energy is separated into two main beams, with an adjustable energy balance. The first part is directly compressed under vacuum into a picosecond pulse and is used for TCE soft X-ray laser plasma pumping. The second beam can be used directly in order to get a high energy uncompressed (500 ps) pulse, mainly to form the plasma of the TCE soft X-ray laser. A variable part of this uncompressed beam can be sent into an auxiliary compressor and redirected in the high order harmonic beamline or as an auxiliary beam for pump-probe experiments. The harmonic beamline and the soft X-ray laser beamline are in a large extend independent but can also be used together in a same experiment. Since 2009, the Laserix facility provides XUV beamtime for users interested in soft X-ray sources development or in their use for application experiments. As a member of the European Laserlab network, Laserix shall provide four weeks every years of dedicated European beamtime.

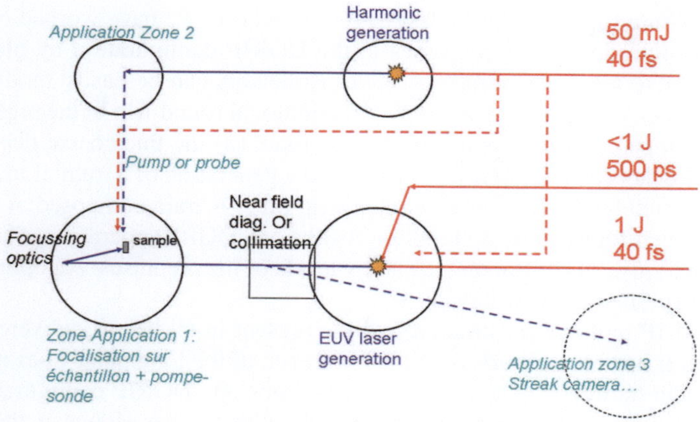

Fig. 1 General view of the XUV beamlines and their possible combinations

3 The Soft X-Ray Laser Beamline

The soft X-ray laser beamline requires most of the infrared laser energy. The soft X-ray laser source relies on the Transient Collisional Excitation (TCE) scheme [1] implemented in a GRIP (Grazing Incidence Pumping) irradiation geometry [2]. In the nominal configuration, the preforming long pulse is focused by a combination of cylindrical lenses in normal incidence compared to the target plane. The short (picosecond) pumping pulse is line-focused on the target by an off-axis spherical mirror with a grazing angle of 17 to 23 degrees on target. The delay between the long and short pulses can be adjusted with a delay line. Efficient lasing has been achieved with Ni-like Molybdenum (18.9 nm) and Ni-like Silver (13.9 nm). However, it has been observed that this scheme is sensitive to beam pointing instability which leads to an unperfect overlap of the two line-foci when the pointing is drifting. Recently a variant of the GRIP scheme has been proposed to overcome this drawback [3, 4]. In the DGRIP configuration, only the beam travelling through the compressor and focused by the spherical mirror is used and supports both long and short pulses. This train of pulses is synthesized between the stretcher and the regenerative amplifier cavity by a Mach-Zendher type device, with adjustable delay and energy balance. A Martinez Stretcher installed in one of the arms of the Mach-Zendher adds an extra chirp to the corresponding pulse. When recompressed by the final compressor, this pulse has a duration of 200 ps and will act as the long pulse of TCE X-ray laser. Beam pointing differences between the two pulses are strongly reduced by the regenerative amplifier cavity leading to an always perfect overlap of the two line-foci on target. Successful operation with Molybdenum, Silver and Ne-like Titanium (32.6 nm) have been achieved. The implementation of low energy prepulses prior to the two main pulses is known to improve the soft X-ray laser output. This has been checked in the case of DGRIP configuration. Moreover, these prepulses also lead to a significant extend of the time a soft X-ray laser can be generated from a same target position [5]. This property is useful for experiments requiring high photon exposure [4, 6]. Since the last SPIE conference, we have worked on different ways to introduce these prepulses in the DGRIP configuration by playing on the regenerative amplifier properties. Two parameters can be easily modified and lead to dramatic changes: (i) the number of additional round-trip in the regenerative amplifier compared to the optimal extraction time, (ii) the fine coarse delay of the extracting Pockels cell. Parameter (i) enables the generation of a train of low energy pulses separated by the regenerative round-trip time. A train composed of two prepulses has been found to be optimal for Ni-like Mo DGRIP operation. The second parameter (ii) gives a control on the energy levels of the prepulses. An optimum has also been experimentally observed.

The DGRIP multipulses structure will be present in all beams delivered by the laser chain. It has been observed that the presence of the long pulse does not affect the high order harmonic generation in Argon. Thus, the DGRIP configuration may contribute to make seeding experiment with solid target easier. However, the DGRIP approach is not well suited for pump-probe experiments requiring a high contrast in the auxiliary beam. If only single shot operation is required, GRIP configuration will be choosen instead. For experiment requiring both pump-probe approach and important shot accumulation, DGRIP can be used in conjunction with an auxiliary beam frequency-doubled with a LBO crystal and filtered by 400 nm coated mirrors.

4 The High Order Harmonics Beamline

In parallel to the soft X-ray laser, a high order harmonic beamline has been developed. Up to 30 mJ of infrared energy, compressed to 35 fs, are available for high order harmonic generation. A typical setup involves a one meter focal length lens and a gas cell, followed by a transmission grating spectrometer. Absolute measurement with a calibrated photodiode indicates XUV radiation in the nJ range per pulse and per harmonic when argon is used as a generation medium. This standard setup has been recently completed with a Mach-Zendher system placed before the focussing optics. This device is represented in Fig. 2. It enables the obtention of two infrared pulses and, after conversion, two harmonic pulses with variable delay, energy balance and directions. When the two output beams are collinear such a device can be used to probe the gain dynamics of the soft X-ray laser gain in the context of seeding experiments. When a small angle is introduce and the delay is set to zero, harmonic generation in the bissectrice direction of the two IR beams can be observed. This type of experiment may give physical insight in processes involved in some intracavity harmonic generation scheme. The device is designed to be easily switched to two-color harmonic generation experiments.

5 Some Pump-Probe Experiments Examples

The infrared beam feeding the harmonic beamline can also be used as an auxiliary beam for pump-probe experiments and can enter an interaction chamber placed

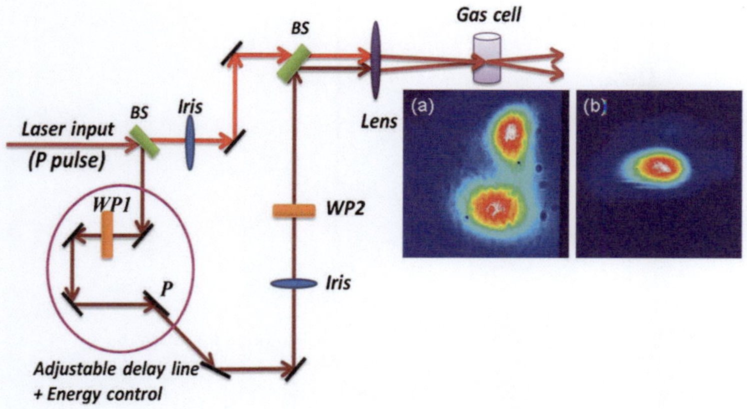

Fig. 2 Principle of the two harmonic pulses generation setup. *Inset*: farfield image of the harmonic beam. The two infrared pulses are quasi-collinear (1 mrad of angular separation) and delayed by 10 ps (*inset* (**a**)). The angular separation is also observed in the XUV range. For a 0 ps delay (*inset* (**b**)), on-axis harmonic generation can be observed

after the X-ray laser chamber. The soft X-ray laser is focussed onto a sample by an $f = 300$ mm multilayer spherical mirror, whereas the auxiliary beam is focussed by a short focal lens. The sample holder presented in Fig. 3 is equipped with a BBO crystal for precise timing between the pump and the probe. Besides, a YAG-Cerium crystal is mounted on the holder for XUV focal spot characterisation and spatial overlap checking. The soft X-ray laser beam has been used either as a probe or as a pump. Considering the soft X-ray laser as a pump, the XUV beam can induce permanent or transient changes in a polymer sample (polyvinyl-alcohol). The photoelectrons produced by the XUV photons will recombine and generate free radicals or new chemical species on time scales ranging from picosecond to permanent. This evolution can be followed with a visible light shining through the X-ray laser focal volume and absorbed by the induced species. For a proof-of-principle experiment, a CW He-Ne laser was used as a probing beam. Calculations showed that a one thousands variation of the He-Ne signal can be expected in the tens of nanosecond range following the interaction. A fast photodiode and an oscilloscope are therefore sufficient to record the evolution after the interaction, provided that the signal to noise ratio is high. The first experimental run was unsuccessful but helped us to identify the sources of noise and significant reduction has been achieves.

The soft X-ray laser is also recognized as an interesting probe in the context of hot dense plasmas. Recently we performed a plasma probing experiment in collaboration with the University of York. A sample composed of thin layer (50 nm) of Iron buried in a plastic (Parylene) foil was irradiated and heated by the compressed auxiliary beam in a non-normal incidence geometry and with a P polarisation. The change in opacity at 13.9 nm was followed with a Ni-like Silver X-ray laser shining through the sample. The size of the beam was greater than the heating pulse focal spot in order to get on a same image the transmission in the perturbed and

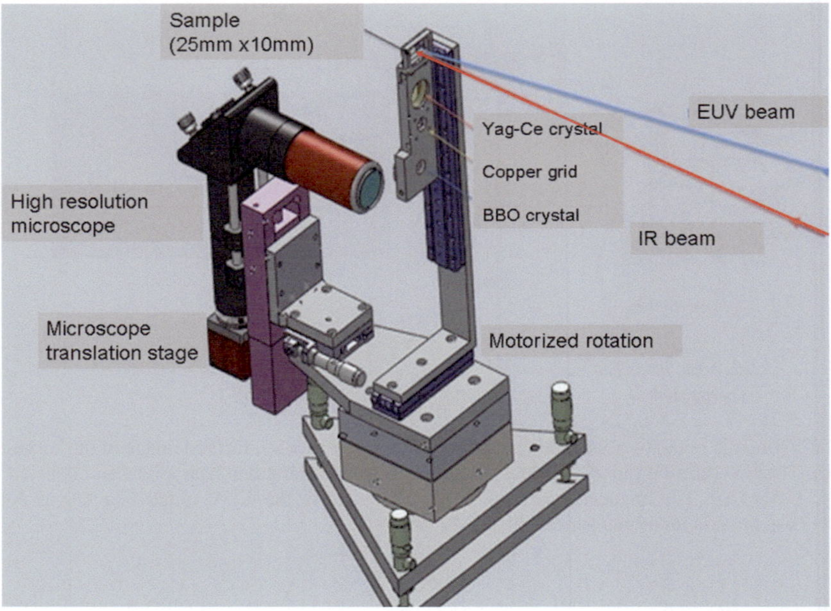

Fig. 3 Schematic of the pump-probe sample manipulator

unperturbed material. This type of experiment may provide important insight in suprathermal electron generation and propagation, heating by return current, and non-local thermal conduction in solid density plasmas.

6 Towards Advanced Soft X-Rays Diagnostics

The determination of soft X-ray laser temporal profile is an important but delicate task. Two techniques have been so far experimented. XUV streak cameras have already provided important results on soft X-ray lasers pulse durations. But their dynamical range is small, their resolution limited to the picosecond range and they are affected by important jitter. The other techniques relies on two color (XUV-IR) ionization of gas, and is in essence a cross-correlation measurement between the IR field (known) and the XUV field (unknown). This techniques potentially offers femtosecond resolution but requires electron spectroscopy in a magnetic bottle spectrometer where ultrahigh vacuum is mandatory. An ideal approach would be to get some sample that, when exposed to a short laser pulse, would experience a fast change in XUV transmission or reflectivity. The measurement of the transmitted/reflected XUV signal has a function of the delay between the short pulse and the XUV pulse would give, after deconvolution, the temporal profile of the XUV pulse. This measurement would be jitter-free, and no ultrahigh vacuum would be required. Finally, by exploiting focussing geometry generating travelling waves, in

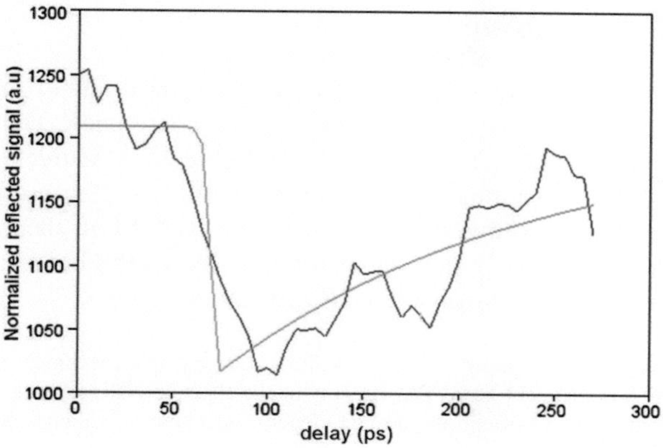

Fig. 4 Normalized reflected XUV signal on the magnetic sample as a function of the delay between the heating pulse and the X-ray laser pulse. A fitting curve by a simple convolution model corresponding to a 5 ps X-ray laser duration is superimposed on experimental data

association with XUV imaging optics, it should be possible to get the correlation function in single shot.

It has been demonstrated recently that the XUV reflectivity of a polarised HHG beam on a magnetic sample can change of 20 % on a 100 fs time scale after irradiation by a low energy, femtosecond IR laser pulse [7]. For IR fluence of the order of 1 mJ/cm^2, this fast change is only transient and the same sample can be used an infinite number of time. In collaboration with LOA, we performed an experiment to assess the interest of such a technique for soft X-ray laser pulse duration estimation. The sample was placed in the pump-probe setup presented in the previous section. The soft X-ray laser was P polarised before the focussing optics by two multilayer mirrors. The angle of incidence on the sample was set to 45° in order to maximise the magneto-optical Kerr effect (MOK). A XUV camera observed the specular reflexion of the XUV beam. A live measurement of the incident XUV energy is given by the photocurrent generated by the XUV beam into the focussing optics. The heating pulse is incident on the sample with a normal angle in order to minimise the time mismatch between the two pulses. It has been frequency doubled in order to remove all the prepulses introduced by the DGRIP setup. A delay scan of the reflectivity is presented in Fig. 4, where each point results from the accumulation of 100 X-ray laser shots. Despite important fluctuations, a deep in reflectivity can be observed at a delay corresponding to the temporal overlap of the two pulses. The experimental trend has been fitted by a model resulting from the convolution of a reflectivity response similar to the one presented in [7] by a XUV pulse of single-sided exponential shape. The best fit leads to a X-ray laser pulse duration of 5 ps (at $1/e$). This results are very preliminary and important work will be needed to improve the signal quality. However, the total signal level experienced during this work might be compatible with the idea of a single shot experiment involving a travelling wave irradiation.

7 Conclusion

In conclusion, we gave in this paper an overview of the work performed with the Laserix facility since the last ICXRL conference. The laser driver is now highly reliable increasing the availability of the XUV sources. Both harmonic and soft X-ray laser beamlines are now in operation, and their use in pump-probe experiments is underway. Investigations of a large variety of phenomena using the soft X-ray sources not only raise hopes of new discoveries but also will catalyse new idea for XUV radiation measurement and manipulation.

Acknowledgements We thank the LPGP and LOA technical and administrative staff for their always efficient help. The support of the Agence Nationale de la Recherche (ANR) is also acknowledged through the project "jeunes chercheuses et jeunes chercheurs" ASOURIX ANR-09-JCJC-0056.

References

1. Nickles, P.V., Shlyaptsev, V.N., Kalachnikov, M., Schnürer, M., Will, I., Sandner, W.: Short pulse x-ray laser at 32.6 nm based on transient gain in Ne-like titanium. Phys. Rev. Lett. **78**, 2748–2751 (1997)
2. Keenan, R., Dunn, J., Patel, P.K., Price, D.F., Smith, R.F., Shlyaptsev, V.N.: High-repetition-rate grazing-incidence pumped soft X-ray laser operating at 18.9 nm. Phys. Rev. Lett. **94**, 103901 (2005)
3. Zimmer, D., Zielbauer, B., Pittman, M., Guilbaud, O., Habib, J., Kazamias, S., Ros, D., Bagnoud, V., Kuehl, T.: Optimization of a tabletop high-repetition-rate soft X-ray laser pumped in double-pulse single-beam grazing incidence. Opt. Lett. **35**, 450–452 (2010)
4. Zielbauer, B., Zimmer, D., Habib, J., Guilbaud, O., Kazamias, S., Pittman, M., Ros, D.: Stable and fully controlled long-time operation of a soft X-ray laser for user application experiments. Appl. Phys. B **100**, 731–736 (2010)
5. Habib, J., Guilbaud, O., Zielbauer, B., Zimmer, D., Pittman, M., Kazamias, S., Montet, C., Kuehl, T., Ros, D.: Low energy prepulse for 10 Hz operation of a soft-x-ray laser. Opt. Express **20**, 10128–10137 (2012)
6. Capeluto, M.G., Vaschenko, G., Grisham, M., Marconi, M.C., Luduena, S., Pietrasanta, L., Lu, Y., Parkinson, B., Menoni, C.S., Rocca, J.J.: Nanopatterning with interferometric lithography using a compact $\lambda = 46.9$ nm laser. IEEE Trans. Nanotechnol. **5**, 3–7 (2006)
7. La-O-Vorakiat, C., Siemens, M., Murnane, M., Kapteyn, H.C., et al.: Ultrafast demagnetization dynamics at the *M* edges of magnetic elements observed using a tabletop high-harmonic soft x-ray source. Phys. Rev. Lett. **103**, 257402 (2009)

Chapter 19
Observation of the Laser-Induced Surface Dynamics Using the Single-Shot Soft X-Ray Laser Probe

N. Hasegawa, Y. Ochi, T. Kawachi, M. Nishikino, M. Ishino, T. Imazono, T. Kaihori, T. Morita, A. Sasaki, K. Terakawa, Y. Minami, M. Baba, T. Tomita, M. Yamamoto, M. Yamagiwa, and T. Suemoto

Abstract We constructed optical laser pump and soft x-ray laser probe system and succeed to obtain the temporal evolution of the soft x-ray images on the laser irradiated Pt surface during the femtosecond laser ablation. The temporal evolution of the reflectivity of the soft x-ray strongly depends on the pump laser fluence, and the apparent ablation threshold was observed. The results were well explained by the molecular dynamics simulations of femtosecond laser ablation process.

1 Introduction

The unique sub-micron scaled structures formed by the irradiation of the femtosecond lasers, such as the ripple and bubble structures [1], are expected to have new material functionalities with low processing cost. In order to precisely control the laser ablation, the detailed observation of the laser ablation dynamics is required. The researches for the ablation dynamics by using the visible laser probe have been reported by several groups. The expansion speed of the ablation front was estimated to be 900 m/s from the result of the interference pattern (Newton ring) between the ablation front and melting front [2]. While the visible laser probe enables the measurement below and around the ablation threshold, it is difficult to observe the dynamics of the laser ablation with generation of the surface plasmas that reflect the visible light. For the direct observation of the femtosecond laser ablation dynamics,

N. Hasegawa · Y. Ochi · T. Kawachi (✉) · M. Nishikino · M. Ishino · T. Imazono · T. Kaihori · T. Morita · A. Sasaki · M. Yamagiwa
Quantum Beam Science Directorate, Japan Atomic Energy Agency, 8-1-7 Umemidai, Kizugawa, Kyoto 619-0215, Japan
e-mail: kawachi.tetsuya@jaea.go.jp

K. Terakawa · Y. Minami · M. Baba · T. Suemoto
Institute of Solid State Physics, The University of Tokyo, 5-1-5 Kashiwanoha, Kashiwa, Chiba 277-8581, Japan

T. Tomita · M. Yamamoto
Department of Ecosystem Engineering, The University of Tokushima, 2-1 Minamijo-Sanjima, Tokushima 770-8506, Japan

S. Sebban et al. (eds.), *X-Ray Lasers 2012*, Springer Proceedings in Physics 147,
DOI 10.1007/978-3-319-00696-3_19,

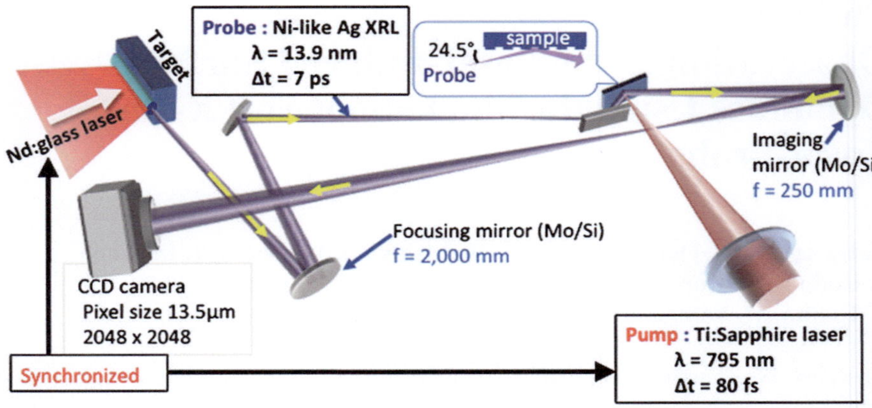

Fig. 1 Single shot x-ray laser probe with optical pump laser system

we propose to use the soft x-ray laser (SXRL) probe that can penetrate the surface plasmas and not penetrate the surface layer. In this study, we observed the temporal evolution of the soft x-ray reflective image of the laser induced metal surface, because the reflectivity of the soft x-ray is not sensitive to the electric state of the surface but quite sensitive to the surface morphology (see Fig. 3).

2 Observation of the Laser-Induced Surface Dynamics Using the Single-Shot Soft X-Ray Laser Probe

The experimental setup is shown in Fig. 1. The Ni-like silver SXRL at the wavelength of 13.9 nm generated by the CPA Nd:glass laser was used for the probe beam. The output energy and duration were 1 μJ and 7 ps, respectively. The SXRL was loosely focused on the sample with the oblique incidence angle of $\theta = 24.5$ deg. The SXRL beam size on the sample was about 1.2 mm (horizontal) and 0.5 mm (vertical). The spatial resolution of the soft x-ray image on the sample surface was measured to be 1.8 μm [3]. Ti:Sapphire laser at the central wavelength of 795 nm with the duration of 80 fs was used for the pump source. The spatial profile of the pump laser was Gaussian, and this enables us to discuss the pump fluence dependence at each radial position. The timing jitter between Ti:Sapphire laser and SXRL was measured to be 10 ps in RMS.

We observed the soft x-ray reflective images of the laser induced Pt film (300 nm thickness) for the study of the ablation dynamics. The surface roughness of Pt was measured to be better than 1 nm by atomic force microscope. Figure 2 shows the temporal evolution of the soft x-ray reflective images and the cross sections of the relative reflectivity at various times. Relative reflectivity was obtained under the condition that reflectivity at the non-irradiation area by the pump laser was 1.0. The typical pump laser energy, peak fluence and intensity on the sample surface

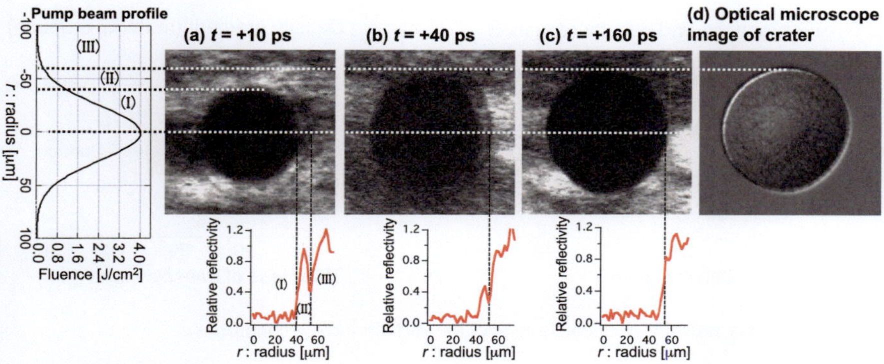

Fig. 2 The soft x-ray reflective images of femtosecond laser induced Pt surface

were 200 μJ, 4 J/cm^2 and 5×10^{13} W/cm^2, respectively. In Fig. 2(a), the dark disk appeared at the central part (region I: $r < 40$ μm, $F > 1.0$ J/cm^2) of irradiation area at $t = +10$ ps. Here, F and r show the pump laser local fluence and radial position, respectively. Around the dark disk (region II: 40 μm $< r <$ 55 μm, $0.4 < F < 1.0$ J/cm^2), the reflectivity slightly decreased. Outside of region II (region III: $r >$ 55 μm, $F < 0.4$ J/cm^2), reduction of the reflectivity was not observed. At $t = +40$ ps, reduction of the reflectivity was observed in region II (40 μm $< r <$ 55 μm) compared with that of $t = +10$ ps. At $t = +160$ ps, the size of the dark disk was equal to that of the bright ring in Fig. 2(d). The bright ring shows the rim structures. From these results, temporal evolution of the reflectivity was not linear function of the pump laser fluence, and the pump fluence at the boundary between the regions II and III leads to the ablation threshold fluence, $F = 0.4$ J/cm^2. More interestingly, thin dark ring was observed at the boundary between the regions II and III at $t = +10$ ps and $+40$ ps.

Next, we discuss the origin of the reduction of the soft x-ray reflectivity. The penetration depth of the soft x-ray with the oblique incidence angle of $\theta = 24.5$ deg for the solid Pt is less than 10 nm. Therefore, the reflectivity of the soft x-ray depends on the nanometer scaled surface morphology. We calculated the soft x-ray reflectivity on the several surface conditions. From Fig. 3(a), the reflectivity is quite sensitive to the surface roughness. But the density reduction by the phase transition from solid to liquid is not critically to explain the results. As shown in Fig. 3(b), 20 nm-density transition layer decreases the reflectivity to less than one-third. From these calculations, soft x-ray reflectivity drops occurred mainly due to the nanometer scaled surface roughness and density gradient.

The several ablation processes, the spallation ($F = F_{th}$), homogeneous nucleation ($F = 1.2F_{th}$), fragmentation ($F = 2.8F_{th}$), and vaporization ($F > 2.8F_{th}$), are predicted by the molecular dynamics simulations [4]. Here, F_{th} is the ablation threshold fluence. In vaporization and fragmentation processes, the solid surface is destroyed into gas and clusters just after the laser irradiation. In homogeneous nucleation process, the gas-bubbles are formed in the homogeneous liquid under the

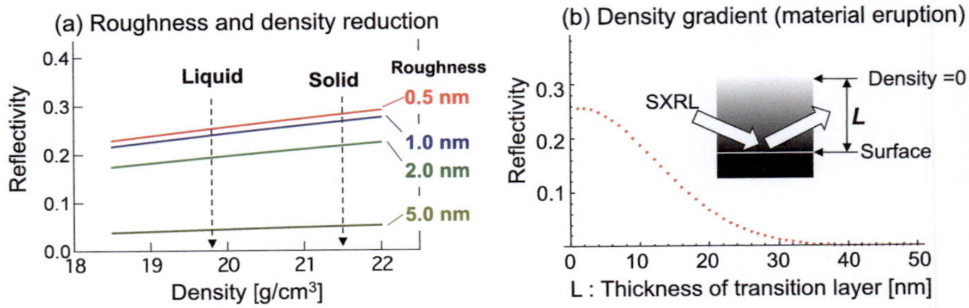

Fig. 3 Soft x-ray reflectivity calculated under the various surface conditions

condition that the pressure of liquid is lower than the saturation vapor pressure. In this case, the smooth surface is retained during dozens of picoseconds. The dependence of the transient surface conditions on the pump fluence well accorded with that of the experimental results in regions I∼III. Therefore, the experimental results in the regions I and II are supposed to be equivalent to the vaporization (or fragmentation) and homogeneous nucleation, respectively. Spallation process is the rapid peeling of the solid surface layer induced by the tensile strain. The tensile strain is caused by the compressive pressure wave reflected at the free surface, therefore the spallation is required the solid surface and observed only under the condition that the pump laser fluence close to the ablation threshold. It just explains the thin dark ring observed at the boundary between regions II and III.

3 Summary

We observed that the temporal evolution of the soft x-ray reflectivity after the femtosecond laser irradiation on Pt surface. The results showed strong dependence on the pump laser fluence. In addition, the experimental results well accorded with the ablation process predicted by the molecular dynamics simulations.

References

1. Tomita, T., et al.: Effect of surface roughening on femtosecond laser-induced ripple structures. Appl. Phys. Lett. **90**, 153115 (2007)
2. Sokolowski-Tinten, K., et al.: Transient states of matter during short pulse laser ablation. Phys. Rev. Lett. **81**, 224–227 (1998)
3. Suemoto, T., et al.: Single-shot picosecond interferometry with one-nanometer resolution for dynamical surface morphology using a soft x-ray laser. Opt. Express **18**, 14114–14122 (2010)
4. Perez, D., Lewis, L.J.: Molecular-dynamics study of ablation of solids under femtosecond laser pulses. Phys. Rev. B **67**, 184102 (2003)

Chapter 20
Nano-meter Size Modification of Metal Surfaces Induced by Soft X-Ray Laser Single Pulse

Masahiko Ishino, Anatoly Faenov, Momoko Tanaka, Tatiana Pikuz, Satoshi Tamotsu, Noboru Hasegawa, Masaharu Nishikino, Sergei Starikov, Vladimir Stegailov, Genry Norman, Vladimir Fortov, Igor Skobelev, Takeshi Kaihori, and Tetsuya Kawachi

Abstract We show experimentally the possibility of the precise nanometer size surface structuring of metal surfaces induced by ultra low fluencies of picosecond soft x-ray laser single pulse. After irradiation processes, we observed the modified surfaces to understand the interactions between the soft x-ray laser pulses and various materials by a scanning electron microscope. The formations of unique modified structures caused by irradiations of the soft x-ray laser pulses were seen. On Al surface, the formations of conical structures were observed. It was found that on Au surface, the ripple-like structures were formed. The atomistic model of ablation is developed that reveals the ultra-low threshold fluence values of this process. Calculated ablation depth as a function of fluence is in good agreement with the experimental data.

1 Introduction

Laser pulses having short durations have abilities to make high temperature and pressure, excite states of electrons in materials, emit rays, and so on, as the results of interaction with matters. Soft x-ray laser (SXRL) beam is one of attractive x-ray source for scientific studies, because of its features of short wavelength, short pulse duration, and highly spatial coherence. At the Japan Atomic Energy Agency, we have generated the SXRL beam, which has a fully spatial coherence, a short

M. Ishino (✉) · A. Faenov · M. Tanaka · T. Pikuz · N. Hasegawa · M. Nishikino · T. Kaihori · T. Kawachi
Quantum Beam Science Directorate, Japan Atomic Energy Agency, Kyoto, Japan
e-mail: ishino.masahiko@jaea.go.jp

A. Faenov · T. Pikuz · S. Starikov · V. Stegailov · G. Norman · V. Fortov · I. Skobelev
Joint Institute for High Temperatures, Russian Academy of Sciences, Moscow, Russia

S. Tamotsu
Division of Natural Science, Faculty, Nara Women's University, Nara, Japan

S. Starikov · V. Stegailov · G. Norman
Moscow Institute of Physics and Technology, Moscow, Russia

S. Sebban et al. (eds.), *X-Ray Lasers 2012*, Springer Proceedings in Physics 147,
DOI 10.1007/978-3-319-00696-3_20,

wavelength of 13.9 nm and the short duration of 7 ps. When we investigated a characteristic of the focal patterns of SXRL beam by means of color center formation in lithium fluoride (LiF) crystal [1], we found the ablation structures in the central parts of color center [2]. The ablation property appearing on LiF was very unique, because the ablation threshold was much smaller than those obtained with other lasers having longer wavelengths [2, 3].

It is quite interesting in the investigations of surface modifications induced by SXRL pulse irradiations, because understanding of surface modifications induced by laser irradiations is both fundamental and technological applications. The low ablation threshold of materials and formation of peculiar surface structure have possibilities of efficient and unique surface machining. There are still a lot of gaps in the understanding of the ablation mechanisms, especially for ultra short pulse lasers, because the existing experimental data are inconsistent with theoretical predictions. To confirm the appearance of surface modification, we irradiated the focused SXRL pulses to the target surfaces, and to describe the observed modification process, we proposed to use the atomistic model.

2 Experimental Results and Discussion

The SXRL irradiation experiment was carried out by use of the SXRL facility at Japan Atomic Energy Agency. The SXRL pulse was generated from the silver (Ag) plasma mediums using an oscillator-amplifier configuration. Characteristics of the generated SXRL pulse had a wavelength of 13.9 nm, bandwidth of narrower than 10^{-4}, and duration of 7 ps. The SXRL pulse was focused on the target surfaces by using a Mo/Si multilayer coated spherical mirror. A zirconium filter having 0.2 μm thick was placed in front of the spherical mirror to reduce the scattered optical radiations from the laser produced Ag plasmas. The total energy of the SXRL beam on the target surface was calculated to be 48 nJ. Using the LiF detector technique for estimation of the focused SXRL beam properties, we found that the SXRL energy concentration in focusing areas reached about 60 %. And around 40 % of the focusing SXRL beam energy was concentrated in the center of the best focusing spots.

After the SXRL irradiation processes, target surfaces were observed by use of various techniques, such as visible microscope, scanning electron microscope (SEM), and atomic force microscope (AFM). Details of experimental setup are described in previous papers [4].

Figure 1 shows SEM images of the irradiated aluminum (Al) and gold (Au) surfaces after a single shot exposure of the SXRL pulse around the focal points. The modified structures caused by the SXRL pulse irradiation are clearly observed on both targets, and we can confirm that the induced structures have quite different topology. On Al surfaces, it can be confirmed that the conical structures, which observed previous work [4], are formed around the ablated holes, and the relatively area of the conical structures increases by the reduction of the SXRL fluence. On

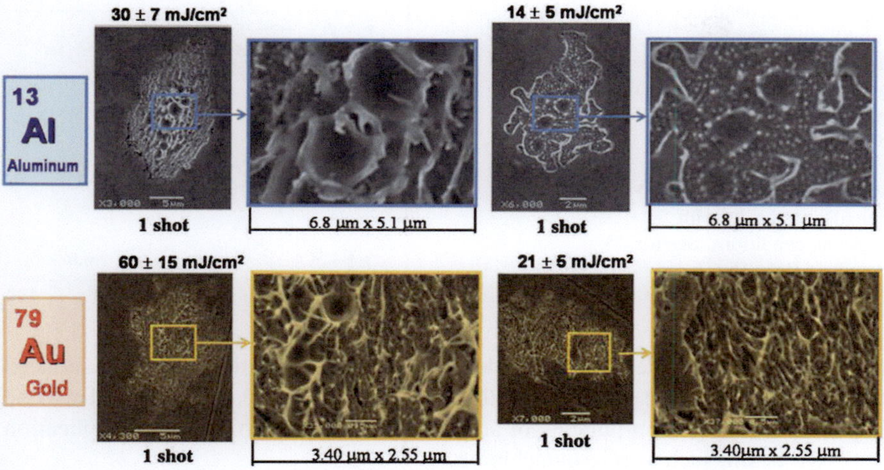

Fig. 1 SEM images of Al and Au surfaces after the SXRL single pulse irradiation

Au surfaces, on the other hand, the ripple-like shallow structures are seen. Areas of modified parts of Au surfaces are essentially smaller than those of Al cases.

The obtained depths of surface irregularities are 400 nm and 170 nm for Al surfaces with the fluences of 30 mJ/cm^2 and 14 mJ/cm^2, respectively [5]. For Au cases, 80 nm and 20 nm of surface irregularities are obtained by the fluences of 60 mJ/cm^2 and 21 mJ/cm^2 [6]. With increasing of SXRL fluence on the target surfaces, the depths and height of surface modifications are growth. As the experimental results of interactions between SXRL pulses and metal surfaces with various fluences, the difference type of surface modifications are distinguished, such as (1) at relatively low fluence ($\sim$10–20 mJ/cm^2), modified surfaces with nanometer size structures are formed, and (2) at relatively high fluence ($>$30–60 mJ/cm^2), ablated holes deeper than the attenuation lengths of SXRL beam for metals are confirmed.

To explain the experimental results obtained by the laser pulses with ultra short duration, the atomistic model of ablation was developed [6, 7]. To describe the ablation process near the threshold fluence, we propose the molecular dynamic (MD) model of two-temperature system with electron-temperature-dependent interatomic potential. Figure 2 shows the dependence of modification depth on absorbed laser fluence (F_{abs}). For SXRL pulses, the ablation depth smoothly grows together with F_{abs}, and it is clearly seen that the calculated ablation depth is in a rather good agreement with the experimental data. Our model calculation reveals that the first changes of surface structure are most probable due to the splash of molten gold.

3 Conclusion

We show the possibility of the precise nanometer size modified structures on Al and Au surfaces induced by the irradiation of the focused SXRL single pulse (experi-

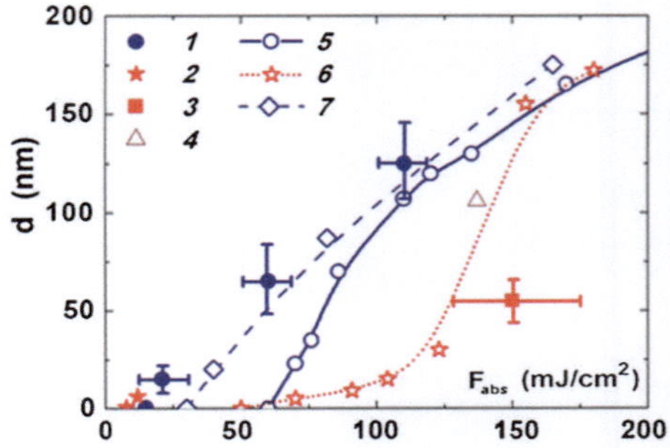

Fig. 2 The dependence of modification depth on absorbed fluence (F_{abs}) for gold target: *1*—experimental results (this work); *2* and *3*—optical pulses; *4*—MD calculation result (previous work); *5*, *6* and *7*—our MD calculation results with different conditions. See for details in Refs. [6, 7]

ment). We also provide the MD model to describe the ablation processes induced by SXRL single pulse irradiations (theory). The results we demonstrated will be important not only for a pursuite of ablation process but also for future application of SXRL beam, such as micromachining.

Acknowledgements This work was partly supported by the Grant-in-Aid for Scientific Research (B) from MEXT, Japan (No. 21360364), the Russian Foundation for Basic Research (Nos. 12-02-0947, 12-08-00666), and the Russian Academy of Science Presidium Program of Basic Research (No. 2). The presentation at conference was supported by Grant-in-Aid for Young Scientists (B) from MEXT, Japan (No. 24710096).

References

1. Faenov, A.Ya., et al.: Opt. Lett. **34**, 941–943 (2009)
2. Faenov, A.Ya., et al.: Appl. Phys. Lett. **94**, 231107 (2009)
3. Inogamov, N.A., et al.: Appl. Phys. A **101**, 87–96 (2010)
4. Ishino, M., et al.: J. Appl. Phys. **109**, 013504 (2011)
5. Ishino, M., et al.: J. Laser Micro Nanoeng. **7**, 147–151 (2012)
6. Norman, G., et al.: J. Appl. Phys. **112**, 013104 (2012)
7. Starikov, S.V., et al.: JETP Lett. **93**, 642–647 (2011)

Chapter 21
Speckle Statistics, Coherence and Polarization of a Collisional Soft X-Ray Laser

K.A. Janulewicz, C.M. Kim, P.V. Nickles, H. Stiel, M. Nishikino, N. Hasegawa, and T. Kawachi

Abstract A simple model of dynamics of a collisional X-ray laser is given applying Maxwell-Bloch equations in a uniform approximation. The main considered parameters include polarization and coherence of the emitted radiation. Speckle pattern, typical for a radiation source of partial coherence, especially, if involving process of the amplified spontaneous emission (ASE), are used as an alternative source of information on the processes occurring in the active medium of an X-ray laser. A level of partial coherence is deduced by statistical analysis of the speckle pattern generated in the output beam of the laser. It is shown that plasma fluctuation and non-Gaussian character of the statistics are needed to reproduce the transverse coherence values reported in the experiment.

1 Introduction

Plasma-based, transient-inversion X-ray lasers work nearly exclusively in the mirrorless arrangement. The main reason for that is short-lived gain with a typical duration between 10 and 15 ps. As a consequence, the amplified spontaneous emission (ASE) is used as the energy extraction mechanism [1]. Present-day X-ray lasers excited by the collisional processes in the grazing incidence pumping geometry (GRIP) or by electron residual energy of the above-threshold-ionization (ATI) in an

K.A. Janulewicz (✉) · C.M. Kim
Advanced Photonics Research Institute, Gwangju Institute of Science and Technology, Gwangju, 500-712, Rep. of Korea
e-mail: kaj@gist.ac.kr

P.V. Nickles
WCU Dept. of Nanobio Materials and Electronics, Gwangju Institute of Science and Technology, Gwangju, 500-712, Rep. of Korea

H. Stiel
Max Born Institute, 12489 Berlin, Germany

M. Nishikino · N. Hasegawa · T. Kawachi
Japan Atomic Energy Agency, 8-1-7 Umemidai, Kizugawa, Kyoto 619-215, Japan

S. Sebban et al. (eds.), *X-Ray Lasers 2012*, Springer Proceedings in Physics 147,
DOI 10.1007/978-3-319-00696-3_21,

optical field, show exceptionally high (50–80 cm^{-1}) small-signal-gain coefficient [2–4]. This causes them to work even more readily in the ASE regime.

One of the distinguishing effects of ASE-based lasers is presence of speckle (deep and random modulation of the intensity distribution giving a kind of granular structure) in the output beam. This is caused mainly by lack of a mode selecting cavity [5]. For these reasons X-ray lasers seem to be an inherent source of the speckle effect and indeed, randomly distributed hot spots were observed in the experiment [4, 6].

Application perspective was always important in the development of plasma-based X-ray lasers (XRLs). One of the fundamental parameters for using them in practice is coherence. While plasma-based XRLs proved excellent temporal coherence (probably the best among the short-wavelength sources), their spatial coherence is limited. Partial coherence in a plane perpendicular to the propagation direction is frequently quantified by a ratio of the coherence area A_{coh} and the beam cross-section A. This value is typically, for the systems without injection of an external signal and not pumped by fast electric discharge, about one percent [7].

The Young's interferometry is commonly used method of spatial coherence measurement. This is a well-established and simple method but it has some drawbacks when applied to XRLs. The Young's interferometry applied to the plasma-based sources requires removal of a noticeable background of the plasma thermal emission to avoid reduction in the fringes visibility. The principal practical assumption of this measurement scheme is equality of the signals at each of two distant slits. If this is not fulfilled, the correction factor $2\sqrt{I(1)I(2)}/[I(1)I(2)]$ has to be introduced in the formula for the fringes visibility, where $I(1)$ and $I(2)$ denote the intensities at each of the slits. The speckle effect causing randomness of the intensity distribution in the beam pattern precludes exact knowledge of the correction factor. Influence of this effect on the experiment was analyzed and clearly demonstrated in [8]. As a consequence, the Young's interferometry should typically underestimate a spatial coherence level of the plasma-based X-ray lasers. On the other hand, the saturation intensity level of X-ray lasers is about 10^{10} W/cm^2 and at this level of the optical field plasmonic effects on the slit edges can also affect fringes visibility by their modulation [9].

Radiation polarization is another parameter crucial for applications. It is commonly assumed that the beam originating from a spontaneous noise is naturally polarized, i.e. it does not show any preferential direction in the field oscillations. There were totally contradictory reports regarding this aspect. One showed strong s-component [10] and another one natural (undefined) polarization [11].

Recently, a simple model for the origin of the speckle structure in the X-ray laser output beam has been proposed [12]. This structure was termed as a speckle-like one and modeled by standard dependencies from coherence and diffraction theories. Description of the effect was done in terms of so-called classic or full developed speckle [13, 14]. It was concluded that the length of the output pulse exceeding the coherence time was responsible for creation of this structure. It was also assumed that the ratio of the pulse length t_p and coherence time t_{coh} was a measure of the modal structure of the emitted radiation. This is in our opinion, taking into account a very high level of medium inhomogeneity, unnecessarily strong limitation. It tacitly assumes, that an X-ray laser beam shows full spatial coherence and this is not the case in the experimental practice. In fact, in the most general case, separation of the effects of the spatial and temporal coherences is possible, subject to the mutual spectral purity.

In the present contribution we are analyzing, whether the observed and discussed earlier intensity inhomogeneity in the X-ray laser output beam is really a classic speckle structure and undergoes well established statistical rules of the speckle-related phenomena [13–15]. No assumption regarding the coherence state of the beam is introduced. It was attempted to determine whether the speckle structure could be used as a source of reasonable quantitative information on the X-ray laser beam coherence level.

2 The Method

2.1 X-Ray Laser Dynamics

We applied the X-ray laser dynamics model developed and presented in detail in [17]. The model formulated in a uniform approximation describes emission and amplification of the polarized radiation in terms of Maxwell-Bloch equations. Observing build-up of polarization in a single mode, it was found that saturation of the amplification process is responsible for decreasing in the polarization built-up. Polarization achieves nearly completeness before saturation onset, to decrease in the saturated regime to a value slightly above 80 %. It should be stressed that here the well-defined polarization has been considered (one of the P-states different from the natural polarization) and not some specific polarization state. In other words, the polarization state could be different from shot to shot, but it would be always well defined. The build-up of a random, but defined, polarization state with the length of the medium is shown in Fig. 1. In the model it was assumed that the seeding process by spontaneous noise is perfectly random and of uniform distribution. It is seen in Fig. 1 that (in this specific case) elliptic polarization achieves maximum for 3 mm and then is reduced. At a length of 3 mm the gain saturation effect begins. The horizontal axis describes the polarization state by the phase difference between the perpendicular field components. It was tried to verify experimentally the effect observed in simulations. A double-target arrangement was used to generate coherent X-ray radiation at 13.9 nm and those was subsequently analyzed by a membrane/multilayer beam-splitter allowing quantitative distinguishing between the "amount" of s- and p-polarized components in the output radiation. More detailed description of the experiment is given in these proceedings (see [16]). It was found that the p-component of the wave electric field dominated in the emitted radiation. There are some shot-to-shot fluctuations but on a very strong background of the p-component. This clearly suggests elliptical polarization of radiation emitted by the X-ray laser under investigation.

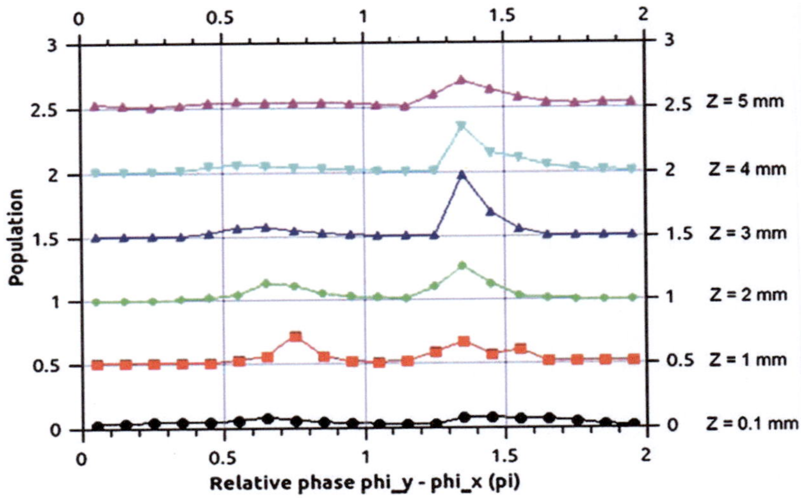

Fig. 1 Development of a defined polarization state with a length of the active medium in the output radiation of a nickel-like silver X-ray laser

Analysis of the amplification process showed also interesting dependence of the gain-narrowing effect on the medium length. The rate of the bandwidth reduction with the length is in the medium beginning noticeably lower than that in the strong amplification (see Fig. 2). The probable reason for that is strong influence of the "noisy" character of the signal within the first few millimeter of propagation. This could explain difficulty faced by Koch [18] to fit its experimental data of short length amplifier with a bandwidth reduction curve based on the rate equations.

2.2 Data Acquisition and Reproducibility

The experimental data processed here were obtained from the far-field images of the output beam of a Ni-like silver soft X-ray laser working at 13.9 nm. The images were recorded during the experiment described in [4]. The observation plane was placed 55 cm from the laser exit. A plasma column of 7 mm in length was obtained by irradiation of a slab target in the typical form of the GRIP geometry (double-pulse scheme with a long pulse incident normally to the target). The total pump laser energy contained in the irradiating laser pulse was equal to 1.5 J. The small-signal-gain coefficient was estimated to be 61 cm^{-1}. More detailed description of the experimental layout as well as measurements of the transverse coherence done in the classic Young's double-slit arrangement are given in [4]. Visibility of the fringes determined as a function of the slits separation was fitted under the assumption of the uniform circular source by using van Cittert-Zernike (vCZ) theorem. The measurement, with all mentioned earlier limitations, gave an effective incoherent source

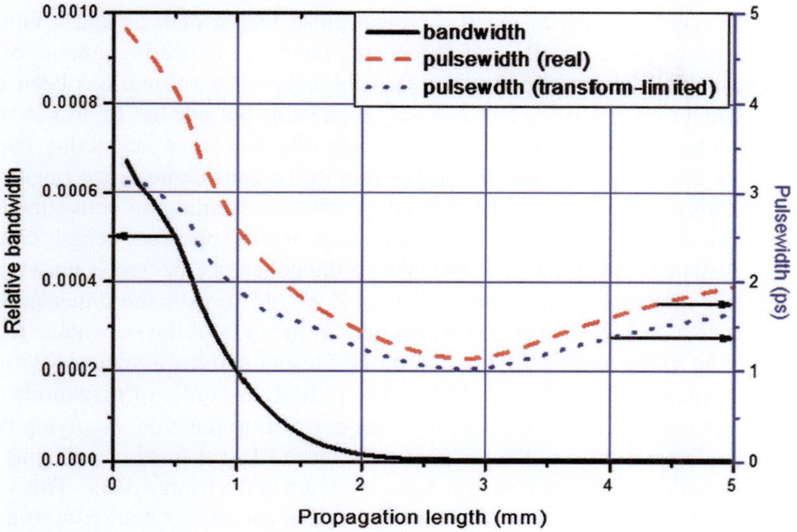

Fig. 2 Band- and pulse-width of the output pulse. Characteristic is that close to the origin (very short medium) the decrease in the bandwidth and the pulse-width are slightly slower in the beginning of the amplification process, in contradiction to the standard rate-equation result [18]

of 6.6 µm and a coherence length of L_{coh} of 57 µm at a position of 8.4 cm behind the laser exit [4]. Following this, it was estimated that the double-pulse irradiation delivered 3 % of fully coherent photons in a beam of 4 mrad divergence.

X-ray lasers are, due to the character of the processes involved in the generation scheme, very sensitive to small and accidental fluctuation in the laser pulse energy, accuracy of the pump beams overlapping and the focal line parameters. These depend partly on stability of the driving laser system and partly on quality of the medium/target itself. Beam pointing of a pump laser (spatial stability of the emitted beam) is one the most serious deteriorating effects. Generally, a typical reproducibility of XRL output determined by a standard deviation from the mean usually exceeds slightly 20 % and is considered as a good one [19].

On the other hand, it was assumed—what frequently used to be done—that the source is circular and uniform with a very narrow field correlation function justifying the use of the standard analytic formulae resulting from the vCZ theorem [20]. In fact, the beam profile is better described by more realistic distribution function $I(r) = I_0 \cosh^{-2}(r/a)$, approximating well a parabolic function close to its maximum [8, 21].

2.3 Data Processing and Simple Estimates

Using the coherent length value experimentally determined by the Young's interferometry, one can get a coherence volume of $V_{coh} = A_{coh} \times t_{coh} \times c =$

1.97×10^{-6} cm^3. Having assumed a typical pulse length of 6 ps and a number of photons in the pulse of 10^{11}, the degeneracy factor δ_c, indicating amount of the indistinguishable (coherent) photons within the coherence volume has been estimated as equal to 8.5×10^{10}. This is a very high value but one has to bear in mind that it was obtained for strongly idealized case of plane wave neglecting dependence of the coherence volume on the position in the considered space (medium). A simple estimate method of XRL spatial coherence assuming an active medium of a cylindrical shape and a less strict definition of the coherence length considered as a distance from the origin over which the coherence degree is not smaller than 0.5 give a simple analytic formula $L_{coh} \cong \lambda L/4a$ to estimate dimensions of the coherence area [22]. It was concluded that, in the case of the coherence length corresponding to the plasma/XRL width a, the Fresnel number associated with the plasma column should be $N_F = a^2/\lambda L \cong 1/4$ [22]. These order of magnitude estimates are related only to the main peak of the correlation function. Applying these approximated expressions to the parameters of our XRL (assumed $a = 32$ μm) and a medium length of $L = 7$ mm we got $L_{coh} \cong 0.5$ μm at the source plane. This value is related to the main peak of the correlation function and differs markedly from the value of 2 μm obtained from the experimental far-field (FF) intensity profile by the Fourier transform.

It is well known that in the ideal case of full coherence a laser emits photons in a single mode with a statistics ruled by the Poisson distribution and the relevant probability density (probability density to obtain a given intensity level I within the intensity gap of I, $I + dI$) has negative exponential distribution [13, 14]. The speckle pattern governed by this distribution is called the fully developed speckle. Corresponding distribution for a partially coherent light is given in the form of the gamma distribution, sometimes also called m-distribution [13, 14]

$$p_M(E/\overline{E}) = M^M (E/\overline{E})^{M-1} \exp(-ME/\overline{E})/\overline{E}\Gamma(M) \quad (1)$$

where M is a number of statistically independent but equally intense (average value) coherence modes in the output radiation. The experimental practice (detectors and the time scale) forces one to use temporally integrated intensity as a variable. Thus, for the sake of consistence we can use $E/\overline{E}$—the normalized pulse integrated intensity instead of $I/\overline{I}$ for the normalized peak intensity. The beam images recorded in the far-field zone presented in Fig. 3 have been processed to obtain dependence of the probability density on the normalized integrated intensity $E/\overline{E}$. The histograms of the image intensity level were created with binning including 50 bins. The part of the beam used for the histogram was selected by a rectangular window (Fig. 3), drawn to cover the maximum possible area of the irregular beam spot and minimize contribution from the non-irradiated but noisy vicinity.

The nature of the lasing process in X-ray lasers makes analysis of the speckle structure difficult. Due to mentioned limited reproducibility of the plasma dynamics, X-ray lasers show a significant shot-to-shot fluctuation of the beam position caused by variable deflection on not fully relaxed density gradients. As a consequence, averaging the speckle images as it is recommended in the scattering experiments at a synchrotron [23, 24] is impossible as it leads to artificial beam homogenization and

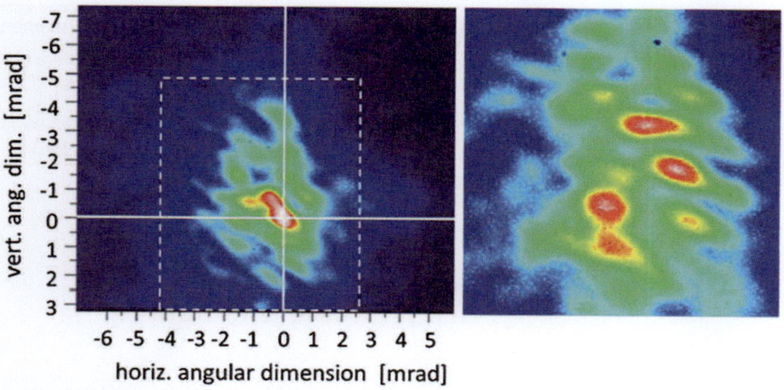

Fig. 3 Far-field image of X-ray laser output recorded at a distance 55 cm from its exit and with the marked are taken to the statistical analysis. The image of the speckle pattern on the *right side* is already equal to the analysis area

the speckle statistics tends towards the negative exponential distribution equivalent to reduction in the effective number of the contributing modes. By averaging in this way five images from our experiment with the M value between 2.4 and 6.5 we got an effective number of coherence modes $M = 1.6$, that is definitely incorrect. Another procedure realized by averaging statistics of the individual beam images was found, during speckle analysis of a synchrotron beam, to deviate too much from the realistic values [23]. Moreover, the gain area of X-ray lasers is not exactly stable in time [25] and this can result in additional smearing and blurring effects. The detectors integrate the signal within the time gap equal to the pulse length that is typically longer than the coherence time. That means, the integrated intensity (pulse energy) includes a few overlapped (incoherently added) speckle patterns, each corresponding to the time gap equal to the coherence time. All this makes the picture to be very complex and the question arises if the standard statistical tools allow for fast, simple and reasonable estimate of the radiometric features of the observed radiation.

3 Results and Discussion

The far-field images of the beam were recorded independently in five consecutive shots and analyzed using the standard software of the back illuminated CCD camera of Princeton Instruments. The average intensity and corresponding standard deviation were the main output parameters of the statistics. The mean value from this processing was put in Eq. (1) with M as a fitting parameter—the procedure suggested by Goodman as the reasonable one [13].

Dimension of the selected analysis window (area) had limited influence on the output of the statistical procedure. Usually, the M value varied by much less than 10 % as a function of the window area choice. Figure 3 shows an example of such a recorded image with indication of the angular distribution of the intensity and the area from which the data were used. In spite of the local deviations, the best fit gave

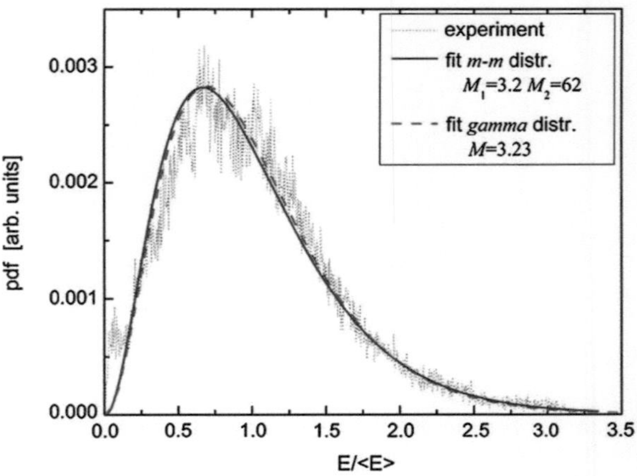

Fig. 4 Experimental histogram of the right-side image in Fig. 3 with two possible fits: m distribution and m-m distribution. m-Distribution is a synonym of gamma distribution

$M = 2.4$, which is a very low value. The poor shot-to-shot reproducibility of the X-ray laser was confirmed as within the registered series of shots the derived number of the coherence modes varied between 2.4 and 6.5.

All the above discussed characteristics and experiments confirm that a collisional X-ray laser is a device of limited (partial) spatial coherence. The histograms of the recorded far-field patterns, the obtained values of the mean intensity and the standard deviation allowed for determination of the speckle contrast K_{sp} and the number of modes $1/M = \sigma_E^2/\overline{E}^2$ involved in creation of the partially developed speckle pattern ($K_{sp} < 1$). It is not surprising that these values are very close to those derived from the fitting process as the experimental value of the mean was used in the fitting process as a fixed parameter. This confirms, however, a high goodness of the fit. It has to be stressed here, that the obtained M values have to be halved when the polarization state of the XRL output radiation is unknown and commonly assumed as natural one (undefined polarization).

From here on, we concentrate on the speckle pattern presented in Fig. 3 (right image) delivering the best fit and relatively low value of M. A contrast value of the speckle $K_{sp} = \sigma_E/\overline{E} = 1/\sqrt{M}$ equal to 0.56 was estimated from the statistics when M estimated in the fitting procedure was equal to 3.23. This confirms, that the photon statistics is definitely non-Poisonian (in this case the contrast is equal to 1) and one can guess that the presented in Fig. 3 speckle patterns show structures resulting from intensity summing of some finite number of spots (coherent modes). This guess is justified even if Eq. (1) is, in fact, obtained under the assumption that all the contributing statistically independent spots (modes) have the same mean intensity and an uniform phase distribution.

While the second condition is fulfilled in the X-ray laser medium the former one is actually not fulfilled. The active medium is inhomogeneous and the radial density gradient is sufficient to cause radiation to be refracted (bent off the active area in the extreme case) from the axial direction and the modes can partially overlap each other, especially at the end of the medium where the saturation effect is strongest. On the other hand, the gain itself is sufficiently high to cause significant differences in the group velocities of the photon fluxes traversing the medium areas of different gain [26]. The absolute gain values are extremely high and one can consider two realistic values of the gain coefficient, namely 80 cm^{-1} and 50 cm^{-1}. The corresponding group velocities υ_{gr} in a medium with a Gaussian spectral line profile ($\Delta\nu_D = 1 \times 10^{12}$ 1/s) are equal to $0.735c$ and $0.816c$, respectively. The time delay for two groups of photons amplified in these areas will achieve a value of ≈2.3 ps. This is much more than the coherence time of the radiation. Hence the assumption about the constant average intensity of all correlation (coherence) areas is definitely not fulfilled.

All these effects cannot, unfortunately, account for the large discrepancy between the experimental results and the estimates of the transverse coherence lengthy derived from the speckle data. The latter suggest existing up to 30 % of fully coherent photons in the output beam, at least one order more than the experimental data. The field correlation function at the XRL exit indicates multi-mode distribution. Most of the area is weakly correlated but the number of the hot spots is large. The resulting speckle has to be described by other distribution function. It was found, that a bi-statistical process described by gamma-gamma (m-m) distribution fits favorably the experimental speckle statistics (see Fig. 4) and differs negligibly from gamma distribution. However, the number of modes is much larger. Plasma fluctuations cause mode mixing and movement of the source and it results in non-Gaussian character of the beam propagation through the medium. This process is overlapped (modulated) by formation of large-scale modes determined by the medium geometry.

Taking into account that M can be treated as a measure of the ratio A/A_{coh} we can introduce two separate numbers M_1, M_2 characterizing the multiplicative component processes of the mode formation. Hence one can estimate the value of spatial coherence length $L_{coh} \cong \sqrt{A_{coh}}$ or $L_{coh} \cong \sqrt{A/M}$, where $M = M_1 \times M_2$. Using obtained from the fitting process values $M_1 = 3.2$ and $M_2 = 62$ we obtained a coherence length in the observation plane equal to 263.5 μm. This value corresponding to 1/14 of the beam diameter is in reasonable agreement with 1/18–1/20 reported in the experiment [8].

4 Conclusion

In summary, it was found that polarization of a slab target collisional X-ray laser is not undefined (natural polarization), but it has a strong p-component fluctuating from shot-to-shot, what indicates tendency to elliptical polarization. This finding is not in perfect agreement with 1D modeling suggesting that the output beam can have well defined, but random polarization state.

Analysis of the speckle pattern in a beam of an X-ray laser stated that these undergo the usual statistics of partially coherent source, i.e. the probability density corresponds to a gamma distribution. However, more exact inspection of the results shows that in general, the coherence level is very seriously overestimated. This results from the plasma inhomogeneity and reasonable agreement with the experimental reports requires implementation of plasma fluctuations and non-Gaussian statistics of the speckle. The random inhomogeneities of the medium give the probability density described in an accurate way by gamma-gamma (or m–m) distribution. As a consequence, the speckle structure has a bi-statistical character and the derived in this way coherence length compares favorably with the values reported in the experiment. Taking into account the drawbacks of the Young's interferometry the speckle analysis can be considered as a simple tool of reasonable estimate of the coherence level from the far-field beam patterns.

Acknowledgements We would like to pay credit to Dr. Hyung Taek Kim for his help in the experimental effort. The work was supported partially by TBP GIST, NCRC and DASAN.

References

1. Pert, G.J.: Output characteristics of amplified-spontaneous-emission lasers. J. Opt. Soc. Am. B **11**, 1425 (1994)
2. Keenan, R., et al.: High-repetition-rate grazing-incidence pumped X-ray laser operating at 18.9 nm. Phys. Rev. Lett. **94**, 103901 (2005)
3. Luther, B.M., et al.: Saturated high-repetition-rate 18.9-nm tabletop laser in nickellike molybdenum. Opt. Lett. **30**, 165–167 (2005)
4. Kim, H.T., et al.: Characteristics of Ni-like silver X-ray laser pumped by single profiled laser pulse. J. Opt. Soc. Am. B **25**, B76–B84 (2008)
5. Kravis, S.P., Allen, L.: Measurement of the statistics and spatial distribution of pulsed spontaneous emission. Opt. Commun. **23**, 289 (1977)
6. Nishikino, M., et al.: Characterization of a high-brilliance soft x-ray laser at 13.9 nm by use of an oscillator-amplifier configuration. Appl. Opt. **47**, 1129 (2008)
7. Lu, P., et al.: Spatial coherence of prepulse-induced neon-like x-ray lasers. Phys. Rev. A **58**, 628 (1998)
8. Liu, Y., et al.: Spatial coherence measurements of a 13.2 nm transient nickel-like cadmium soft X-ray laser pumped at grazing incidence. Opt. Express **14**, 12872–12879 (2006)
9. Gan, C.H., Gbur, G., Visser, T.D.: Surface plasmons modulate the spatial coherence of light in Young's interference experiment. Phys. Rev. Lett. **98**, 043908 (2007)
10. Kawachi, T., et al.: Observation of polarization of the soft X-ray laser line in neonlike germanium ions. Phys. Rev. Lett. **75**, 3826 (1995)
11. Rus, B., et al.: Demonstration of amplification of a polarized soft-x-ray laser beam in a neon-like germanium plasma. Phys. Rev. A **51**, 2316–2327 (1995)
12. Gilbaud, O., et al.: Origin of microstructures in picosecond X-ray laser beam. Europhys. Lett. **74**, 823–829 (2006)
13. Goodman, J.W.: Statistical properties of laser speckle pattern. In: Dainty, J.C. (ed.) Laser Speckle and Related Phenomena, 2nd edn. Springer, Berlin (1984)
14. Goodman, J.W.: Speckle Phenomena in Optics: Theory and Applications. Roberts & Company, Englewood (2006)
15. Goodman, J.W.: Speckle with a finite number of steps. Appl. Opt. **47**, A111 (2008)
16. Kawachi, T., et al.: Source development of novel coherent x-rays and their applications in JAEA. In: Sebban, S., et al. (eds.) X-Ray Lasers 2012. Springer Proceedings in Physics, vol. 147. Springer, Cham (2013). Chapter 23 in this book
17. Kim, C.M., Lee, J., Janulewicz, K.A.: Coherent amplification of an ultrashort pulse in a high- and swept-gain medium with level degeneracy. Phys. Rev. Lett. **104**, 053901 (2010)
18. Koch, J.A., et al.: Experimental and theoretical investigation of neonlike selenium x-ray laser spectral linewidths and their variation with amplification. Phys. Rev. A **50**, 1877 (1994)

19. Rocca, J.J., et al.: Saturated 13.2 nm high-repetition rate laser in nickel-like cadmium. Opt. Lett. **30**, 2581–2583 (2005)
20. Goodman, J.W.: Statistical Optics, Chap. 9.1. A Wiley-Interscience Publication. Wiley, New York (1985)
21. London, R.A., Strauss, M., Rosen, M.D.: Modal analysis of X-ray laser coherence. Phys. Rev. Lett. **65**, 563 (1990)
22. Feit, M.D., Fleck, J.A. Jr.: Spatial coherence of laboratory soft x-ray lasers. Opt. Lett. **16**, 76 (1991)
23. Abernathy, D.L., et al.: Small-angle X-ray scattering using coherent undulator radiation at the ESRF. J. Synchrotron Radiat. **5**, 37–47 (1998)
24. Tsui, O.K.C., Mochrie, S.G.J., Berman, L.E.: Statistical analysis of X-ray speckle at the NSLS. J. Synchrotron Radiat. **5**, 30–36 (1998)
25. Pert, G.J.: Optimizing the performance of nickel-like collisionally pumped x-ray lasers. Phys. Rev. A **73**, 033809 (2006)
26. Casperson, L., Yariv, A.: Pulse propagation in a high-gain medium. Phys. Rev. Lett. **26**, 293 (1971)

Chapter 22
Upscaling of X-Ray Laser Repetition Rate Using an OPCPA Architecture

Fei Jia, Felix Staub, and Jürg Balmer

Abstract We present a novel design of a hybrid amplification chain for pumping a Sn target producing ∼μJ coherent soft-X-ray pulses at few-Hz repetition rate. The driver aims to generate multi-joule energy at the central wavelength of 1053 nm and consists of optical-parametric chirped-pulse amplification (OPCPA) combined with diode-pumped Nd:YLF amplifiers. The soft-X-ray laser at the wavelength of 11.9 nm is produced by focusing the infrared beam into a line focus on a solid Sn target. Benefiting from this hybrid pump scheme and grazing-incidence pump geometry (GRIP), such a system has the potential to upscale the repetition rate of soft-X-ray lasers from single shot to the few-Hz region.

1 Introduction

Single-shot X-ray lasers based on collisional-excitation-pumped nickel-like ions have proved to deliver coherent photons with high brilliance and narrow bandwidth. Inspired by recent work on soft-X-ray lasers at few-Hz repetition rate [1] and driven by demands of versatile applications, novel approaches to increase the duty cycle with a more compact and affordable system are currently under study. The optical-parametric chirped-pulse amplification (OPCPA), a nonlinear technique, benefits from high gain ($\sim 10^6$), wide gain bandwidth and absence of issues due to thermal load. Combined with diode-pumped Nd:YLF rods, this hybrid amplification system provides access to deliver pulse energies up to a few Joules at repetition rates of a few Hz with pulse durations of a few ps.

In this work, numerical simulations have been performed to optimize major parameters in OPCPA and Nd:YLF amplification chains. With the pulse energy and duration from the simulations, we intend to exploit the Ni-like Sn 4d-4p laser line at 11.9 nm approaching saturation at a repetition rate of sub-10 Hz.

F. Jia (✉) · F. Staub · J. Balmer
Institute of Applied Physics, University of Bern, Sidlerstrasse 5, 3012 Bern, Switzerland
e-mail: fei.jia@iap.unibe.ch

S. Sebban et al. (eds.), *X-Ray Lasers 2012*, Springer Proceedings in Physics 147, DOI 10.1007/978-3-319-00696-3_22,

2 Experimental Setup and Numerical Model

A schematic drawing of the OPCPA/Nd:YLF hybrid driver setup is shown in Fig. 1. A two-stage OPA, consisting of two lithium triborate (LBO) crystals in the pre-amplifier stage and a single crystal in the power-amplifier stage, is pumped by a commercial Q-switched Nd:YAG laser with a total pulse energy of 200 mJ and a pulse duration of 4.1 ns (FWHM) at the central wavelength of 532 nm. The spatial profile of the pump laser is nearly top-hat, whereas the temporal profile is Gaussian-like as naturally obtained from the Q-switched resonator. The pump beam is then separated by a beamsplitter into a ratio of 25 %/75 % for each stage, in order to maximize the pump energy extraction as well as to optimize the energy stability. Collimated with a telescope, the diameter for the two stages is 1.5 mm and 2.25 mm respectively, which corresponds to a peak irradiance of 0.7–0.8 GW/cm^2. These pump pulses are required to spatio-temporally overlap the stretched signal pulse. Therefore, a double-folded 4-pass stretcher is designed to stretch the signal pulse (initial pulse duration of 180 fs with 7 nm bandwidth (Origami-10)) to 1 ns with a pass-bandwidth of 2.5 nm. The purpose of over-stretching and subsequent major clipping of the signal spectrum is to reduce the gain narrowing effect in Nd:YLF. The output spectrum of the OPCPA has to be adapted into similar gain bandwidth of Nd:YLF (FWHM 1.4 nm). In this case, energy losses due to gain narrowing of an overfilled spectrum can be reduced by the pre-selection of bandwidth before the OPCPA. The few-mJ pulse from the OPA is further amplified by a 2-pass Nd:YLF amplifier and a 4-pass booster amplifier, resulting in a pulse energy of $\sim$2 J.

The major parameters of this hybrid amplification chain are roughly pre-determined by simulations. For OPA simulation, a numerical integration of the three coupled wave equations for difference-frequency generation (DFG) in the slowly-varying-envelope approximation is used [2].

3 Results and Discussion

Electronic trigger synchronization and pulse build-up timing of Q-switch commercial pump laser are the main sources of temporal jitter in an OPCPA system. Figure 2(a) shows the measurements of jitter of the pump laser. We propose an alternative compensation method and compare it with the method proposed by Zhang et al. [3] in the same frame of simulation. The major difference in our method is the shifting of one pump pulse a few hundred ps away from the peak of the signal pulse (the dashed line in Fig. 2(b) shows the temporal shift of the pump pulse). Owing to the symmetric profile of the pump pulse, a jitter compensation mechanism is introduced naturally. By tailoring the crystal length in both stages, an optimized stability of signal gain will be achieved, shown in Fig. 2(d) where every line meets at the exit of the crystal. The reconversion-induced minor spatial profile modification will bring benefits in terms of a more homogeneous line focus. The spatio-temporal profile at the OPCPA output is introduced into the Nd:YLF amplifiers, resulting in the final

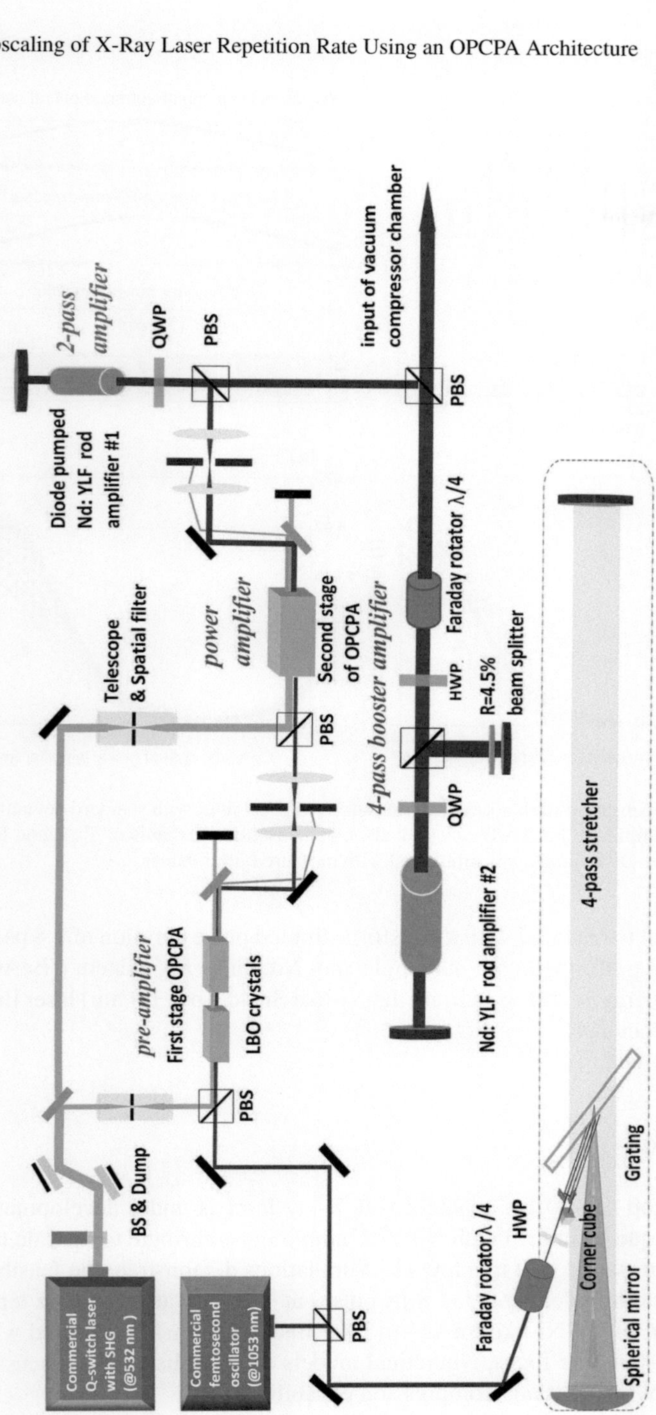

Fig. 1 Experimental setup of the OPCPA/Nd:YLF hybrid amplification system

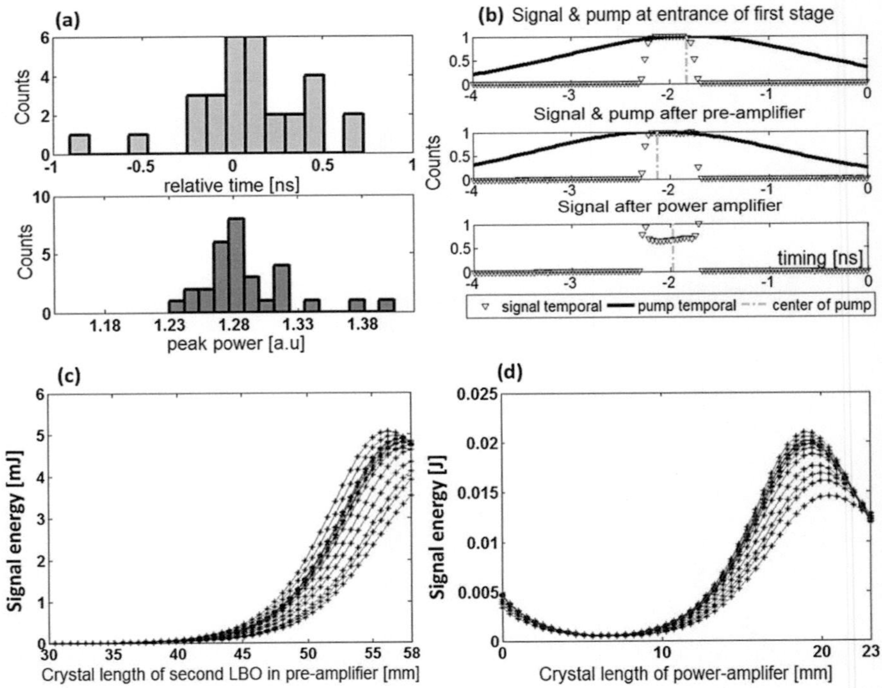

Fig. 2 Jitter compensation simulations. (**a**) Measured temporal jitter with standard deviation (StD) of 0.32 ns, and amplitude jitter StD = 0.036. (**b**) Compensation mechanism, ∇: signal temporal profile, (**c**), (**d**) the OPA signal gain simulation with measured jitter values

amplified signal to reach 2 J with a transform-limited pulse duration of 1.8 ps. Based on experiments performed with our single-shot Nd:Glass TW system ('BeAGLE'), such a pump energy is able to saturate the Ni-like Sn 4d-4p (11.9 nm) laser line with a 12-mm line focus [4].

4 Conclusion

A novel tabletop sub-10 Hz coherent soft X-ray laser is under development. The OPCPA technique combined with Nd:YLF amplifiers contribute to upscale the repetition rate from single shot to a few Hz. Simulations demonstrate the feasibility of this hybrid system to deliver 2-J, 1.8-ps pulses at 1053 nm at sub-10-Hz repetition rate. Saturation of the Ni-like Sn 4d–4p laser line (11.9 nm) is expected with this energy in a 12-mm line focus. Numerical models are established for discussion on stability control methods and compression capability.

Acknowledgements This project is funded in part by the Swiss National Science Foundation (Swiss NSF).

References

1. Rocca, J.J., Wang, Y., Larotonda, M.A., Luther, B.M., Berrill, M., Alessi, D.: Opt. Lett. **30**, 2581 (2005)
2. Smith, A.V., Bowers, M.S.: J. Opt. Soc. Am. B **12** (1995)
3. Zhang, S.K., Fujita, M., Yamanaka, M., Nakatsuka, M., Izawa, Y., Yamanaka, C.: Opt. Commun. **184**, 451–455 (2000)
4. Grünig, M., Imesch, C., Staub, F., Balmer, J.E.: Opt. Commun. **282**, 267–271 (2009)

Chapter 23
Source Development of Novel Coherent X-Rays and Their Applications in JAEA

T. Kawachi, A. Sasaki, M. Nishikino, M. Ishino, N. Hasegawa, T. Imazono, Y. Ochi, M. Tanaka, A.Y. Faenov, T.A. Pikuz, A. Pirozhkov, T. Esirkepov, T. Nakamura, M. Kando, S.V. Bulanov, K. Kondo, K. Janulewicz, C.M. Kim, H. Stiel, G. Norman, T. Suemoto, T. Tomita, K. Namikawa, M. Yamagiwa, and Y. Kato

Abstract This paper reviews recent progress in the source development of intense coherent x-rays and the applications in the research fields such as material science and laser processing in Japan Atomic Energy Agency (JAEA). In the source development, the polarization of the fully spatial coherent 13.9 nm soft x-ray laser (SXRL) in the transient collisional excitation (TCE) scheme was investigated experimentally. The result indicated that the SXRL beam was strongly polarized in the direction perpendicular to the target surface. In the alternative schemes toward the

---

T. Kawachi (✉) · A. Sasaki · M. Nishikino · M. Ishino · N. Hasegawa · T. Imazono · Y. Ochi · M. Tanaka · A.Y. Faenov · T.A. Pikuz · A. Pirozhkov · T. Esirkepov · T. Nakamura · M. Kando · S.V. Bulanov · K. Kondo · M. Yamagiwa · Y. Kato
Quantum Beam Science Directorate, JAEA, 8-1-7 Umemidai, Kizugawa, Kyoto 619-0215, Japan
e-mail: kawachi.tetsuya@jaea.go.jp

A.Y. Faenov · G. Norman
Joint Institute for High Temperatures, Russian Academy of Science, Izhorskaya st. 13 Bd. 2, Moscow, Russia

K. Janulewicz · C.M. Kim
Advanced Photon Research Institute, GIST, Cheomdan-gwagiro, Buk-gu, Gwangju 500-712, Republic of Korea

H. Stiel
Max-Born Institute for Nonlinear Optics and Short Pulse Spectroscopy, Max-Born-Strasse 2a, 12489 Berlin, Germany

T. Suemoto
Institute of Solid State Physics (ISSP), University of Tokyo, 5-1-5 Kashiwanoha, Kashiwa, Chiba 277-8581, Japan

T. Tomita
Faculty of Engineering, University of Tokushima, 2-1 Minamijo-Sanjima, Tokushima 770-8506, Japan

K. Namikawa
Research Institute for Science & Technology, Tokyo University of Science, 2641 Yamazaki, Noda, Chiba 278-8510, Japan

S. Sebban et al. (eds.), *X-Ray Lasers 2012*, Springer Proceedings in Physics 147,
DOI 10.1007/978-3-319-00696-3_23,

short wavelength region, intense higher-order harmonics with the photon energy of more than 300 eV was observed from a relativistic plasma generated with the laser intensity higher than 10^{18} W/cm^2. In the applications of the 13.9 nm TCE laser, we observed temporal evolution of surface distortion of Pt sample pumped by a 80 fs optical pulse by use of x-ray laser interferometer using double Lloyd's mirrors and reflectometer. In the study of SXRL ablation, the melting depth of the substances using single shot SXRL exposure was measured, and the result was quantitatively consistent with the calculated result based upon the spallative ablation model by molecular-dynamics code.

1 Introduction

Several years ago, we have firstly demonstrated fully spatial coherent x-ray laser beam at the wavelength of 13.9 nm by the method of double targets geometry, in which the first gain medium works as the soft x-ray oscillator and the second gain medium works as the soft x-ray amplifier [1]. Now this novel light source with pico-second duration and more than 10^9 coherent photons/pulse is routinely used for the wide variety of research fields such as material science [2], atom and molecular science [3], soft x-ray imaging [4], laser ablation physics and laser processing [5].

In a view point of the source development, one of the most serious limitations so far is the lasing wavelength, and the extension of the lasing wavelength to shorter region is urgent subject. Another important issue is to have deeper understanding of characteristics of x-ray amplifier plasmas, because under the coupling of x-ray seeder with x-ray amplifier, the final beam quality is determined by plasma parameters such as density gradient, homogeneity etc., of the x-ray amplifier plasma. Polarization characteristics is one of the parameters, and the polarization transport in x-ray amplifier is important in the applications of SXRL beam with higher-order harmonics x-ray seeder and double target geometry.

In the applications of the SXRLs, observation of *non-periodic* or *unrepeatable* ultra-fast phenomena such as nano-meter scale deformation of domain structure of substances under the phase transition or surface distortion in laser ablation are quite attractive application in terms of full use of the advantages of plasma x-ray lasers. In order to apply the SXRL beam to this field, we have developed x-ray laser interferometer, which enable us to obtain surface information with 1 nm depth-resolution and 7 ps temporal resolution. On the other hand, the study of SXRL ablation (EUV ablation) has created feasible opportunities for technological breakthrough in the research field of laser surface nano-processing and for fundamental investigations of materials in warm dense matter state. The experimental data obtained through this study can provide good benchmarks for theoretical codes and contribute to understand the laser matter interaction.

In this paper, we describe recent progress in the source development of coherent soft x-rays and the applications in Japan Atomic Energy Agency (JAEA).

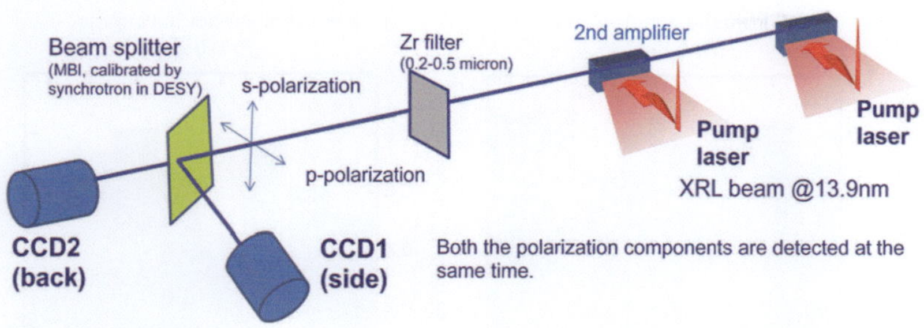

Fig. 1 Experimental-set up for measurement of the polarization

2 Source Development of Novel Coherent X-Rays

2.1 Observation of Polarization of SXRL in Transient Collisional Excitation Scheme

The polarization of plasma x-ray lasers is determined by the summation of plasma anisotropy as the polarization processes such as anisotropic collision, radiation trapping and depolarization processes such as collisional magnetic-sublevel-mixing and localized plasma electric-field. Through polarization measurement, we may obtain the information about the anisotropic feature in plasma and about the polarization transport in x-ray amplifier. The polarization of collisional excitation lasers has been measured for quasi-steady state scheme [6], however there are few reports for transient excitation scheme. Recently concerning with the x-ray amplifier coupled with higher-order harmonics as x-ray seeder, theoretical investigation of polarization transport has been studied, which implies the importance of the information of polarization characteristics of the x-ray amplifier [7]. In order to measure the polarization characteristics of TCE gain medium plasma, we conducted an experiment under the collaboration with APRI-GIST and MBI. Figure 1 is the experimental set-up. The SXRL beam at the wavelength of 13.9 nm passes through a Zr filter to cut visible light and is divided into two polarization components by a Mo/Si multilayer free-stand soft x-ray beam splitter (BS) fabricated in MBI. The reflectance and transmittance, together with the polarization separation ratio of this BS has been calibrated using synchrotron radiation source. Here s- and p-polarization component means that the direction of electric field is parallel and perpendicular to the target surface, respectively. We observed the two polarization components at the same time by use of two charged couple devices (CCDs).

We measured the polarization degree, P, $P = (I_p - I_s)/(I_p + I_s)$, for three cases, i.e., XRL from the first target, the second target, and the double target geometry.

Figure 2 shows the experimental result. It was found that the SXRL in the TCE scheme was strongly polarized, $P \sim 0.8$, in the direction perpendicular to the target surface. For the double target case, the tendency was the same however, the polarization degree was slightly lower ($P \sim 0.65$), which might attribute to the gain

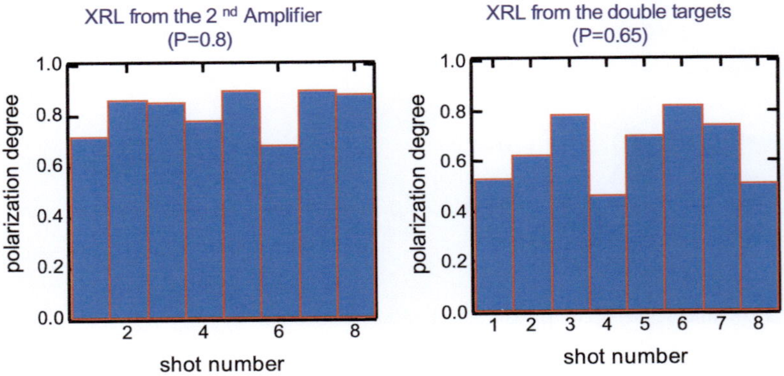

Fig. 2 Measured polarization degree for the second target (*left*), and double target (*right*)

saturation effect. Detailed analysis of the result using ray-trace code is now going on.

2.2 *Observation of Higher-Order Harmonics Emission from Relativistic Plasma*

In JAEA, several attempts toward shorter wavelength lasing have been studied [8, 9]. Recently higher-order harmonics generation experiment using ultra-intense laser pulse was conducted. A He gas target with the gas density of around 10^{19} cm^{-3} was irradiated by J-KAREN laser which provided 0.4–1 J, 27–40 fs pulse with the focusing intensity of around 4×10^{18} W/cm^2. The charged particles from the He plasma were ejected by the bending magnet, and the x-ray emission from the plasma was observed by grazing incidence spectrometers. Figure 3 shows the spectrum obtained in this experiment. Higher-order harmonics signal up to the order of $N = 180$ were clearly observed. One of the interesting features was that we had even-number order of the harmonics together with the odd-number order. This means this emission is not originated from the atomic HHG. Observation using another spectrometer which covered shorter wavelength region showed that the signal extended to the water-window region.

Present higher-order harmonics is quite different from other coherent sources such as atomic higher-order harmonics, nonlinear Thomson scattering and betatron radiation. We performed 3D PIC (particle-in-cell) simulation to explain the present results. The result of the calculation shows due to the ponder-motive force and Lorentz force induced wake wave and bow wave together with laser-induced modulation generates high density spikes in the plasma. This localized high density regions are collectively oscillated under the laser electric field, resulting in the generation of the higher-order harmonics. Following this model, as the laser intensity is increased, the expected photo energy of higher-order harmonics becomes larger,

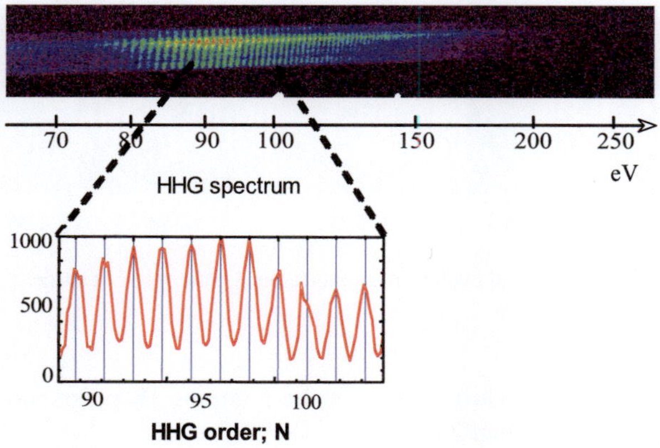

Fig. 3 Obtained HHG spectrum. Up to N = 180, HHG signal are clearly resolved. The decrease of the signal for the photon energy larger than 200 eV is due to poor efficiency of the spectrograph

and the upper limit of the laser intensity may be determined by radiation damping regime with the laser intensity of around 10^{23} W/cm^2. This means this scheme can become one of the feasible candidates for generating intense coherent x-ray in 10 keV range.

3 Application of SXRLs

The 13.9 nm laser in JAEA has been used for wide variety of application researches, such as material science, plasma physics, atom and molecular physics, ultrafast soft x-ray imaging and laser processing under the collaborations with universities and other institutes. In these applications, we choose two examples and describe the recent progress: The first one is the development of x-ray laser interferometer for probing surface dynamics, and the second is the study of SXRL ablation of substances.

3.1 X-Ray Laser Interferometer for Probing Surface Dynamics

Schematic figure of the x-ray interferometer is described in Ref. [10]. A Pt sample was pumped by an IR pulse with the Gaussian spatial profile of the spot size of 60 μm, duration of 80 fs, and the peak intensity of 5×10^{13} W/cm^2. The SXRL probe beam illuminated the pumped region, and this probe beam with having the information of the surface deformation, was reflected by the imaging mirror and went to the double Lloyd's mirrors. Here the probe beam was divided into two parts, one had the information of surface deformation and the other was used as the reference, and the two components were merged at the detector (CCD) at the distance of about 5 m. The resolution of this interferometer was tested using a Si sample with several 6 nm-depth grooves on the surface, and the depth and lateral resolution was determined to be 1 nm and 1.5 μm, respectively. Using double time fiducial technique,

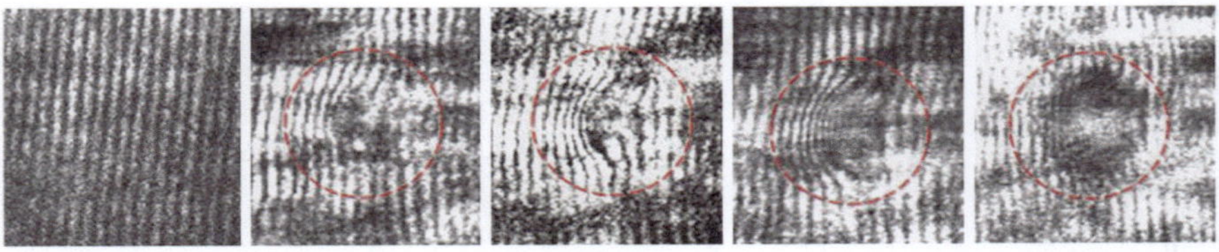

Fig. 4 Temporal evolution of interference patterns of Pt surface pumped by 80 fs-duration Gaussian optical pulse

temporal synchronization of the pumping IR pulse and SXRL probe was achieved to be better than 2 ps.

Figure 4 shows the temporal evolution of the interference pattern as the function of the delay time of the SXRL probe. Our observation shows even in the time of 10 ps, fringe shift can be seen. The absence of the fringe shift in the central portion is due to that the higher fluence of the pump laser (more than 1 J/cm^2) destroys and ionizes the surface in the early time region. This is consistent with observation using soft x-ray reflectometer.

The pumping area may be categorized into 3 areas: (a) The central portion with the fluence larger than 1 J/cm^2, (b) middle region with 0.4 to 1 J/cm^2, and (c) outer region with the fluence of around laser ablation threshold. We could see a ring structure in region (c), whose reflectivity was low even in early time region. This temporal behavior of the reflectivity was similar to that of region (a), whereas for region (b) the reflectivity decreased slowly. Detailed analysis using theoretical code will be published elsewhere.

3.2 The Study of SXRL Ablation of Substances

Laser ablation has many technological applications in material processing and nanostructure fabrication. The laser-induce damage of the materials has been intensively studied and its dependence upon the pulse width, the photon energy, and the fluence has come to be understood. Furthermore novel light source in soft x-ray region, i.e., laser-based SXRL and X-FEL, provides new possibility of laser processing, and it is urgent subject to have understanding of the interaction between intense soft x-ray and substances. Recently we found that the threshold fluence value for the ablation of Si and LiF crystal using pico-second duration SXRL pulse was smaller by 2–3 orders magnitude than that of IR and visible laser case. This was qualitatively explained by spallative ablation model: Molecular dynamics simulation shows that SXRL absorption increases temperature of surface layer of the material, and if the heating time is shorter than the heat transport (sonic) time then the temperature growth increases the tensile stress in the material which induces the spallation of the surface layer.

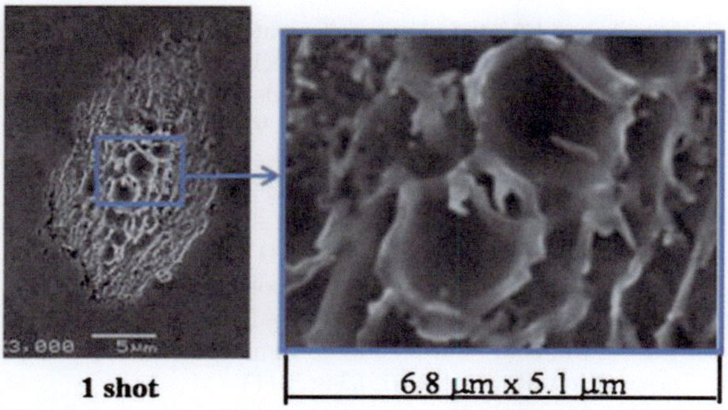

Fig. 5 SEM image of the irradiation area of Aluminum sample under the laser fluence of 30 mJ/cm^2

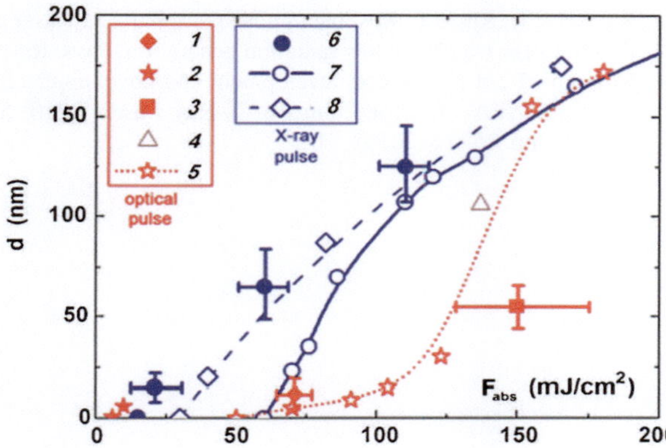

Fig. 6 Fluence dependence of the melting depth of the Au sample. *Solid circle* is the experimental result and *dashed line* is the result by molecular dynamics simulation, which is consistent with the experiment

For the quantitative analysis, we measured the depth of melting region of Au and Al in the single-shot SXRL exposure. Figure 5 shows SEM images of typical irradiation area of Al samples. AFM analysis of these irradiation area shows the surface is the mixture of splashed molten metals and hole induced by removed metal from the surface. The measured melting depth of Au is plotted as the function of the fluence in Fig. 6. The blue solid circle and blue dashed line are the present experimental result and the theoretical calculation, respectively. Other experimental result using IR and optical pulse together with the calculation is also added (red marks and lines). Comparison with the present experimental result and the theoretical calculation implies that spallative ablation model is good explanation of the present experimental result.

Acknowledgements This work is partly supported by auspices of Japan Grant-in-Aid for Scientific research, Kiban (B), No. 25289244.

References

1. Tanaka, M., et al.: X-ray laser beam with diffraction-limited divergence generated with two gain media. Opt. Lett. **28**, 1680–1682 (2003)

2. Namikawa, K., et al.: Direct observation of the critical relaxation of polarization clusters in $BaTiO_3$ using a pulsed x-ray laser technique. Phys. Rev. Lett. **103**, 197401 (2009)
3. Namba, S., et al.: Enhancement of double Auger decay probability in Xe clusters irradiated with a SXRL pulse. Phys. Rev. Lett. **99**, 043004 (2007)
4. Suemoto, T., et al.: Single-shot picosecond interferometry with one-nanometer resolution for dynamical surface morphology using a soft x-ray laser. Opt. Express **18**, 14114 (2010)
5. Faenov, A.Ya., et al.: Low-threshold ablation of dielectrics irradiated by picosecond soft x-ray laser pulses. Appl. Phys. Lett. **94**, 231107 (2009)
6. Kawachi, T., et al.: Observation of polarization of the soft x-ray laser line in neonlike germanium ions. Phys. Rev. Lett. **75**, 3826–3829 (1995)
7. Kim, C.M., et al.: Pulse buildup from noise and intrinsic polarization of plasma-based x-ray lasers. Phys. Rev. A **84**, 013834 (2011)
8. Kando, M., et al.: Demonstration of laser-frequency up-shift by electron-density modulation in a plasma wakefield. Phys. Rev. Lett. **99**, 135001 (2007)
9. Kawachi, T., Kato, Y.: Resonant photo-pumping x-ray laser scheme using intense characteristic x-rays for water-window radiation generation. Phys. Rev. A **84**, 063848 (2011)
10. Kawachi, T., et al.: Source development and novel applications of laser driven plasma x-ray lasers in JAEA. In: Proceedings of X-Ray Lasers 2010. Springer Proceedings in Physics, vol. 136, pp. 15–24 (2010)

Chapter 24
Characterization of Zn X-Ray Laser at PALS Centre, Its Applications in Dense Plasma Probing and Astrophysics

M. Kozlova, J. Nejdl, M. Krus, J. Prokupek, J. Dostal, B. Rus, A. Klisnick, L. Meng, F. Tissandier, C. Stehlé, R. Lefevre, U. Chaulagain, N. Champion, P. Barroso, F. Reix, P. Jagourel, J. Larour, F. Delmotte, C. Constancias, F. Suzuki-Vidal, and O. Acef

Abstract This report presents the results from experiments at PALS Centre using a Zn X-ray laser with the pulse length of 0.15 ns and the wavelength of 21.2 nm, working in single or double pass regime with the output energy of 0.4 mJ or 4 mJ per pulse, respectively. The current X-ray laser was experimentally examined to obtain its temporal coherence and spectral width using a path-difference interferometer.

M. Kozlova (✉) · J. Nejdl · M. Krus · J. Prokupek · J. Dostal · B. Rus
IoP/PALS Center, Prague, Czech Republic
e-mail: kozlova@fzu.cz

A. Klisnick · L. Meng
ISMO, Uni. Paris Sud, Paris, France

F. Tissandier
LOA, Ecole Polytechnique, CNRS, Paris, France

C. Stehlé · R. Lefevre · U. Chaulagain · N. Champion
Lerma, Observatoire de Paris, UPMC, CNRS, Paris, France

P. Barroso · F. Reix · P. Jagourel
GEPI, Observatoire de Paris, Univ. Paris-Diderot, CNRS, Paris, France

J. Larour
LPP, Ecole Polytechnique, UPMC, CNRS, Palaiseau, France

F. Delmotte
LCF, Institut d'Optique, CNRS, PXI, Palaiseau, France

C. Constancias
CEA-LETI, Grenoble, France

F. Suzuki-Vidal
Imperial College, London, UK

O. Acef
SYRTE, Observatoire de Paris, UPMC, CNRS, LNE, Paris, France

S. Sebban et al. (eds.), *X-Ray Lasers 2012*, Springer Proceedings in Physics 147,
DOI 10.1007/978-3-319-00696-3_24,

The double pass regime shows that QSS plasma based source-amplifier is promising for "short" fs soft X-ray pulses. The X-ray laser is commonly used for user's experiments. Its advantages can be shown in applications such as probing of dense plasmas (up to 2.5×10^{24} cm^{-3}) or single shot experiments (4×10^{14} photons/pulse). The simple technique based on Talbot effect was used to obtain the gradients of electron densities of line plasmas produced under conditions corresponding to XRL's amplifiers operating in TCE and QSS regime. To investigate radiative shock wave in laboratory is challenging in aspects of the optimization of experimental parameters. Due to the high electron density (10^{22} cm^{-3}) produced in the gas medium propagated by the shock wave, the velocity of the shock wave, and the absorption losses on optical elements, it is necessary to use the energetic single shot probe.

1 Introduction

Coherent and high-brightness radiation sources at very short wavelengths are a tool of significant interest in future studies of dense and/or highly ionized plasmas. Among the prospects offered by these sources, there are both novel plasma-probing techniques able to access extremely high densities and provide nanometer spatial resolution, as well as new ways of generation of extremely dense and highly ionized plasmas, previously inaccessible in laboratory conditions. This report describes the XRL, which was implemented with a goal to produce as great amount of energy as possible near the wavelength of 20 nm, use the energetic XRL for imaging, and to probe dense plasmas.

2 Characteristics of Ne-Like Zn X-Ray Laser at PALS Centre

Over past decade, the most energetic XRL is delivered at Prague Asterix Laser System (PALS) Centre [1]. The pumping iodine laser emitting at 1.3 μm has total energy about 1 kJ. The pumping beam is split into two beams, prepulse and main pulse, and this sequential generation of the XRL (see Fig. 1) can provide output energies up to 10 mJ.

Several pumping schemes were examined to generate the XRL (Ne-like Fe, Zn, Se and Ni-like Ag), but the most robust and the most emitting schema is Ne-like Zinc. The population inversion is produced between $2p^5$ 3p and $2p^5$ 3s fine-structure levels of Ne-like ions, through electron collisional excitation from the Ne-like ground state $2p^6$. The inversion is maintained by strong radiative dipole decay from the 3s levels back to the ground state while the upper 3p levels are metastable with respect to this process. Under appropriate plasma conditions, the strongest population inversion is generated (due to the highest monopole electron excitation rate from the ground) for the transition $(2p^5_{1/2}3p_{1/2})J = 0 \rightarrow (2p^5_{1/2}3s_{1/2})J = 1$, corresponding in Ne-like zinc to the wavelength of 21.22 nm. The intensity and beam profile emitted by this type of XRL strongly depends on the density profile of the

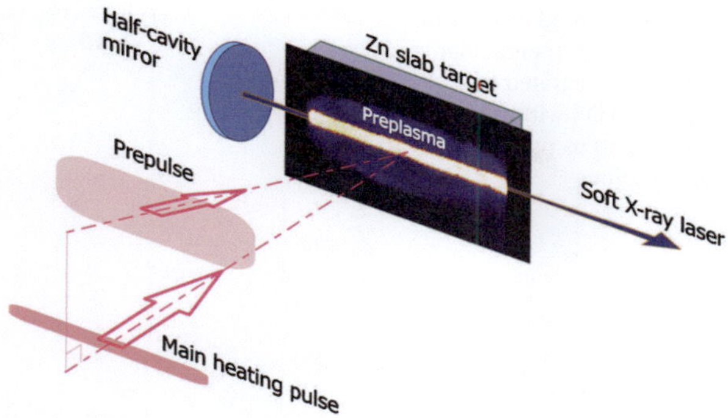

Fig. 1 Sequential pumping of the XRL amplifier, where the typical ratio of the prepulse ($<10^{11}$ W cm^{-2}) and the main pulse ($\sim10^{13}$ W cm^{-2}) is about 10^{-3}. Half-cavity mirror is placed on one side of the Zn slab target to boots output energy

amplifying plasma, as the largest population inversion on the J = 0–1 transition occurs at high electron densities. To reduce density gradients in the plasma ablated by a laser from a massive slab target, a technique, consisting of applying a prepulse typically a few nanoseconds prior the main heating pulse, is employed [2–4]. The parameters of the prepulse relative to the main pulse are one of the key features determining the functionality of the XRL and characteristics of the emitted beam. Two configurations where a weak prepulse is applied 10 ns or 50 ns ahead of the main pulse will be discussed. The prepulse is delivered and focused separately to produce a wide pre-plasma column into which the main pulse, focused much more tightly, is coupled (see Fig. 1). The configuration where the prepulse is 10 ns ahead can be operated in single or double-pass regime. The typical value of energy (double-pass regime) is 4 mJ with peak power 40 MW and with 3×10^{14} photons per pulse. The 10 ns configuration is the most stable and robust, and at PALS is provided as the user beamline. In order to increase the energetic output of the XRL, the longer prepulse delays were investigated. The best results were observed for 50 ns prepulse delay operated in double-pass regime. This was studied by the streak camera and near-field measurements. While the streak camera records indicate comparable kinetics of the exited levels for both regimes, from the near-field data, it is found out that the 50 ns prepulse delay can generate 2× larger (in the direction perpendicular to the target) amplification region than in 10 ns regime (see Fig. 2). Therefore the energy measured for 50 ns prepulse delay is 10 mJ with peak power 100 MW and with 1.2×10^{15} photons per pulse.

3 Further Characterization of the Ne-Like Zn X-Ray Laser

Using the Zn XRL as the user beamline requires a detailed characterization of its output parameters such as the energy, the pulse length etc. In addition to routine monitoring parameters, the spectral profile and linewidth were explored. The bandwidth of the Zn XRL is extremely narrow, about 10^{-4}, therefore direct measurement is beyond capabilities of existing spectrometers in XUV range. Hence, the wavefront-splitting interferometer [5, 6] was used to measure temporal coherence

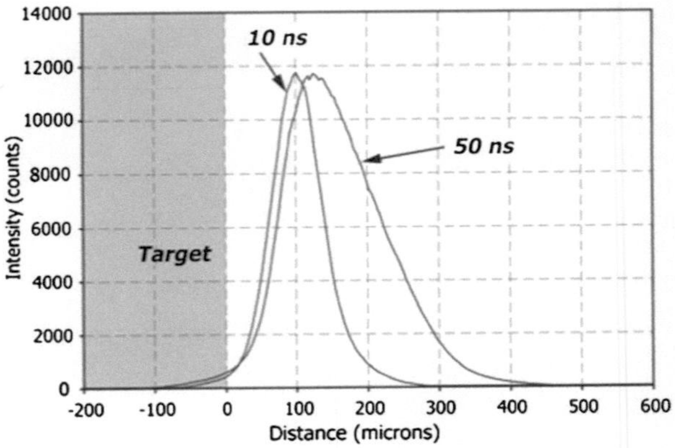

Fig. 2 Near-filed profiles in the direction perpendicular to the target (integrated over lateral FWHM width) for the 10 ns and 50 ns pumping configurations

of the source from which the spectral linewidth is inferred. The experimental setup is shown in Fig. 3. The XRL beam, operated in double-pass regime (better spatial coherence expected), was send either to the footprint to monitor spatial intensity distribution and alignment, or to the interferometer. The direct light from the plasma was blocked by Al filter after multilayer mirror.

Two sets of data were acquired for the distance between the source and the interferometer (5 m and 7 m). The distance of the interferometer and the camera was, in both cases, 1.5 m and the tilt of the camera was 35 degrees.

The spectral linewidth ($\Delta\lambda$) is given by

$$\Delta\lambda = \frac{\lambda^2}{c}\Delta\nu \tag{1}$$

where λ is XRL wavelength. The coherence time ($\Delta\nu$) can be measured by the variation of the path difference between dihedron mirrors. This measurement resulted in 0.7 ps and 1 ps for distance of 5 m and 7 m, respectively. According to Eq. (1), the corresponding spectral linewidths are 11.3 mA and 7.7 mA. The measured data of the coherence time leading to a Fourier transform limit of the duration below 1 ps, which is much smaller that duration of the duration of the SXRL and potentially it indicates the possibility of the amplification of sub-picosecond duration [7].

4 The Zn X-Ray Laser as a Powerful Tool to Probe Dense Plasmas

Because of natural constraint, the electromagnetic radiation can propagate through the plasma only if the electron density does not reach the critical density that scales as λ^{-2}. Therefore, one is obliged to use the short-wavelength probe in order to investigate the region close to the critical density for VIS or UV laser light where a lot of interesting phenomena occur. Besides the wavelength criterion, the probe beam must be bright enough to overcome the thermal radiation of the investigated

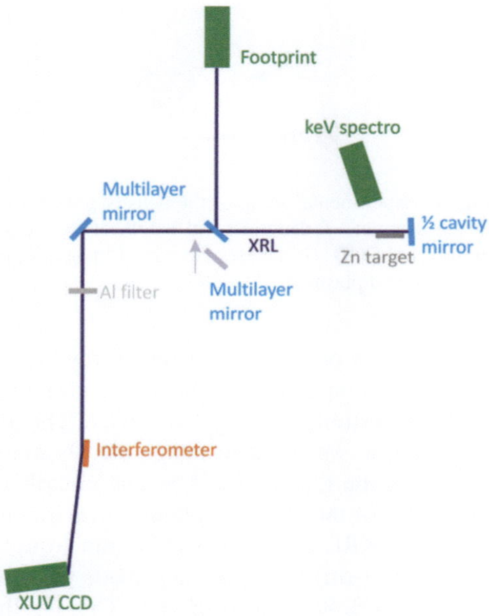

Fig. 3 The sketch of the experimental scheme. The XRL beam is either sent to footprint or to the interferometer

plasma. Therefore the most suitable plasmatic sources for dense plasma probing are X-ray K-alpha radiation sources, or X-ray lasers [8] emitting in soft X-ray or extreme ultra-violet range [9].

4.1 Deflectometry of the Dense Plasmas

This paper introduces a new simple deflectometric technique that enables to resolve the gradients in both directions of the projection plane using the deformation of the Talbot pattern of a single 2D grating. The simplicity and low demands on spatial coherence of the probing beam enables the use of this method in complex experiments where the previous techniques would not be acceptable. The probing technique is based on a modification of the Talbot reconstruction of the 2D grating, caused by density gradients in the probed plasma. The suggested experimental setup is a simple modification of the setup used for X-ray laser backlighting [10]. The plasma is imaged on a CCD camera with sufficiently high magnification, so that good spatial resolution is achieved. The pinhole is placed in the position of the probing source image in order to eliminate most of the plasma self-emission and enhance the signal-to-noise ratio. The size of the pinhole must be set as a compromise between this ratio and maximum detectable wave-front tilt (maximum detectable density gradient). The 2D grating is placed between the pinhole and CCD so it has its first Talbot sub-plane (or plane) on the CCD. The simplified experimental setup is shown in Fig. 4.

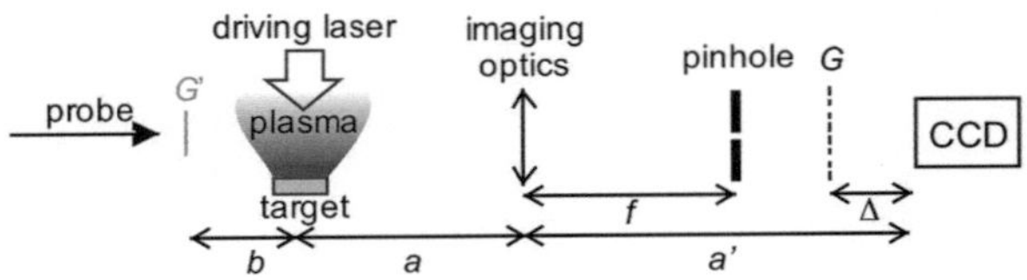

Fig. 4 Experimental setup of the density gradient measurement. a is a object distance, a' is the image distance, f denotes focal length of the imaging element, Δ is the distance between 2D grating and CCD. G' denotes the virtual image of grating G that is placed at a distance b in front of the probed plasma

The beam of the Ne-like zinc X-ray laser emitting at 21.2 nm [1] was used as a probe beam, operating in the single pass regime with the 150 ps (FWHM) pulses of 0.4 mJ estimated energy in a pulse. The Mo-Si multi-layer spherical mirror with f = 300 mm was used to image the plasma on back-illuminated X-ray CCD with magnification factor M = 5.5. The pinhole of the diameter of 0.5 mm was put to the image of the XRL source in order to reduce the signal of plasma self-emission [10]. The XRL source was 2800 mm from the imaging mirror. Then the 100 μm-period laser-drilled 2D grating made of 5 μm thick steel sheet was placed to the distance Δ = 980 mm from the CCD, i.e. the higher sensitivity and worse spatial resolution configuration for the first Talbot sub-plane (q = 1) was used. The plasma was created by the focusing sequence of two pulses in order to create similar plasma conditions on solid target as with XRL sequence. The linear plasma was created by 80 J@250 ps pulse of 1.3 μm laser by pair of cylindrical and spherical lenses with focal lengths of 6 m and 1.2 m, respectively, into the 150 μm-wide and 1.5 mm-long line, overfilling the 1 mm-wide slab in order to produce uniform irradiation on the target with intensity of 4×10^{13} W cm^{-2}. The plasmas created by irradiation of Zn, Fe, and Au slabs were investigated. The recorded pattern $\Phi(x, y)$ is related to density gradient ∇n via the formula

$$\int_0^L \nabla n(x, y, z)\, dz = \Phi(x, y) \tag{2}$$

where L is the length of plasma. If plasma is regarded as uniform in probing direction, Eq. (2) can be simplified to $L\nabla n(x, y) = \Phi(x, y)$ and supposing $n_e \ll n_c$ the relation $\nabla n_e(x, y) = -2n_c\Phi(x, y)/L$ could be employed. The recorded patterns and evaluated electron density gradient maps for Zn are shown in Fig. 5, for Fe in Fig. 6 and for Au in Fig. 7. All presented plasmas were probe 0.3 ns after auxiliary main pulse arrived to the target.

4.2 XUV Probing of the Laser Driven Radiative Shocks

Using high-energy lasers, the radiative shocks can be produced in gases. For instance, the laser intensity about 10^{14} W cm^{-2} permits to generate shocks propagating at ~50 km s^{-1} in gases at low pressure and to produce interesting phenomena

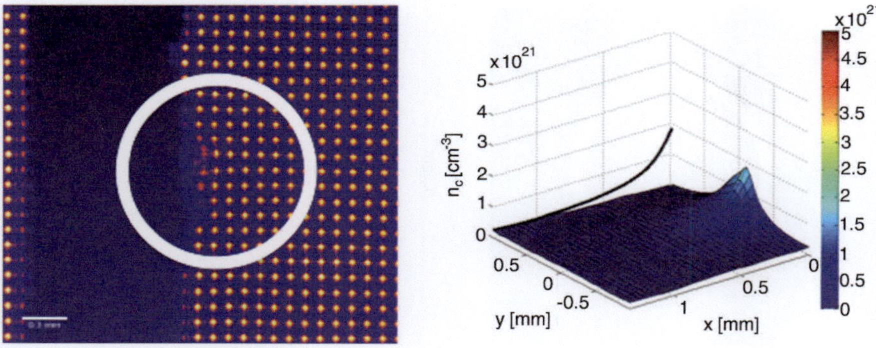

Fig. 5 The example of recorded pattern (*left*) of the Zn plasma created on 1 mm solid target by sequence of prepulse ($I = 4 \times 10^{10}$ W cm^{-2}) and main pulse ($I = 4 \times 10^{13}$ W cm^{-2}), where the prepulse is 10 ns ahead to main pulse, and the corresponding electron density map (*right*)

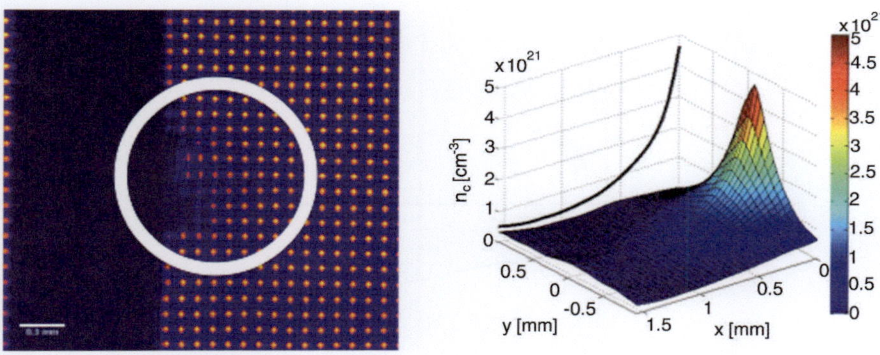

Fig. 6 The example of recorded pattern (*left*) of the Fe plasma created on 1 mm solid target by sequence of prepulse ($I = 4 \times 10^{10}$ W cm^{-2}) and main pulse ($I = 4 \times 10^{13}$ W cm^{-2}), where the prepulse is 10 ns ahead to main pulse, and the corresponding electron density map (*right*)

like ionization wave, also called radiative precursor [11]. In such environment, the electron density may be up to 10^{23} cm^{-3} and thus is difficult to probe it using long wavelength light such as VIS. To be able to probe electron densities close to critical density, the critical density of probe must be 10× higher than the investigated electron density. The critical density of the Zn XRL is about 10^{24} cm^{-3}, which makes it a very promising tool for investigation. Although, all the requirements on the probing beam is fulfilled, the X-ray shadowgraphy is a very delicate and requires a high quality of all used components such as filters, mirrors, windows or membrane of closing the gaseous targets. Recent experiments performed at PALS laser facility in 2010 and 2011 reveal the critical optical elements, which are necessary for such study. In Fig. 8 is schematically shown the experimental setup.

Probe beam enters and exits the gaseous target through membranes which must be transparent enough, to do not absorb all XRL signal, and sturdy enough to with-

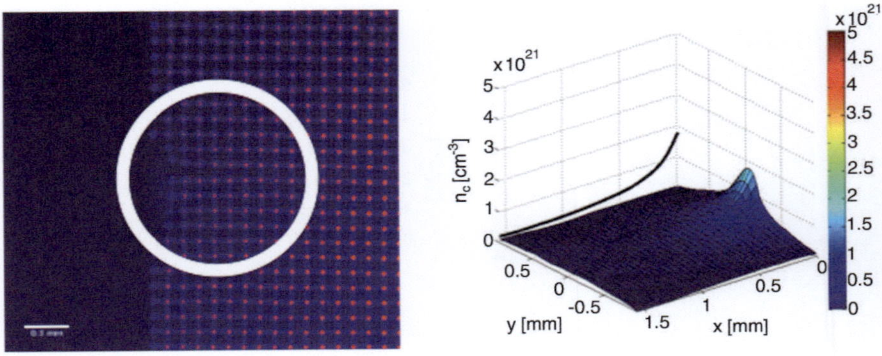

Fig. 7 The example of recorded pattern (*left*) of the Au plasma created on 1 mm solid target by sequence of prepulse ($I = 4 \times 10^{10}$ W cm^{-2}) and main pulse ($I = 4 \times 10^{13}$ W cm^{-2}), where the prepulse is 10 ns ahead to main pulse, and the corresponding electron density map (*right*)

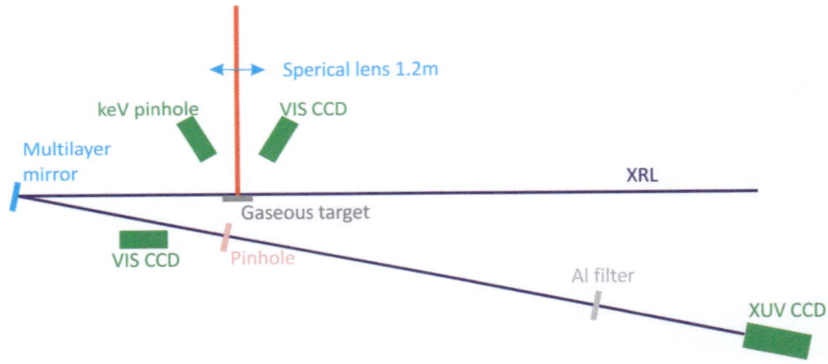

Fig. 8 The experimental schema. The auxiliary beam @1.3 micron is focused on Au/CH piston and launches the shock. 20 ns after the shock start propagating is investigated by XRL. The modifications of transmissivity are imaged on the CCD with magnification 5.5

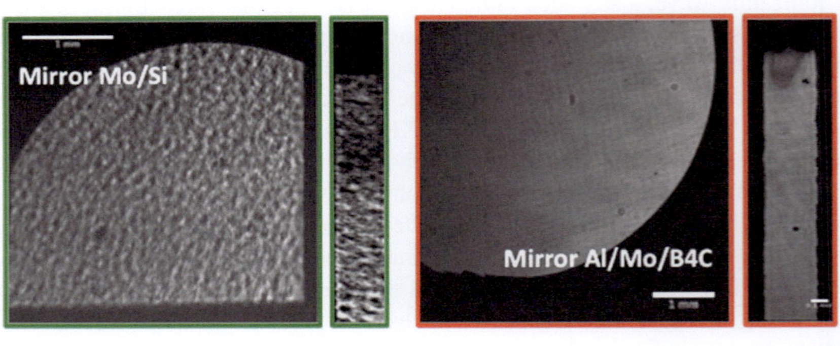

Fig. 9 The image of XRL reflected by Mo/Si multilayer mirror (*left*), Al/Mo/B4C multilayer mirror (*right*) and corresponding images of shock waves

stand a difference of pressure of 0.5 bar (inside the target) and 0 bar outside, without leakage. In order to have the smallest energetic losses, the Si3N4 membrane was used (T = 0.017 @21.2 nm). The imaging mirror must have a good and uniform reflectivity at given wavelength. In Fig. 9, there are shown two examples of imaging mirrors and corresponding shadowgrams of the shocks. On left part (see in Fig. 9) the front of the shock wave is almost not distinguishable due to strong perturbations coming from Mo/Si mirror. To enhance the mirror quality the new Al/Mo/B4C multilayer mirror was used [12]. The difference is obvious, see in Fig. 9 (right).

5 Conclusion

At conclusion, the work at PALS facility over past decade, led us to provide the most energetic X-ray source as standard user beamline, delivering up to 10 mJ. Beside the standard characteristics of the XRL beam, the linewidth measurement was done and it indicates the possibility of amplification of pulses below 1 picosecond. These results are in good agreement with calculations. It was shown that the Zinc X-ray laser is powerful tool to probe dense plasmas. The electron densities of various elements were measured by deflectometry technique. The results will be used as benchmark data for ARWEN code. Also the XRL beam was used to investigate the radiative shock waves. The critical elements of the experimental setup are the optical components. The dramatic improvement in the pattern contrast was presented.

Acknowledgements This work was partially supported by the by the Czech Ministry of Education, Youth and Sports (projects 7E08099 and 7E09092), and from the European Union (EU) Seventh Framework Programme (LASERLAB-EUROPE, grant agreement 228334).

References

1. Rus, B., et al.: Phys. Rev. A **66**, 063806 (2002)
2. Nilsen, J., et al.: Phys. Rev. A **48** (1993)
3. Cairns, G.F., et al.: In: Proc. 4th Int. Coll. on X-Ray Lasers, vol. 332, p. 289 (1994)
4. Rus, B., et al.: In: Proc. 4th Int. Coll. on X-Ray Lasers, vol. 332, p. 152 (1994)
5. Klisnick, A., et al.: JQSRT **99** (2006)
6. Guilbaud, O., et al.: EPJD **40** (2006)
7. Meng, L., et al.: In: These proceedings
8. Tallents, G.J., et al.: Laser Part. Beams **20**, 201–209 (2002)
9. Nejdl, J., et al.: Plasma Phys. **17** (2010)
10. Kozlová, M., et al.: In: Springer Proceedings in Physics, vol. 130, p. 417 (2009)
11. Stehle, Ch., et al.: Opt. Commun. **285** (2012)
12. Meltchakov, et al.: Proc. SPIE **8168** (2011)

Chapter 25
Development of High Reflective Multilayer Mirrors at "Water Window" Wavelengths in IPOE

Haochuan Li, Jingtao Zhu, and Zhanshan Wang

Abstract Some X-ray sources for laboratory near "water window" wavelengths have been developed. High reflective multilayer mirrors are required. In this wavelength region, Cr/Sc, Cr/Ti, Cr/V and Cr/C are promising for high reflective multilayer mirror. The layer thickness is typically about 1.0 nm. We have deposited Cr/C and Cr/Sc multilayers for $\lambda = 4.48$ nm and reflectance of 15.2 % was obtained for Cr/C at near normal incidence. We also show that the interface of Cr/Ti multilayers can be significantly improved by inserting B_4C as diffusion barrier layer. In this report, high-reflective multilayer mirrors with and without barrier layer were deposited by using magnetron sputtering method. The reflective properties of the multilayer mirror were measured by using synchrotron radiation.

1 Introduction

Multilayer-based reflective optics in the extreme ultraviolet and soft x-ray region has tremendously progressed in the past three decades, which has been widely used in EUV lithography, x-ray astronomy, soft x-ray microscopy and synchrotron radiation instruments [1]. In the "water window" wavelength (2.3 nm $< \lambda <$ 4.3 nm), great opportunities emerge in microscopy of biological samples with both high contrast and high resolution [2]. High reflective coatings are needed for laboratory "water window" light source.

In the "water window" wavelengths, C, Sc, Ti, V are suitable spacing layer material near their respective absorption edges. State-of-the-art Cr/C [3] and Cr/Sc [4] multilayers have near-normal reflectance of about 13% and 20% respectively. Ti-based [5] and V-based [6] multilayers are still under-developed with reflectance of 2 %–3 %.

In this paper, we present the development of ultra-thin multilayers near "water window". Cr/C and Cr/Sc multilayers were prepared for $\lambda = 4.48$ nm and measured at BESSY II. Reflectance of 15.2 % was obtained for the Cr/C one. Cr/Ti multilayers

H. Li · J. Zhu (✉) · Z. Wang
Institute of Precision Optical Engineering (IPOE), Physics Department, Tongji University, Shanghai 200092, China
e-mail: jtzhu@tongji.edu.cn

S. Sebban et al. (eds.), *X-Ray Lasers 2012*, Springer Proceedings in Physics 147,
DOI 10.1007/978-3-319-00696-3_25,

for λ = 2.73 nm were studied by introducing B_4C at interfaces. The effect of B_4C thickness and interface symmetry were investigated. The hard x-ray (λ = 0.154 nm) reflectance were found 8 times higher after introducing 0.13 nm B_4C at both interfaces.

2 Experimental Details

All the multilayers were deposited by an ultrahigh vacuum direct current magnetron sputtering system. Base pressure was below 5.0×10^{-5} Pa. The sputter gas was argon at a constant pressure of 1.0 mTorr (0.13 Pa). All multilayers were deposited onto sliced polished silicon (100) wafer.

Grazing-incidence X-ray reflectance (XRR) measurements were made in the θ–2θ geometry using a 4-circle x-ray diffractometer having a sealed-tube Cu source and a Si (220) crystal monochromator tuned to the Cu K–α line (λ = 0.154 nm, E = 8.05 keV). The angular divergence of this system is estimated to be ~0.007°. Near-normal incidence soft x-ray reflectance measurements for Cr/C and Cr/Sc were made on beam line UE56/2-PGM-1 at BESSY-II synchrotron radiation.

3 Results and Discussion

3.1 Cr/C and Cr/Sc Multilayers

Cr/C and Cr/Sc multilayers were designed as near-normal incidence reflective mirror with period D = 2.26 nm, thickness ratio (C layer thickness to period) $\Gamma = 0.55$ and bilayer number N = 300. After deposition, XRR measurements were performed. The results are shown in Fig. 1(a). Both curves show sharp and intense Bragg peaks up to 4° of grazing incidence angle, which indicate sharp interfaces with small roughness or interdiffusion. The reflectance were measured at synchrotron radiation (SR) facility and shown in Fig. 1(b). Cr/C multilayer has a reflectance as high as 15.2 %, while Cr/Sc has only 7.5 %. The less reflectance of Cr/Sc is due to smaller optical contrast at this wavelength than Cr/C.

3.2 Cr/Ti Multilayers

We have deposited 4 sets of Cr/Ti multilayers. Firstly, four pieces of pure Cr/Ti multilayers with $\Gamma = 0.45$, N = 50 were prepared as reference samples and the periods of the multilayers were in the range of 1.37–1.66 nm. Then, other three sets had the same deposition time of Cr and Ti with the 1.37-nm-thick pure Cr/Ti

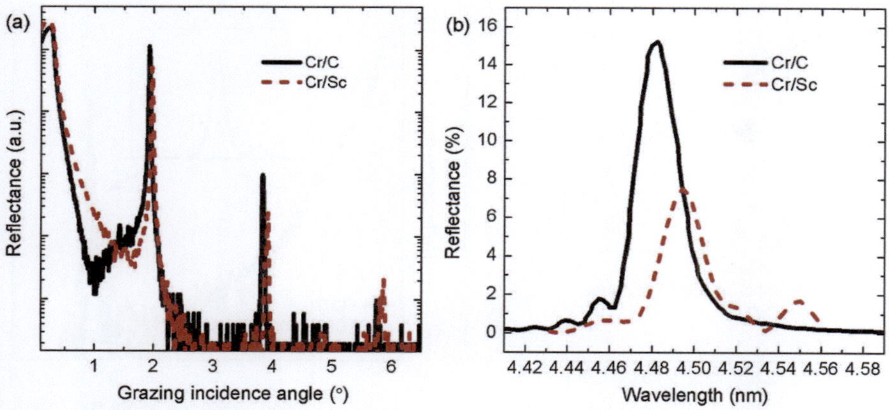

Fig. 1 XRR (**a**) and SR (**b**) measurements of Cr/C and Cr/Sc multilayers

multilayer, but they had B_4C diffusion barrier layer inserting at the Cr-on-Ti or Ti-on-Cr or both interfaces. Each set had five samples with nominal B_4C thickness of 0.08, 0.13, 0.18, 0.22, 0.27 nm.

The four sets of multilayers were measured by XRR. Only the first order peaks can be observed as the periods are ultra-short. Thus we plotted the peak intensities versus peak positions of all the samples in Fig. 2(a). For three sets with B_4C barrier layer, points on the left (low angle) side represent multilayers with thicker B_4C as the Cr and Ti layer thicknesses are identical to each other. Inserting B_4C on either interface would improve multilayer quality, and the best performance was achieved by inserting B_4C at both interfaces. Inserting B_4C at Cr-on-Ti interface gives more reflectance than at Ti-on-Cr interface. Comparing samples with different B_4C thickness, we have found that B_4C of 0.13 nm thick provide reflectance improvement as good as thicker ones. Considering that thicker B_4C will reduce theoretic reflectance, 0.13 nm B_4C layer will be inserted at both interfaces.

Basing on the preliminary results, two pieces of Cr/Ti multilayers with $N = 600$ were deposited for $\lambda = 2.73$ nm, one with 0.13 nm B_4C barrier layer on both interfaces, the other without barrier layer. The XRR measurement results are shown in Fig. 2(b). The peak reflectance of the Cr/Ti multilayer with B_4C barrier is 8 times higher than the one without barrier layer. Such great improvement in hard x-ray reflectance exhibits the effect of B_4C barrier layer, and also predicts improvement in soft x-ray reflectance at $\lambda = 2.73$ nm.

4 Conclusion

Cr/C and Cr/Sc multilayers for $\lambda = 4.48$ nm were deposited and reflectance of 15.2 % was obtained for the Cr/C one. The B_4C diffusion barrier layer inserting into Cr/Ti multilayers were investigated, and the best barrier layer thickness is 0.13 nm.

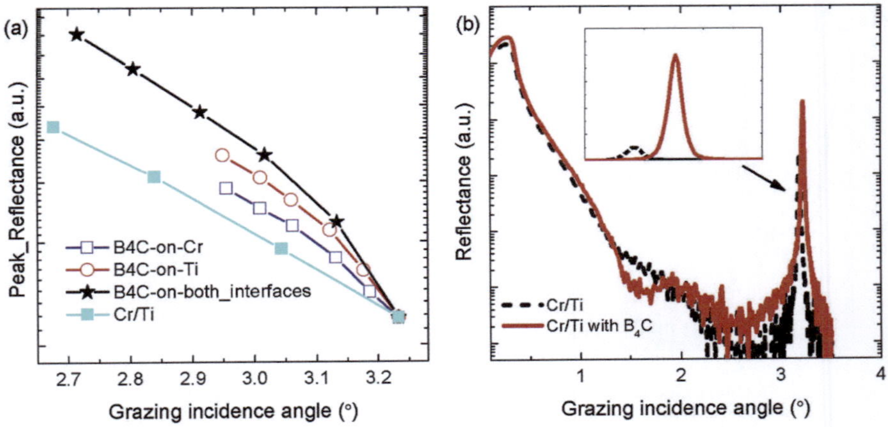

Fig. 2 (**a**) First order peak reflectance versus peak position plot of the four sets of N = 50 Cr/Ti multilayers with or without B_4C barrier layer. (**b**) XRR curves (log scale) of N = 600 Cr/Ti multilayers with or without B_4C barrier layer, the *inset* shows the linear scale zoom of first order peak

After introducing 0.13 nm B_4C in both interfaces, an 8 times improvement of reflectance at 8.05 keV was observed.

Acknowledgements We are indebted to Dr. Franz Schäfers at BESSY-II, and Mingqi Cui at BSRF for their kindly help in synchrotron radiation measurements.

References

1. Schäfers, F.: Multilayers for the EUV/soft X-ray range. Physica B **283**, 119–124 (2000)
2. Hoover, R.B., Shealy, D.L., Brinkley, B.R., Baker, P.C., Barbee, T.W. Jr., Walker, A.B.C. Jr.: Development of the water window imaging x-ray microscope utilizing normal-incidence multilayer optics. Opt. Eng. **30**, 1086–1093 (1991)
3. Andreev, S.S., Mertins, H.C., Platonov, Y.Y., Salashchenko, N.N., Schaefers, F., Shamov, E.A., Shmaenok, L.A.: Multilayer dispersion optics for X-ray radiation. Nucl. Instrum. Methods Phys. Res., Sect. A **448**, 133–141 (2000)
4. Eriksson, F., Ghafoor, N., Schäfers, F., Gullikson, M., Aouadi, S., Rohde, S., Hultman, L., Birch, J.: Atomic scale interface engineering by modulated ion-assisted deposition applied to soft x-ray multilayer optics. Appl. Opt. **47**, 4196–4204 (2008)
5. Ghafoor, N., Persson, O.A., Birch, J., Eriksson, F., Schäfers, F.: Interface engineered ultrashort period Cr-Ti multilayers as high reflectance mirrors and polarizers for soft x rays of $\lambda = 2.74$ nm wavelength. Appl. Opt. **45**, 137–143 (2006)
6. Eriksson, F., Ghafoor, N., Schäfers, F., Gullikson, M., Birch, J.: Interface engineering of short-period Ni/V multilayer X-ray mirrors. Thin Solid Films **500**, 84–95 (2006)

Chapter 26
Time Resolved Holography Scheme Using a Table Top Soft X-Ray Laser

E.B. Malm, N.C. Monserud, P.W. Wachulak, C. Brown, W. Chao, E. Anderson, H. Xu, C.P. Hains, G. Balakrishnan, C.S. Menoni, J.J. Rocca, and M.C. Marconi

Abstract We demonstrate a versatile table-top holography setup capable of acquiring single-shot soft X-ray holograms with a 10–90 % knife edge spatial resolution of 170 ± 26 nm and 1 ns temporal resolution. A Fresnel zone plate is used to create the reference wave as well as to illuminate the sample in a Fourier transform holography scheme. A 100 μm in diameter central opening in the zone plate allows the incident beam to pass through and directly illuminate the object. A pinhole is located in the sample mask allowing the first order from the zone plate to pass while blocking the higher orders. This setup can be used to enhance edges for conventional single-shot soft X-ray holography imaging.

1 Introduction

X-ray imaging promises to provide nanometer spatial and femtosecond temporal resolutions [1]. So far, a major limitation preventing X-ray imaging from becoming more prevalent is that much of X-ray imaging research is conducted at limited-access free electron lasers (XFELs) or synchrotron sources [2–5]. Recently, table-top sources have shown the ability to provide the coherence and intensity required for X-ray imaging [6–10]. As table-top soft X-ray lasers become more available, experimental setups need to be developed in order to image specimens in other areas of science.

E.B. Malm · N.C. Monserud · C. Brown · C.S. Menoni · J.J. Rocca · M.C. Marconi (✉)
Department of Electrical Engineering, Colorado State University and Engineering Research Center for Extreme Ultraviolet Science and Technology, Fort Collins, CO 80523, USA
e-mail: marconi@engr.colostate.edu

H. Xu · C.P. Hains · G. Balakrishnan
Department of Electrical Engineering, University of New Mexico and Center for High Technology Materials, Albuquerque, NM 87106, USA

P.W. Wachulak
Institute of Optoelectronics, Military University of Technology, Warsaw, Poland

W. Chao · E. Anderson
Center for X-Ray Optics, Lawrence Berkeley National Lab, Berkeley, CA 94720, USA

S. Sebban et al. (eds.), *X-Ray Lasers 2012*, Springer Proceedings in Physics 147,
DOI 10.1007/978-3-319-00696-3_26,

In this paper we describe a Fourier transform holography (FTH) setup similar to the one described by McNulty [4]. It utilizes FTH for soft X-ray imaging. A Fresnel zone plate is used to create the reference wave while transmitting the beam directly to illuminate the object creating the object wave. The interference between these two waves allows for direct reconstruction of the sample with a simple two-dimensional Fourier transform applied to the detected intensity. Previous work has used the diffracted field from the zones of a Fresnel zone plate to illuminate the object. This experiment allows the incident beam to pass directly through a central opening in the zone plate. The central opening allows for a more intense and uniform illumination of the sample and provides single-shot nanometer-scale reconstructions.

2 Soft X-Ray Laser

The coherent soft X-ray beam is produced from a table-top discharge-pumped laser emitting at a wavelength of 46.9 nm [11]. The high-gain lasing medium is Ne-like argon plasma. The 46.9 nm wavelength light used for illumination is produced by a table-top discharge pumped capillary Ne-like Ar laser. This compact laser when operated with a 27 cm long capillary produces ~0.3 mJ pulses at a repetition rate of 1 Hz. The laser fits on an optical table occupying a $1 \times 0.5\ \text{m}^2$ footprint. The EUV laser has a high temporal coherence with a fractional spectral bandwidth of $\Delta\lambda/\lambda \approx 3\text{–}4 \times 10^{-5}$ [12], which corresponds to a coherence length of $l_c \approx 750\ \mu\text{m}$. The beam is also highly spatially coherent. The spatial coherence radius in the position where the hologram was recorded was approximately 500 μm for a capillary length of 27 cm and become fully coherent as the capillary was lengthened to 36 cm [13].

3 Experimental Setup

Figure 1 shows a diagram of the experimental setup. The laser pulse reflects off a Mo/Si multilayer mirror at 45° incidence angle (not shown in the figure). The mirror allows a precise alignment of the laser beam in the optical axis of the hologram recording setup and simultaneously provides extra spectral filtering to the illumination. The pulse then illuminates a Fresnel zone plate which produces a reference wave focusing in the 400 nm pinhole made in the sample substrate and at the same time allows the beam to directly illuminate the sample with its 100 μm central opening. A pinhole in the sample mask allows for the first order from the zone plate to pass while filtering out the higher orders. The sample is placed in the first order focal plane of the zone plate which is approximately 1 mm away from the zone plate. The sample is located 5.5 cm from an X-ray back illuminated charge coupled device (CCD) camera from Andor. The CCD has 2048×2048 pixels with a square pixel size of $13.5 \times 13.5\ \mu\text{m}^2$. This setup allows for single-shot time-resolved holograms with real-time detection and reconstruction.

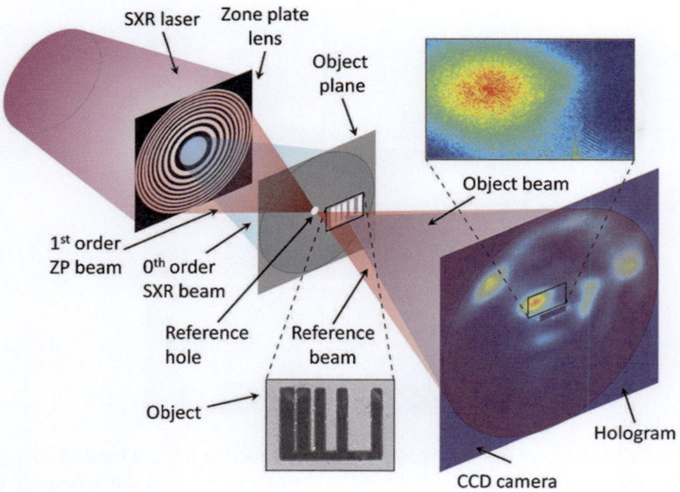

Fig. 1 The Fresnel zone plate creates both the reference and the object waves necessary for FTH. The first order from the zone plate passes through a pinhole in the mask. The 100 μm in diameter central opening in the zone plate allows for the incident light to illuminate the object creating the object wave. The object and reference waves interfere on the CCD

4 Experimental Results

Experiments in the past have used zone plates with all the zones to image samples with X-rays [4]. This previous work used diffracted light through the zones to illuminate the object which might produce a complicated illumination pattern in the sample plane. In the experiment described herein a 100 μm in diameter central opening was created in the zone plate that was used to illuminate the sample. To show the advantage of using a central opening instead of a zone plate with all the zones, we performed a comparison in reconstruction quality for the two cases.

The sample was a series of elbow-shaped apertures made in a 200 nm thick Si_3N_4 membrane with a 200 nm layer of gold deposited on top. The sample was then milled using a focused ion beam (FIB). The slits' widths in this sample vary from 2 μm, 1 μm, 0.5 μm and finally to 0.25 μm in the smallest elbow test pattern. A scanning electron micrograph (SEM) image of the sample is shown in Fig. 2a. Figure 2b was obtained using a zone plate with 100 μm in diameter central block with a zone plate outer zone width of 120 nm.

The reconstruction clearly shows that the sample is illuminated with a distorted wavefront which is less than ideal for imaging purposes. Figure 2c shows the effect of illuminating the object directly with an opening in the zone plate. The bottom section of the sample, indicated with the red dashed line, was illuminated with the opening while the remaining parts were illuminated with the diffracted field from the zone plate. The central opening provides a more intense and uniform field making it more advantageous for imaging. This setup is capable of imaging large samples with

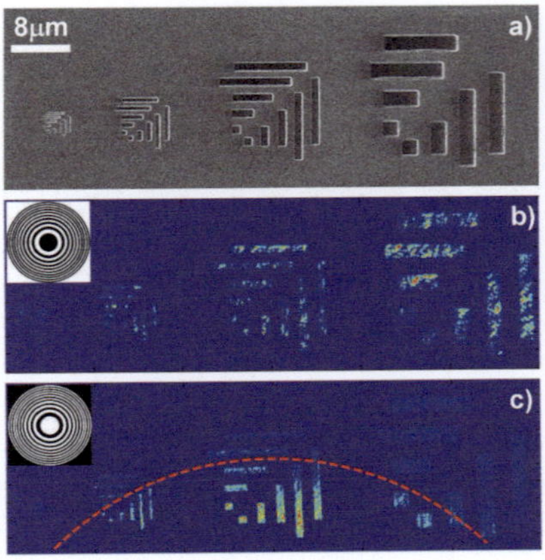

Fig. 2 (**a**) An SEM image of the sample. (**b**) The reconstruction using a Fresnel zone plate with a 100 μm in diameter central block. The *inset* in (**b**) shows a diagram of the zone plate used with a central block. (**c**) A reconstruction using a Fresnel zone plate with a D = 100 μm central opening. The *inset* in (**c**) shows a diagram of a zone plate with a central opening. The *bottom section* of (**c**) is directly illuminated with the incident beam (*red dashed line*), while (**b**) is illuminated by the diffraction through the zone plate entirely

nanometer-scale resolution. The only limitation to the sample size is determined by the central opening of the zone plate which in this case is 100 μm in diameter.

5 Edge Enhancement

To show the versatility of this setup, it is possible to enhance different features of the sample by repositioning the zone plate relatively to the reference hole. In this experiment the sample was fabricated on a 200 nm of Si_3N_4 with 200 nm of gold deposited on top. A FIB was used to create the four pillar design into the membrane. The sample contains four pillars with a height of 10 μm and of varying widths. The pillar widths are 4 μm, 2 μm, 1 μm and 0.5 μm. An SEM of the pillar sample is shown in Fig. 3a.

The image also shows that the sample is contaminated with several small particles. The contamination appears semi-transparent making it difficult to determine its extent with an SEM. In order to see the contamination more clearly, the image contrast was adjusted and is shown in Fig. 3b. As a comparison, Fig. 3c shows the reconstruction of the sample using the optical setup shown above, and the two images agree well with each other. The small differences between the SEM image and the soft X-ray image could be attributed to the different scattering properties between electrons and X-ray photons. While Fig. 3c shows the capability to image samples with a single-shot, Fig. 3d shows the ability of this setup to optically enhance edges. By adjusting the focal spot location it is possible to increase the amount of light

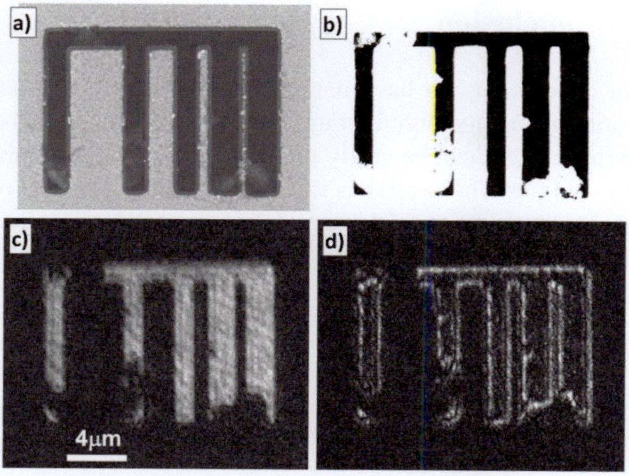

Fig. 3 (**a**) An SEM image of the pillar sample. (**b**) An SEM image with the contrast adjusted to show the contamination more clearly. (**c**) A reconstruction from a single-shot hologram. (**d**) A single-shot reconstruction showing the edge enhancement by readjusting the zone plate focal position

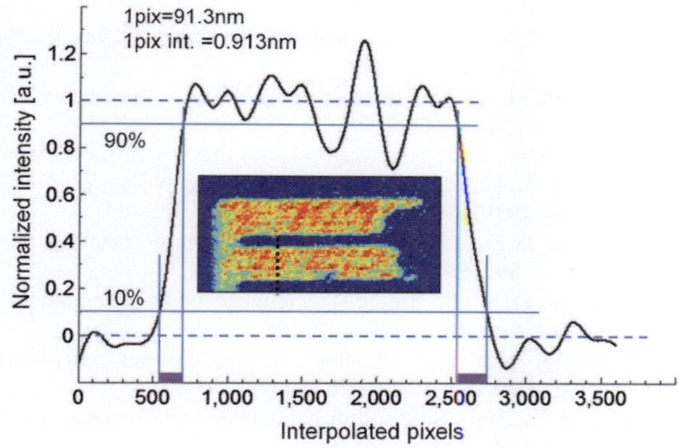

Fig. 4 A knife edge cut from the single-shot reconstruction from Fig. 3c. The resolution obtained is 170 ± 26 nm

at high diffracted angles which would enhance the signal in the hologram corresponding to high spatial frequencies. This annular illumination results in an edge enhanced reconstruction, and the adjustment allows for edge enhancement without any post processing of the reconstruction. A typical example of this illumination effect is shown in the reconstruction in Fig. 3d.

To make a first assessment of the spatial resolution we utilized the knife edge method [14]. A characteristic knife edge cut along the reconstruction in Fig. 3c is shown in Fig. 4. Performing these cuts in different regions of the image gives different values for the 10–90 % intensity transition. We performed a statistical study to obtain a resolution of 170 ± 26 nm from a single-shot hologram based on 12 measurements. The theoretical diffraction limit for the zone plate used in this experiment with a 100 μm in diameter central opening was numerically calculated to be 150 nm, a little larger than for a full zone plate equal to 122 nm.

6 Conclusion

In conclusion, we have demonstrated a soft X-ray optical system capable of acquiring single-shot holograms with a resolution of 170 ± 26 nm. Also, optical edge enhancement can be easily implemented by repositioning the zone plate relative to the sample. This work has shown that having a central opening in the zone plate for direct illumination of the sample is advantageous to illumination with the diffracted field through the zone plate.

Acknowledgements This research was sponsored by the Defense Threat Reduction Agency-Joint Science and Technology Office for Chemical and Biological Defense (Grant no. HDTRA1-10-1-0070) and the NSF ERC Center for Extreme Ultraviolet Science and Technology, NSF Award No. EEC-0310717.

References

1. Gunther, C.M., et al.: Sequential femtosecond X-ray imaging. Nat. Photonics **5**, 99–102 (2011)
2. Chapman, H., et al.: Femtosecond diffractive imaging with a soft-X-ray free electron laser. Nat. Phys. **2**, 839–843 (2006)
3. Chapman, H., et al.: Femtosecond time-delay X-ray holography. Nature **448**, 676–679 (2007)
4. McNulty, I., et al.: High-resolution imaging by Fourier transform X-ray holography. Science **256**, 1009–1012 (1992)
5. Schlotter, W.F., et al.: Multiple reference Fourier transform holography with soft X rays. Appl. Phys. Lett. **89**, 163112 (2006)
6. Gauthier, D., et al.: Single-shot femtosecond X-ray holography using extended references. Phys. Rev. Lett. **105**, 093901 (2010)
7. Sandberg, R.L., et al.: Tabletop soft-x-ray Fourier transform holography with 50 nm resolution. Opt. Lett. **34**, 1618–1620 (2009)
8. Kim, H.T., et al.: Single shot nanometer scale imaging with laser-driven x-ray laser. Appl. Phys. Lett. **98**, 121105 (2011)
9. Carbajo, S., et al.: Sequential single shot imaging of nanoscale dynamic interactions with a table top soft X-ray laser. Opt. Lett. **37**, 2994–2996 (2012)
10. Brewer, C.A., et al.: Single-shot extreme ultraviolet laser imaging of nanostructures with wavelength resolution. Opt. Lett. **33**, 518–520 (2008)
11. Benware, B.R., et al.: Demonstration of a high average power tabletop soft X-ray laser. Phys. Rev. Lett. **81**, 5804–5807 (1998)
12. Urbanski, L., et al.: Spectral linewidth of a Ne-like Ar capillary discharge soft-x-ray laser and its dependence on amplification beyond gain saturation. Phys. Rev. A **85**, 033837 (2012)
13. Liu, Y., et al.: Achievement of essentially full spatial coherence in a high-average-power soft-x-ray laser. Phys. Rev. A **63**, 033802 (2001)
14. Heck, J.M., et al.: Resolution determination in X-ray microscopy: an analysis of the effects of partial coherence and illumination spectrum. J. X-Ray Sci. Technol. (Paris) **8**, 95–104 (1998)

Chapter 27
Plasma Homogenization for Overcoming Refractive Losses in X-Ray Lasers

Leili Masoudnia and Davide Bleiner

Abstract Refractive losses along laser-produced plasmas used as gain media are caused by density gradients. The pump pulse is thus deflected from the high gain region and the short wavelength signal steers-away causing loss of collimation. Hollow cylinders used as targets make the plasma homogeneous and can mitigate refractive losses by means of waveguiding.

1 Introduction

Profiles of the electron-density in laser-produced plasmas used as gain media lead to a deflection of the pump pulse up to a turning point. The generated short-wave also steers-away during propagation along the plasma column causing loss of collimation and illumination quality. The prepulse technique was shown to mitigate the refraction problem [1] as well as original target designs [2–5]. Refraction is thus an important effect limiting the gain. To begin with, the refractive losses in a planar target are discussed. Then at original target design to produce plasma homogenization was studied; besides, a computational study with a hydro-code for planar and hollow cylindrical target is presented which includes a ray tracing modeling based on the Eikonal theory of the trajectory-equation.

2 Refractive Losses in a Planar Target

Figure 1 shows the electron density distribution before and after delivering of the main pump pulse on a planar target. The onset of a density gradient deflects the pump pulse up to a turning point, causing loss of efficiency.

L. Masoudnia (✉) · D. Bleiner
Institute of Applied Physics, University of Bern, Sidlerstrasse 5, 3012, Bern, Switzerland
e-mail: leili.masoudnia@iap.unibe.ch

D. Bleiner
e-mail: bleiner@iap.unibe.ch

S. Sebban et al. (eds.), *X-Ray Lasers 2012*, Springer Proceedings in Physics 147,
DOI 10.1007/978-3-319-00696-3_27,

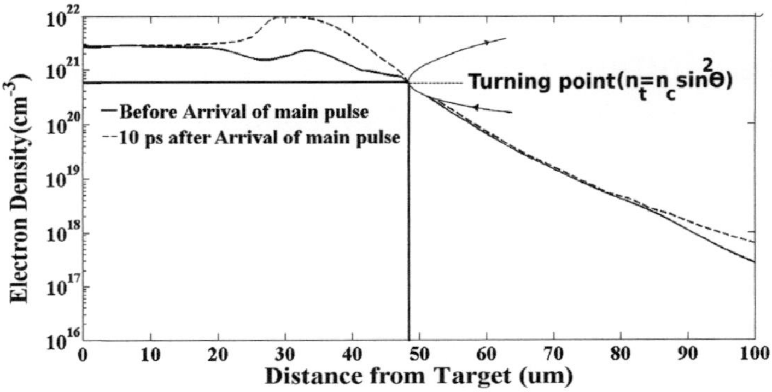

Fig. 1 Electron density distribution before and after delivery of the main pump pulse with the hydro code. The pump is reflected already at ≃56 % of the maximum penetration depth ($n_c = 1.1 \times 10^{21}$)

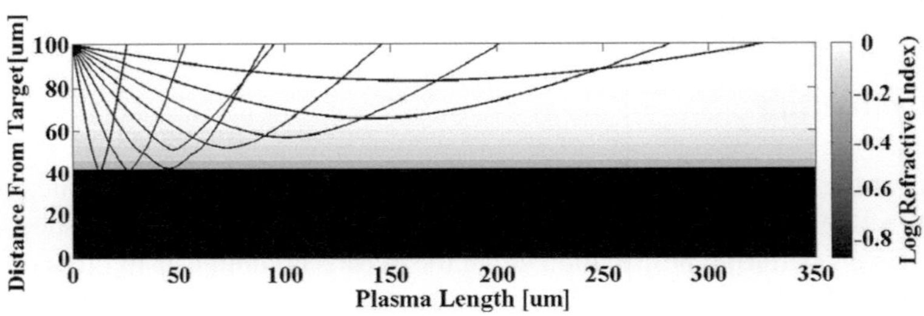

Fig. 2 Trajectory of the pump pulse considering the density profile before the main pulse for the irradiation angle in the range 10° to 80°

Figure 2 shows pump pulse trajectory as a function of irradiation angle. With the smaller angle one notes a longer active volume, that for 49° angle gives a penetration depth of 48 μm (Fig. 1). A longer and shallower active volume has a higher gain-length product for saturation.

The trajectory of the short-wavelength signal is shown on Fig. 3. At a distance larger than 55 μm, the output divergence is approx. 6 mrad, which is in agreement with our experimental studies [6]. The observations point at one common solution; i.e., plasma homogenization is required [7].

3 Plasma Confinement for Signal Waveguide

Using hollow targets confines the plasma and acts as a signal waveguide. Figure 4 shows the two counter-propagating fronts toward the center at 1 ns. These merge at 1.7 ns and create a concave density profile. At 5 ns the profile evolves into a convex. Figure 5 shows that using the hollow cylindrical target can effectively overcome the refraction problem of X-ray lasers.

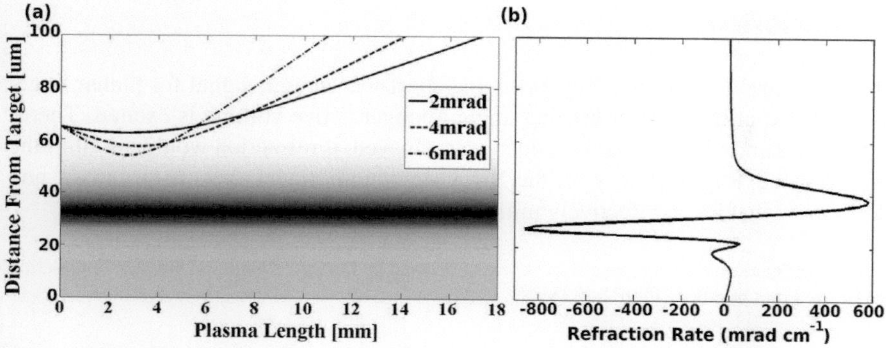

Fig. 3 Short-wavelength signal trajectory along the plasma gain medium (**a**), and refraction rate as a function of distance from the target (**b**)

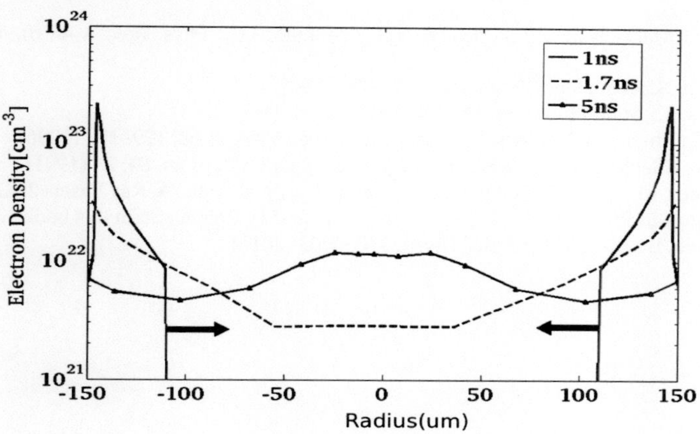

Fig. 4 Temporal and spatial evolution of the electron density profile for hollow cylinder target. The profile tends to self-homogenize

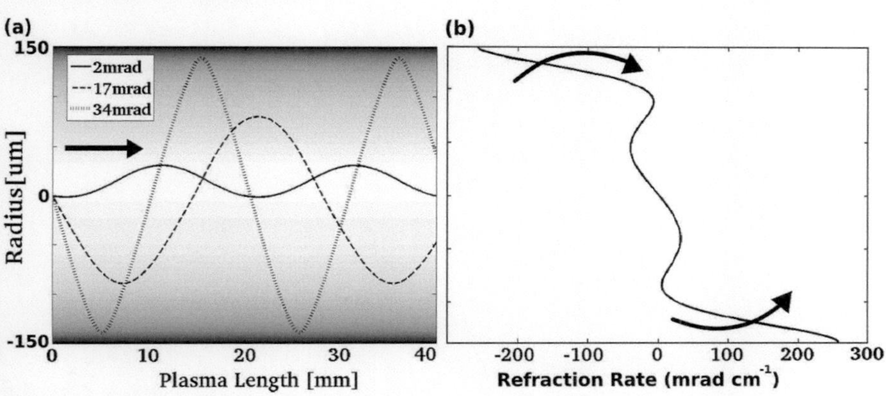

Fig. 5 Short-wavelength signal trajectory inside the hollow cylindrical target (**a**), and refraction rate as a function of radius (**b**)

4 Conclusions

Ray trajectory for the IR pump pulse and short-wavelength signal for planar target showed that at the smaller grazing angle a longer active volume is excited. Therefore, a maximum gain-length product was achieved, if refraction would not limit the performance. Ray trajectory for the X-ray signal in hollow targets is presented, and it is shown that it can effectively mitigate refractive losses.

Acknowledgements The present work was supported by the Swiss National Science Foundation under the grant number PP00P2-133564/1.

References

1. Nilsen, J., MacGowan, B.J., Da Silva, L.B., Moreno, J.C.: Phys. Rev. A **48**(6), 4682–4685 (1993)
2. Lewis, C.L.S., et al.: Opt. Commun. **91**, 71–76 (1992)
3. Lunney, J.G.: Appl. Phys. Lett. **48**, 891 (1986)
4. Daido, H., Murai, K., Kodama, R., Yuan, G.: J. Appl. Phys. B **62**, 129–133 (1996)
5. Soom, B., Ellenberger, U., Weber, R., Balmer, J.E.: Opt. Commun. **84**, 23 (1991)
6. Ruiz-Lopez, M., Staub, F., Bleiner, D.: In: Sebban, S., et al. (eds.) X-Ray Lasers 2012. Springer Proceedings in Physics, vol. 147. Springer, Cham (2013). Chapter 36 in this book
7. Masoudnia, L., Bleiner, D.: Laser Phys. **23**, 056003 (2013)

Chapter 28
Temporal Coherence and Spectral Linewidth of Neon-Like XUV Lasers Pumped in the Quasi-steady State Regime

L. Meng, A. Klisnick, M. Kozlova, K. Bohacek, M. Krus, J. Prokupek, L. Urbanski, M.C. Marconi, M. Berrill, J.J. Rocca, O. Guilbaud, F. Tissandier, S. Sebban, P. Zeitoun, A. Calisti, S. Ferri, C. Mossé, and B. Talin

Abstract We report recent experimental progress in the characterization of the temporal coherence and related spectral linewidth of plasma-based soft X-ray lasers (SXRL). New measurements were carried out with two types of SXRLs pumped in the quasi-steady state (QSS) regime, in a capillary-discharge plasma and in a laser-produced plasma. We describe the main results obtained from both experiments and compare them to dedicated numerical simulations. We discuss the results in the context of the possibility to achieve XUV lasers with pulse duration below 1 picosecond.

1 Introduction

The progress of collisional soft X-ray lasers (SXRL), generated by quasi-steady state [1] or transient pumping in solid target plasmas [2], optical-field-ionization in gaseous targets [3], and capillary discharges [4], have opened the way to compact devices and new prospects for various applications. This requires a detailed

L. Meng (✉) · A. Klisnick
ISMO, CNRS-Université Paris-Sud, Orsay, France
e-mail: limin.meng@u-psud.fr

M. Kozlova · K. Bohacek · M. Krus · J. Prokupek
Institute of Physics of ASCR, Prague, Czech Republic

L. Urbanski · M.C. Marconi · M. Berrill · J.J. Rocca
NSF-ERC for EUV Science and Technology, Colorado State University, Fort Collins, USA

O. Guilbaud
LPGP, CNRS-Université Paris-Sud, Orsay, France

F. Tissandier · S. Sebban · P. Zeitoun
LOA, ENSTA, Palaiseau, France

A. Calisti · S. Ferri · C. Mossé · B. Talin
PIIM, CNRS-Aix-Marseille Université, Marseille, France

S. Sebban et al. (eds.), *X-Ray Lasers 2012*, Springer Proceedings in Physics 147,
DOI 10.1007/978-3-319-00696-3_28,

characterization of their output parameters, like the energy per pulse, duration etc. The spectral linewidth $\Delta\lambda$, or the coherence time τ_c are also very important parameters, since they determine the shortest pulse duration (Fourier-transform limit τ_{FL}) that can be achieved (in a linear amplification regime), through the relation: $\tau_{FL} = \alpha\frac{\lambda^2}{\Delta\lambda} = \beta\tau_c$, where α and β are numerical factors depending on the particular shape of the line profile. The spectral width of the SXRL line results from the complex combination of several processes, namely the contribution of homogeneous (natural and collisional) and inhomogeneous (Doppler) broadenings of the intrinsic profile, followed by gain narrowing with amplification of the line along the plasma length [5]. When inhomogeneous broadening is the dominant cause of broadening the line can be rebroadened when the laser reaches saturation [6].

The purpose of our work is to characterize experimentally the spectral behavior of existing collisional excitation SXRLs, and to evaluate their potential to support the amplification of pulses with duration below 1 ps. Over the recent years, our previous studies were devoted to short pulse (femtosecond to picosecond) pumped systems, in which the duration of the SXRL pulse is of the same order of magnitude as the coherence time. At LOA-Palaiseau (France), a coherence time of ~5 ps was measured in a Ni-like Kr laser generated at 32.8 nm from an optical-field ionized plasma [7]. At Colorado State University (USA), a shorter coherence time of ~1 ps was measured in a Ni-like Mo laser ($\lambda = 18.9$ nm) pumped in the transient regime with a grazing incidence irradiation geometry [8]. Here in the present paper, we present new measurements that were carried out with two types of neon-like SXRLs, pumped in the quasi-steady state (QSS), long pulse (100 ps to nanosecond) regime: the capillary-discharge argon laser generated at CSU (USA) [9, 10], and the Ne-like Zn laser generated at PALS (Czech Republic) [1]. These sources differ from the previous ones by several aspects, among which a higher ionic temperature leading to a larger contribution of Doppler broadening of the laser lines.

We used a variable path-difference, wavefront-division interferometer, which was specifically designed to measure the temporal coherence of the source. This interferometer, described in [11], is composed of two dihedrons, tilted towards each other with a small angle. The incoming beam is separated into two beamlets that slightly converge towards each other after reflection on the dihedrons. In their overlapping region, interference fringes can be recorded using a XUV CCD camera. One of the dihedrons can be accurately translated vertically in order to introduce a path difference between the interfering beams. By following the evolution of the fringe visibility as a function of the delay with successive shots performed in identical conditions, it is possible to measure the coherence time of the pulse, and to infer the spectral width of the line through a Fourier transform.

2 Capillary-Discharge Ne-Like Ar Laser

Capillary discharge XUV lasers are the highest average power tabletop source of coherent soft-x-ray radiation [9, 12] up to date. The capillary discharge Ne-like Ar

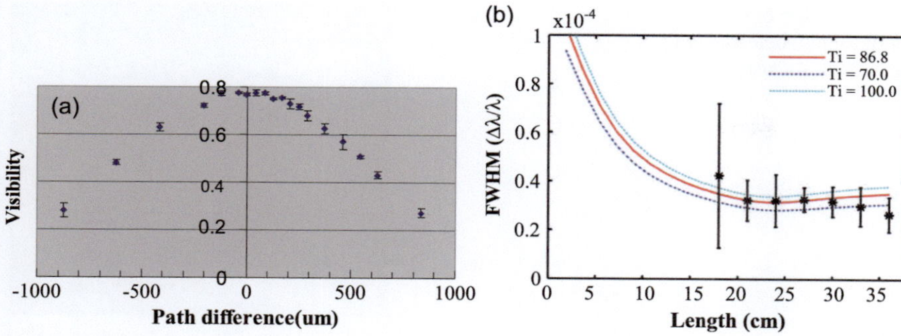

Fig. 1 (**a**) Example of a measured visibility curve; (**b**) Measured spectral width as a function of amplifier length. The *black points* with *error bars* stand for experimental data, and the three *curves* show the simulation results for different temperatures [10]

laser operating at $\lambda = 46.9$ nm (3p-3s J = 0–1 line), was first demonstrated in 1994 [4] and has been widely characterized [13–17]. However, the spectral linewidth and the temporal coherence, which are important parameters in applications such as interferometry [18] and large area nanopatterning [19], remained to be measured. Capillary discharge plasmas are characterized by a low electron density ($N_e \sim 2 \times 10^{18}$ cm^{-3}) and a relatively high ionic temperature ($kT_i \sim 100$ eV). The Doppler effect is thus expected to be the dominant cause of broadening of the laser line. The Ne-like Ar laser is thus an appropriate system to study the possible existence of saturation rebroadening predicted in highly inhomogeneous line profiles [20]. This is achieved by measuring the linewidth evolution with the plasma column length.

The experiment was performed at CSU (USA). The experimental setup and the results obtained are presented in more detail in [10] as well as in a companion paper (see [21]). Figure 1(a) shows an example of the measured variation of the fringe visibility with the path difference. The coherence time is defined as the delay at which the visibility is decreased by $1/e$ of its maximum. We find that the experimental data fit very well to a Gaussian curve. The corresponding spectral profile, inferred by a Fourier transform of the visibility curve, is also Gaussian. The linewidth is defined as the FWHM of the spectral profile.

The measured variation of the spectral bandwidth as a function of the plasma length is shown in Fig. 1(b). One can see that the bandwidth decreases slowly, while no rebroadening of the line is apparent for the longest lengths, although the laser is operated at saturation. The experimental data are compared to the predictions of numerical simulations, performed at CSU, that compute the line propagation along the amplifier axis taking into account gain saturation and refraction losses. One can see that these calculations predict a weak rebroadening of the line beyond $L \sim 24$ cm. However this effect is small compared to our experimental error bars. For the longest plasma length, the typical coherence time τ_c is 2.3 ps, corresponding to a Fourier limit pulse duration τ_{FL} of 1.9 ps, which are both much shorter than the duration of the SXRL pulse (1.2–1.8 ns).

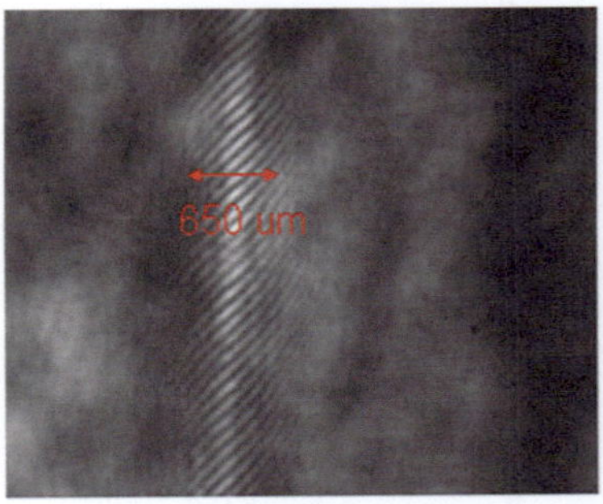

Fig. 2 Portion of the interferogram detected with the Ne-like Zn laser at 21.2 nm

3 Laser-Based Ne-Like Zn Laser

More recently we carried out a temporal coherence measurement of the Ne-like Zn laser, emitting at 21.2 nm (3p-3s J = 0–1 line, as above). This SXRL is generated at the PALS (Prague, Czech Republic) laser facility and was used for several applications [22]. The lasing plasma is pumped by a succession of a low-energy prepulse and high-energy main pulse, both ~300 ps in duration, delivered by an iodine pump laser ($\lambda = 1.315$ μm). After double-pass amplification in a 3-cm zinc plasma column, the 21.2 nm pulse reaches saturation with a typical energy of ~1 mJ per pulse or more [1].

Although lasing is obtained on the same Ne-like 3p-3s transition as for the capillary discharge laser discussed above, the plasma conditions in the gain zone are markedly different. Here the electron density is typically ×100 times higher ($N_e \sim 2\text{–}5 \times 10^{20}$ cm^{-3}), while the ionic temperature is only twice as high ($kT_i \sim 150\text{–}200$ eV). As a result, in Ne-like Zn, both collisional and Doppler broadenings contribute at the same level, as will be shown below.

The interferometer was implemented along the 21.2 nm beam path, at a distance of ~5.5 meters from the source. Figure 2 shows a typical example of a single-shot interferogram, when the path-difference between the interfering beams was close to zero. The size of the overlapping region, where fringes are apparent, is ~650 μm. This is slightly larger than the spatial coherence of the beam at the interferometer position, so that the fringe visibility is relatively low, of the order of 30 %. The distance from the source was increased by ~1.5 m in a second series of measurements, but the fringe visibility was not significantly improved. Another difficulty was the limited number of shots available in a data set, due to the low repetition rate of the laser (2 shots per hour). Two series of data, where the path difference was varied from −300 μm to +300 μm, were obtained for the two different distances from the source mentioned above. They led to slightly different, although consistent values for the coherence time of the 21.2 nm laser, namely $\tau_c = 1$ ps and 0.7 ps for the 5.5 m and 7 m distance respectively. The corresponding spectral linewidth are 11.3 mÅ and 7.7 mÅ respectively.

The measured $\Delta\lambda$ values were compared to the predictions of numerical simulations, using the PPP lineshape code [23] to calculate the intrinsic broadening, and a 1D-radiative transfer code [24] that computes the amplified profile, taking into account saturation. The plasma conditions in the gain zone are not known precisely, but estimated as $N_e \sim 2\text{–}5 \times 10^{20}$ cm^{-3}, $kT_e = 200\text{–}300$ eV; $kT_i = 150\text{–}200$ eV. A small-signal gain coefficient of 7 cm^{-1} was considered. We find that within this parameter range, the amplified linewidth of the 21.2 nm laser varies between 7.4 and 10.6 mÅ, in good agreement with our measurements. The contribution of the collisional broadening to the intrinsic line profile ranges from 11 to 24 mÅ, whereas the Doppler one is of the order of 25 mÅ.

Finally the measured τ_c values were used to estimate the corresponding Fourier-transform limit τ_{FL}, which is the shortest duration that could be reached if the 21.2 nm was fully temporally coherent. From the definition recalled in Sect. 1, and considering a Gaussian shape, we find that τ_{FL} is below 1 ps for both measurements, namely 0.6 and 0.7 ps. This is the shortest value ever measured for a plasma-based SXRL and this confirms the high potential of the Zn quasi-steady state laser for the generation of femtosecond amplified pulses, through seeding with high-harmonic radiation.

4 Conclusion

We have measured the temporal coherence and spectral width of two Ne-like soft X-ray lasers pumped in the quasi-steady state regime with two distinct techniques, leading to different plasma conditions and spectral characteristics.

The Ne-like Ar, capillary-discharge pumped laser exhibits a spectral width that decreases slowly as the plasma length increases, as a result of gain narrowing. The small saturation rebroadening, predicted by numerical simulations, was not observed and could be beyond our experimental accuracy. The measured coherence time is ~2 ps, much shorter than the pulse duration of 1.2–1.8 ns. For the laser-pumped Ne-like Zn laser, the measured coherence time is shorter (0.7–1 ps), leading for the first time to a Fourier-transform limit duration that lies below 1 ps.

In both cases, the inferred values for the spectral width of the laser line are in good quantitative agreement with the predictions of numerical simulations. Finally since the measured coherence times are in both cases much smaller than the duration of the SXRL pulse, picosecond (or potentially sub picosecond) durations could be reached through seeding with femtosecond high-order harmonic radiation.

References

1. Rus, B., et al.: Multimillijoule, highly coherent x-ray laser at 21 nm operating in deep saturation through double-pass amplification. Phys. Rev. A **66**, 063806 (2002)
2. Wang, Y., et al.: High-brightness injection-seeded soft-X-ray-laser amplifier using a solid target. Phys. Rev. Lett. **97**, 123901 (2006)
3. Zeitoun, Ph., et al.: A high-intensity highly coherent soft X-ray femtosecond laser seeded by a high harmonic beam. Nature **431**, 426 (2004)

4. Rocca, J.J., et al.: Demonstration of a discharge pumped table-top soft-X-ray laser. Phys. Rev. Lett. **73**, 2192 (1994)
5. Koch, J.A., et al.: Experimental and theoretical investigation of neonlike selenium x-ray laser spectral linewidths and their variation with amplification. Phys. Rev. A **50**, 1877 (1994)
6. Pert, G.J.: Output characteristics of amplified-stimulated-emission lasers. J. Opt. Soc. Am. B **11**, 1425 (1994)
7. Tissandier, F., et al.: Observation of spectral gain narrowing in a high-order harmonic seeded soft-x-ray amplifier. Phys. Rev. A **81**, 063833 (2010)
8. Meng, L., et al.: Temporal coherence and spectral linewidth of an injection-seeded transient collisional soft x-ray laser. Opt. Express **19**, 12087 (2011)
9. Benware, B., et al.: Demonstration of a high average power tabletop soft X-ray laser. Phys. Rev. Lett. **81**, 5804 (1998)
10. Urbanski, L., et al.: Spectral linewidth of a Ne-like Ar capillary discharge soft-x-ray laser and its dependence on amplification beyond gain saturation. Phys. Rev. A **85**, 033837 (2012)
11. Guilbaud, O., et al.: Longitudinal coherence and spectral profile of a nickel-like silver transient soft X-ray laser. Eur. Phys. J. D **40**, 125 (2006)
12. Macchietto, C.D., et al.: Generation of millijoule-level soft-x-ray laser pulses at a 4-Hz repetition rate in a highly saturated tabletop capillary discharge amplifier. Opt. Lett. **24**, 1115 (1999)
13. Tomassetti, G.A., et al.: Toward a full optimization of a highly saturated soft-X-ray laser beam produced in extremely long capillary discharge amplifiers. Opt. Commun. **231**, 403 (2004)
14. Heinbuch, S., et al.: Demonstration of a desk-top size high repetition rate soft x-ray laser. Opt. Express **13**, 4050 (2005)
15. Le Pape, S., et al.: Electromagnetic-field distribution measurements in the soft X-ray range: full characterization of a soft X-ray laser beam. Phys. Rev. Lett. **88**, 183901 (2002)
16. Marconi, M.C., et al.: Measurement of the spatial coherence buildup in a discharge pumped table-top soft X-ray laser. Phys. Rev. Lett. **79**, 2799 (1997)
17. Liu, Y., et al.: Achievement of essentially full spatial coherence in a high-average-power soft-x-ray laser. Phys. Rev. A **63**, 033802 (2001)
18. Brewer, C.A., et al.: Single-shot extreme ultraviolet laser imaging of nanostructures with wavelength resolution. Opt. Lett. **33**, 518 (2008)
19. Wachulak, P.W., et al.: Patterning of nano-scale arrays by table-top extreme ultraviolet laser interferometric lithography. Opt. Express **15**, 3465 (2007)
20. Koch, J.A., et al.: Collisional redistribution effects on x-ray laser saturation behavior. In: AIP Conf. Proc., vol. 332, pp. 574–578. AIP, New York (1994)
21. Urbanski, L., et al.: Spectral linewidth measurement of a Ne-like Ar capillary discharge soft x-ray laser. In: Sebban, S., et al. (eds.) X-Ray Lasers 2012. Springer Proceedings in Physics, vol. 147. Springer, Cham (2013). Chapter 41 in this book
22. Mocek, T., et al.: Plasma-based X-ray laser at 21 nm for multidisciplinary applications. Eur. Phys. J. D **54**, 439–444 (2009)
23. Calisti, A., et al.: Model for the line shapes of complex ions in hot and dense plasmas. Phys. Rev. A **42**, 5433 (1990)
24. Guilbaud, O., et al.: Fourier-limited seeded soft x-ray laser pulse. Opt. Lett. **35**, 1326 (2010)

Chapter 29
Spectral Broadening of Ni-Like XUV Laser Lines

L. Meng, A. Calisti, S. Ferri, C. Mossé, B. Talin, D. Benredjem, O. Guilbaud, and A. Klisnick

Abstract We have used the PPP lineshape code to calculate the intrinsic (i.e. before amplification) line profile of the Mo XUV laser over an extended range of plasma densities and temperatures, chosen to cover conditions for collisional excitation pumping in the transient and quasi-steady state regimes. The calculated profiles were then used to simulate the amplified line profile, using a detailed 1D-radiative transfer code, taking into account the effect of saturation. We discuss the possibility to achieve a gain bandwidth that would support pulse amplification below 1 ps.

1 Introduction

The demonstration of seeded, plasma-based XUV lasers with high-order harmonic radiation has opened new and promising prospects for both these sources towards higher peak brightness and output energies. Compared to the usual amplified spontaneous emission (ASE) mode, the optical quality of the seeded output beams is dramatically improved, with a smaller divergence and a higher degree of spatial coherence [1, 2]. The degree of temporal coherence was also observed to be improved by the seeding operation [3], but the measured coherence time was of ~ 1 ps thereby limiting the shortest duration accessible to a similar value [4]. A challenging goal is to extend the pulse duration of seeded XUV lasers to the femtosecond range, since this would considerably enhance the potential of these sources to application fields

L. Meng (✉) · A. Klisnick
ISMO, CNRS-Université Paris-Sud, Orsay, France
e-mail: limin.meng@u-psud.fr

A. Calisti · S. Ferri · C. Mossé · B. Talin
PIIM, CNRS-Aix-Marseille Université, Marseille, France

D. Benredjem
LAC, CNRS-Université Paris-Sud, Orsay, France

O. Guilbaud
LPGP, CNRS-Université Paris-Sud, Orsay, France

S. Sebban et al. (eds.), *X-Ray Lasers 2012*, Springer Proceedings in Physics 147,
DOI 10.1007/978-3-319-00696-3_29,

like ultrafast coherent imaging or interaction with matter at high XUV intensities. This will require significant enlargement of the XUV laser spectral bandwidth.

The main purpose of the work described in this paper was to determine the main plasma parameters that could be used in the experiment to reach the desired bandwidth. We have used the PPP lineshape code [5] to calculate the respective contributions of the broadening processes that affect the intrinsic spectral profile of the 4d–4p (J = 0–1) lasing line in Ni-like Mo. Our study cover an extended range of plasma temperatures and densities over which collisional excitation pumping of these ions can be induced. We consider two different pumping regimes: (i) the transient pumping regime, which relies on a picosecond pump laser pulse, leading to plasmas with relatively low ionic temperatures; (ii) the quasi-steady state (QSS) regime, in which the ionic temperature is higher due to a longer (few 100 ps) pump pulse and more efficient electron-ion thermalization. The calculated intrinsic line profiles were then used to simulate the amplified line profile, using a detailed 1D-radiative transfer code [6], taking into account the effect of saturation. We discuss the possibility to achieve a gain bandwidth that would support pulse amplification below 1 ps.

2 Calculation of the Intrinsic Line Profile

The PPP code is a multi-electron radiator line broadening code [5, 7], which includes natural and Doppler broadening, as well as (electronic) collisional broadening using the classical impact approximation. The effect of ions was treated using the static approximation. Figure 1 shows the computed overall FWHM linewidth $\Delta\nu$ plotted as a function of the electron density, for ionic temperatures of 20 eV (transient pumping, left graph) and 200 eV (QSS pumping, right graph) respectively. In both cases the electron temperature was taken as 200 eV. The overall linewidth was inferred from the calculated spectral profile, which was first fitted with a Voigt profile. The individual contributions of the thermal Doppler component (independent of N_e) and of the (electron) collisional component are also shown in Fig. 1.

It can be seen that the respective weight of collisional versus Doppler broadening varies considerably over the considered N_e range. At high N_e the collisional broadening component dominates in both cases, while the linewidth increases with N_e, although less than proportionally. In the transient pumping case the collisional broadening is always larger than the Doppler one, except for $N_e < 8 \times 10^{19}$ cm^{-3}. In the QSS pumping case the Doppler broadening starts to dominate for densities below $\sim 3 \times 10^{20}$ cm^{-3}.

It should be noted that the Doppler broadening is calculated here in the free-particle limit, assuming that the velocity of ion emitters does not change over the radiative lifetime. A more recent investigation, involving molecular dynamics simulations, shows that this approximation can be inappropriate in our conditions, where ionic correlation effects cannot be neglected. This aspect is discussed in a companion paper [7]. The inclusion of these effects in a line-transfer model used to calculate the amplified line profile is not straightforward. Thus for the discussion below, we consider the usual free-particle limit Doppler broadening only.

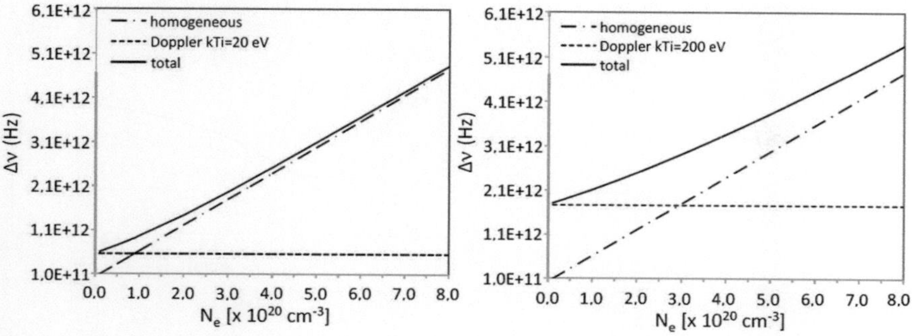

Fig. 1 Calculated intrinsic linewidth (FWHM) as a function of N_e, for 2 pumping regimes: (*left*) transient pumping ($kT_i = 20$ eV); (*right*) QSS pumping ($kT_i = 200$ eV). In both cases $kT_e = 200$ eV. Individual Doppler and collisional broadening components are also shown

3 Amplified Line Profile and Fourier-Transform Limit Duration

The calculated intrinsic linewidths were then used to calculate the amplified profile, using a 1D-line transfer code [6]. This code describes the frequency-dependent amplification of the laser line intensity, taking into account the homogeneous (natural and collisional broadening) and inhomogeneous (thermal Doppler frequency shifts) components of the spectral profile, while allowing for gain saturation. A small-signal gain coefficient of $70\ \mathrm{cm}^{-1}$ and a plasma length of 7 mm were considered.

Figure 2 shows the evolution of the spectral linewidth as a function of amplification length, calculated in three distinct cases, corresponding to different plasma conditions: (a) $N_e = 8 \times 10^{20}\ \mathrm{cm}^{-3}$, $kT_i = 20$ eV (homogeneous profile); (b) $N_e = 2 \times 10^{19}\ \mathrm{cm}^{-3}$, $kT_i = 200$ eV (inhomogeneous profile); (c) $N_e = 3 \times 10^{20}\ \mathrm{cm}^{-3}$, $kT_i = 200$ eV (mixed profile). As expected, saturation rebroadening is observed only in the inhomogeneous case (b), while the output linewidth recovers $\sim 2/3$ of its intrinsic value for $L = 7$ mm. It should be noted that in all cases (even in the homogeneous case (a)), the amplified line profile tends to a Gaussian shape.

The Fourier-transform limit duration τ_{FL}, i.e. the shortest pulse that could be extracted from the plasma (in a linear amplification regime), is related to the spectral linewidth $\Delta\lambda$ as: $\tau_{FL} = \kappa \frac{\lambda^2}{\Delta\lambda}$, where κ is a numerical factor depending on the particular shape of the line profile (here $\kappa = 1.5 \times 10^{-9}$ for a Gaussian profile). In the three above cases we find that τ_{FL} is below 1 ps, namely (a) $\tau_{FL} = 450$ fs; (b) $\tau_{FL} = 380$ fs; (c) $\tau_{FL} = 650$ fs. The shortest duration is obtained for the strongly inhomogeneous case (b), due to the linewidth rebroadening at saturation. However ionic correlation effects, discussed in [7], were found to be significant in these plasma conditions. As a result, collisional cross-relaxation is efficient and tend to homogenize the profile, preventing the rebroadening behavior.

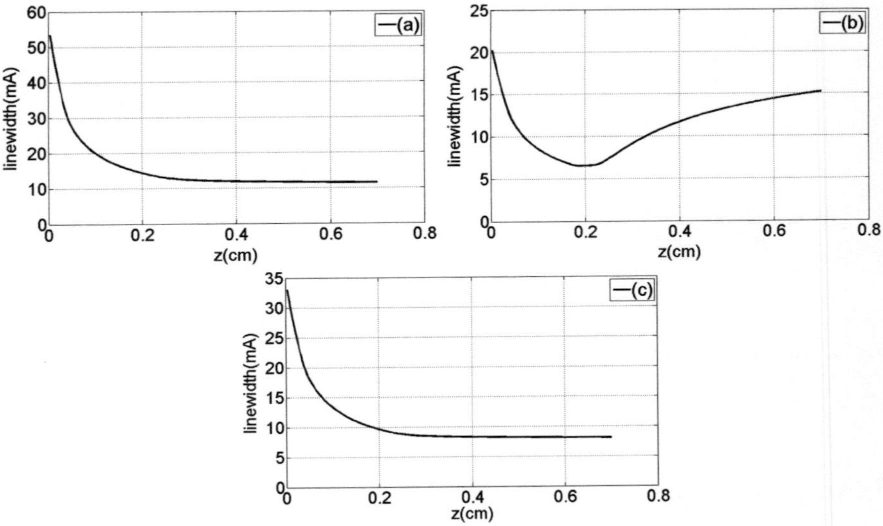

Fig. 2 Calculated amplified linewidth (FWHM) as a function of plasma length for three different plasma conditions, corresponding to different types of profile (see text): (**a**) homogeneous; (**b**) inhomogeneous; (**c**) mixed

References

1. Tissandier, F., et al.: Phys. Rev. A **81**, 063833 (2010)
2. Berrill, M., et al.: Opt. Lett. **35**, 2317 (2010)
3. Meng, L.M., et al.: Opt. Express **19**, 12087 (2011)
4. Wang, Y., et al.: Phys. Rev. A **79**, 023810 (2009)
5. Calisti, A., et al.: Phys. Rev. A **42**, 5433 (1990)
6. Guilbaud, O., et al.: Opt. Lett. **35**, 1326 (2010)
7. Calisti, A., et al.: Study of particle correlation effects on line profiles of Ni-like collisional XUV laser amplifiers. In: Sebban, S., et al. (eds.) X-Ray Lasers 2012. Springer Proceedings in Physics, vol. 147. Springer, Cham (2013). Chapter 9 in this book

Chapter 30
Visualization of Rapid Dynamic Interactions by Flash Soft X-Ray Microscopy

C.S. Menoni, S. Carbajo, I.D. Howlett, W. Chao, E.H. Anderson, A.V. Vinogradov, I.A. Artyukov, K. Buchanan, M.C. Marconi, and J.J. Rocca

Abstract We demonstrate sequential single-shot real space imaging using a compact soft x-ray (SXR) microscope. A sequence of real space flash images acquired with a table-top SXR laser was used to follow and record the motion of a rapidly oscillating magnetic nanoprobe tip and detect nanoscale dynamic interactions. Changes of tens of nm in the oscillation amplitude were detected when the nanoprobe was made to interact with stray fields from a magnetic sample. Modeling of this interaction confirms the results of the experiments. The ability of compact SXR microscopes to freeze dynamics open up the pathway to investigate the dynamics of a variety of processes that include the motion on nanodevices and potentially nonrecurrent phenomena.

1 Introduction

Time resolved x-ray imaging is a powerful technique to record dynamic interactions at the nanoscale. Stroboscopic pump and probe imaging techniques using synchrotron illumination have been extensively used to monitor dynamics of magnetic materials [1–3]. These experiments have been successful at probing dynamic processes that can be held unaltered over minutes or hours, the time it takes to average

C.S. Menoni (✉) · S. Carbajo · I.D. Howlett · W. Chao · E.H. Anderson · M.C. Marconi · J.J. Rocca
NSF ERC for Extreme Ultraviolet Science and Technology, Department of Electrical & Computer Engineering, Colorado State University, Fort Collins, CO 80523, USA
e-mail: carmen.menoni@colostate.edu

W. Chao · E.H. Anderson
NSF ERC for Extreme Ultraviolet Science and Technology and Center for X-ray Optics, LBNL, Berkeley, CA 94720, USA

A.V. Vinogradov · I.A. Artyukov
P.N. Lebedev Physical Institute, Moscow, Russia

K. Buchanan
Physics Department, Colorado State University, Fort Collins, CO 80523, USA

S. Sebban et al. (eds.), *X-Ray Lasers 2012*, Springer Proceedings in Physics 147,
DOI 10.1007/978-3-319-00696-3_30,

several million of events before moving to the next time delay. There are, however, other material processes such as twinning [4] and nanoscale ablation [5] that are statistically reproducible but individually nonrecurring which can only be imaged using flash illumination. SXR single shot imaging has been demonstrated [6–10]. As of today dynamic SXR imaging has only been realized in holographic and coherent diffractive imaging geometries which require extensive post-image processing [11, 12] using free electron lasers. Electron microscopes have also the ability to follow dynamics at the nanoscale [13–16]. Femtosecond temporal and picometer spatial resolution can be obtained using a scanning imaging configuration similar to photon-based stroboscopic pump and probe, that uses a few electrons per pulse at the expense of long image acquisition times [13, 14]. Single-shot full field imaging instead uses a large number of electrons per pulse, thus the spatial and temporal resolution are limited by space charge effects to 10's of nanometers and 10's of nanoseconds respectively [15, 16]. Electron beams are also affected by the presence of electric and magnetic fields. These constraints do not apply to x-ray microscopes.

Soft x-ray microscopes owe their ability to render real-space images with nanoscale resolution to the use of tailored diffractive optics. Advances in the fabrication of zone plate nanostructures have been instrumental in achieving 12 nm spatial resolution [17]. In turn, the use of bright table-top SXR lasers for illumination give SXR microscopes the capability to acquire single-shot real-space images in a non-destructive way [6]. We demonstrate the recording of nanoscale dynamic interactions using a compact SXR microscope. A sequence of single-shot real-space SXR captured the interaction of a magnetic cantilever tip and a magnetic sample. Analysis of the time-sequenced images revealed an increase in the amplitude of the tip oscillation of a few tens of nm due to the magnetic interaction. These results open new opportunities in material science and nanotechnology, such as the characterization of rapid motion in micro/nano-electro-mechanical devices, and the study of fast non-periodic nanoscale processes in materials, such as the propagation of twins and nanoablation related events. Both these and other laser-initiated phenomena can be accurately synchronized with SXR laser probe pulses by using the same laser driver.

2 Experimental Details and Results

Sequential table-top SXR imaging was implemented using a compact microscope that combines illumination from a desk-top size SXR capillary discharge laser with reflective and diffractive optics [18]. The laser emits pulses of 1.5 ns duration at a wavelength of 46.9 nm (26.5 eV), with a spectral bandwidth of $\Delta\lambda/\lambda = 3.5 \times 10^{-5}$ [19, 20]. A capillary length of 22 cm was selected to obtain pulses of 10 μJ energy, equivalent to 2.4×10^{12} photons/pulse and to maintain the spatial coherence of the laser beam relatively low to minimize coherence effects on the image. The microscope uses a Schwarzschild condenser and a zone plate objective. The Schwarzschild primary convex mirror, 10.8 mm in diameter, and the secondary concave mirror, 50 mm in diameter, are coated with Sc/Si multilayers that result in

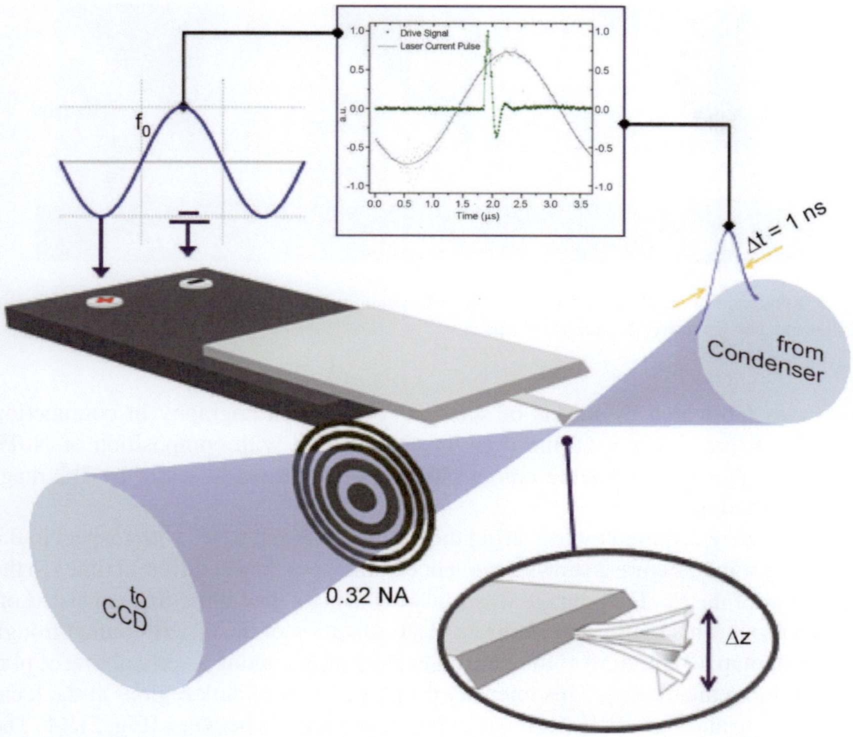

Fig. 1 Schematic of the set up used to implement single shot imaging with a transmission SXR microscope. The object is a commercially available cantilever tip. The tip was driven by a sinusoidal voltage signal of frequency equal to the tip's resonance frequency

a throughput of 16 % at $\lambda = 46.9$ nm and a numerical aperture of 0.18 NA [21]. The objective is a freestanding zone plate with 500 μm diameter, numerical aperture 0.32 NA, and a focal length $f = 740$ μm at $\lambda = 46.9$ nm. Its efficiency is 10 % in first order [22]. Sequential single shot images were captured by a back-illuminated array detector with 2048×2048 pixels of 13.5 μm size. The detector was cooled to -15 °C to minimize thermal noise. We have previously shown through static imaging of gratings with different period that the combination of a 0.32 NA objective zone plate with illumination from the 46.9 nm laser yields images with a spatial resolution of 54 nm [6].

Two commercially available nanoprobe tips were used in the experiments: a high-aspect ratio Si tip and a Co-alloy coated standard magnetic tip that was magnetized along the tip axis prior to use. The former consisted of a cone shaped tip with a nominal radius of 20 nm, a nominal spring constant of 40 N/m, and a resonance frequency of 318.6 kHz. The magnetic nanoprobe had a cone shaped tip radius of 40 nm, a nominal spring constant of 0.7 N/m, and a resonance frequency of 65.2 kHz. Each tip was clamped onto a rectangular bimorph actuator driven by a sinusoidal voltage signal (Fig. 1). The 1:1 array of 10 μm × 2 μm × 40 nm magnetic

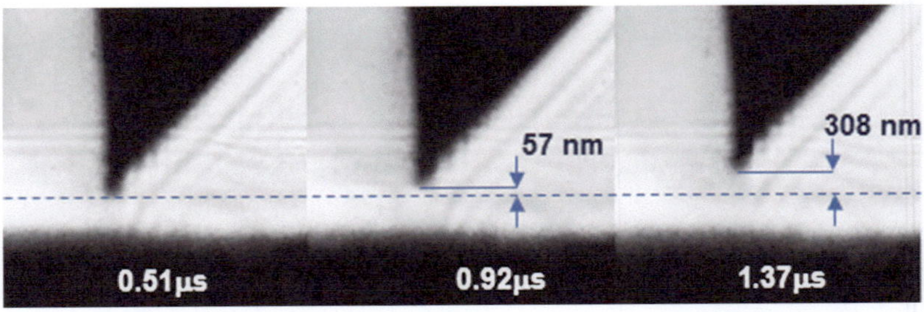

Fig. 2 Sequence of single-shot SXR images of the nanoprobe tip freeze the dynamics of the tip oscillation. The tip moves 57 nm in 0.41 μs

strips was fabricated by means of standard UV photolithography in conjunction with lift-off process in a 40 nm thick Permalloy film with composition of Ni/Fe 81:19 wt% that was deposited onto a photoresist-patterned Si wafer by DC magnetron sputtering.

Images were acquired by triggering the laser at different times with respect to the tip driving signal. Figure 2 shows a sequence of images taken at different times in the oscillation of the tip. Each image was analyzed using a procedure that consisted of: (1) background-histogram equalization; (2) localization of the tip extremum through a gradient map; (3) plotting of the image intensity profile along a vertical row of pixels containing the tip axis. This intensity profile consists of dark regions at the location of the tip and surface respectively and a clear region in between (Fig. 2). (4) The tip-to-surface separation was determined as the full-width-at-half-maximum of the intensity profile. A set of over 100 sequential images were captured to map the entire period of the tip oscillation. Averaging of five independent sets of tip trajectories traces over a period allowed us to calculate the mean tip-to-sample distance with an uncertainty of ± 3 nm corresponding to one standard deviation. Finally, sequences of images of the nanoprobe tip dynamics were assembled into motion picture by sequencing the SXR images in temporal order [23].

We have used the SXR flash imaging capabilities to record nanoscale dynamic interactions. We exploited the immunity of photons to electrostatic and magnetostatic fields to record the interaction of the magnetic nanoprobe tip with the stray magnetic fields created by the magnetized Permalloy microstrip (Fig. 3a). This type of interaction is what occurs in a magnetic force microscope. From a sequence of single shot SXR images, the motion of the magnetic nanoprobe tip was reconstructed when placed in a region between the strips and above the edge of a strip (Fig. 3). For a given sign of the Permalloy strip remnant field, an increase in the amplitude of the oscillation of (40 ± 3) nm and (30 ± 3) nm were measured. Reversing the direction of the remnant field caused the tip amplitude to decrease (Fig. 3c).

The amplitude changes in the magnetic tip oscillation caused by the interaction with the Permalloy microstrips were modeled assuming the tip to be a perturbed harmonic oscillator in the presence of a magnetic dipole-dipole force. The force gradients were calculated using established values of magnetization of 8×10^5 A/m

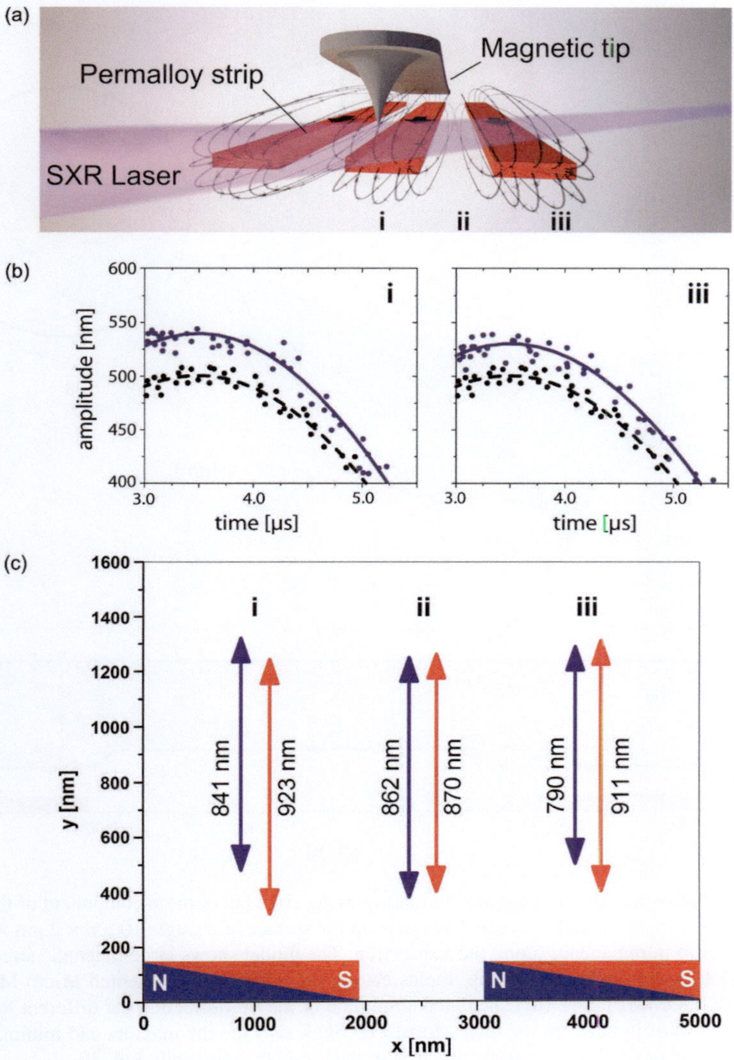

Fig. 3 (**a**) Schematic rendering of the geometry used to interrogate the magnetic interaction of the magnetized nanoprobe tip and magnetized patterned Permalloy sample. (**b**) Tip motion reconstructed from the single shot SXR images for locations of the tip above the edge of a microstrip (*i* and *iii*) and in between two microstrips (*ii*). (**c**) The amplitude of the tip corresponding to the locations indicated by *i*, *ii*, and *iii* is shown for two possible orientations of the remnant field, showing an increase in the amplitude and a decrease compared to the unperturbed oscillation when the direction of the magnetization is changed

and 14×10^5 for Permalloy and Co-alloy which, for the volumes under consideration, result in magnetic moments of 4.1×10^{-17} A m^2 and 6.4×10^{-13} A m^2 for the tip and microstrips respectively. The magnetic force fields effectively change

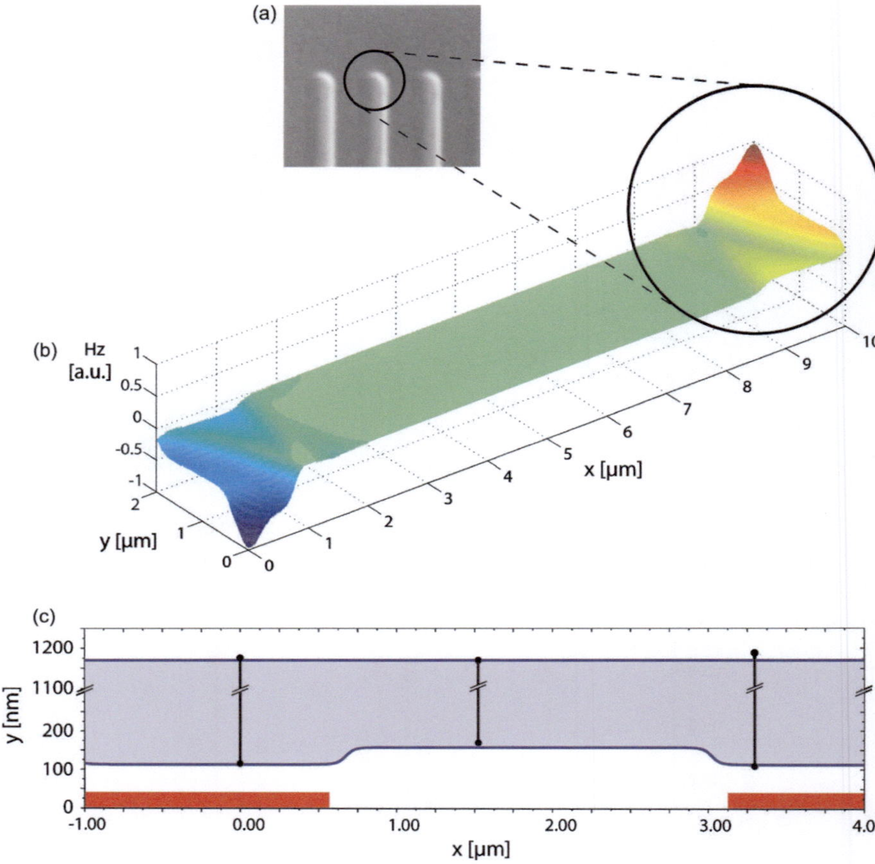

Fig. 4 (**a**) Photograph of the patterned Permalloy strips. (**b**) Map of the z-component of the magnetic field strength (Hz) present at z = 150 nm from the surface of a single 10 μm × 2 μm × 40 nm Permalloy strip in remanence along the x direction. The model shows large magnetic strength on the edges of the Permalloy microstrip. Fields calculated using Object Oriented Micro Magnetic Framework (OOMMF) [24]. (**c**) Calculated amplitude of the tip oscillation for different locations of the tip above or in between the microstrips. The *black dots* are the maxima and minima of the tip oscillation obtained from the reconstruction of the tip motion shown in Fig. 3b

the restoring force of the oscillating magnetic tip through the addition of a term that is proportional to the magnetic force gradient. A change in the restoring force constant modifies the tip's resonant frequency. When driving the tip at a fixed frequency, a small change in the restoring force varies the amplitude and phase of the tip oscillation. Figure 4 shows the variation of the z-component of the magnetic field calculated at 150 nm above the surface of the microstrip where the tip is located are significant at the edges of the microstrip. The results of the model are shown in Fig. 4c. The model predicts changes in the amplitude of the tip oscillation of tens of nanometers, which is in very good agreement with the experimental results.

3 Conclusions and Outlook

In conclusion we have directly visualized nanoscale phenomena with a table-top microscope using bright flashes from a compact SXR laser. We have recorded single-shot real-space images that mapped the motion of a magnetic nanoprobe tip and its interaction with a magnetic sample. The ability to visualize nanoscale motion on a table top will impact the development of nano-electro-mechanical devices. Moreover, the use of a similar SXR microscope in a reflection configuration [25] will allow to visualize nanoscale material surface dynamics such as laser ablation, nanomachining and the study of material dislocation dynamics, nucleation and growth. Furthermore, our table-top SXR laser dynamic imaging technique is readily scalable in wavelength and temporal resolution. The recent advances in saturated table-top SXR laser sources, that now provide picosecond pulses with energies $>10\,\mu J$ per pulse at 13.9 nm [26], will open the possibility to visualize ultrafast dynamics with a few tens of nanometer spatial resolution and picosecond time resolution. Further advances in injection-seeding table-top SXR lasers that are expected to produce high energy femtosecond pulses will make it possible to extend this technique to the femtosecond range [27].

Acknowledgements Work supported by the NSF ERC for Extreme Ultraviolet Science and Technology under NSF Award EEC-0310717 using equipment developed under NSF Award MRI-ARRA 09-561, and by the Chemical Sciences, Geosciences and Biosciences Division, Office of Basic Energy Sciences, U.S. Department of Energy. K.S.B. was also supported by NSF award number 0907706.

References

1. Ade, H., et al.: Near-edge X-ray absorption fine-structure microscopy of organic and magnetic materials. Nat. Mater. **8**(4), 281–290 (2009)
2. Stoll, H., et al.: High-resolution imaging of fast magnetization dynamics in magnetic nanostructures. Appl. Phys. Lett. **84**(17), 3328–3330 (2004)
3. Kammerer, M., et al.: Magnetic vortex core reversal by excitation of spin waves. Nat. Commun. **2**, 279–284 (2011)
4. Capolungo, L., et al.: Nucleation and growth of twins in Zr: a statistical study. Acta Mater. **57**(20), 6047–6056 (2009)
5. Vaschenko, G., et al.: Nanometer-scale ablation with a table-top soft x-ray laser. Opt. Lett. **31**(24), 3615–3617 (2006)
6. Brewer, C.A., et al.: Single-shot extreme ultraviolet laser imaging of nanostructures with wavelength resolution. Opt. Lett. **33**(5), 518–520 (2008)
7. Chapman, H.N., et al.: Femtosecond diffractive imaging with a soft-X-ray free-electron laser. Nat. Phys. **2**(12), 839–843 (2006)
8. Chou, M.-C., et al.: Single-shot soft-x-ray digital holographic microscopy with an adjustable field of view and magnification. Opt. Lett. **34**(5), 623–625 (2009)
9. Kim, H.T., et al.: Single-shot nanometer-scale holographic imaging with laser-driven x-ray laser. Appl. Phys. Lett. **98**(12), 121105 (2011)
10. Ravasio, A., et al.: Single-shot diffractive imaging with a table-top femtosecond soft X-ray laser-harmonics source. Phys. Rev. Lett. **103**(2), 028104 (2009)

11. Barty, A., et al.: Ultrafast single-shot diffraction imaging of nanoscale dynamics. Nat. Photonics **2**(7), 415–419 (2008)
12. Gunther, C.M., et al.: Sequential femtosecond X-ray imaging. Nat. Photonics **5**(2), 99–102 (2011)
13. Zewail, A.H.: Four-dimensional electron microscopy. Science **328**(5975), 187–193 (2010)
14. Flannigan, D.J., et al.: Nanomechanical motions of cantilevers: direct imaging in real space and time with 4D electron microscopy. Nano Lett. **9**(2), 875–881 (2009)
15. Taheri, M.L., et al.: Laser-based in situ techniques: novel methods for generating extreme conditions in TEM samples. Microsc. Res. Tech. **72**(3), 122–130 (2009)
16. Kim, J.S., et al.: Imaging of transient structures using nanosecond in situ TEM. Science **321**(5895), 1472–1475 (2008)
17. Chao, W., et al.: Demonstration of 12 nm resolution Fresnel zone plate lens based soft X-ray microscopy. Opt. Express **17**(20), 17669–17677 (2009)
18. Vaschenko, G., et al.: Nanoimaging with a compact extreme-ultraviolet laser. Opt. Lett. **30**(16), 2095–2097 (2005)
19. Heinbuch, S., et al.: Demonstration of a desk-top size high repetition rate soft x-ray laser. Opt. Express **13**(11), 4050–4055 (2005)
20. Urbanski, L., et al.: Spectral linewidth of a Ne-like Ar capillary discharge soft-x-ray laser and its dependence on amplification beyond gain saturation. Phys. Rev. A **85**(3), 033837 (2012)
21. Artioukov, I.A., et al.: Schwarzschild soft-x-ray microscope for imaging of nonradiating objects. Opt. Lett. **20**(24), 2451–2453 (1995)
22. Anderson, E.H.: Specialized electron beam nanolithography for EUV and X-ray diffractive optics. IEEE J. Quantum Electron. **42**(1), 27–35 (2006)
23. Carbajo, S., et al.: Sequential single-shot imaging of nanoscale dynamic interactions with a table-top soft x-ray laser. Opt. Lett. **37**(14), 2994–2996 (2012)
24. NIST: Available from http://math.nist.gov/oommf/
25. Brizuela, F., et al.: Reflection mode imaging with nanoscale resolution using a compact extreme ultraviolet laser. Opt. Express **13**(11), 3983–3988 (2005)
26. Martz, D.H., et al.: High-energy 13.9 nm table-top soft-x-ray laser at 2.5 Hz repetition rate excited by a slab-pumped Ti:sapphire laser. Opt. Lett. **35**(10), 1632–1634 (2010)
27. Wang, Y., et al.: Phase-coherent, injection-seeded, table-top soft-X-ray lasers at 18.9 nm and 13.9 nm. Nat. Photonics **2**(2), 94–98 (2008)

Chapter 31
Using the X-FEL to Understand X-Ray Thomson Scattering for Partially Ionized Plasmas

Joseph Nilsen, Walter R. Johnson, and K.T. Cheng

Abstract For the last decade numerous researchers have been trying to develop experimental techniques to use X-ray Thomson scattering as a method to measure the temperature, electron density, and ionization state of high energy density plasmas such as those used in inertial confinement fusion. X-ray laser sources have always been of interest because of the need to have a bright monochromatic X-ray source to overcome plasma emission and eliminate other lines in the background that complicate the analysis. With the advent of the X-ray free electron laser (X-FEL) at the SLAC Linac Coherent Light Source (LCLS) we now have such a source available in the keV regime. Other X-FEL sources are being built in Germany and around the world. One challenge with X-ray Thomson scattering experiments is understanding how to model the scattering for partially ionized plasmas. Most Thomson scattering codes used to model experimental data greatly simplify or neglect the contributions of the bound electrons to the scattered intensity. In this work we take the existing models of Thomson scattering that include elastic ion-ion scattering and the electron-electron plasmon scattering and add the contribution of the bound electrons in the partially ionized plasmas. Except for hydrogen plasmas almost every plasma that is studied today has bound electrons and it is important to understand their contribution to the Thomson scattering, especially as new X-ray sources such as the X-FEL will allow us to study much higher Z plasmas. Currently most experiments have looked at hydrogen or beryllium. We will first look at the bound electron contributions to beryllium by analyzing existing experimental data. We then consider several higher Z materials such as Cr and predict the existence of additional peaks in the scattering spectrum that requires new computational tools to understand. For a Sn plasma we show that the bound contributions changes the shape of the scattered spectrum in a way that would change the plasma temperature and density inferred by the experiment.

J. Nilsen (✉) · K.T. Cheng
Lawrence Livermore National Laboratory, Livermore, CA 94551, USA
e-mail: nilsen1@llnl.gov

W.R. Johnson
University of Notre Dame, Notre Dame, IN 46556, USA

S. Sebban et al. (eds.), *X-Ray Lasers 2012*, Springer Proceedings in Physics 147,
DOI 10.1007/978-3-319-00696-3_31,

1 Introduction

Thomson scattering of X-rays is being developed as an important diagnostic technique to measure temperatures, densities, and ionization balance in warm dense plasmas. Glenzer and Redmer [1] have reviewed the underlying theory of Thomson scattering being used in experiments.

In this work we start with the theoretical model proposed by Gregori et al. [2] but with one important difference. We evaluate the Thomson-scattering dynamic structure function using parameters taken from our own average-atom code [3, 4]. The average-atom model is a quantum mechanical version of the temperature-dependent Thomas-Fermi model of plasma developed years ago by Feynman et al. and is described in Ref. [5]. It consists of a single ion of charge Z with a total of Z free and bound electrons in a Wigner-Seitz cell that is embedded in a uniform "jellium sea" of free electrons whose charge is balanced by a uniform positive background. This model enables us to consider the contributions from the bound electrons in a self-consistent way for any ion. Other approaches such as Gregori's includes hydrogenic wavefunctions with screening factors to approximate the contribution from the bound electrons for a limited number of materials.

In this paper we look first look at the bound electron contributions to beryllium by analyzing existing experimental data. We then consider several higher Z materials such as Cr and predict the existence of additional peaks in the scattering spectrum that requires new computational tools to understand. For a Sn plasma we show that the contributions from the bound electrons can change the shape of the scattered spectrum in a way that would change the plasma temperature and density inferred by the experiment.

2 Validating the Average-Atom Code for Be Experiments

Numerous experiments have been done to look at Thomson scattering for low-Z materials such as hydrogen and beryllium. We examine one particular experiment [6] done at the Omega laser facility that looked at Thomson scattering in the forward scattering direction (40 degrees) off solid Be with a Cl Ly-α source at 2963 eV. Figure 1 shows the measured spectrum as the noisy dotted line. An electron temperature of 18 eV, ion temperature of 2.1 eV, and density of 1.647 g/cc is used in the average-atom model to give an electron density of 1.8×10^{23} per cc, in agreement with the analysis in Ref. [6]. The dashed line shows the experimental source function from the Cl Ly-$\boldsymbol{\alpha}$ line. Because of satellite structure we approximate the source by 3 lines: a Cl Ly-$\boldsymbol{\alpha}$ line at 2963 eV with amplitude 1 and two satellites at 2934 and 2946 eV with relative amplitudes of 0.075 and 0.037 respectively. Using the 3 weighted lines to do the Thomson scattering calculation, we calculate the scattering amplitude for Thomson scattering (solid line) and compare against the experimental data (dotted line). We observe excellent agreement within the experimental noise. Contributions from the bound 1**s** electrons have a threshold at 2876 eV that are beyond the range of the data shown in the figure.

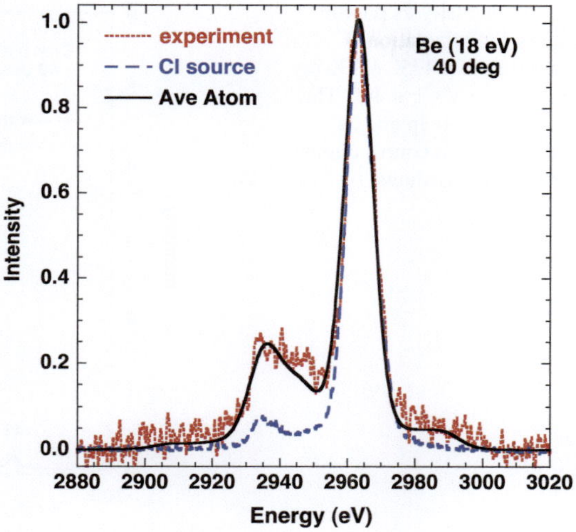

Fig. 1 Intensity vs photon energy for scattering of a Cl Ly-α X-ray off Be at 40°. *Dotted line* is experimental data (see Ref. [6]) while *solid line* is calculation for an electron temperature of 18 eV. The *dashed line* shows the Cl X-ray source

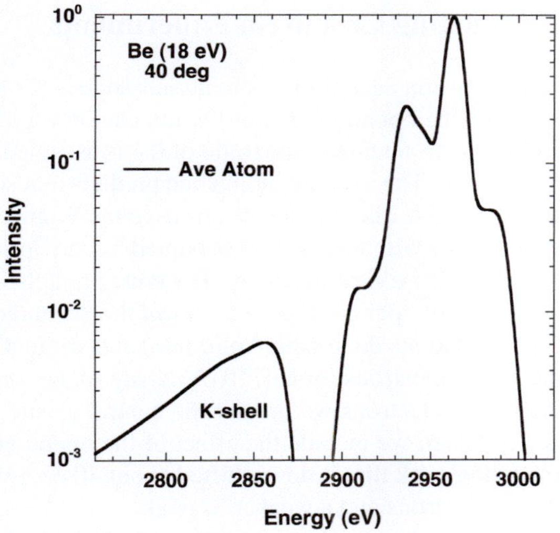

Fig. 2 Intensity vs photon energy for calculation of scattering of a Cl Ly-α X-ray off Be at 40° for an electron temperature of 18 eV. The contribution from the K-shell bound electrons is shown

For Be under the conditions just described the L-shell is completely stripped but the K-shell is 97 % occupied. The average-atom code has $Z_f = 1.647$ for the number of free electrons in the jellium sea outside the Wigner-Seitz cell. The binding energy of the K-shell electron is 86.8 eV which means the threshold for seeing the bound state contribution should begin at $2963 - 86.8 = 2876.2$ eV. To understand the contributions of the bound K-shell electrons to the Thomson scattering we expand the energy scale in Fig. 1 and re-plot it on a log-scale looking just at the average-atom calculation. In Fig. 2 we observe a K-shell contribution that is about 40 times weaker than the low-energy plasmon peak so it would be very difficult to observe in a laser-plasma experiment given the experimental noise.

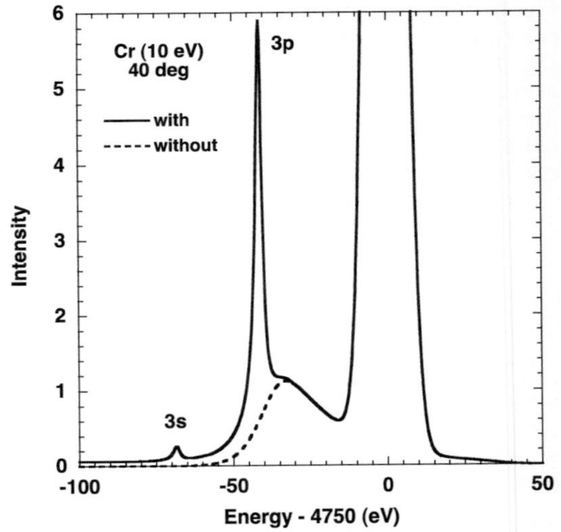

Fig. 3 Intensity vs photon energy for calculation of scattering of a 4750 eV X-ray line off 10 eV Cr at 40°. The case where the 3p and 3s bound electron contributions are included is shown by *solid line*

3 Modeling Cr and Sn Experiments

If we now consider higher-Z materials such as Cr and Sn at an electron temperature of 10 eV the bound state contribution can be very important. Following Refs. [4, 6] an ion/electron temperature ratio of 0.1 is assumed to reduce the central elastic scattering peak. The average-atom code predicts that solid density Cr (7.19 g/cc) heated to 10 eV has 6.2 continuum electrons in the Wigner-Seitz cell. This makes it a closed Ar-like core with nearly fully occupied 3s and 3p subshells with binding energies of 57.7 and 30.7 eV, respectively. The code predicts $Z_f = 2.92$ and an electron density of 2.4×10^{23} per cc. Figure 3 shows the scattered intensity versus photon energy for calculations done with (solid line) and without (dashed line) the contribution of the bound electrons for a 4750 eV X-ray source scattered at 40° off the Cr. Without the bound electrons we predict the central elastic scattering peak and the plasmon peaks. When we include the effect of the bound electrons we predict a very strong scattering peak that is downshifted about 40 eV from the central elastic peak due to the 3p electrons and a weaker 3s peak.

For solid density Sn (7.3 g/cc) at 10 eV we look at the case of the 2960 eV X-ray source scattering in the backward direction at 130°. The average-atom code predicts that warm dense Sn has 4.4 continuum electrons with an average occupation of 8.8 for the 4d electrons and 0.8 for the 5s electrons outside a closed Kr-like core with all the lower orbital subshells fully occupied. We predict $Z_f = 3.37$ which gives an electron density of 1.25×10^{23} per cc. The binding energies of the 4d and 5s electrons are 22.4 and 2.14 eV, respectively, which means the effect of the 5s electron will be hidden under the elastic scattering peak. Figure 4 plots the scattering intensity versus photon energy for calculations done with (solid line) and without (dashed line) the contribution of the bound electrons. For backward scattering the scattering is no longer in the collective regime and the distinct plasmon peaks are replaced by a broader scattering structure. The contribution from the bound 4d electrons is now an

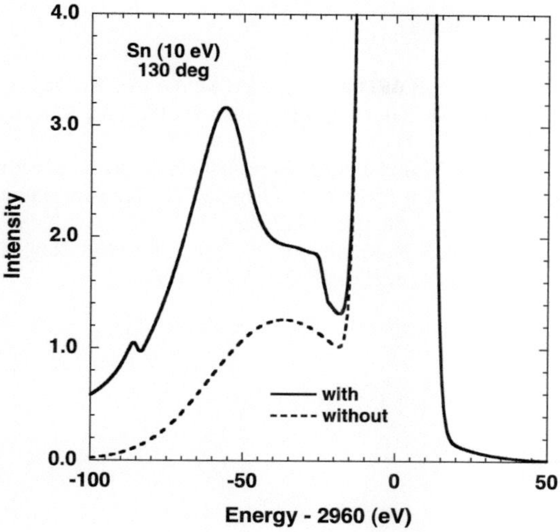

Fig. 4 Intensity vs photon energy for calculation of scattering of a 2960 eV X-ray line off Sn at 130° for an electron temperature of 10 eV. The strong contribution from the 4d bound electrons are apparent for the case where the bound electrons are included (*solid line*) when compared with case without bound electrons (*dashed line*)

additional large broad feature that lies on top of the Thomson scattering and would make it easy to misinterpret the spectrum if one did not understand the contribution of the bound electrons.

4 Conclusions

The availability of bright monochromatic X-ray line sources from X-ray free electron laser facilities opens up many new possibilities to use Thomson scattering as an important diagnostics technique to measure the temperatures, densities, and ionization balance in warm dense plasmas.

Current attempts to model Thomson scattering tend to use very simplified models, if anything, to model the effect of the bound electrons on the measured scattered intensity. Our approach here is to evaluate the Thomson-scattering dynamic structure function using parameters taken from our own average-atom code [3, 4]. This model enables us to consider contributions from the bound electrons in a self-consistent way for any ion.

In this paper we validate our average-atom based Thomson scattering code by comparing our model against existing experimental data for Be near solid density and temperature near 18 eV. For solid density Cr at 10 eV we predict the existence of additional peaks in the scattering spectrum that requires new computational tools to understand. We also analyze solid density Sn at 10 eV and show that the contributions from the bound electrons can change the shape of the scattered spectrum in a way that would change the plasma temperature and density inferred by the experiment.

Acknowledgements This work was performed under the auspices of the U.S. Department of Energy by Lawrence Livermore National Laboratory under Contract DE-AC52-07NA27344.

References

1. Glenzer, S.H., Redmer, R.: Rev. Mod. Phys. **81**, 1625 (2009)
2. Gregori, G., Glenzer, S.H., Rozmus, W., Lee, R.W., Landen, O.L.: Phys. Rev. E **67**, 026412 (2003)
3. Johnson, W.R., Guet, C., Bertsch, G.F.: J. Quant. Spectrosc. Radiat. Transf. **99**, 327–340 (2006)
4. Johnson, W.R., Nilsen, J., Cheng, K.T.: Thomson scattering in the average-atom approximation. Phys. Rev. E **86**, 036410 (2012)
5. Feynman, R.P., Metropolis, N., Teller, E.: Phys. Rev. **75**, 1561 (1949)
6. Döppner, T., et al.: HEDP **5**, 182 (2009)

Chapter 32
Development of Soft X-Ray Microscopy Using Fresnel Zone Plate for Observation of Laser-Induce Surface Dynamics

M. Nishikino, M. Baba, T. Suemoto, N. Hasegawa, M. Ishino, T. Kaihori, T. Kawachi, and M. Yamagiwa

Abstract X-ray reflective microscope in EUV region has been tried with a Fresnel zone plate (FZP). FZP with an outermost zone width of 0.28 μm and thickness of 250 nm is used as an objective lens. The focal length of FZP is 10 mm at the wavelength of 13.9 nm. Image of a test object has been taken, and the spatial resolution of the measured image is estimated to be less than 1 μm in the case of the magnification of 20.

1 Introduction

The dynamical processes of the ultra-short pulse laser induced surface modifications such as laser ablation come to attract much attention for the micro processing [1, 2]. The understanding of the dynamics of the laser ablation process is important for the accuracy and control of the femtosecond laser processing technique. However, the dynamics of the femtosecond laser ablation is still not clear. Several studies on the time-resolved imaging of femtosecond laser ablation process have been performed on various materials. We have developed the femto-second laser pump and soft x-ray laser interferometer system in order to observe the dynamical processes of the femto-second laser ablation. By using this system, we succeed to obtain surface expansion of the laser induced Platinum (Pt) surface in the early stage [3–5]. It is necessary for the observation of detailed laser ablation to develop the measurement technique with high spatial resolution. An x-ray microscopy technique has expanded in the various fields due to its capability of nondestructive, high resolution and highly sensitivity analysis [6, 7]. The x-ray microscopy with high spatial resolution has been developed for mainly x-ray transmitted samples. In the case of

M. Nishikino · N. Hasegawa · M. Ishino · T. Kaihori · T. Kawachi (✉) · M. Yamagiwa
Quantum Beam Science Directorate, Japan Atomic Energy Agency, 8-1-7 Umemidai, Kizugawa, Kyoto 619-0215, Japan
e-mail: kawachi.tetsuya@jaea.go.jp

M. Baba · T. Suemoto
The Institute for Solid State Physics, University of Tokyo, 5-1-5 Kashiwanoha, Kashiwa, Chiba 277-8581, Japan

S. Sebban et al. (eds.), *X-Ray Lasers 2012*, Springer Proceedings in Physics 147,
DOI 10.1007/978-3-319-00696-3_32,

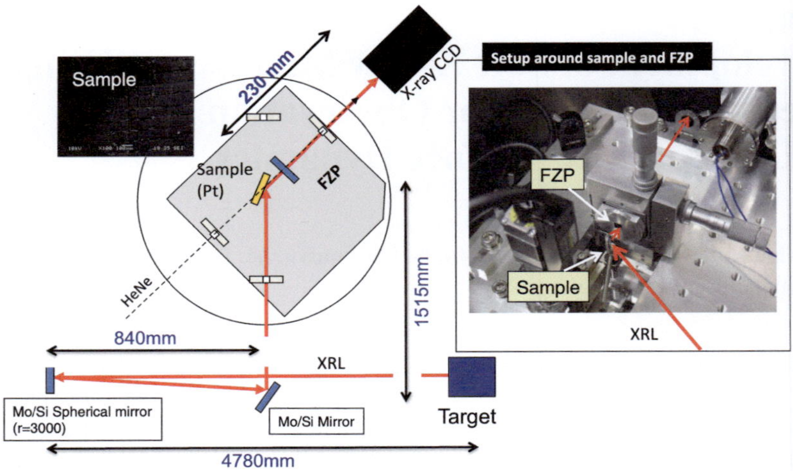

Fig. 1 Schematic view of the experimental setup

the measurement of the opaque sample surface, we can use the refraction type of the x-ray microscopy. Recently, the refraction type of x-ray microscopy with high resolution is developed for the technique of the mask defects in the EUV lithography [8]. However, the decrease in the spatial resolution caused by the deference of focal depth is seemed to be a problem. In this study we have developed the proto-type reflected soft x-ray microscopy with a practical level resolution using the Fresnel zone plate.

2 Experimental Setup and Results

A schematic view of the experimental setup for proto-type refracted x-ray microscopy is shown in Fig. 1. The experiment has been performed at JAEA plasma x-ray laser (XRL) beam lines. A highly coherent XRL at a wavelength of 13.9 nm was generated by the oscillator-amplifier configuration with two targets. The output energy of the XRL beam was about 0.1 μJ, corresponding to a flux of about 10^{10} photons/pulse. The pulse width of the XRL was about 7 ps. The distance between the FZP and the CCD camera is 230 mm in the case of magnification of 20. The image of the test pattern was measured with the CCD camera. The FZP contains a poly(methylmethacrylate) (PMMA) zone on a 0.75×0.75 mm^2 silicon nitride (SiN) membrane with a thickness of 50 nm. The diameter of the FZP was 0.5 mm, and the total zone number and outermost zone width were 450 and 280 nm, respectively. The focal length was 10 mm for a λ of 13.9 nm. The theoretical resolution was 680 nm. We have obtained the single shot image of the grid on the sample surface.

Figure 2(a) shows single-shot images and spatial profiles of the test pattern in the different focal position. The spatial resolution is about 1 μm and keeps for about 160 μm on the sample in the horizontal direction. In the case of few photon numbers

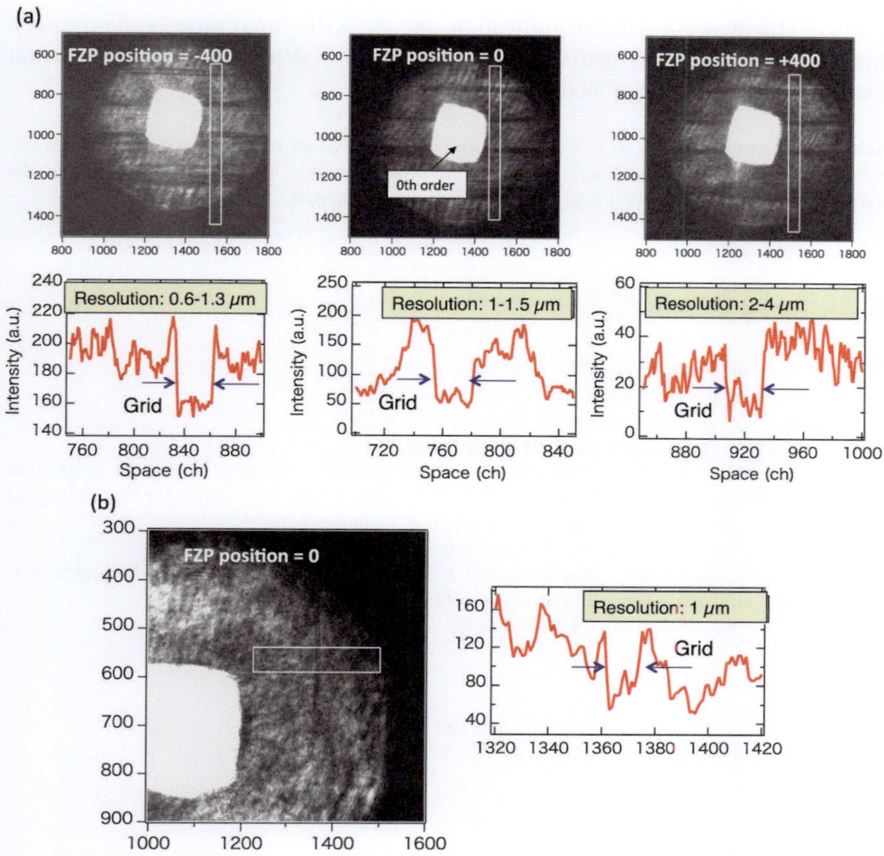

Fig. 2 Single shot x-ray image of test pattern. (**a**) The horizontal grid image. (**b**) The vertical grid image

in the single-shot image, it becomes a high spatial resolution by the data accumulation of x-ray images. Figure 2(b) shows single-shot images and spatial profiles of the test pattern in the vertical direction. The spatial resolution on sample is about 2.6 μm, because of a grazing incidence in the reflected x-ray microscopy system. In this experiment, the spatial resolution by the limitation of the CCD pixel size and the magnification was demonstrated over whole region of the sample surface.

3 Summary

We have developed the reflective x-ray microscopy and observed the surface groove of test pattern. The spatial resolution is evaluated about 1 μm in vertical and about 3 μm in horizontal direction, respectively. In future, the spatial resolution will be improved by increasing the magnification, and the XRL pump-probe system can be upgraded to have higher spatial resolution of 1 nm in depth and 100 nm in lateral space by combining the reflective x-ray microscopy.

Acknowledgements This work was partly supported by a Grant-in-Aid for Young Scientist (B) No. 24710096 from the Japan Society for the Promotion of Science. The authors are grateful to the staff of the x-ray laser application group for their technical support.

References

1. Korte, F., et al.: Appl. Phys. A **79**, 879 (2004) (2010)
2. Perieira, A., et al.: Appl. Phys. A **79**, 1443 (2004)
3. Suemoto, T., et al.: Opt. Express **18**, 14114 (2010)
4. Ochi, Y., et al.: Jpn. J. Appl. Phys. **51**, 016601 (2012)
5. Hasegawa, N., et al.: Observation of the laser-induced surface dynamics using the single-shot soft x-ray laser probe. In: Sebban, S., et al. (eds.) X-Ray Lasers 2012. Springer Proceedings in Physics, vol. 147. Springer, Cham (2013). Chapter 19 in this book
6. Schmahl, G., Rudolph, D. (eds.): X-Ray Microscopy. Springer, Berlin (1984)
7. Atwood, D.T.: Soft X-Rays and Extreme Ultraviolet Radiation: Principles and Applications. Cambridge University Press, Cambridge (1999). Chapter 9
8. Brizuela, F., et al.: Opt. Lett. **34**, 271 (2009)

Chapter 33
0.27 GW Soft X-Ray Pulse Using a Plasma-Based Amplification Chain

E. Oliva, M. Fajardo, P. Velarde, D. Ros, S. Sebban, and P. Zeitoun

Abstract Seeding plasma-based soft-x-ray lasers (PBSXRL) with high order harmonics (HOH) has been demonstrated in plasmas created from gas targets (Zeitoun et al. in Nature 431:426, 2004) and solid targets (Wang et al. in Nat. Photonics 2:94, 2008), obtaining 1 μJ, 1 ps pulses. Reaching multi-microJoule, hundreds of fs regime is the ultimate goal. Recent papers (Oliva et al. in Opt. Lett. 34(17):2640–2642, 2009; Phys. Rev. E 82(5):056408, 2010) showed that increasing the width (up to 1 mm) of the plasma increases the amplification surface and improves the gain zone properties. Up to 20 μJ could be extracted when seeding but the temporal duration and profile was not studied. Simulations show that the HOH is weakly amplified whereas most of the energy is within a long (several picoseconds) wake induced by the HOH (Al'miev et al. in Phys. Rev. Lett. 99(12):123902, 2007; Kim et al. in Phys. Rev. Lett. 104:053901, 2010). Amplified Spontaneous Emission (ASE) is also present in the output beam. Using the 1D Maxwell-Bloch code *DeepOne* (Oliva et al. in Phys. Rev. A 84(1):013811, 2011) we will show that fully coherent, wake and ASE-suppressed, 21.6 μJ, 80 fs pulse can be obtained when optimizing at the same time both the seed and the plasma conditions.

E. Oliva (✉) · S. Sebban · P. Zeitoun
Laboratoire d'Optique Appliquée, ENSTA ParisTech, École Polytechnique ParisTech, CNRS, UMR 7639, Palaiseau, France
e-mail: eduardo.oliva@ensta-paristech.fr

M. Fajardo
Centro de Física dos Plasmas, Instituto Superior Técnico, Lisbon, Portugal

P. Velarde
Instituto de Fusión Nuclear, Universidad Politécnica de Madrid, Madrid, Spain

D. Ros
CLUPS, EA4127, Bat 106, Université Paris-Sud, 91405 Orsay, France

S. Sebban et al. (eds.), *X-Ray Lasers 2012*, Springer Proceedings in Physics 147,
DOI 10.1007/978-3-319-00696-3_33,

1 Introduction

The emergence of free-electron lasers (FEL) emitting in soft X-rays (30 eV–1 keV) and hard X-rays (energy greater than 1 keV) [9] is making astonishing advances in time-resolved X-ray Science [3, 8]. Plasma based seeded soft X-ray lasers [14, 15] are table-top sources that deliver several hundreds nanoJoules with optimal spatial properties as they are fully coherent [14] and aberration-free [4]. However, current and future applications need pulses of hundred femtosecond or shorter. In addition to this, time-dependent Maxwell-Bloch models show that the output beam consists of amplified spontaneous emission (ASE), weakly amplified HOH and a wake field with a complex structure (Rabi oscillations and coherent decay) [1, 6].

In this article, using our time-dependent Maxwell-Bloch code *DeepOne*, we will study the seeding and amplification of high order harmonics (HOH) inside plasma-based amplifiers. We will show that adjusting HOH duration and energy to some values given by plasma hydrodynamics leads to a single component pulse of 1.5 μJ, 140 fs. The amplifier required is a 2.5 mm long, 100 μm wide plasma. The energy of the amplified pulse can be enhanced by increasing the width (i.e. the amplification surface) of the plasma [10, 11]. Nevertheless, to keep the duration of the HOH along the amplifier and avoid the creation of strong ASE and wake, it is necessary to seed at high intensity. Among other possible sources, we selected as seeding source a 100 μm wide plasma amplifier seeded by HOH as described above. We simulated the full pre-amplifier main-amplifier chain, demonstrating that it is possible to obtain a 21.6 μJ, 80 fs single pulse when seeding carefully a 1 mm wide, 2 mm long plasma with a 100 μm wide, 2.5 mm long plasma.

2 Maxwell-Bloch Equations

Amplification of the HOH through a plasma is a multiscale phenomenon involving different physical processes: temporal variation of the envelop of the HOH electric field (tens of femtoseconds), plasma response to this electric field (tens of picoseconds) and plasma hydrodynamics (hundreds of picoseconds to nanoseconds). Maxwell-Bloch type equations (Eqs. (1), (2), (3) and (4)) take into account the first two processes whereas hydrodynamic data is used as an input [12]:

$$\frac{\partial E}{\partial \zeta} = \frac{i\omega_0}{2c}\left[\mu_0 c^2 P - \left(\frac{\omega_p}{\omega_0}\right)^2 E\right] \tag{1}$$

$$\frac{\partial P}{\partial \tau} = \Gamma - \gamma P - \frac{i z_{21}^2}{\hbar} E(N_2 - N_1) \tag{2}$$

$$\frac{\partial N_2}{\partial \tau} = \sum_k C_{k2} N_k + Imag\big(E^* P\big)\frac{1}{2\hbar} \tag{3}$$

$$\frac{\partial N_1}{\partial \tau} = \sum_k C_{k1} N_k - Imag\big(E^* P\big)\frac{1}{2\hbar} \tag{4}$$

Equation (1) is deduced from Maxwell wave equation for the electric field. It describes the propagation, amplification by the polarization density P and damping by the free electron current of the envelope of the electric field. ω_p and ω_0 are respectively the plasma frequency and the electric field frequency. To simplify the equations we use the coordinate change $\tau = t - \frac{z}{c}$, the so-called reduced time (i.e. the referential frame of the photon), and define $\zeta = c\tau$.

The rest of equations are deduced from Bloch equations. Equation (2) gives the temporal evolution of the polarization [7, 13], where γ is the depolarization rate, which we take to be proportional to the electron-ion collision frequency [7] and N_2, N_1 are respectively the population of the upper and lower level of the transition. z_{21} is the dipole matrix element. To model the spontaneous emission, we have added a stochastic source term, Γ. This term is a random variable with vanishing correlation function, properly normalized to obtain the expected Lorenzian lasing line profile [2, 7].

Finally, the populations of the levels involved on the lasing transition are computed using the standard rate equations coupled with the electric field (Eqs. (3), (4)). The populations of the levels involved on the transition are coupled with collisional (de)excitation and radiative deexcitation rates included in the C_{ij} coefficients and with the soft X-ray laser electric field through polarization density. We have modeled the $2p^5 3s\ J = 1 \rightarrow 2p^5 3p\ J = 0$ transition in Ne-like Zinc, situated at 21.2 nm.

3 Polarization Dynamics and Amplification

The amplification of a short (tens of femtoseconds) HOH pulse inside a plasma has been studied recently in several papers [1, 5, 6, 12]. The HOH immediately polarizes the medium and creates a trailing pulse. The harmonic itself is barely amplified and is the trailing pulse (the wake) who extracts the energy stored in the amplifier, but during several picoseconds. The pulse shape is given by the plasma response to the HOH electric field, i.e. by polarization dynamics.

An exact solution of the polarization induced by a Gaussian HOH pulse, assuming that the HOH pulse does not modify the population inversion (i.e. far from the saturation regime) can be written in as a product of three terms $P(t) = P_0 \cdot P_1 \cdot P_2$ given by:

$$P_0(t) = -\frac{i z_{21}^2}{\hbar} E_0 (N_2 - N_1) e^{\frac{\gamma^2 \sigma^2}{2}} \sigma \sqrt{2\pi} \tag{5}$$

$$P_1(t) = \frac{1}{\sigma\sqrt{2\pi}} \int_{-\infty}^{t} e^{-\frac{(\tau - t_{cent} - \gamma\sigma^2)^2}{2\sigma^2}} d\tau \tag{6}$$

$$P_2(t) = \begin{cases} e^{-\gamma(t - t_{cent})} & t \geq t_{cent} \\ 1 & \text{otherwise} \end{cases} \tag{7}$$

where E_0 is the maximum seeded electric field, σ is the standard deviation of the Gaussian profile of the field and t_{cent} is the time of the maximum seed intensity. P_1 and P_2 define the temporal shape of the polarization. P_1 controls the rising time (which depends on σ) and P_2 is the decay time, given by γ, the electron-ion collision frequency. Since both terms are normalized to unity, the numerical value of the polarization is given by P_0. Thus, the temporal shape of the amplified beam is given by the last two factors and controlled by σ and γ. In the case of 20 fs HOH seeded in a plasma with $n_e = 2.2 \times 10^{20}$ cm^{-3} and $T_e = 531$ eV, the two characteristic times are $\sigma = 9$ fs and $\gamma^{-1} = 176$ fs. The small value of σ induces a sudden rise of the polarization, which decays slowly, due to the higher value of γ. This slow decay is the origin of the picosecond wake. Then, if the decay time is shorter, the duration of the output beam would be also shortened. This is achieved increasing the electron density of the plasma. In addition to this, increasing the duration of the pulse will slow the rising of the polarization. This will improve the matching between HOH and polarization, enhancing the HOH amplification and reducing the wake. Electron density is greater in wider plasmas. Then, using wide plasmas will not only benefit from greater gain zone but also from higher electron density and thus, shorter pulse duration. A 200 fs, 1.5 nJ HOH seeded in a 100 μm wide, 2.5 mm long plasma will make an output beam of 1.5 μJ 140 fs, keeping ASE negligible. The 10 MW power attained by this source is 100 times greater than the experimental values using solid target [14] and its spatial and temporal coherences makes it suitable for many applications.

4 Multi-stage Amplifiers

As explained, wider plasmas have a greater amplification zone and thus, more energy (tens of microJoules) can be extracted from then. Nevertheless, in order to keep ASE and wake negligibles and maintain the duration of the amplified pulse below 100 fs, it is necessary to seed at high energies (greater than 1 μJ) while having a duration between 100 fs and 200 fs depending on plasma electron-ion collision [12]. State-of-the-art HOH deliver this energy but, since the bandwidth is between 30 and 100 greater than the bandwidth of the plasma amplifier, only a small part of the energy is coupled, being nowadays not possible to reach the in-band energy requirements. On the other hand, a plasma amplifier seeded with HOH as explained in Sect. 3 can be used as a pre-amplifier. Since both plasmas have similar electron density and temperature, the bandwidth of the pre-amplifier will match that of the main amplifier.

In Fig. 1 the modeled amplification chain is depicted. The pre-amplifier is a 100 μm × 2.5 mm plasma seeded with 50 nJ, 160 fs HOH, similar to the scheme described in Sect. 3. The output is a 1.6 μJ 130 fs beam which is seeded in the 1 mm × 2 mm main amplifier, taking into account some loses on the focusing optics. This beam is strongly amplified while conserving the short duration till 21.6 μJ and 80 fs. As shown in Fig. 1 the HOH completely dominates the output beam. The high power attained by this source makes it perfect for many challenging experiences.

In conclusion, we have shown that a careful matching of the HOH characteristic with plasma properties and the use of pre-amplifiers to seed wide (1 mm) plasmas

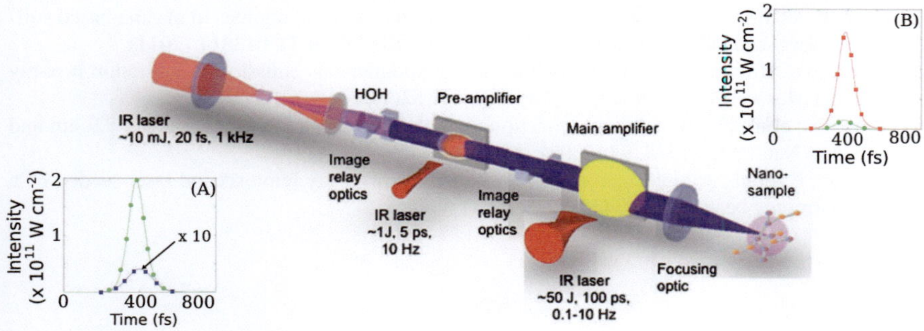

Fig. 1 Artistic view of an amplification chain. Output beam of a 1 mm × 2 mm main amplifier seeded with 1.6 μJ, 130 fs beam (sub-figure **A** *blue*, sub-figure **B** *green*) coming from a 2.5 mm × 100 μm pre-amplifier plasma. The output beam has an energy of 21.6 μJ and a duration of 80 fs (sub-figure **B** *red*)

is the key to obtain tens of microJoules in less than 100 fs. We have simulated a full pre-amplifier main-amplifier chain capable of delivering beams of 21.6 μJ and 80 fs.

Acknowledgements The authors would like to acknowledge the financial support provided by the LASERLAB II 228334 (SFINX) European Project and RTRA "Triangle de la Physique" project (SHYLAX).

References

1. Al'miev, I.R., et al.: Dynamical description of transient x-ray lasers seeded with high-order harmonic radiation through Maxwell-Bloch numerical simulations. Phys. Rev. Lett. **99**(12), 123902 (2007)
2. Chandrasekhar, S.: Stochastic problems in physics and astronomy. Rev. Mod. Phys. **15**(1), 1–89 (1943)
3. Chapman, H.N.: X-ray imaging beyond the limits. Nat. Mater. **8**, 299–301 (2009)
4. Goddet, J.P., et al.: Aberration-free laser beam in the soft x-ray range. Opt. Lett. **34**(16), 2438–2440 (2009)
5. Kim, C.M., Janulewicz, K.A., Kim, H.T., Lee, J.: Amplification of a high-order harmonic pulse in an active medium of a plasma-based x-ray laser. Phys. Rev. A **80**, 053811 (2009)
6. Kim, C.M., Lee, J., Janulewicz, K.A.: Coherent amplification of an ultrashort pulse in a high- and swept-gain medium with level degeneracy. Phys. Rev. Lett. **104**, 053901 (2010)
7. Larroche, O., et al.: Maxwell-Bloch modeling of x-ray-laser-signal buildup in single- and double-pass configurations. Phys. Rev. A **62**(4), 043815 (2000)
8. Mancuso, A., Yefanov, O., Vartanyants, I.: Coherent diffractive imaging of biological samples at synchrotron and free electron laser facilities. J. Biotechnol. **149**(4), 229–237 (2010)
9. McNeil, B.W.J., Thomson, N.R.: X-ray free-electron lasers. Nat. Photonics **2**, 814–821 (2010)
10. Oliva, E., et al.: Optimization of soft x-ray amplifier by tailoring plasma hydrodynamics. Opt. Lett. **34**(17), 2640–2642 (2009)
11. Oliva, E., et al.: Hydrodynamic study of plasma amplifiers for soft-x-ray lasers: a transition in hydrodynamic behavior for plasma columns with widths ranging from 20 μm to 2 mm. Phys. Rev. E **82**(5), 056408 (2010)

12. Oliva, E., et al.: Comparison of natural and forced amplification regimes in plasma-based soft-x-ray lasers seeded by high-order harmonics. Phys. Rev. A **84**(1), 013811 (2011)
13. Sureau, A., Holden, P.B.: From amplification of spontaneous emission to saturation in x-ray lasers: a Maxwell-Bloch treatment. Phys. Rev. A **52**(4), 3110–3125 (1995)
14. Wang, Y., et al.: Phase-coherent, injection-seeded, table-top soft-x-ray lasers at 18.9 nm and 13.9 nm. Nat. Photonics **2**, 94 (2008)
15. Zeitoun, P., et al.: A high-intensity highly coherent soft x-ray femtosecond laser seeded by a high harmonic beam. Nature **431**, 426 (2004)

Chapter 34
Harmonic Generation in Argon by Femtosecond Ti:Sapphire Laser

Rabia Qindeel, Ricardo Elgul Samad, Anderson Zanardi de Freitas, Paulo Sergio Fabris de Matos, Edilson Lucena Falcão, and Nilson Dias Vieira Junior

Abstract Generation of harmonics using a gas nozzle has remarkable feature in various applications. Pulses from a Ti:Sapphire laser, centered at 785 nm, in a 4 kHz train, with 25 femtoseconds and 800 μJ of maximum energy were employed to generate harmonics in an argon gas nozzle. We present the current results on the focusability of the nozzle, harmonic radiation, measurement of the influence of laser power and laser focus position on the divergence of gas nozzle. We have successfully generated 3rd, 5th and 7th harmonics in Argon at different laser powers. The results show that the harmonic signals are almost same for laser average powers over 1.0 W and variation always appears below 1.0 W. It means that there is saturation in the physical phenomenon happening inside the gas nozzle at high laser powers and the variation is non-linear below 1.0 W. These results are embedded in an effort towards x-ray generation in the water window.

1 Introduction

Generation of coherent extreme ultraviolet (EUV) by focusing femtosecond laser into gasses is a well-known method [1–4]. High pulse intensities, greater than $10^{13}\ \mathrm{W\,cm^{-2}}$ and above, are required for the High Harmonic Generation (HHG) nonlinear conversion process. These high pulse intensities are not directly attainable using only the output power of a femtosecond oscillator. Chirped Pulse Amplification (CPA) enables the pulse intensity to exceed this threshold by incorporating several regenerative and/or multi-pass amplifier cavities in tandem [5]. When a very intense laser pulse is focused into a gas at reduced pressures, a strong nonlinear interaction can lead to the generation of odd harmonics of the pulse optical frequency, as an extreme form of nonlinear frequency conversion. Although only a small fraction of the laser power is converted into harmonics, this output can still be useful for

R. Qindeel (✉) · R.E. Samad · A.Z. de Freitas · P.S.F. de Matos · N.D. Vieira Junior
Centro de Lasers e Aplicações, IPEN, CNEN/SP, Sao Paulo, Brazil
e-mail: plasma_qindeel@yahoo.com

E.L. Falcão
Physics Department, UFPE, Recife, Brazil

S. Sebban et al. (eds.), *X-Ray Lasers 2012*, Springer Proceedings in Physics 147,
DOI 10.1007/978-3-319-00696-3_34,

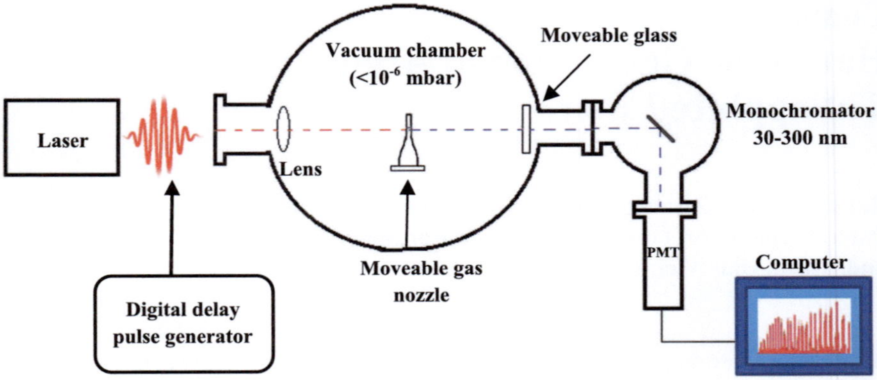

Fig. 1 Experimental setup for High Harmonic Generation in a gas nozzle

measurements down to wavelengths in the ultraviolet or even the soft X-ray spectral region.

When femtosecond laser pulses propagate in a transparent medium, the light field is localized in a narrow filament, whose formation is the result of the nonlinear optical interaction of the short-duration radiation with a medium [6–9]. Depending on the laser intensity, the creation of free electrons can be described by multiphoton or tunneling ionization. The applied laser intensity exhibits a highly nonlinear behavior, resulting in a saturation of the nonlinear refractive index. The mechanism of harmonic generation and saturation with respect to laser power is still not understood in detail. Therefore we have carried out this research to understand and generate harmonics. In this work we have studied the laser peak power effect on harmonic generation in argon gas flowing through a nozzle. We have discussed the saturation in harmonic signals with respect to increase in laser power.

2 Experimental Set up

The laser facility used in the experiment of high-harmonics generation is a Ti:Sapphire (CPA) system at the Centro de Lasers e Aplicações (CLA), IPEN, CNEN/SP, generating pulses centered at 785 nm, in a 4 kHz train, with 29 femtosecond and 800 μJ of maximum energy. These pulses were focused by a 50 cm focal length lens to generate harmonics in an argon nozzle with back pressure of 140 mbar, placed just after the lens focal point for the short trajectory. The nozzle used was 0.5 mm thick, to minimize the possibility of phase mismatch and reabsorption, and placed inside a vacuum chamber. A turbomolecular vacuum pump was used to evacuate the chamber and an attached monochromator (McPherson 234/302, 30–300 nm range) to a background pressure of 10^6 mbar. Harmonic radiation produced in the nozzle passed into the monochromator where the beam was

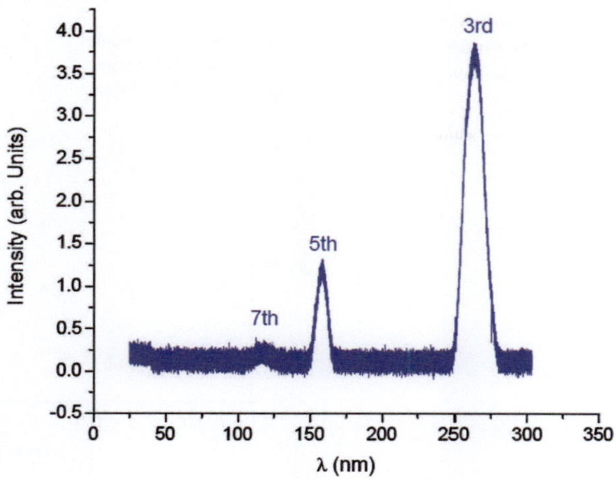

Fig. 2 Harmonic generation at after focus at 140 mbar pressure and 2.0 W of laser power, 4 kHz

spectrally dispersed by grating and then imaged onto a PMT with a Sodium Salicylate scintillator. Only odd harmonics were generated in atomic gases due to the interaction selection rules. When gas was injected into the target chamber, the pressure increases to about 10^{-3} mbar. Pulses with focused intensities in the range from 10^{14} W/cm^2 to 10^{15} W/cm^2 were needed to generate the harmonics. The resulting spectra were recorded in a digital oscilloscope and then processed in a computer. The experiment was carried out with different laser powers. A moveable glass plate was placed to confirm the generation of harmonics. The experimental setup is depicted in Fig. 1.

3 Results and Discussion

The 3rd, 5th and 7th harmonics were successfully generated in the argon filled gas nozzle at various laser powers, under the same gas pressure. Figure 2 shows the harmonics at 2.0 W, 4 kHz repetition rate and 140 mbar of pressure. The phase mismatch due to ionization of the argon is negligible due to the short 0.5 mm gas nozzle.

The observed quadratic increase in harmonic yield with different laser power, shown in Fig. 3, proves that we are operating in a regime that is free of significant phase mismatch or re-absorption. Figure 3 also represents the focusability of nozzle harmonic radiation, measurement of the influence of laser power and laser focus position on the divergence of gas nozzle harmonics. When highly intense laser pulses are focused into the gas target, the pulses propagate over few millimeters, where it can ionize, distort, refract and accelerate particles. The rapid ionization occurs by the fact that the laser field perturbs the Coulomb barrier of the atom, allowing electrons to tunnel free. The kinetic energy of the electrons pulled out from the atom by the laser field is typically smaller than both the ionization potential and quiver

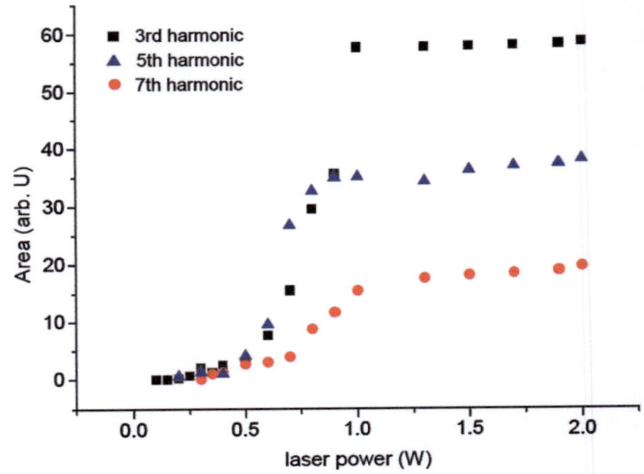

Fig. 3 Area versus laser power of graph for 3rd, 5th and 7th harmonics

energy. This low residual energy arises because within an optical cycle, ionization occurs at the peak of the electric field. The released electron energy is increased by the electric field forced oscillation, and ultimately rises the final temperature of the plasma created and only a tiny fraction of the laser power can be converted into harmonics [10].

The results show that the harmonics signals quickly increase with the laser power, until 1.0 W, over which they become almost constant. It means that there is saturation in the physical phenomenon happening inside the gas nozzle at high laser power and the variation is non-linear below 1.0 Watt. The observed saturation of the harmonic in the gas jet is the result of a combination of neutral atom depletion and free-electron contribution to phase mismatch. The depletion of the ground state occurs when the laser intensity is so high that all the neutral atoms near the focus are ionized. The depletion of the neutral atoms in the gas jet is responsible for the existence of an optimum laser intensity at which the harmonics are generated with maximum efficiency. This is closely related to the saturation intensity of the gas, since it does not seem to depend on the harmonic order, ruling out other possible explanations based on the phase matching with the excitation beam.

The other possibility of saturation is the intensity clamping effect [11], which sets an upper limit to the laser intensity due to the dynamic balance between Kerr self-focusing and plasma defocusing resulting in a stabilized harmonic intensity. We can also say that the intensity clamping effect should be responsible for the saturation in harmonic generation, occurring at high excitation intensities and saturating the conversion efficiencies of the 3rd, 5th, and 7th harmonic processes.

4 Conclusions

In summary, we have performed systematic studies of the harmonic generation in argon using a gas nozzle. We have successfully generated 3rd, 5th and 7th harmonics using ultrashort laser pulses with intensities in the range from 10^{14} W/cm^2 to higher than 10^{15} W/cm^2. We have carefully examined the behavior of the obtained signals in argon with different laser power observed that the saturation in signals are due to depletion of the neutral atoms.

Acknowledgements The authors thank FAPESP, CNPq and FINEP for supporting this work.

References

1. Seungchul, K., Jonghan, J., Young, J.K., In-Yong, P., Yunseok, K., Seung, W.K.: High-harmonic generation by resonant Plasmon field enhancement. Nature Lett. **453**(5), 757–760 (2008)
2. Corkum, P.B.: Plasma perspective on strong-field multiphoton ionization. Phys. Rev. Lett. **71**, 1994–1997 (1993)
3. Lewenstein, M., Balcou, P., Ivanov, M., L'Huillier, A., Corkum, P.B.: Theory of high-harmonic generation by low-frequency laser fields. Phys. Rev. A **49**, 2117–2132 (1994)
4. Chang, Z., Rundquist, A., Wang, H., Murnane, M.M., Kapteyn, H.C.: Generation of coherent soft X rays at 2.7 nm using high harmonics. Phys. Rev. Lett. **79**, 2967–2970 (1997)
5. Seres, J., Seres, E., Verhoef, A.J., Tempea, G., Streli, C., Wobrauschek, P., Yakovlev, V., Scrinzi, A., Spielmann, C., Krausz, F.: Laser technology: source of coherent kiloelectrovolt X-rays. Nature **433**, 596 (2005)
6. Chin, S.L., Hosseini, S.A., Liu, W., Luo, Q., Theberge, F., Akozbek, N., Becker, A., Kandidov, V., Kosareva, O., Schroeder, H.: Can. J. Phys. **83**, 863 (2005)
7. Kandidov, V.P., Shlenov, S.V., Kosareva, O.G.: Quantum Electron. **39**, 205 (2009)
8. Couairon, A., Mysyrowicz, A.: Phys. Rep. **441**, 47 (2007)
9. Berge, L., Skupin, S., Nuter, R., Kasparian, J., Wolf, J.P.: Rep. Prog. Phys. **70**, 1633 (2007)
10. Gibbon, P., Forster, E.: Short-pulse laser-plasma interactions. Plasma Phys. Control. Fusion **38**, 769–793 (1996)
11. Ni, J., Yao, J., Zeng, B., Chu, W., Li, G., Zhang, H., Jing, C., Chin, S.L., Cheng, Y., Xu, Z.: Comparative investigation of third- and fifth-harmonic generation in atomic and molecular gases driven by midinfrared ultrafast laser pulses. Phys. Rev. A **84**, 063846 (2011)

Chapter 35
Demonstration of a 100 Hz Repetition Rate Soft X-Ray Laser and Gain-Saturated Sub-10 nm Table-Top Lasers

J.J. Rocca, B.A. Reagan, Y. Wang, D. Alessi, K. Wernsing, B.M. Luther, M.A. Curtis, M. Berrill, D. Martz, S. Wang, L. Yin, F. Furch, M. Woolston, D. Patel, V.N. Shlyaptsev, and C.S. Menoni

Abstract We report the first operation of a table-top soft x-ray laser at 100 Hz repetition rate. This gain saturated laser produces 0.15 mW average power in the 18.9 nm line of nickel-like molybdenum in the form of 1.5 μJ pulses. This is the highest average power reported to date from a compact coherent soft x-ray laser source operating at wavelengths shorter than 20 nm. The soft x-ray laser is excited by a diode-pumped chirped pulse amplification Yb:YAG laser that produces 1 J pulses of 5 ps duration. We have also demonstrated the efficient generation of sub-9 nm wavelength laser pulses of microjoule energy at 1 Hz repetition rate with a table-top laser. Gain-saturated lasing was obtained at 8.85 nm in nickel-like lanthanum ions. Isoelectronic scaling along the lanthanide series resulted in lasing at wavelengths as short as 7.36 nm. Simulations show that the collisionally broadened atomic transitions in these dense plasmas can support the amplification of sub-picosecond soft x-ray laser pulses.

1 Introduction

The generation of high average power coherent soft x-ray radiation on a table-top is of high interest for numerous applications. However, this is a challenging task, with both high harmonics and plasma-based soft x-ray lasers presently producing average powers in the microwatt range. In the case of plasma-based soft x-ray lasers the

J.J. Rocca · B.A. Reagan · Y. Wang · D. Alessi · K. Wernsing · B.M. Luther · M.A. Curtis · M. Berrill · D. Martz · S. Wang · L. Yin · F. Furch · M. Woolston · D. Patel · V.N. Shlyaptsev · C.S. Menoni
National Science Foundation ERC for Extreme Ultraviolet Science and Technology, Fort Collins, USA

J.J. Rocca (✉) · B.A. Reagan · Y. Wang · D. Alessi · K. Wernsing · B.M. Luther · M.A. Curtis · M. Berrill · D. Martz · S. Wang · L. Yin · M. Woolston · D. Patel · V.N. Shlyaptsev · C.S. Menoni
Department of Electrical and Computer Engineering, Colorado State University, Fort Collins, CO 80523, USA
e-mail: jorge.rocca@colostate.edu

J.J. Rocca · F. Furch
Department of Physics, Colorado State University, Fort Collins, CO 80523, USA

S. Sebban et al. (eds.), *X-Ray Lasers 2012*, Springer Proceedings in Physics 147,
DOI 10.1007/978-3-319-00696-3_35,

limitation has been their low repetition rate, with the first soft x-ray lasers operating at repetition rates ranging from a few shots per day to one shot per minute [1, 2]. In spite of significant progress that has greatly reduced the size of plasma-based soft x-ray lasers their average power has remained insufficient for applications requiring a high photon flux.

We discuss the operation of a *table-top* soft x-ray laser at 100 Hz repetition rate with a record-breaking average power of 0.15 mW at $\lambda = 18.9$ nm [3]. Gain-saturated, picosecond soft x-ray laser pulses are produced by amplification in a molybdenum plasma rapidly heated by ultra-short laser pulses with a tailored time profile from a diode-pumped chirped pulse amplification Yb:YAG laser that generates picosecond optical pulses of 1 J energy at a record 100 Hz repetition rate. Soft x-ray lasing was also obtained in the $\lambda = 13.9$ nm line of nickel-like silver. In a related effort we have worked in increasing the repetition rate of compact sub-10 nm lasers. We have generated gain-saturated picosecond SXRL pulses at 8.85 nm wavelength at 1 Hz repetition rate. Also, isoelectronic scaling along the lanthanide series has resulted in lasing at wavelengths as short as 7.36 nm [4].

The results increase by an order of magnitude the repetition rate of plasma-based soft x-ray lasers, demonstrating the feasibility of multi-milliwatt average power soft x-ray beams on a table-top. This increase in soft x-ray laser power will open up new scientific research and make possible advanced technology applications requiring high photon flux.

2 Demonstration of a 100 Hz Repetition Rate, Diode-Pumped, Gain-Saturated Soft X-Ray Laser

The average power of plasma-based, sub-20 nm soft x-ray lasers is currently limited by the repetition rate of the solid state optical lasers that drive them. State of the art chirped pulse amplification (CPA) laser systems producing picosecond duration pulses of the energy required to pump table-top soft x-ray lasers based on either neodymium-doped glass or titanium sapphire are limited to about 10 Hz repetition rate by thermal effects originating in the flashlamp-pumped gain medium of these driver lasers. The need for higher average power has motivated the development of high repetition rate directly diode-pumped CPA laser systems [5, 6]. We have developed a compact, directly diode-pumped CPA laser based on cryogenically-cooled Yb:YAG that produces 1 Joule, 5 ps FWHM duration, $\lambda = 1.03$ μm laser pulses at a record high repetition rate of 100 Hz. This laser was employed to drive a gain-saturated 18.9 nm soft x-ray laser in a nickel-like molybdenum plasma producing an average power of ~0.15 mW. Strong lasing at 13.9 nm was also achieved at 100 Hz repetition rate from a silver plasma. These results increase the repetition rate and average power of compact, sub-20 nm lasers by about an order of magnitude [5, 7] and are expected to extend the use of these lasers to new applications such as nanoscale surface imaging and nano-scale materials processing [8, 9].

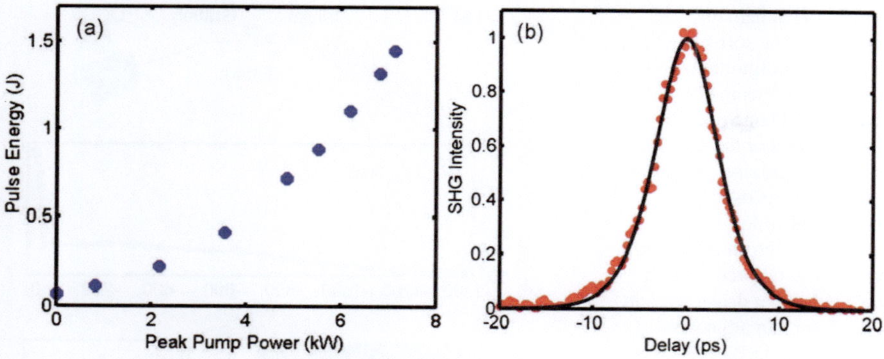

Fig. 1 (**a**) Output pulse energy prior to compression as a function of total peak pump power at 100 Hz repetition rate. The pump pulse duration was 1.5 ms. (**b**) Second-order autocorrelation of 1 J pulses after compression at 100 Hz. The *solid curve* is a sech^2 fit of the data, assuming this pulse shape the pulses have durations as short as 4.9 ps FWHM

The all diode-pumped CPA laser that drives the drives the soft x-ray laser consists of a Yb:KYW mode-locked oscillator, a grating stretcher, three stages of amplification, and a grating compressor. This pump laser occupies a single standard optical table, with vacuum chambers for pulse compression, focusing optics, and the soft x-ray laser target on an adjacent table. Stretched pulses from the oscillator are first amplified to the millijoule-level at 100 Hz repetition rate by a diode-pumped regenerative amplifier based on Yb:YAG at room temperature. The pulses are then amplified to about 100 mJ by a four-pass liquid nitrogen-cooled Yb:YAG amplifier pumped by a 500 W peak power fiber-coupled laser diode. The combination of a room temperature preamplifier with cryogenic power amplifiers allows us to exploit the excellent thermal properties and high gain of Yb:YAG at cryogenic temperature while avoiding excessive loss of bandwidth in the preamplifier [10, 11].

The third stage of amplification consists of two Yb:YAG crystals used in the active mirror configuration mounted on a single cryo-cooling head. Each crystal is pumped with ~4 kW peak power, 1.5 ms duration laser pulses at 100 Hz repetition rate from laser diode arrays. The 100 mJ level pulses are directed to make three passes through one crystal and two through the other and are amplified to 1.5 J at the maximum pump power. We define a pass as traveling through the crystal, reflecting off the high reflecting face and traveling back through the crystal. Figure 1(a) shows the pulse energy obtained prior to compression as a function of pump power in the 3rd stage. At 100 Hz repetition rate the beam quality remains high, with no observable thermal depolarization and only slight thermal lensing. Following amplification, the laser pulses are compressed with approximately 70 % efficiency by a dielectric grating pair to durations as short as 5 ps FWHM as can be seen from the second-order autocorrelation of Fig. 1(b). The system therefore produces 1 Joule, 5 ps laser pulses at 100 Hz repetition rate.

In order to achieve lasing on the $4d^1S_0$–$4p^1P_1$ transition of Ni-like Mo the output pulses of the pump laser were focused at a grazing incidence angle of 29° onto

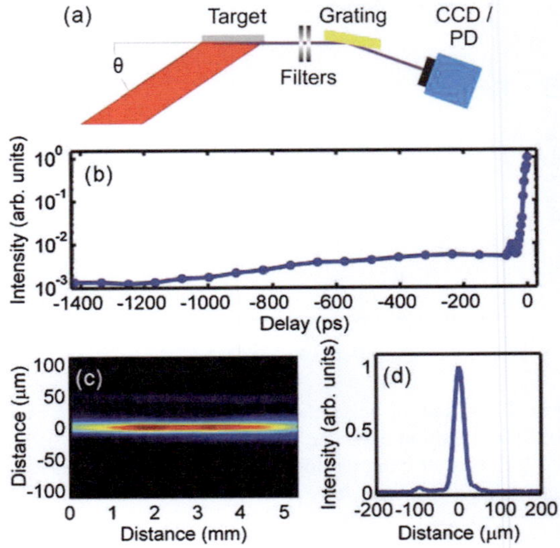

Fig. 2 (**a**) Schematic diagram of the soft x-ray laser pumping configuration and detection spectrometer with either a CCD or Si photodiode. For the 18.9 nm Mo laser a grazing incidence angle of 29° is used. (**b**) Second-order autocorrelation showing the temporally-tailored single laser pulse that drives the soft x-ray laser. An adjustable nanosecond ASE pedestal is followed by a 6 ps FWHM laser pulse. (**c**) and (**d**) Image and lineout of the high quality line focus used to generate the plasma amplifier

a 10 mm wide, polished Mo target as illustrated in Fig. 2(a). This inherently quasi-traveling wave geometry takes advantage of the refraction of the pump beam in the electron density gradient of the plasma to efficiently deposit the energy in the plasma region with optimum density for amplification [12, 13]. A single laser pulse with the tailored temporal profile shown in the autocorrelation of Fig. 2(b) was employed to create and heat the Mo plasma gain medium. Single laser pulses at grazing incidence have been previously demonstrated [14], but with different characteristics. The main laser pulse is preceded by an intentionally-created amplified spontaneous emission (ASE) pedestal with intensity that ramps from $\sim 10^{-3}$ to $\sim 2.5 \times 10^{-3}$ of the peak value in the 1.5 ns prior to the time of the peak intensity. This ASE originates in the regenerative amplifier, and we tuned both its intensity and duration to optimize the 18.9 nm laser output energy. The relative intensity of the ASE was adjusted by controlling the strength of the seed to the first stage amplifier, and its duration was controlled by adjusting the timing of a Pockels cell between the first and second amplifiers. Following this pedestal, a 6 ps duration pulse rapidly heats the plasma generating a transient population inversion and gain. We found that under the optimal conditions the ASE pedestal contained about 25 % of the total pulse energy. These temporally-tailored laser pulses were focused into 5.5 mm FWHM long by ~ 30 μm wide line on the Mo target. Figure 2(c) and (d) show the high quality line focus which is required for strong lasing. On-axis soft x-ray spectra of the plasma emission were measured using a flat-field, grazing incidence diffraction grating and an x-ray CCD with aluminum and zirconium foils used to reject visible plasma emission and scattered driver laser light. For high repetition rate operation of this laser, an x-ray sensitive photodiode was placed in the image plane of the spectrometer to monitor the emission at 18.9 nm.

Figure 3 shows the measured soft x-ray laser pulse energy as a function of plasma length. The laser pulse energy is seen to increase exponentially with a gain coeffi-

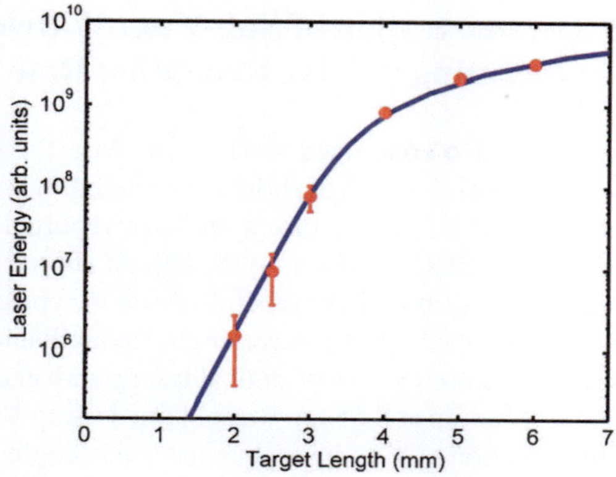

Fig. 3 Energy of the 18.9 nm laser pulses as a function of plasma length. Each *point* represents the mean of 10 shots. The *line* is a fit of the data with the Linford formula corrected for gain saturation [15] that yields a small signal gain coefficient of 43 cm^{-1}

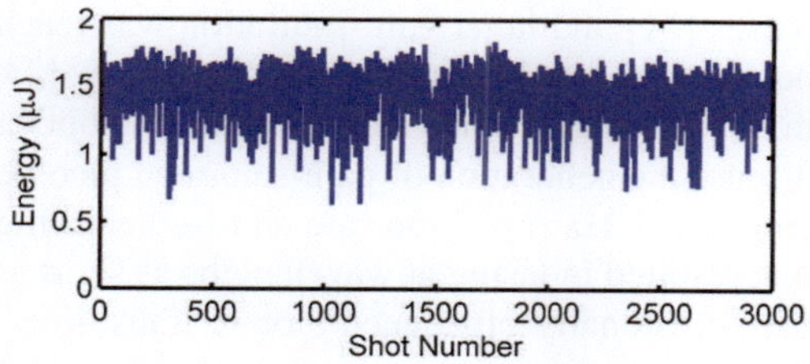

Fig. 4 Sequence of 3000 consecutive $\lambda = 18.9$ nm soft x-ray laser pulses at 100 Hz repetition rate

cient of 43 cm^{-1} until it saturates at a plasma column length of about 3.5 mm. A fit of the data with the Linford formula corrected for gain saturation [15] results in a gain-length product of 16.8, suggesting the rollover is due to gain saturation. Gain-saturated operation was verified by the measurement of a soft x-ray laser output pulse energy of ~1.5 μJ that, for a pulse duration of ~5 ps and near field beam spot size of a few tens of μm typical of these plasma amplifiers [16], corresponds to an intensity that comfortably exceeds the computed gain-saturation intensity [13]. Figure 4 shows 3000 consecutive $\lambda = 18.9$ nm laser pulses obtained operating the laser at 100 Hz repetition rate. For this series a 1 cm wide Mo slab target was translated at a constant velocity of 200 μm/s (2 μm per shot). Prolonged laser operation at this high repetition rate will require the use of a renewable target with a larger useful surface area, such as that demonstrated in [17]. The mean laser pulse energy in this series was 1.5 μJ with a shot-to-shot stability defined by a standard deviation of 11.5 %. The corresponding laser average power amounts to 0.15 mW, the highest value reported to date for a sub-20 nm coherent table-top source. Strong lasing was also obtained on the 13.9 nm line of Ni-like Ag by irradiating a 10 mm wide Ag target with a similar laser pump pulse, but impinging on target at a grazing incidence angle of 33°.

3 Demonstration of Sub-9 nm Wavelength Table-Top Laser Operating at 1 Hz Repetition Rate

We have also conducted work to increase the repetition the repetition rate of soft x-ray lasers operating at shorter wavelengths, below 10 nm. At the last International Conference on X-Ray lasers we have reported 1 Hz operation of a gain-saturated table-top SXRL at 10.9 nm [18, 19]. At shorter wavelengths the large pump energy required has limited the repetition rate to typically a shot per hour [20–24]. Soft x-ray lasing at sub-10 nm wavelengths in lanthanide ions was first demonstrated using several hundred joules of optical laser pump energy [21, 22]. Latter lasing in nickel-like lanthanum at 8.85 nm was obtained using 18 J pump pulses from a chirped pulse amplification (CPA) laser, but the gain-length product ($g \times l = 7.7$) obtained remained insufficient to reach gain-saturation [24]. Progress towards saturated lasing in this transition was reported at the last conference of this series [18, 25]. In turn, lasing at 7.36 nm in nickel-like Sm was initially demonstrated using 130 J pump pulses [21] and latter gain-saturation was reached using a ~40 J picosecond pump pulse added to a pre-pulse of similar energy [23]. More recently the lasing threshold in this line was reached using 36 J of total optical pump pulse energy [26]. Below we discuss the generation of gain-saturated picosecond SXRL pulses at 8.85 nm wavelength at 1 Hz repetition rate and isoelectronic scaling along the lanthanide series that resulted in lasing at wavelengths as short as 7.36 nm [4]. Simulations show that the collisionally broadened atomic transitions in these dense plasmas can support the amplification of sub-picosecond soft x-ray laser pulses.

The experiments were conducted irradiating solid slab targets with a sequence of two laser pulses from a $\lambda = 800$ nm Ti:sapphire CPA laser consisting of a normal incidence pre-pulse followed by a main picosecond pulse impinging at a grazing incidence angle of 35° with a traveling wave velocity of $\sim(1.04 \pm 0.03)$ c. The pump laser has four stages of amplification, of which the last two are pumped with the frequency doubled output of Nd-glass slab amplifiers designed to operate at repetition rates of up to 4 Hz [7]. The plasmas were created by normal incidence irradiation at $I = 6 \times 10^{12}\ \mathrm{W\,cm^{-2}}$ with a pre-pulse of 210 ps duration that is focused onto the target to form a ~30 μm × 6.4 mm FWHM line focus using the combination of a spherical and a cylindrical lens. The plasma created by this pre-pulse is allowed to expand to reduce the density gradient and is subsequently rapidly heated by irradiation at $I = 6 \times 10^{14}\ \mathrm{W\,cm^{-2}}$ with a 4 J pulse of 1–3 ps duration shaped into a line-focus of the same dimension. The target surface was tilted with respect to the axis of the short pulse to define a grazing incidence angle of $\theta = 35°$ for efficient heating [12, 13]. At this angle of incidence refraction couples the pump beam energy into the plasma region where $n_e = 5.7 \times 10^{20}\ \mathrm{cm^{-3}}$. To assist in achieving efficient pumping we developed a new focusing geometry designed to create a plasma column of constant width along the target (Fig. 5a). In the conventional implementations of grazing incidence pumping in which a parabolic or spherical mirror is used to focus the beam [12, 13, 27–29], the tilted target intercepts the short pulse pump beam at different distances from the beam waist. This creates a "butterfly" shape line-focus on target that for small f-number systems can deposit

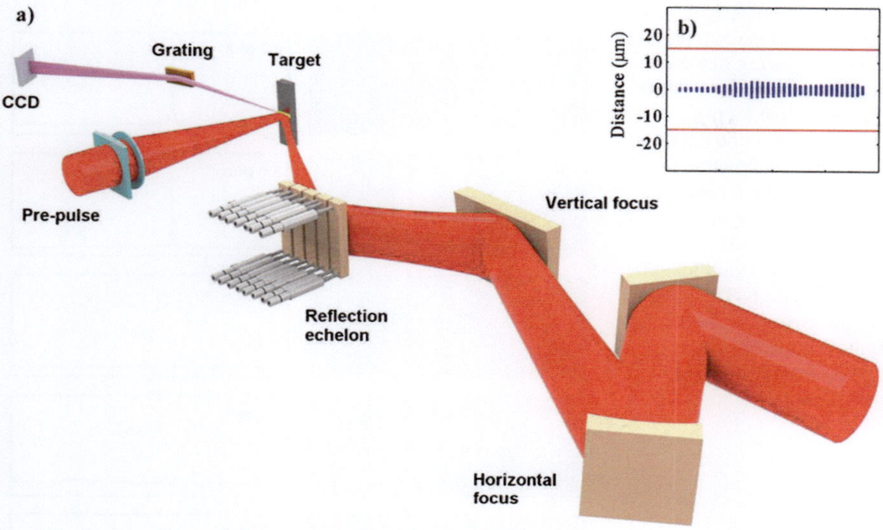

Fig. 5 Schematic representation of the set up used in the sub-9 nm SXRL experiments. (**a**) A pair of cylindrical mirrors focuses the short pulse beam at 35° grazing incidence on target creating a line-focus of uniform width. A reflection echelon is used to obtain quasi-traveling wave excitation. (**b**) Raytrace simulation of the line-focus on target. All rays fall within the line width defined by diffraction (*red lines*)

part of the pump energy outside the useful gain column, decreasing the gain. To overcome this limitation we focused the short pulse beam with a pair of cylindrical mirrors designed to create a line-focus of constant width along a target irradiated at grazing incidence [4]. A reflection echelon composed of five mirror segments was used to obtain a quasi-traveling wave excitation velocity of $\sim(1.04 \pm 0.03)$ c. Measurements at 1-Hz repetition rate were obtained continuously moving the target at a speed of 200 μm/s to renew the surface after each shot. Zirconium filters with thickness up to 2.1 μm were used to avoid saturating the detector and to eliminate visible plasma light.

Figure 6a shows a series of on-axis spectra as a function of plasma column length for a lanthanum plasma column created depositing 4 J of short pulse energy and a total of 7.5 J of optical pump energy on target at the irradiation conditions described above. Strong amplification is observed in the 8.85-nm $4d^1S_0 \rightarrow 4p^1P_1$ transition of Ni-like La. Figure 6b shows that the intensity grows by nearly four orders of magnitude as the plasma length increases from 1.7 mm to 6.2 mm. Saturation of the intensity is observed to have an onset at a plasma column length of ~4 mm. A fit of the data with an expression that takes into account gain saturation [15] yields a gain coefficient of 33 cm^{-1} and a gain-length product of 14.6. The energy of the most intense laser pulses was estimated to be 2.7 μJ from the CCD counts. Gain-saturated operation was confirmed by measuring the SXRL flux at the exit of the amplifier. Near field imaging of the 8.85 nm laser amplifier output determined a laser fluence of ~0.6 J cm^{-2} for the most intense laser shots. Assuming the laser pulse duration of

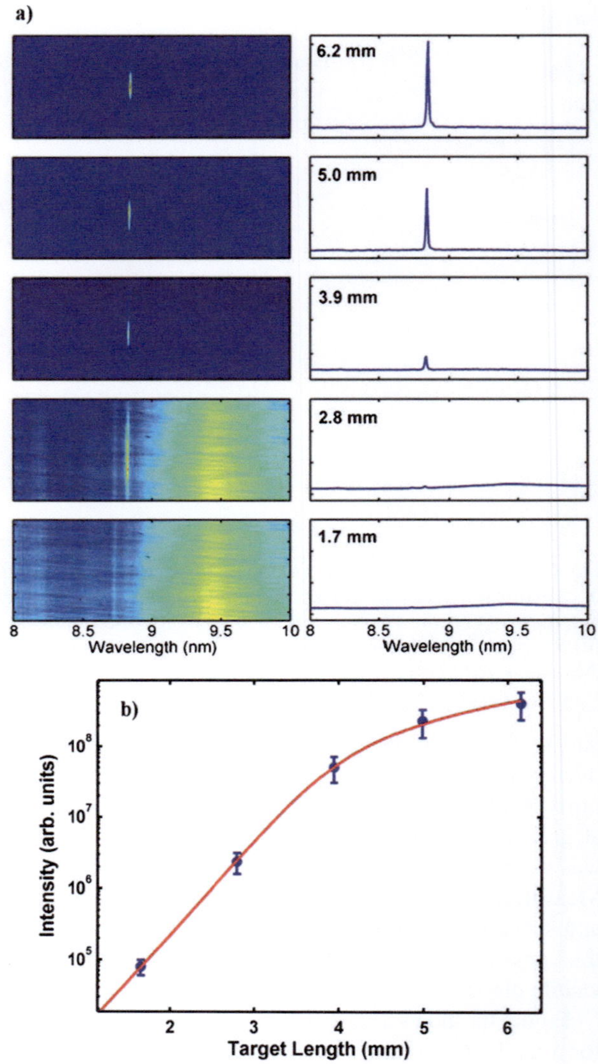

Fig. 6 End-on spectra of a line focus lanthanum plasma column showing saturated amplification in the 8.85-nm line of Ni-like La (**a**). (**b**) Intensity of the 8.85-nm laser line as a function of plasma column length. The *line* is a fit of the data that yields a gain coefficient of 33-cm^{-1} and a gain-length product of 14.6. The *points* are the average of 10 laser shots and the *error bar* represents one standard deviation

2.5 ps from the simulations, the laser intensity at the amplifier exit is estimated to be 2.4×10^{11} W cm^{-2}, which significantly exceeds the computed saturation intensity of 1.5–3 $\times 10^{10}$ W cm^{-2} [4]. The use of similar irradiation conditions resulted in lasing at 8.5 nm in Ni-like Ce, 8.2 nm in Ni-like Pr, and 7.9 nm in Ni-like Nd (Fig. 7). Lasing was also observed at 7.36 nm in Ni-like Sm by reducing the main pump pulse duration to 1.1 ps. The laser emission at 8.5 nm in Ni-like Ce is very strong, with an output energy >1 μJ. Future advances in diode-pumped CPA pump lasers promise to lead to the realization of high average power table-top lasers at sub 10-nm wavelengths.

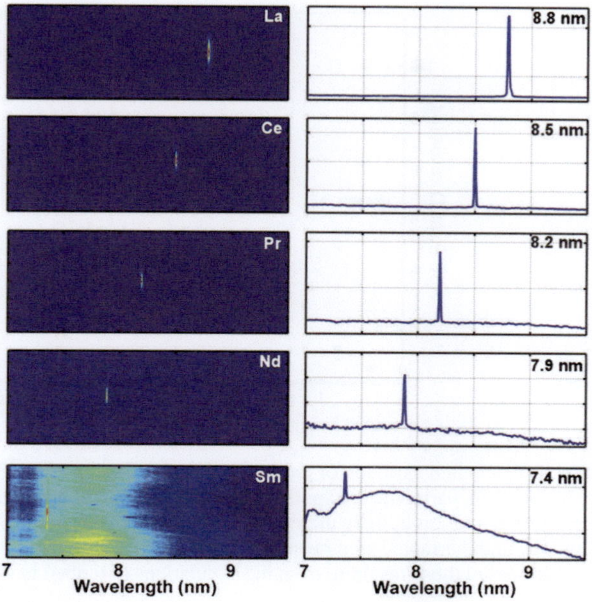

Fig. 7 End-on spectra showing lasing at progressively shorter wavelengths in the $4d^1S_0 \rightarrow 4p^1P_1$ line of nickel-like lanthanide ions, down to 7.36-nm in nickel-like samarium

Another aspect of interest in these amplifiers is the fact that they can support the amplification of sub-picosecond seed XRL pulses. Model simulations indicate amplification at 8.85 nm occurs in a plasma region where the plasma density is 5–9 $\times$ 10^{20} cm^{-3} and the ion temperature ~150 eV. At the plasma conditions for peak gain the laser line profile is defined by collisional broadening ($\Delta\nu_L = 1.7 \times 10^{12}$ Hz) and Doppler contributions ($\Delta\nu_D = 2.8 \times 10^{12}$ Hz) that define a Voigt linewidth of $\Delta\nu = 3.8 \times 10^{12}$ Hz that can support the amplification of sub-picosecond pulses. The variation of the bandwidth and pulse duration of an injected high harmonic seed pulse as it propagates thought the lanthanum plasma amplifier was computed with a 3D ray-trace post processor that couples the amplification and ray propagation equations with a time and space depended atomic physics model that includes all relevant levels and stimulated emission. The ray trace code is capable of fully resolving the temporal duration and bandwidth of both the amplified stimulated emission and the amplified seed beam. Pulse broadening of a seed due to line-narrowing is included using the Fourier transform of the amplified bandwidth accounting for the group velocity delay caused by the gain profile. Figure 8 shows the computed pulsewidth variation of an injected 8.85 nm, 30 fs duration high harmonic seed pulse as it propagates thought the lanthanum plasma amplifier for plasma conditions that result in an ASE laser output energy similar to that obtained in the experiment. Simulations conducted for a range of different pumping parameters predict output pulse durations in the range of 500–700 fs. The computed variation of the pulse duration of the ASE laser pulse as a function of plasma column length is also shown.

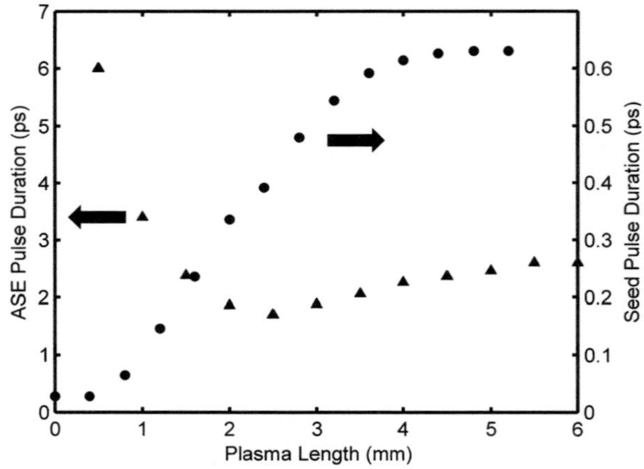

Fig. 8 Computed variation of the FWHM pulse duration of the 8.85 nm ASE laser (*circles*), and amplified high harmonic seed pulse (*triangles*) as a function lanthanum plasma column length. The seed pulse duration initially increases due to linewidth narrowing and later by saturation broadening. The high harmonic seed pulse is assumed to have an energy of 0.1-nJ and the gain and plasma conditions are those resulting from the simulations described in the text

4 Summary

In summary, we have demonstrated 100 Hz generation of soft x-ray laser pulses of microjoule energy at $\lambda = 18.9$ nm, generating a record high average power 0.15 mW soft X-ray laser beam on a table-top. This was accomplished by developing an all diode-pumped CPA laser system based on cryogenic Yb:YAG amplifiers that produce 1 J, 5 picosecond duration pulses at 100 Hz repetition rate. We have also demonstrated bright gain-saturated sub-9 nm wavelengths soft x-ray laser pulses at 1 Hz repetition rate for the first time and obtained laser amplification at wavelengths as short as 7.36 nm. This increase in soft x-ray laser power will open up new scientific research and make possible advanced technology applications requiring high photon flux.

Acknowledgements This work was supported by the NSF ERC for Extreme Ultraviolet Science and Technology under NSF Award EEC 0310717 and by the Chemical Sciences, Geosciences and Biosciences Division, Office of Basic Energy Sciences, U.S. Department of Energy using equipment developed under NSF Award MRI-ARRA 09-561, MB acknowledges the support of an Eugene P. Wigner Fellow for staff members Oak Ridge National Laboratory managed by UT-Battelle, LLC, for the U.S. Department of Energy under Contract DE-AC05-00OR22725.

References

1. Suckewer, S., et al.: Phys. Rev. Lett. **55**, 1753 (1985)
2. Mathews, D.L., et al.: Phys. Rev. Lett. **54**, 110 (1985)

3. Reagan, B.A., et al.: Opt. Lett. (in press)
4. Alessi, D., et al.: Phys. Rev. X **1**, 021023 (2011)
5. Furch, F.J., et al.: Opt. Lett. **34**, 3352 (2009)
6. Tümmler, J., et al.: Opt. Lett. **34**, 1378 (2009)
7. Martz, D., et al.: Opt. Lett. **35**, 1632–1634 (2010)
8. Urbanski, L., et al.: Opt. Lett. (in press)
9. Bravo, H., et al.: IEEE J. Sel. Top. Quantum Electron. **18**, 443 (2012)
10. Curtis, A.H., et al.: Opt. Lett. **36**, 2164 (2011)
11. Reagan, B.A., et al.: IEEE J. Quantum Electron. **48**, 827 (2012)
12. Keenan, R., et al.: Phys. Rev. Lett. **94**, 103901 (2005)
13. Luther, B.M., et al.: Opt. Lett. **30**, 165 (2005)
14. Kim, H.T., et al.: Phys. Rev. A **77**, 023807 (2008)
15. Lin, J.Y., et al.: Opt. Commun. **158**, 55 (1998)
16. Berrill, M., et al.: Opt. Lett. **35**, 2317 (2010)
17. Weith, A., et al.: Opt. Lett. **31**, 1994 (2006)
18. Rocca, J.J., et al.: In: 12th International Conference on X-Ray Lasers, Guanju, Korea (2010)
19. Alessi, D., et al.: Opt. Lett. **35**, 414–416 (2010)
20. MacGowan, B.J., et al.: Phys. Fluids B **4**, 2326 (1992)
21. Daido, H., et al.: Phys. Rev. Lett. **75**, 1074 (1995)
22. Daido, H., et al.: Opt. Lett. **21**, 958–960 (1996)
23. Zhang, J., et al.: Science **276**, 1097 (1997)
24. Kawachi, T., et al.: Phys. Rev. A **69**, 033805 (2004)
25. Balmer, J.E., et al.: In: 12th International Conference on X-Ray Lasers, Guanju, Korea (2010)
26. Zimmer, D., et al.: Phys. Rev. A **82**, 013803 (2010)
27. Wang, Y., et al.: Phys. Rev. A **72**, 053807 (2005)
28. Grunig, M., et al.: Opt. Commun. **282**, 267–271 (2009)
29. Alessi, D., et al.: Opt. Lett. **35**, 414 (2010)

Chapter 36
Optical Correction of X-Ray Laser Illumination for Short-Wavelength Microscopy

Mabel Ruiz-Lopez, Felix Staub, and Davide Bleiner

Abstract A critical aspect in the design of a short-wavelength microscope is the illumination profile and stability. Experimental measurements of the pointing stability and divergence and optical corrections of X-ray laser source are shown. The optical corrections provided one order of magnitude improvement in the illumination performance.

1 Introduction

The continuing progress of the nano-technology has increased the interest on nano-imaging tools. Short-wavelength microscopy allows resolving tens of nanometers by pushing down the diffraction limit. Numerous groups have developed imaging-systems for lithography and nano-inspection applications [1], but the most of them are not for laboratory scale operation. Compact sources from plasma light are interesting to became lab-scale enabling tools. The work in the design of a microscope begins with the understanding of the light's source properties: beam size, divergence, pointing stability, etc. [2]. Aim of this work was to characterize the X-ray laser illumination for the construction of a short-wavelength microscope.

2 Set-ups for Source Characterization

The transmission microscope that is being developed is composed of three optical elements: (i) condenser, (ii) sample and (iii) objective. The illumination should match the geometry of the condenser and provide stable and homogeneous filling. For diagnosing the beam stability and divergence of the pulse at the position of

M. Ruiz-Lopez (✉) · F. Staub · D. Bleiner
Institute of Applied Physics, University of Bern, Sidlerstrasse 5, 3012, Bern, Switzerland
e-mail: maria.ruiz-lopez@iap.unibe.ch

D. Bleiner
e-mail: bleiner@iap.unibe.ch

S. Sebban et al. (eds.), *X-Ray Lasers 2012*, Springer Proceedings in Physics 147,
DOI 10.1007/978-3-319-00696-3_36,

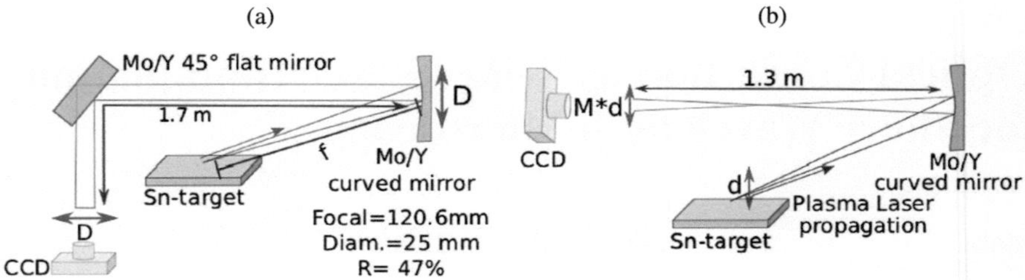

Fig. 1 (**a**) Experimental set up using a micro-collector for correction of pointing stability and divergence. The curved mirror is at a distance equal to the focal length to collimate the beam. (**b**) Experimental set up for a relay-imaging. The CCD captures an image of the source magnified 10-fold. Drawings not to scale

Fig. 2 Illumination profile in the image plane obtained with different TCE pre-pulse schemes: (**a**) single orthogonal long pre-pulse [3], (**b**) dual collinear short pre-pulses [4]

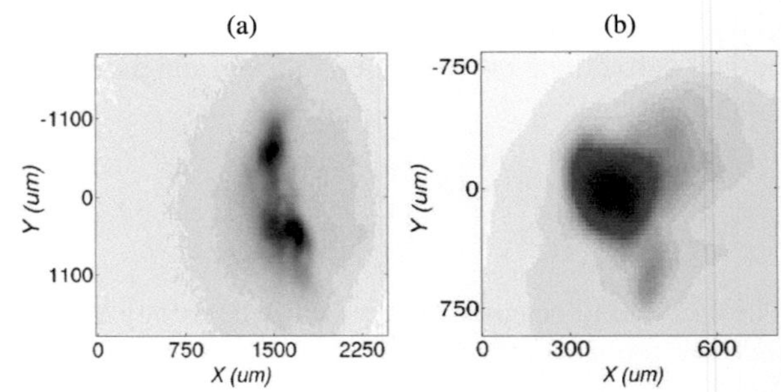

the condenser, two different set-ups were used. In Fig. 1a the set-up employed to compare the stability of a non-collimated and collimated pulse is shown. The setup in Fig. 1b was used for investigations on relay-imaging (near field). The curved mirror images the plasma source on a phosphor screen which converts the short-wavelength to the visible range. Before the CCD detector, a 150 nm-thick layer of zirconium is used to block optical background light. The CCD camera has 9 μm pixel size.

3 Characterization of Illumination and Optical Correction

Two transient collisional-excitation (TCE) schemes of pre-pulse delivery were compared, i.e. a single orthogonal long pre-pulse (Fig. 2a) versus a GRIP dual collinear short pre-pulse (Fig. 2b). One observes that the illumination profile improves with the latter scheme. Pointing stability and divergence were determined and optical correction setups were implemented. The first setup utilized was a telecentric-collector consisting in a concave normal-incidence Mo/Y mirror positioned at the focal plane distance from the source. Figure 3 shows that the improved pointing stability is a factor 4 (Fig. 3a) and the divergence decreases a factor of 10 (Fig. 3b).

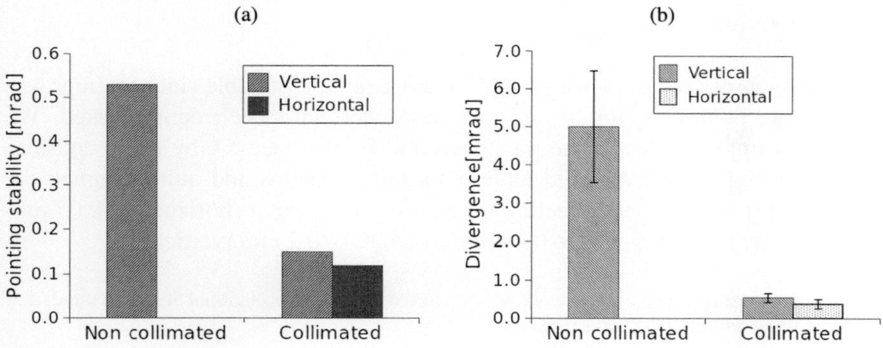

Fig. 3 (**a**) Pointing stability before and after collimation. (**b**) Divergence before and after collimation

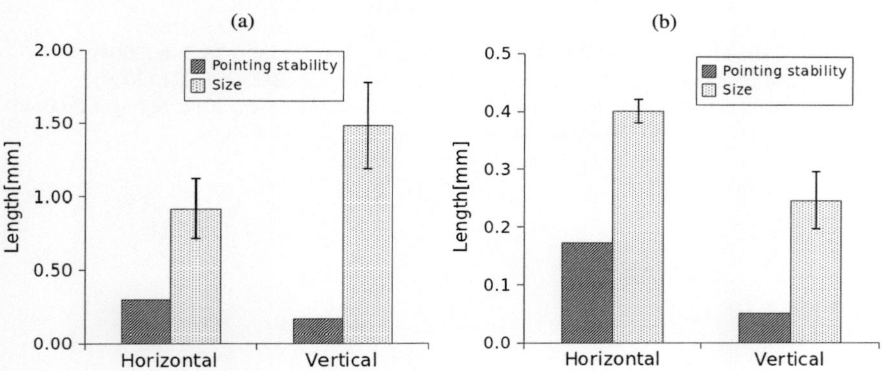

Fig. 4 (**a**) Size and pointing stability for a single orthogonal long pre-pulse. (**b**) Size and pointing stability in the current dual collinear pre-pulse. A factor of 1.5 (horizontal)–4 (vertical) improved stability is obtained as well as a decrease of factor of 2 (horizontal)–6 (vertical) in the divergence, when using the scheme with dual collinear pre-pulse

The second optical correction setup deployed relay-imaging. On the detector the near-field image of the X-ray source, magnified 10-fold, was acquired. Comparison of the two pre-pulse schemes showed that the present dual collinear pre-pulse set-up gives factor of 1.5 (horizontal)–4 (vertical) more stable illumination. The divergence was corrected a factor of 2 (horizontal)–6 (vertical). The smaller divergence (overall in the vertical dimension) implies an increment in the illumination brightness (Fig. 4). Pointing stability defines the reproducibility of the pulse position and must be the most precise possible. Both stability and divergence, contribute to the source etendue, which is important for the brightness of the source. All the results are in agreement with our computational study [5].

4 Conclusions

Lab-scale sources for short-wavelength microscopy are possible since an improvement in the pointing stability and the divergence have been demonstrated. We showed an improvement in the pointing stability of a factor 4 by using collimation of the light. The near-field indicates a homogeneous and quite symmetrical source profile and allows correcting the stability to 0.2 mm (horizontal)–0.05 mm (vertical), and the divergence to 0.4 mm (horizontal)–0.2 mm (vertical).

Acknowledgements The present work was supported by the Swiss National Science Foundation under the grant number PP00P2-133564/1.

References

1. Goldberg, K.A., Mochi, I.: J. Vac. Sci. Technol. B **28**(6), C6E1(2010)
2. Präg, A.R., Kozlovà, M., Rus, B., Rohlena, K.: Eur. Phys. J. B **26**, 59–65 (2003)
3. Staub, F., Braud, M., Balmer, J.E., Nielsen, J., Bajt, S.: Appl. Phys. B **78**, 7–8 (2004)
4. Grünig, M., Imesch, C., Staub, F., Balmer, J.E.: Opt. Commun. **282**, 267–271 (2009)
5. Masoudnia, L., Bleiner, D.: In: Sebban, S., et al. (eds.) X-Ray Lasers 2012. Springer Proceedings in Physics, vol. 147. Springer, Cham (2013). Chapter 27 in this book

Chapter 37
Repetitive XUV Discharge-Pumped Laser at 46.9 nm

Jiri Schmidt, Karel Kolacek, Oleksandr Frolov, Vaclav Prukner, and Jaroslav Straus

Abstract For testing and application purposes we built a new small Marx generator capable to run in a repetitive regime. Its repeating frequency is currently up to 1 Hz. The generator is covered by metal sheets and feeds the capillary discharge device. The first experimental measurements were performed for a 400 mm long capillary and these results are presented.

1 Introduction

Capillary-discharge-based XUV lasers are sources that appeared about twenty years ago. These lasers, due to their current compactness, good beam quality, and stable operation have become useful tool for applications (e.g. for XUV lithography, for high-resolution microscopy, and for high-plasma-density measurements).

The first demonstration of XUV Ne-like argon laser working on 46.9 nm was published in [1]. Rocca's experiment opened a new way to the development of simpler and compact XUV lasers based on excitation pumping by capillary discharge.

Our research group was inspired by many applications of coherent XUV radiation and began to build the fast capillary discharge device (CAPEX) just about fifteen years ago. The experimental device is currently capable to reach current pulse amplitude of ~70 kA (~240 ns period) in ~230 mm long gas-filled capillary. The short (less than 2 ns) intense spike of the soft X-ray signal has been detected by a vacuum photodiode. This short pulse evidently corresponds to amplified spontaneous emission of neon-like argon line. The time-resolved (~30–50 ns exposition) XUV spectra recorded the strong spectral line, which dominates the whole observed spectrum in the range of 10–110 nm [2]. The wavelength of this dominating spectral line is just 46.9 nm. According to our experiences with CAPEX the new apparatus CAPEX-U (-Upgrade) was assembled approximately seven years ago [3] with intention to find lasing at shorter wavelength on discharge-pumped installations. Both apparatuses the CAPEX and the CAPEX-U [4] are fully operational.

J. Schmidt (✉) · K. Kolacek · O. Frolov · V. Prukner · J. Straus
Pulse Plasma Systems Department, Institute of Plasma Physics AS CR, v.v.i., Za Slovankou 1782/3, 182 00 Prague, Czech Republic
e-mail: schmidt@ipp.cas.cz

S. Sebban et al. (eds.), *X-Ray Lasers 2012*, Springer Proceedings in Physics 147,
DOI 10.1007/978-3-319-00696-3_37,

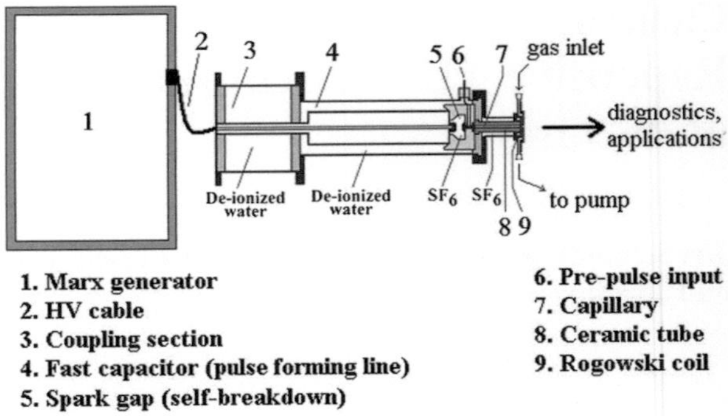

Fig. 1 Schematic drawing of the CAPEX device

2 Experimental Set-up

The CAPEX apparatus consists of a Marx generator, a coupling section, a pulse forming line (fast cylindrical capacitor), a main spark gap and a capillary. Experimental setup of the device is shown in Fig. 1.

As a power supply for the capillary discharge experiment the fully screened Marx generator is used. It consists of two parts—two bodies of commercial Marx generators GIN 400-0.06/5 of the soviet origin, which were equipped with a new system of spark-gaps. Spark-gaps are filled with nitrogen (SF_6 alternatively). The whole generator was successfully tested up to the charging voltage 50 kV (erected voltage 0.5 MV), i.e. up to the stored energy 0.75 kJ.

Significant part of the CAPEX device is the capillary. At present we use ceramic capillaries made from alumina (Al_2O_3). The 400 mm (232 mm alternatively) long capillary is directly attached to the main spark gap (having one common electrode). The gas filling and pumping part is attached to the outer end of the capillary (through the orifice in the outer electrode). The soft X-ray measurements were performed in the argon-filled-capillary with inner diameter of 3.2 mm. The capillary is placed in a shielding and circuit closing metallic cylinder of the diameter of 80 mm. The generated radiation was brought out through $\sim$0.5 mm orifice in the outer electrode. Typical values of gas (argon) pressure inside the capillary prior to breakdown are in the range from 10 Pa to 100 Pa. More details of the experimental setup can be found in [2, 5].

3 XUV Radiation

The first experimental measurements were performed for a 400 mm long capillary. The XUV radiation emitted during the radial compression of the plasma column was measured using a vacuum diode.

The vacuum photodiode with Au photocathode was powered by 500 V. HP 54542C oscilloscope (0–500 MHz analog bandwidth, 700 ps rise-time, 2 GSa/s maximum sampling rate in each of four channels) was used for recording signals. The diode was placed at a distance of $\sim$2 m from capillary output.

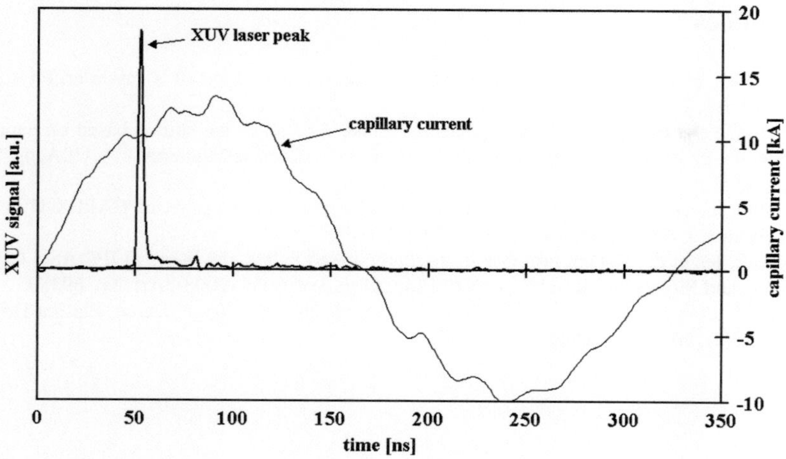

Fig. 2 Temporal evolution of XUV emission in the correlation with the main capillary current

The dependence of XUV pulse shape in the correlation with the capillary current is shown in Fig. 2. The full width at half maximum (FWHM) of the large short intense spike of the soft X-ray signal is approximately ∼1.5 ns and it appears ∼30 ns prior to the capillary current maximum (the current quarter-period is ∼80 ns). This short pulse detected by vacuum diode corresponds to amplified spontaneous emission of neon-like argon line ($\lambda = 46.9$ nm). The amplitude and period of the capillary current pulse are approximately 15 kA and 320 ns. The main capillary current pulse was preceded by a pre-ionization current pulse of the amplitude ∼160 A and of the duration 20 μs to produce a uniform pre-plasma. The capillary pre-pulse was generated by an external driver [5].

4 Conclusion

The paper reports on some technological modification (the new repetitive high voltage Marx generator) and the recent experimental measurement (400 mm long capillary). The new small Marx generator is capable to run in the repetitive regime with frequency up to a few Hz.

The optimal conditions (initial pressure of argon, capillary pre-pulse amplitude and duration, etc.) for lasing were found for 400 mm long capillary. The short intense spike was detected by vacuum diode in XUV region. This pulse corresponds to amplified spontaneous emission of neon-like argon line ($\lambda = 46.9$ nm).

Acknowledgements This work was performed under auspices and with the support of the Ministry of Education, Youth, and Sports of the Czech Republic (contract LA08024).

References

1. Rocca, J.J., et al.: Demonstration of a discharge pumped table-top soft X-ray laser. Phys. Rev. Lett. **73**(16), 1236 (1994)
2. Schmidt, J., et al.: Strong amplification of Ne-like Ar line in the source based on capillary discharge. In: IEEE 15th International Pulsed Power Conference, Monterey, CA, USA, pp. 370–373 (2005)
3. Schmidt, J., et al.: Comparing of calculated and experimental results of CAPEX-U device. Czechoslov. J. Phys., Suppl. B **56**, 371–376 (2006)
4. Kolacek, K., et al.: Recent progress in discharge-based soft X-ray lasers at IPP ASci CR. In: Atomic and Molecular Pulsed Laser VII. Proc. SPIE, vol. 6938 (2008), Art. No. 693805
5. Schmidt, J., et al.: Soft X-ray emission of a fast-capillary-discharge device. Plasma Devices Oper. **13**(2), 105–109 (2005)

Chapter 38
Line-Focus Generation for X-Ray Laser Pumping

F. Staub, D. Bleiner, F. Jia, and J.E. Balmer

Abstract Down-scaling of X-ray lasers towards shorter wavelengths imposes the need for higher irradiance of the pump laser systems. Typically, for a factor of 2 reduction in wavelength, a factor of ~10 increase in irradiance was so far required. This discourages the use of refractive optics after the compressor stage of the high-power CPA laser systems commonly used for pumping X-ray lasers due to increasing nonlinear phase distortion (B-Integral). We present ray-tracing calculations of alternative focusing schemes applying reflective optics for grazing-incidence line-focus generation. The proposed mirrors deliver a narrow and shortened line focus from a high-aperture input beam, yielding an increase of irradiance on target.

1 Introduction

Grazing-incidence pumping (GRIP) combined with pico-second pulses puts high demands on the optics forming a line focus from the beam exiting the pulse compressor stage of a chirped-pulse amplifier (CPA) laser system. The beam diameter is usually determined by the damage threshold of the last compressor grating. This results in a large-aperture, high-irradiance beam, which has to be converted to a high-aspect-ratio line focus. The development of collisionally excited X-ray lasers towards shorter wavelengths asks for increasing irradiance on target. For a given laser energy (and power) this can be addressed by optimizing the length and the quality of the pump line focus intensity distribution.

2 Focusing Schemes

The simplest focusing optics consists of a tilted spherical mirror, which geometrically forms a perfect line parallel to the collimated incident beam. However, the length of the focal line can not be adjusted independently of the GRIP angle and has

F. Staub (✉) · D. Bleiner · F. Jia · J.E. Balmer
Institute of Applied Physics, University of Bern, Sidlerstrasse 5, 3012 Bern, Switzerland
e-mail: felix.staub@iap.unibe.ch

S. Sebban et al. (eds.), *X-Ray Lasers 2012*, Springer Proceedings in Physics 147,
DOI 10.1007/978-3-319-00696-3_38,

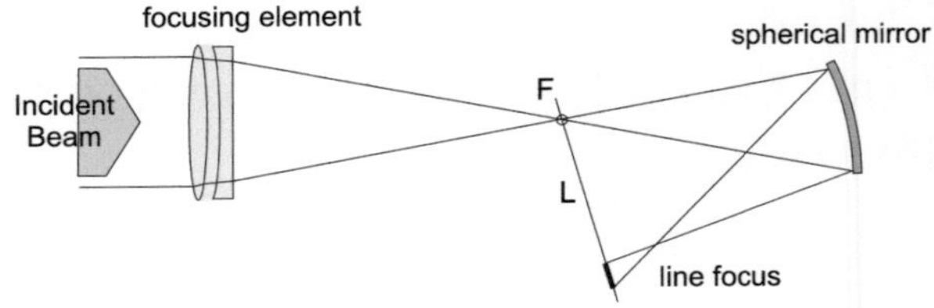

Fig. 1 Line focus generation by imaging an intermediate point focus F [1, 2]

Table 1 B-integrals of 10 TW in a 132-mm beam

Nonlinear medium	B
1:1.7 reduction telescope	17.5
focusing objective	4.6
1-inch vacuum window	3.8
5 m air	1

to be adapted by the numerical aperture (NA) of the sphere. The NA can be reduced by means of a reduction telescope. More versatile is the method of imaging a point source in finite distance from the spherical mirror to adjust the length of the line focus and the GRIP angle via the location of the point source with respect to the line focus [1, 2], as shown in Fig. 1. Both methods need additional optics.

The high irradiance of the beam exiting the compressor of a CPA system can lead to severe nonlinear phase distortion in subsequent transmission optics (B-integral), given by

$$B = 2\frac{\pi}{\lambda} \int \gamma I(z)\, dz$$

where λ is the wavelength, γ the nonlinear index of refraction in SI units and I the irradiance. Examples of the B-integral for typical transmission optics used in our setup are given in Table 1 for a 10-TW beam having a diameter of 132 mm.

As can be seen, the B-Integral in glass dominates over propagation in air. The propagation in glass after pulse compression limits the quality of the (line) focus, as the focus breaks up for B between 4 and 10 [3]. In order to keep the B-integral as low as possible, transmission optics should be avoided.

As an alternative to refractive optics, the intermediate focus can be formed by an off-axis parabola, which is expensive. A further possibility is to use a tilted on-axis parabola, forming a distorted intermediate focus. A spherical mirror is used to image the distorted spot, resulting in a widened propeller-shaped line-focus. Ray-tracing calculations showed that the focal line is rotated with respect to the setup shown in Fig. 1, as indicated in Fig. 2. As a consequence, the line L, indicating the direction of the narrowest line focus, does not intersect the point F, but is tilted in the direction of the tilt φ of the parabolic mirror. In Fig. 2a the spherical mirror is tilted to form a down-folded line focus, while it is up-folded in Fig. 2b. The distance between the two mirrors is roughly the sum of their focal lengths. The narrowest focus is obtained by optimizing the distance between the parabolic and spherical mirror for each GRIP angle.

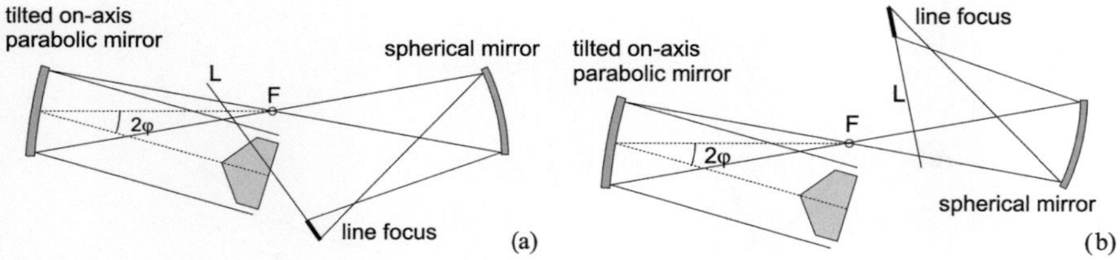

Fig. 2 Line focus generation by imaging a distorted intermediate focus F formed by an on-axis parabola tilted by the angle φ. On the *left* (**a**), the spherical mirror is tilted to form a down-folded line focus, while the setup on the *right* (**b**) forms an up-folded line focus

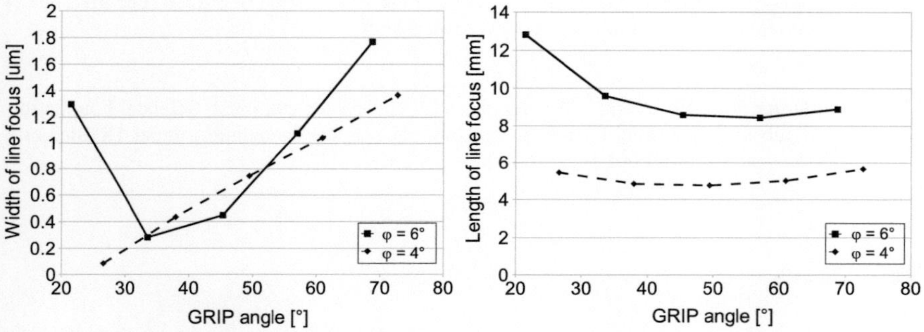

Fig. 3 Calculated geometrical size of the line focus of the setup shown in Fig. 2a. The tilt angle φ of the parabolic mirror is 6° (*solid line*) and 4° (*dashed line*)

Calculations were performed for the focusing schemes of Fig. 2 using a diameter of the incident beam of 132 mm in a top hat profile. The focal length of the parabolic mirror is $f_1 = 609.6$ mm, the focal length of the spherical mirror $f_2 = 304.8$ mm, respectively. The ratio $f_1/f_2 = 2$ was chosen in order to achieve identical beam diameters on both mirrors. The pump laser wavelength is 1054 nm, yielding a diffraction limit of the line-focus width of 6 μm.

Figures 3 and 4 show the calculated length and width of the line focus as a function of GRIP angle for the setups of Figs. 2a and 2b, respectively. Each data point is optimized for minimum width of the focal line. The geometrical line width is seen to stay well below the diffraction limit over a wide range of GRIP angles. The line foci of Figs. 3 and 4 are not identical, as the intensity distribution at F is not symmetric. Consequently, the spherical mirror forms different line foci for the up or down-folded configuration of Fig. 2. The length of the line focus shows less variation for Fig. 2a, while Fig. 2b offers shorter lines for GRIP angles below 60°.

In conclusion, a scheme for line-focus generation is presented using a standard parabolic mirror and a spherical mirror. Ray-tracing calculations imply that it is possible to keep the distortions well below the diffraction limit without the need for an expensive off-axis parabolic mirror. Due to the absence of transmission optics, there is no contribution to the B-Integral of the focusing optics.

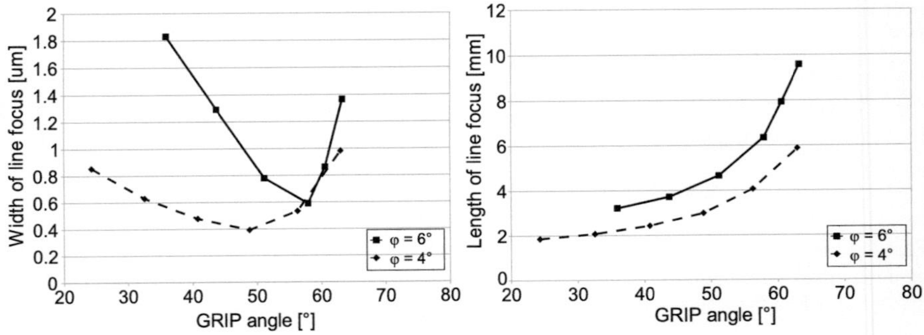

Fig. 4 Calculated geometrical size of the line focus of the setup shown in Fig. 2b. The tilt angle φ of the parabolic mirror is 6° (*solid line*) and 4° (*dashed line*)

Acknowledgements This work was supported by the Swiss National Science Foundation (SNSF). F. Staub was supported by the Swiss National Science Foundation under D. Bleiner's Förderungsprofessur (grant PP00P2-133564/1).

References

1. Ross, I.N., et al.: Some optical designs for the generation of high quality line foci. J. Phys. E, Sci. Instrum. **18**, 169 (1985)
2. Ross, I.N., et al.: Design and performance of a new line focus geometry for x-ray laser experiments. Appl. Opt. **26**(9), 1584 (1987)
3. Seznec, S., et al.: Opt. Commun. **87**, 331–339 (1992)

Chapter 39
Critical Components for XUV Probing of Laser Driven Shocks

C. Stehlé, R. Lefèvre, U. Chaulagain, N. Champion, P. Barroso, F. Reix, P. Jagourel, J. Larour, E. Meltchakov, R. Mercier, F. Delmotte, M. Kozlova, J. Nejdl, M. Krus, J. Dostal, J. Prokupek, C. Constancias, F. Suzuki-Vidal, and O. Acef

Abstract Radiative shocks can be produced in gases using high-energy lasers. As the electron density may be higher than the critical density for visible light, radiography with soft x-ray radiation becomes very promising technique to probe these shocks. Feasibility of this method has been proven employing zinc soft x-ray laser at 21.2 nm as backlighter. The experiment has shown high requirements on quality of the imaging optics and windows of the gas filled cell.

1 Introduction

Radiative shocks (RS) are strongly hypersonic shocks, where the high temperature generates a strong radiation structuring the shock. For instance, these shocks

---

C. Stehlé (✉) · R. Lefèvre · U. Chaulagain · N. Champion
Lerma, Observatoire de Paris, UPMC, CNRS, Meudon, France
e-mail: chantal.stehle@obspm.fr

P. Barroso · F. Reix · P. Jagourel
GEPI, Observatoire de Paris, CNRS, Paris, France

J. Larour
LPP, Ecole Polytechnique, UPMC, Palaiseau, France

E. Meltchakov · R. Mercier · F. Delmotte
LCF, Institut d'Optique, CNRS, PXI, Palaiseau, France

M. Kozlova · J. Nejdl · M. Krus · J. Dostal · J. Prokupek
Institute of Physics, Prague, Czech Republic

C. Constancias
CEA-LETI, Grenoble, France

F. Suzuki-Vidal
Imperial College, London, UK

O. Acef
SYRTE, Observatoire de Paris, UPMC, CNRS, Paris, France

S. Sebban et al. (eds.), *X-Ray Lasers 2012*, Springer Proceedings in Physics 147,
DOI 10.1007/978-3-319-00696-3_39,

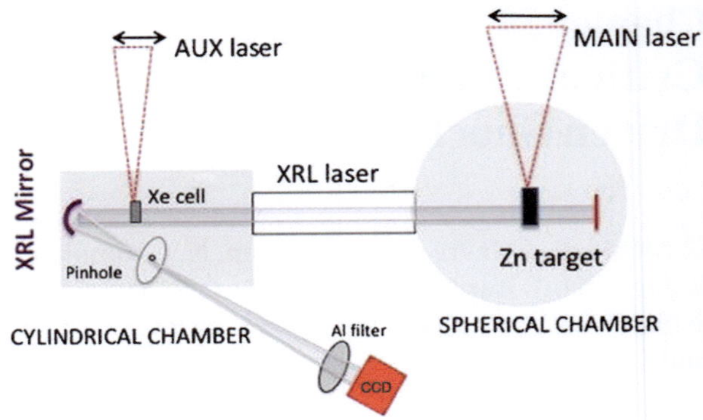

Fig. 1 Experimental setup with the 3 laser beams and the imaging diagnostic

may present a radiative precursor. Radiative shocks can be found in astrophysical media and in laboratory. Producing such shocks on large scale lasers requires an intensity in the range of 10^{14} W/cm^2 at least. Several experiments have been performed at PALS installation on Xenon at 0.2–0.3 bar, either using visible (0.4 μm) or the infrared (1.3 μm) laser beam (0.3 ns) and producing shocks at ∼50 km/s [1]. Visible interferometry demonstrated the presence of a radiative precursor and an electron density up to 10^{19} cm^3 was measured. As the plasma is opaque for electron density exceeding the critical density: n_c (cm^{-3}) $\sim 10^{21}/\lambda$ (μm)^2 (where λ is the laser wavelength), and strongly absorbing one decade below, it is difficult to probe the shocked part of the plasma, where 1D simulations predicts a density of 10^{20}–10^{21} cm^{-3}. We recently showed that XRL laser beam at 21 nm may be used to probe these shocks [2]. New results are presented below.

2 Experimental Setup and Critical Components

The PALS laser beam (operating at 1.3 μm) is split into two beams: one (AUX, 60 J) is focalized on a target located in the "cylindrical" vacuum chamber. It is used to generate the shock wave. The second beam (MAIN, 500 J) is focused on a Zn target, located in the "spherical" chamber, to generate a laser beam at 21.2 nm (3 mJ, 0.2 ns) [3], which is sent towards the cylindrical chamber, shining the cell perpendicularly to the direction of AUX, with a precise delay of 20 ns. The XRL beam then reaches a spherical mirror and a CCD camera. A pinhole, placed at the focus of XRL beam, and an Aluminum foil placed in front of the CCD, perform a filtering of the plasma stray light. The XRL mirror also images the target on the CCD, with a magnification G of 8 or 4 (see Fig. 1 for the experimental setup).

The targets are miniaturized shocks tubes, with an internal canal of $6 \times 0.4 \times 0.4$ mm^3, filled *in situ* with Xe. They are closed on one end by a gilt foil of polystyrene (10 μm), on which the AUX laser is focalized and which acts as a piston. They are closed laterally by two membranes (see below) through which the XRL beam propagates (Fig. 2). These membranes have not only to be transparent at 21.2 nm (i.e. they have to be thin) but also to withdraw a difference of pressure

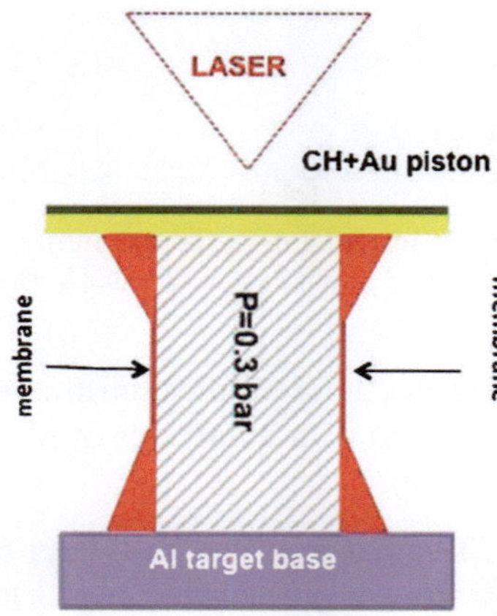

Fig. 2 Principle of the target with the gilt piston and the 2 membranes

of ~0.3 bar between the two faces (pressure inside the target, vacuum outside). A crude approximation of the minimum thickness h required to withdraw a difference of pressure ΔP is, for a long window of section a, given by:

$$\Delta P = \frac{h}{a}\sqrt{\frac{\sigma^3}{0.167E}},$$

where σ and E are the breaking point stress and the Young modulus respectively. Two types of membranes were used: (i) Si_3N_4 ($h = 100$ nm, 0.4×3 mm^2) and (ii) Si ($h = 100$ nm, 0.6×5 mm^2), respectively manufactured by SILSON and CEA-LETI.

The Si_3N_4 membranes are manufactured using hydroxide-based anisotropic etching of silicon wafers coated with LPCVD silicon nitride [4]. Using the values $\sigma = 2 \times 10^9$ N/m^2 and $E = 2.6 \times 10^{11}$ N/m^2, one obtains ΔP (bar) $= 4.3 \times 10^3\, h/a$ (here $a = 0.4$ mm), in qualitative agreement with the measured value of ± 0.5 bar at least. The transmission of each membrane is ~0.13 and their optical properties are excellent. The only difficulty was the roughness of the external section of the Si frame which was presenting small defaults of ~0.1 mm, causing difficulties in the gluing of the polystyrene foil on the membranes.

Few Si membranes were also tested. They were manufactured starting from SOI substrate. The deep etching is: RIE on 500 μm then completed by a wet etching with TMAH to be stopped on SiO_2, finally removed by HF acid [5]. The transmission (0.75) at 21 nm is excellent. The membranes withdraw the required difference of pressure. Our prototypes presented winkles. However, flat high stress membranes can be provided and these membranes are very promising.

Spherical XUV mirrors ($f = 300$ mm, used of axis, with a total deviation of 11 degrees) required special attention. Two mirrors were used:

- Mirror 1: fused silica substrate (roughness 0.2 nm rms, pitch polished at Institut d'Optique). The layer deposition consisted of 20 periods of Al/Mo/B4C with B4C capping layer, made at CEMOX platform of Institut d'Optique [6]. The expected reflectivity at 21.2 nm is 45 %.

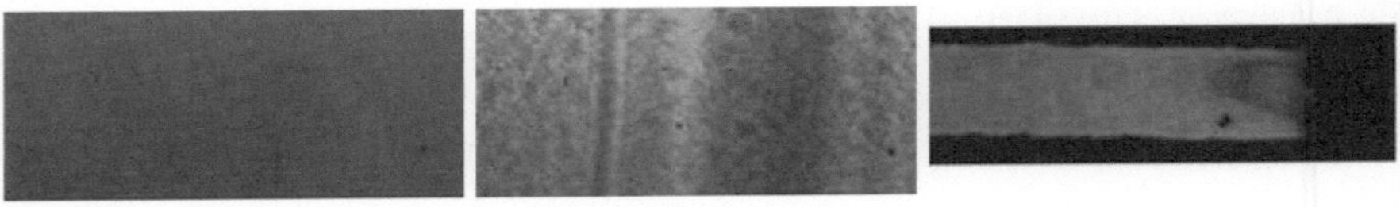

Fig. 3 Image of the laser with mirror 1 (*left*; G = 8; ∼5.5 × 13 mm^2), mirror 2 (*center*; G = 8; ∼5.5 × 13 mm^2), image of a shock wave (*right*; G = 4; ∼9 × 3 mm^2)

– Mirror 2: fused silica substrate (roughness 0.2 nm rms, shape $\lambda/50$, from Winlight Optics) with 30 periods of Mo/Si deposited at CEMOX (expected reflectivity = 35 %).

The images of the XRL laser beam at the position of the target, taken with these mirrors are given in the Fig. 3 together with a sharp image of a RS shock using mirror 1. Due to several defaults, the mirror 2 is less adequate for XUV radiography. Indeed, a similar mirror (EAGLE substrate, MoSi deposition performed at Brno) was used in previous experiment. Although the image of the shock wave was visible, it was difficult to analyze.

3 Conclusions

To conclude, XRL radiography is a powerful diagnostic for probing laser generated shock waves. This requires an energetic XRL laser, good quality windows for the shock tube and a XUV mirror, with a well polished substrate and an adequate multilayer deposition. Additive tests have to be made to understand the underachievement of our MoSi mirrors.

Acknowledgements The authors thank the CEA-LETI and the Basic Technological Research French program for providing the silicon membranes. The work was supported by LASERLAB access program, French ANR (grant 08-BLAN-0263-07), Observatoire de Paris, PICS4343 of CNRS, Academy of Sciences of the Czech Republic (Project No M100100911), Czech Ministry of Education Youth and Sports (Project Nos. 7E08099 and 7E09092), and by the Czech Science Foundation (Grant No. 202/08/1734).

References

1. Stehlé, C., et al.: Laser Part. Beams **28**, 253 (2010)
2. Stehlé, C., et al.: Opt. Commun. **285**, 64 (2012)
3. Kozlova, M., et al.: In: Sebban, S., et al. (eds.) X-Ray Lasers 2012. Springer Proceedings in Physics, vol. 147. Springer, Cham (2013). Chapter 24 in this book
4. Anastasi, P., et al.: In: Michette, A., et al. (eds.) X Ray Microscopy III. Springer Series in Optical Sciences, vol. 67, p. 441 (1992)
5. Constancias, C., et al.: J. Vac. Sci. Technol. B **28**, 194 (2010)
6. Meltchakov, E., et al.: Proc. SPIE **8168**, 816819 (2011)

Chapter 40
High Density Optical-Field-Ionization Soft X-Ray Lasers

F. Tissandier, S. Sebban, M. Kozlova, J. Gautier, P. Zeitoun, A. Klisnick, and G. Maynard

Abstract Generating an optical-field-ionization (OFI) collisional soft X-ray laser amplifier in a high density plasma would allow a significant increase of the saturation intensity, laser gain, and reduction of the emitted pulse duration. Due to strong refraction of the pump infrared beam in the plasma, guiding techniques have to be implemented. Using an optically preformed plasma waveguide, we successfully achieved guiding of J-energy pulses in a near-critical density krypton plasma with 50 % total transmission over 5 to 10 mm. X-ray lasing at 32.8 nm has also been demonstrated. This technique is a step towards OFI recombination soft X-ray lasers. This pumping scheme requires the production of a cold and dense plasma. We propose to pump a high density neon VIII plasma by a frequency doubled intense Ti:Sa laser. Guiding of the 405 nm pump pulse has already been achieved at LOA.

1 Introduction

In the past few years, there has been noticeable progress in the development of high-brightness coherent radiation beams in the X-ray and soft X-ray range. The X-ray free electron lasers have demonstrated unprecedented peak powers in the nanometer and angström regions with femtosecond time resolution and have been used for numerous applications since available. However, due to the relative lack of these

F. Tissandier (✉) · S. Sebban · J. Gautier · P. Zeitoun
Laboratoire d'Optique Appliquée, ENSTA, 104 Bvd des Maréchaux, 91120 Palaiseau, France
e-mail: fabien.tissandier@ensta-paristech.fr

M. Kozlova
Institute of Physics, AVČR, Na Slovance 2, 18221 Prague 8, Czech Republic

A. Klisnick
Institut des Sciences Moléculaires d'Orsay, Université Paris Sud (Bât. 210-350-351), 91405 Orsay, France

G. Maynard
Laboratoire de Physique des Gaz et des Plasmas, Université Paris Sud (Bât. 210), 91405 Orsay, France

S. Sebban et al. (eds.), *X-Ray Lasers 2012*, Springer Proceedings in Physics 147,
DOI 10.1007/978-3-319-00696-3_40,

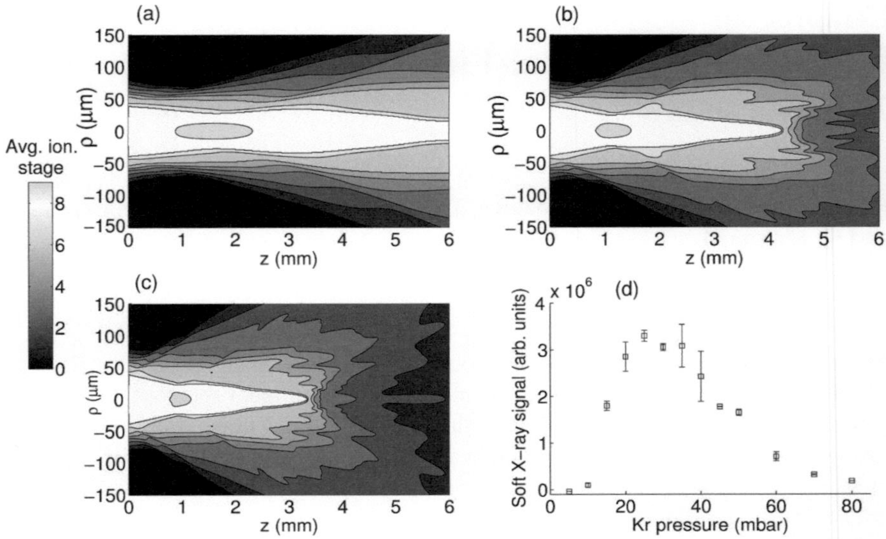

Fig. 1 Calculated ionization map of the OFI plasma for a pump intensity in vacuum of 10^{18} W/cm^2 and a krypton pressure of (**a**) 10 mbar, (**b**) 30 mbar and (**c**) 50 mbar. The laser propagates from left to right. The lasing Kr^{8+} is represented in *white*. (**d**) Measured and calculated output intensity of the soft X-ray laser as a function of krypton pressure

large and costly facilities, there is a need to develop more compact and less expensive sources. Soft X-ray lasers (SXRLs) generated through laser amplification in a collisional plasma are a promising alternative. Since their first experimental demonstration in 1985, dramatic progress has been made worldwide in the understanding and performances of these sources. However, they remain strongly limited in terms of focusable intensity or temporal resolution (ps pulse duration at best).

The work we report here mostly deals with an optical-field-ionization (OFI) collisional SXRL at 32.8 nm generated in a Ni-like krypton plasma [1]. We will first summarize the performances of that SXRL in the amplification of spontaneous emission (ASE) regime and when seeded by a high-order harmonic (HOH) of an intense infrared laser, in both cases with a maximum gas pressure of a few tenths of mbar (atom density around 10^{18} cm^{-3}). We will then discuss the advantages and drawbacks of drastically (2 orders of magnitude) increasing the gas density and report experimental measurements—including soft X-ray lasing—in a near-critical density plasma. Before concluding, we will discuss the possibility of generating a plasma suitable to the SXRL recombination pumping scheme.

2 Low-Density Performances

In the OFI scheme, both the creation of the lasing Kr^{8+} ion and the pumping of the population inversion are driven by the pump laser pulse. In its simplest form,

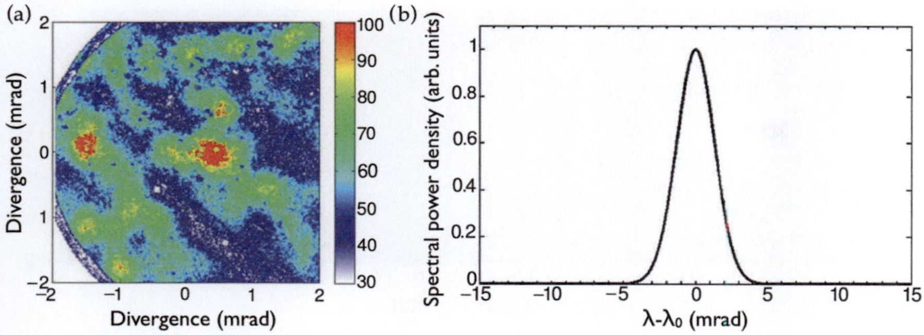

Fig. 2 (**a**) Far-field profile of the ASE SXRL beam. The beam was clipped by the Al filter mounting. (**b**) Spectrum of the ASE radiation, deducted from longitudinal coherence measurements

it consists in focusing an ultrashort and intense laser pulse into a cell filled with krypton at a pressure of a few tenths of mbar. The SXRL is then generated along the propagation of the pump beam in the plasma. We remind here the low-density soft X-ray amplifier performances obtained using the 10 Hz multi-terawatt "Salle Jaune" Ti:Sapphire laser installation at LOA. The 1 J, 35 fs pulses at a central wavelength of 815 nm were focused into a variable-length krypton cell. A quarter-wave plate was used to ensure the pump field has a circular polarization, and an aluminum filter was used to separate the SXRL beam from the remaining infrared pump beam. The best performances of the 32.8 nm SXRL were usually obtained for a gas pressure of 30 mbar [2] as can be seen on Fig. 1(d). The reason they were not obtained for the highest pressure is the main issue of the OFI scheme: this is a longitudinal pumping scheme, so the pump laser propagates in the plasma it created. This plasma exhibits an refraction index radial profile such that the pump beam suffers strong defocusing along its propagation. This defocusing gets stronger for higher plasma densities, as can be seen on Fig. 1(a–c) giving a 2D (ρ, z) map of the ionization stage in the plasma. It was calculated in a cylindrical geometry using the code OFI_PROP. While the amplifier is more than 6 mm long for a gas pressure of 10 mbar, its length drops to 3 mm at 50 mbar. This effect indirectly seriously limits the output pulse energy, which may vary from 1 to 100 nJ.

In the ASE regime, the divergence of the emitted SXRL beam is determined by the geometry of the amplifier. It is usually set between 10 and 15 mrad. The spatial profile also exhibits speckle (Fig. 2(a)) patterns. They are a consequence of the fact that the source behaves as a collection of micro sources mutually temporally coherent but spatially incoherent [3]. These modulations are characteristic of "short" pulse duration SXRLs (OFI and transient pumping); in the case of a long pulse duration SXRL (quasi-steady state), time averaging contributes to a drastic dampening of these speckle patterns. The spatial coherence of this source was also measured using variable-spacing Young slits and was found to be almost nil at the source, yielding a beam with a coherent zone diameter accounting for only of few percent of the beam diameter at 4 m from the source.

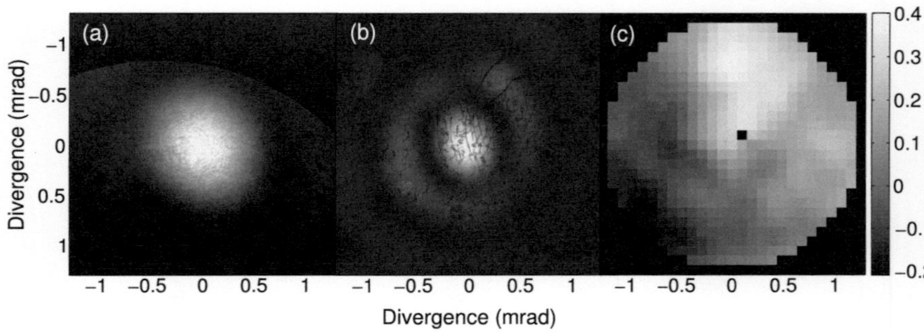

Fig. 3 (**a**) Usual Gaussian far-field profile of the seeded SXRL beam. (**b**) Bessel far-field profile of the seeded SXRL beam, due to a hole the amplifier caused by over ionization of krypton. (**c**) Wavefront of the seeded SXRL beam. The RMS distortions are of the order of $\lambda/11$

At last, the temporal coherence of the ASE source was measured by longitudinal variable-path difference interferometry using a Fresnel bi-dihedron interferometer. Using Wiener-Khinchin theorem, we could calculate the power density spectrum of the pulse (equal to the Fourier transform of the fringe visibility as a function of optical path difference curve), given in Fig. 2(b). The bandwidth is extremely narrow, around 3 mÅ, corresponding to $\Delta\lambda/\lambda = 10^{-5}$. The $1/e$ coherence time was found to be 6 ps for a krypton pressure of 30 mbar, and since this duration is the same as the gain duration at the same gas pressure, we concluded that the soft X-ray pulse is Fourier limited, with a duration around 6 ps [4].

The OFI ASE SXRL source offers then poor spatial characteristics, a low pulse energy and a long pulse duration. Seeding the amplifier by a high-order harmonic allowed a drastic improvement of the soft X-ray beam [5]. In this configuration, a second Ti:Sa beam (a few tenths of mJ, 35 fs) is used to generate in particular the 25th harmonic of the infrared frequency. This harmonic can be accurately tuned to wavelength of 32.8 nm by adjusting the chirp of the driving laser. The HOH are generated in a argon gas cell at a pressure of 20 mbar, and the source is imaged onto the amplifier entrance plane using a grazing incidence toroidal mirror. Figure 3(a–b) shows two typical far-field profile of the seeded SXRL beam. Due to spatial filtering of the harmonic by the plasma [6], the output amplified beam profile exhibits a symmetric Gaussian (a) or Bessel (b) profile [7], depending on the shape of the amplifier. The beam is also well collimated, with a divergence of the order of 1 mrad. Another interesting new feature of this source is that it offers a high spatial coherence, and an almost diffraction-limited wavefront (Fig. 3(c)) with RMS distortions of $\lambda/11$ [8].

On the other side, the output pulses still have a duration in the ps range. Since the HOH duration (fs) is much shorter than the amplifier characteristic time scale (ps), the amplified field grows as a coherent wake behind the harmonic peak, with a duration of the order of the phase relaxation time of the dipoles constituting the amplifier medium. The amplification is well saturated, but Rabi oscillations of the fields were neither observed, nor expected [9]. Since that saturation is quickly reached (after

1 or 2 mm of propagation in the amplifier), the output pulse energy is not very high. Pulses with energy up to 1 μJ were produced.

On the whole, the seeded SXRL source is promising but suffers from a lack of focusable intensity of temporal resolution. This could be taken care of by increasing the amplifier density. When increasing the density, collisional processes are faster. In particular, collisional excitation (thus pumping) is faster, but also collisional ionization—that destroys the lasing ion—is also faster. The gain lifetime mechanically decreases. On Fig. 4 are given the SXRL intensity as a function of the harmonic injection delay into the amplifier, for different krypton pressures. These are basically gain temporal evolution measurements. Time-dependent only collisional-radiative calculations of the gain dynamics are also reported on the same graphics.

As expected, the gain duration decreases when we increase the plasma density. That decrease is however not of great importance: the gain duration (FWHM) is around 10 ps at 15 mbar, and 5 ps at 40 mbar. The gain peak value is also slightly higher at 40 mbar (180 cm^{-1}) than at 15 mbar (150 cm^{-1}). It could nonetheless be interesting to see the effects of an increase of density by 2 orders of magnitude, up to 10^{20} at/cm^3. Using the same-as-before time-dependent OFI_0D collisional-radiative code, we have been able to calculate the resulting gain maximum value, and saturation intensity. They are reported on Fig. 5 for a krypton density ranging from 10^{18} cm^{-3} (40 mbar)) to 2×10^{20} cm^{-3}.

It can be seen that at a density of 10^{20} cm^{-3}, the peak gain value is only twice that obtained for a density of 10^{18} cm^{-3}. The most important improvement we can expect is the saturation intensity. Over the same range of density, it gains 4 orders of magnitude. If we are able to let the pump pulse propagate over a significant distance in the plasma (a few mm), we can expect a very strong increase of the emitted pulse energy. The gain duration is also expected to decrease drastically.

3 High-Density Waveguiding and Soft X-Ray Lasing

As we have already seen, the main issue of this OFI scheme is the pump pulse propagation. Using a neutral density of 10^{20} cm^{-3}, some waveguiding technique must be implemented to allow the pump laser to propagate over more than a few hundreds of microns. Guiding of the pump pulse of an OFI SXRL has already been achieved using a capillary discharge [10, 11], hollow capillary tubes [12, 13], or an laser-preformed plasma waveguide [14, 15]. Although the results obtained with capillary discharge or tubes were convincing, both these methods are not suited for a high density plasma. Using a laser-produced plasma waveguide however gave very interesting and promising results at neutral densities around 10^{20} cm^{-3}.

The principle of plasma guiding is to generate a plasma channel with an electron density profile that will compensate the effects of diffraction, i.e. lower along the propagation axis than further away from that axis. An effective method to generate such a plasma channel is the so-called "ignitor-heater" method [16–18]. It basically follows three steps. The first step is the creation of primary electrons. A first

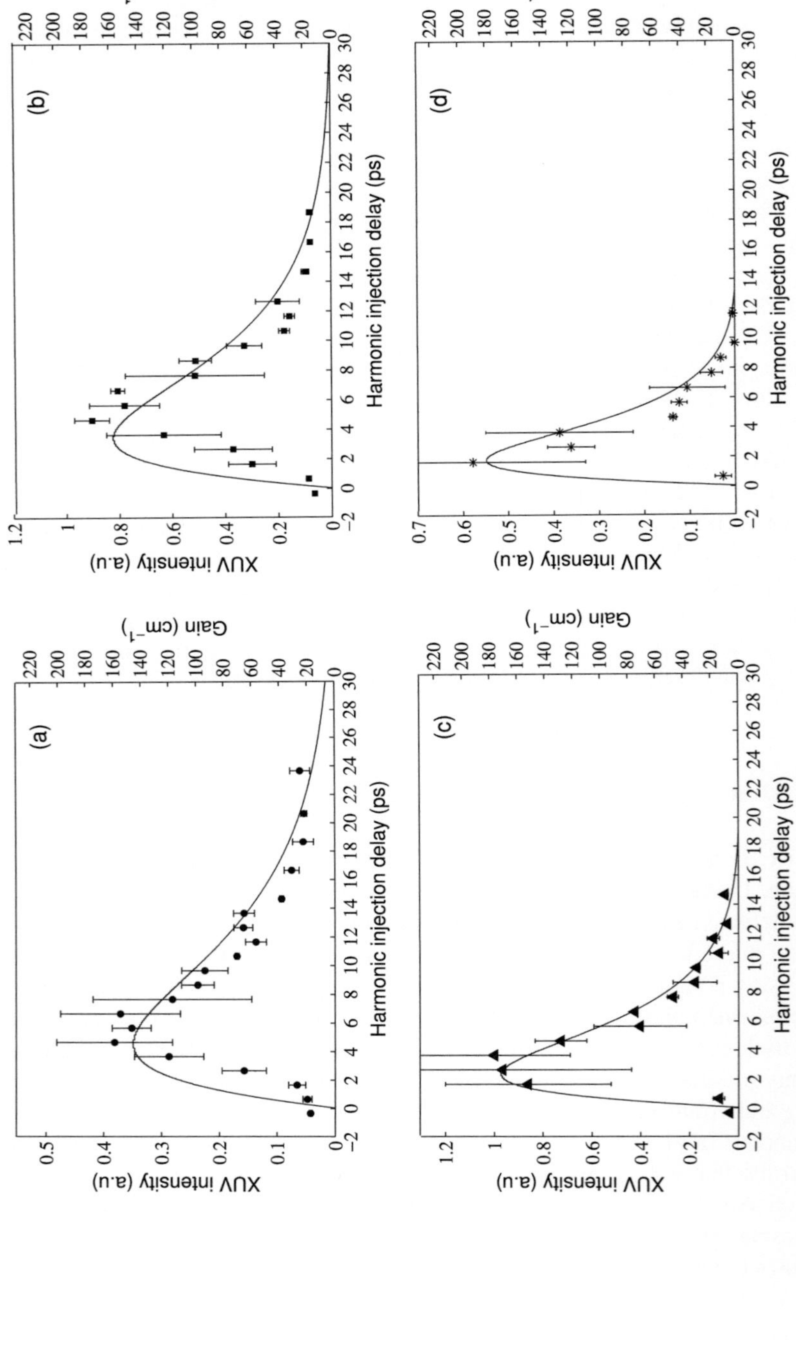

Fig. 4 Soft X-ray output intensity as a function of harmonic injection time into the amplifier for a pressure of (**a**) 15 mbar, (**b**) 20 mbar, (**c**) 30 mbar and (**d**) 40 mbar. Gain calculations are also reported on each graph

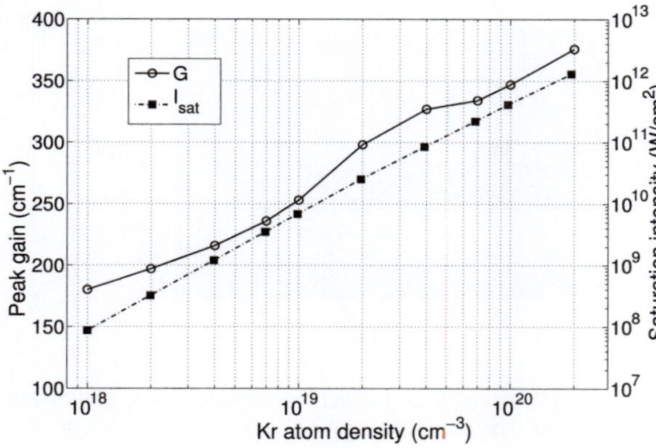

Fig. 5 Calculated peak gain and saturation intensity for a krypton density up to 2×10^{20} cm^{-3}

short (fs) and intense pulse, called the ignitor, is then focused into the gas. Via tunnel ionization, a weakly-ionize plasma is generated. A second long duration (100's ps) and energetic (100's mJ) pulse, called the heater, will then heat the free electrons and trigger collisional avalanche ionization that will lead to a plasma of ionization stage around 3–4 and thus greatly increase the electron density. The last step is the hydrodynamic expansion of that plasma at the local sound speed. This expansion will result in a decrease of the electron density along the axis.

We implemented that technique for our Kr^{8+} amplifier using an axicon [18], that focuses the ignitor and heater pulses along a longitudinal line. The experimental setup is detailed on Fig. 6. The plasma channel is created using a 100 mJ, 35 fs ignitor pulse and a cross-polarized 250 mJ, 400 ps heater pulse delayed by 350 ps. These pulses are focused by the axicon along a line 2–3 cm-long into the gas jet. The gas jet is a prototype that is able to deliver a neutral density up to 2×10^{20} cm^{-3} with a

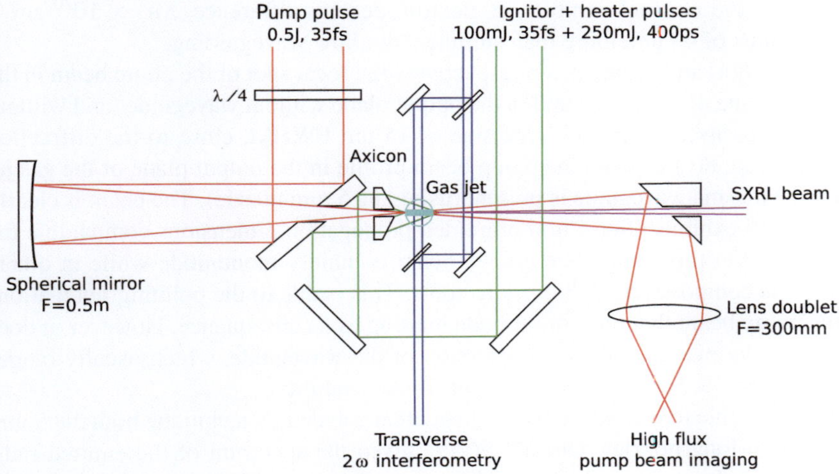

Fig. 6 High density plasma waveguiding and soft X-ray laser: experimental setup

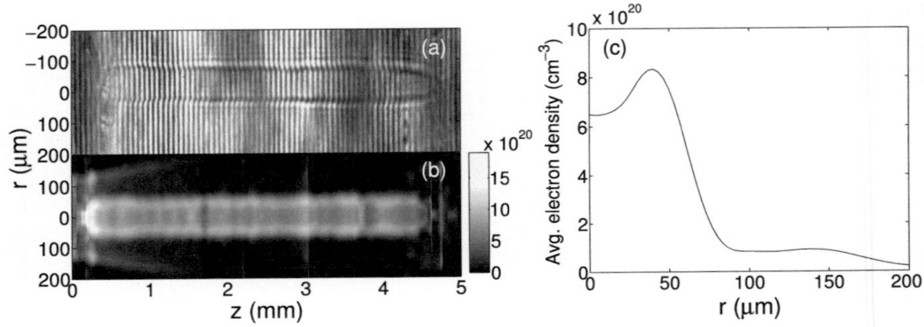

Fig. 7 (**a**) Typical plasma channel interferogram and (**b**) electron density map calculated from that interferogram. The pump laser propagates from left to right. (**c**) Averaged electron density profile along the channel

slit nozzle 0.2 × 5 mm or 0.2 × 10 mm. After a delay of 3.5 ps, the main pulse (0.5 J, 35 fs) is focused into the plasma channel by a 0.5 m focal length spherical mirror. Our main diagnostics are plasma transverse Mach-Zehnder interferometry using a frequency-doubled Ti:Sa pulse, high flux pump beam imaging in the entrance and exit planes of the channel, and a flat-field soft X-ray spectrometer (not showed here) using a grazing incidence toroidal mirror coupled to a 1000 mm^{-1} reflexion grating.

In Fig. 7(a) is given a typical plasma channel interferogram obtained with the 5 mm-long nozzle. The channel clearly runs over most of the 5 mm of gas and has a diameter around 100 μm. This interferogram can be used to calculate the electron density map in the channel. It is given on Fig. 7(b) and the longitudinally-averaged profile is given on Fig. 7(c). The density is very high at the entrance of the channel, more than 10^{21} cm^{-3}, and quickly decreases to a value slightly below 10^{21} cm^{-3}. On average, the density inside the channel is around 7×10^{20} cm^{-3}. We can also notice that the channel exhibits an electron density difference $\Delta n_e > 10^{20}$ cm^{-3} over a radius of 40 μm, more than sufficient to allow waveguiding.

On Fig. 8(a) and (b) are given respectively the focal spot of the pump beam in the entrance plane of the gas jet and in the output plane without waveguide (and without gas). It is focused down to a spot size of 15 μm FWHM, close to the diffraction limit. On Fig. 8(c) is given the pump beam profile in the output plane of the gas jet when the plasma waveguide is present (using the 5 mm nozzle). The beam is clearly guided, and exhibits after 5 mm of guided propagation dimensions comparable the focal spot. On this image, the guided beam is mainly monomode while in others cases, it is composed by 2, 3 or more spots. This is due to the pointing fluctuations of the pump beam that have multimode guiding as a consequence. However, it does not affect the overall energy transmission of the waveguide, which usually ranges between 35 % et 50 % with a good shot-to-shot stability.

We were able to generate a SXRL using that waveguide technique both the 5 mm and 10 mm-long nozzles. On Fig. 9(a) is given the spectrum of the emitted radiation. It is only composed by the 32.8 nm SXRL line. We have not been able to calibrate our sensor, but we estimate the energy contained in the SXRL pulses to

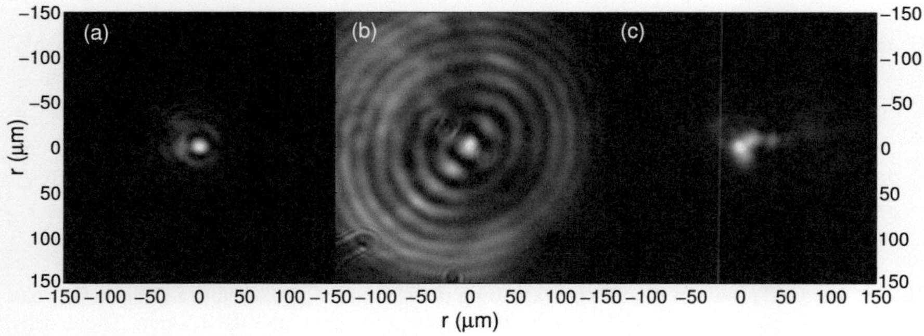

Fig. 8 High flux images of the pump beam (**a**) in the entrance plane of the jet, (**b**) in the output plane, without waveguide, and (**c**) in the output plane, guided in the 5 mm-long plasma channel

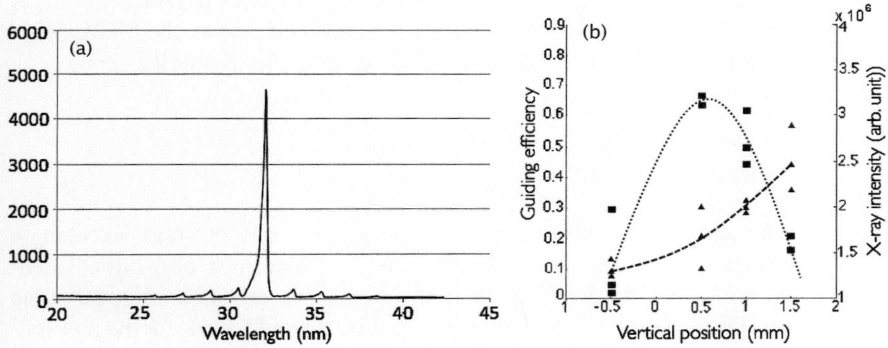

Fig. 9 (**a**) Spectrum of the emitted soft X-ray radiation. (**b**) Guiding efficiency (*dotted line*) and SXRL intensity (*dashed line*) as a function of the relative spacing between the jet nozzle and the plasma channel. The density rises when the vertical position decreases

several μJ. On Fig. 9(b) is given the dependence of the guiding efficiency of the waveguide and the output soft X-ray signal on the gas density. It can be seen that the guiding efficiency increases while the density decreases, because propagation of the preforming beams is easier at lower densities. The resulting channel is also slightly smaller when the density is increased, so that the pump beam is harder to couple to the waveguide. On the other hand, there is a density optimum in terms of soft X-ray generation. This is due to some competition between the high gain and saturation intensity, and the less efficient guiding that are all consequences of a higher density.

At last, the far-field pattern of that SRL beam has been recorded using a spherical mirror to ensure all the light is collected on our soft X-ray CCD camera. A typical footprint of the beam, generated using the 1 cm-long nozzle, is presented on Fig. 10. It exhibits the speckle patterns common to short-duration ASE SXRLs, and a dough-nut shape due to steep density gradients inside the waveguide that deflect the soft X-ray beams away from the axis. Its divergence, between 10 and 15 mrad, is deter-

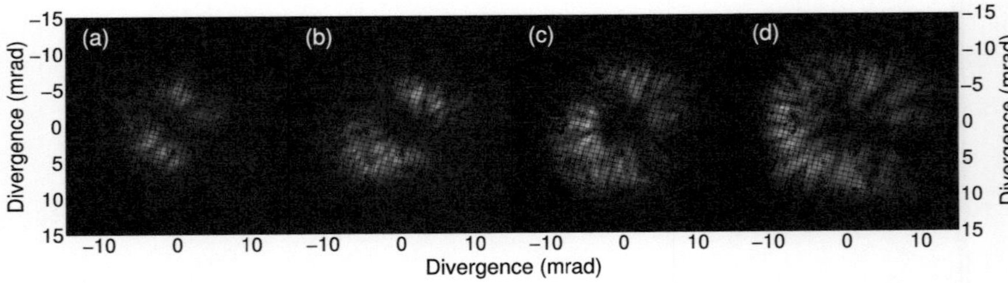

Fig. 10 Far-field profile of the SXRL beam using the 10 mm-long nozzle and a distance between the jet and the plasma channel of (**a**) ≃0, (**b**) 0.5 mm, (**c**) 1 mm, and (**d**) 1.5 mm. The density decreases from left to right

mined by the geometry of the amplifier column. For the same reason, we notice that when the gas density decreases (Fig. 10(a–d)), the beam divergence increases (larger and potentially longer amplifier). When using a 5 mm-long nozzle, the SXRL beam is usually a little more divergent (20 mrad) since he amplifier is shorter.

4 Recombination Scheme

These results are very promising, but still limited in terms of short wavelength SXRL generation. The ability to guide an intense pulse in a near-critical density plasma could be the first step to reinvestigate the recombination pumping scheme [19]. This scheme is very interesting because it would allow for the possibility of lasing from an excited state of an ion down to the ground state. The advantage of lasing down to the ground state is the higher energy (thus shorter wavelength) of the lasing transition as compared with lasing between excited states. The plasma conditions that are required are a very complete emptying of the ground state and an electron temperature which is small compared to the ionization energy in order to favor collisional recombination (over radiative recombination). Since the collisional recombination rate varies as n_e^3/T_e, the plasma needs also to be as dense as possible. If these requirements are met, the upper energy levels of the ion of charge n will be populated from recombination of the ion of charge $n + 1$. Collisional-radiative cascades will occur and will populate the lower energy levels before the ground state, allowing a population inversion between the ground state and the first energy level (among others). This population inversion is expected to be very short, below 1 ps [20]. Good candidates for this scheme are H-like or Li-like ions. In particular, an intensity of only 2×10^{17} W/cm^2 is needed to ionize neon to He-like Ne which may recombinate to Li-like Ne [21]. In Li-like Ne, the transition between the first excited level (3d) and the ground state (2p) corresponds to a wavelength of 9.8 nm.

We propose to use the preformed plasma waveguide technique in a near-critical density Ne plasma to investigate the OFI recombination scheme. To limit electron heating by the pump pulse, the pump field has to be linearly polarized. The smaller pump field wavelength, the less heating it provides, so we decided to use a

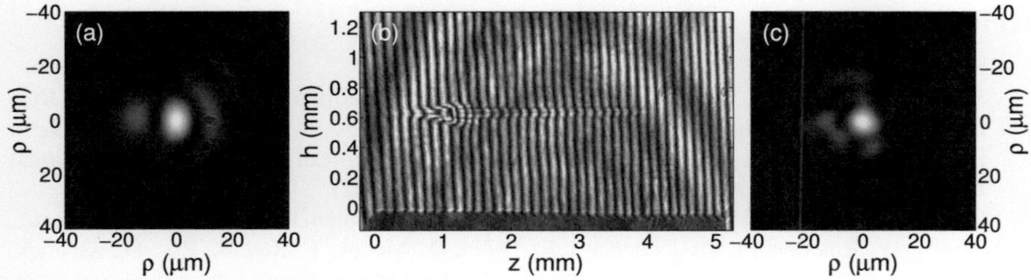

Fig. 11 (**a**) Focal spot of the 2ω beam, in the entrance plane of the plasma channel. (**b**) Interferogram of the plasma channel guiding the 2ω pulse over 5 mm of neon plasma. (**c**) Guided 2ω beam output profile in the exit plane of the channel

frequency-doubled pump pulse. We used a 120 mm-diameter, 0.8 mm-thick type I KDP crystal to generate the second harmonic of the main pulse of our Ti:Sa laser system. Up to 40 % of the infrared was converted to 405 nm. Figure 11(a) shows the 2ω focal spot focused by a 0.5 m-focal length spherical mirror. The ω beam is focused down to a spot size of 16 μm FWHM encircling 50 % of the energy while the 2ω focal spot is twice smaller (8 μm) and contains 57 % of the energy. We have not been able to measure the spectrum of the 2ω pulse, but manufacturer data shows no significant temporal broadening of that pulse.

Figure 11(b) is a typical interferogram of the Ne plasma using a 5 mm-long nozzle. We can notice a plasma channel running over most of the 5 mm of neon, and the guided 2ω pulse in the channel. On Fig. 11(c) is given the beam profile of the guided 2ω beam in the exit plane of the channel. It is comparable to the input beam, with the same spot diameter of 8 μm. We measured an energy transmission of the waveguide of 60 %. This is even better than when guiding the ω beam, and probably due to the smaller size of the 2ω beam at the entrance of the channel. We have not observed any soft X-ray signal, but this was only a preliminary experiment.

5 Conclusion

We have seen that—although they offer interesting features—low density OFI collisional SXRL amplifiers are limited in terms of pulse energy, duration, and wavelength of emission. We successfully increased the density of the amplifier by 2 orders of magnitude, with prospects of multi-10th μJ soft X-ray pulses also expected to have a shorter duration. To achieve that, an ignitor+heater optically preformed plasma waveguide with a transmission of 50 % has been implemented. The ability to guide ultra-intense pulses in near-critical plasmas is a good start to a reinvestigation of the OFI recombination pumping scheme. As a preliminary experiment, we have been able to frequency-double the main IR pump pulse and guide this pulse in a neon plasma at a density around 10^{21} cm^{-3}. If successful, this experiment would allow the production of sub-ps pulses below 10 nm.

References

1. Sebban, S., Mocek, T., Ros, D., Upcraft, L., Balcou, Ph., Haroutunian, R., Grillon, G., Rus, B., Klisnick, A., Carillon, A., Jamelot, G., Valentin, C., Rousse, A., Rousseau, J.P., Notebaert, L., Pittman, M., Hulin, D.: Demonstration of a Ni-like Kr optical-field-ionization collisional soft x-ray laser at 32.8 nm. Phys. Rev. Lett. **89**, 253901 (2002)
2. Sebban, S., Upcraft, L.M., Balcou, Ph., Pittman, M., Haroutunian, R., Grillon, G., Valentin, C., Rousse, A., Rousseau, J.-Ph., Notebaert, L., Hulin, D., Mocek, T., Rus, B., Ros, D., Klisnick, A., Carillon, A., Jamelot, G.: Investigations of collisionally pumped optical field ionization soft-x-ray lasers. J. Opt. Soc. Am. B **20**(1), 195–202 (2003)
3. Guilbaud, O., Klisnick, A., Cassou, K., Kazamias, S., Ros, D., Jamelot, G., Joyeux, D., Phalippou, D.: Origin of microstructures in picosecond x-ray laser beams. EPL **74**(5), 823 (2006)
4. Guilbaud, O., Tissandier, F., Goddet, J.-P., Ribière, M., Sebban, S., Gautier, J., Joyeux, D., Ros, D., Cassou, K., Kazamias, S., Klisnick, A., Habib, J., Zeitoun, P., Benredjem, D., Mocek, T., Nedjl, J., de Rossi, S., Maynard, G., Cros, B., Boudaa, A., Calisti, A.: Fourier-limited seeded soft x-ray laser pulse. Opt. Lett. **35**(9), 1326–1328 (2010)
5. Zeitoun, Ph., Faivre, G., Sebban, S., Mocek, T., Hallou, A., Fajardo, M., Aubert, D., Balcou, Ph., Burgy, F., Douillet, D., Kazamias, S., de Lacheze-Murel, G., Lefrou, T., le Pape, S., Mercere, P., Merdji, H., Morlens, A.S., Rousseau, J.P., Valentin, C.: A high-intensity highly coherent soft x-ray femtosecond laser seeded by a high harmonic beam. Nature **431**(7007), 426–429 (2004)
6. Goddet, J.Ph., Sebban, S., Morlens, A.S., Gautier, J., Rousseau, J.Ph., Burgy, F., Zeitoun, Ph., Valentin, C., Hauri, C., Maynard, G., Boudaa, A., Caumes, J.P., Merdji, H., Mocek, T., Kozlova, M., Jakubczak, K.: Demonstration of a spatial filtering amplifier for high-order harmonics. Opt. Lett. **32**(11), 1498–1500 (2007)
7. Tissandier, F., Sebban, S., Ribière, M., Gautier, J., Zeitoun, Ph., Lambert, G., Goddet, J.-Ph., Burgy, F., Valentin, C., Rousse, A., Nejdl, J., Mocek, T., Maynard, G.: Bessel spatial profile of a soft x-ray laser beam. Appl. Phys. Lett. **97**(23), 231106 (2010)
8. Goddet, J.Ph., Sebban, S., Gautier, J., Zeitoun, Ph., Valentin, C., Tissandier, F., Marchenko, T., Lambert, G., Ribière, M., Douillet, D., Lefrou, T., Iaquaniello, G., Burgy, F., Maynard, G., Cros, B., Robillard, B., Mocek, T., Nejdl, J., Kozlova, M., Jakubczak, K.: Aberration-free laser beam in the soft x-ray range. Opt. Lett. **34**(16), 2438–2440 (2009)
9. Tissandier, F., Sebban, S., Gautier, J., Zeitoun, Ph., Oliva, E., Rousse, A., Maynard, G.: 3d Maxwell-Bloch calculation of the temporal profile of a seeded soft x-ray laser pulse. Appl. Phys. Lett. **101**, 251112 (2012). doi:10.1063/1.4773208
10. Butler, A., Gonsalves, A.J., McKenna, C.M., Spence, D.J., Hooker, S.M., Sebban, S., Mocek, T., Bettaibi, I., Cros, B.: Demonstration of a collisionally excited optical-field-ionization XUV laser driven in a plasma waveguide. Phys. Rev. Lett. **91**, 205001 (2003)
11. Butler, A., Gonsalves, A.J., McKenna, C.M., Spence, D.J., Hooker, S.M., Sebban, S., Mocek, T., Betttaibi, I., Cros, B.: 41.8nm Xe^{8+} laser driven in a plasma waveguide. Phys. Rev. A **70**, 023821 (2004)
12. Mocek, T., McKenna, C.M., Cros, B., Sebban, S., Spence, D.J., Maynard, G., Bettaibi, I., Vorontsov, V., Gonsavles, A.J., Hooker, S.M.: Dramatic enhancement of XUV laser output using a multimode gas-filled capillary waveguide. Phys. Rev. A **71**, 013804 (2005)
13. Cros, B., Mocek, T., Bettaibi, I., Vieux, G., Farinet, M., Dubau, J., Sebban, S., Maynard, G.: Characterization of the collisionally pumped optical-field-ionized soft-x-ray laser at 41.8 nm driven in capillary tubes. Phys. Rev. A **73**, 033801 (2006)
14. Chou, M.-C., Lin, P.-H., Lin, C.-A., Lin, J.-Y., Wang, J., Chen, S.-Y.: Dramatic enhancement of optical-field-ionization collisional-excitation x-ray lasing by an optically preformed plasma waveguide. Phys. Rev. Lett. **99**, 063904 (2007)
15. Lin, P.-H., Chou, M.-C., Lin, C.-A., Chu, H.-H., Lin, J.-Y., Wang, J., Chen, S.-Y.: Optical-field-ionization collisional-excitation x-ray lasers with an optically preformed plasma waveguide. Phys. Rev. A **76**, 053817 (2007)

16. Durfee, C.G., Milchberg, H.M.: Light pipe for high intensity laser pulses. Phys. Rev. Lett. **71**, 2409–2412 (1993)
17. Durfee, C.G., Lynch, J., Milchberg, H.M.: Mode properties of a plasma waveguide for intense laser pulses. Opt. Lett. **19**(23), 1937–1939 (1994)
18. Durfee, C.G., Lynch, J., Milchberg, H.M.: Development of a plasma waveguide for high-intensity laser pulses. Phys. Rev. E **51**, 2368–2389 (1995)
19. Nagata, Y., Midorikawa, K., Kubodera, S., Obara, M., Tashiro, H., Toyoda, K.: Soft-x-ray amplification of the Lyman-α transition by optical-field-induced ionization. Phys. Rev. Lett. **71**, 3774–3777 (1993)
20. Eder, D.C., Amendt, P., Wilks, S.C.: Optical-field-ionized plasma x-ray lasers. Phys. Rev. A **45**, 6761–6772 (1992)
21. Amendt, P., Eder, D.C., Wilks, S.C.: X-ray lasing by optical-field-induced ionization. Phys. Rev. Lett. **66**, 2589–2592 (1991)

Chapter 41
Spectral Linewidth Measurement of a Ne-Like Ar Capillary Discharge Soft X-Ray Laser

L. Urbanski, M.C. Marconi, L.M. Meng, M. Berrill, O. Guilbaud, A. Klisnick, and J.J. Rocca

Abstract We report on the measurement of spectral linewidth of a Ne-like Ar table-top capillary discharge laser. The linewidth was measured as a function of the gain medium length. Due to inhomogeneous character of the linewidth, saturation re-broadening is expected. However no such behavior was observed while the amplifier length was extended beyond saturation. This situation is compared with a numerical model, identifying that even a small amount of collisional broadening effectively homogenizes the line profile.

1 Introduction

The table top capillary discharge Ne-like Ar laser has been widely used in many applications requiring combination of light coherence and high photon flux [1]. In past experiments the spatial coherence of the Ne-like Ar laser was measured using a Young's interferometer. The measurement was performed for different amplifier lengths yielding to the conclusion that a fully spatially coherent beam is obtained for 36 cm long plasma column [2]. However there was no direct measurement of the temporal coherence and thus the linewidth of the capillary discharge laser. The measurement of the linewidth can give insight on the plasma conditions in the amplifier and provide a better understanding of the laser dynamics. In a capillary discharge laser, the electron density is of the order of 10^{18} cm^{-3} and the ion temperature ex-

L. Urbanski · M.C. Marconi (✉) · J.J. Rocca
NSF ERC for Extreme Ultraviolet Science and Technology, Colorado State University, Fort Collins, CO 80523, USA
e-mail: marconi@engr.colostate.edu

L.M. Meng · A. Klisnick
ISMO, Bât. 350, CNRS, Université Paris-Sud 11, Orsay, France

M. Berrill
Oak Ridge National Laboratory, Oak Ridge, TN 37831, USA

O. Guilbaud
LPGP, Bât. 210, CNRS, Université Paris-Sud 11, Orsay, France

S. Sebban et al. (eds.), *X-Ray Lasers 2012*, Springer Proceedings in Physics 147,
DOI 10.1007/978-3-319-00696-3_41,

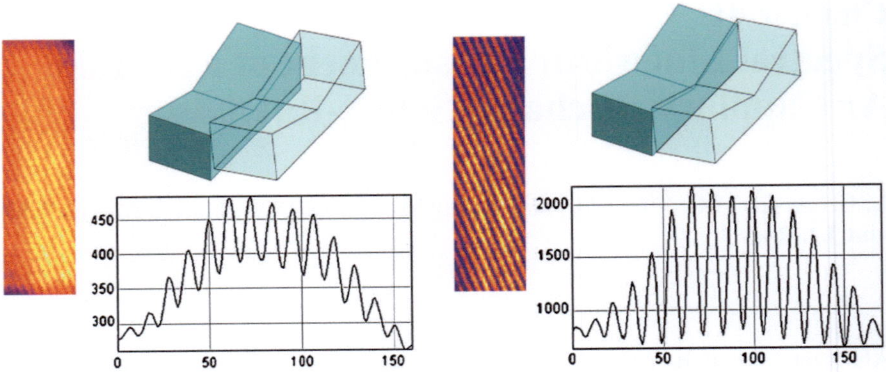

Fig. 1 Different position of the scanning mirror resulting in different interference contrast. Insets showing relative mirror placement, interferogram and a corresponding line-out

ceeds the electron temperature. In such conditions, Doppler broadening dominates the line profile and the collisional component that homogenizes it should be less significant.

This situation gives a justified premise to expect linewidth re-broadening after saturation. Several attempts to measure the linewidth of different types of optically pumped soft-x-ray lasers have been reported in the past [3, 4]. Due to extremely narrow line, they had to be conducted using very high resolution spectrometers [3]. In the experiment described herein, we measured the laser linewidth using a Fresnel bi-mirror interferometer [4]. The line profile was inferred from the visibility profile obtained by measuring the fringe visibility of the interferograms as a function of the path difference in the interferometer. We measured the laser linewidth for different amplifier lengths between 18 and 36 cm and compare the results to the numerical simulations.

2 Measurement Principle

The capillary discharge laser illuminates the Fresnel bi-mirror interferometer after reflecting in a pair of multilayer mirrors (Sc/SiO), with peak reflectivity at 46.9 nm. This arrangement serves in the experiment twofold. First, it allows aligning the laser beam with the optical axis of the interferometer. Second, due to the narrow reflectivity bandwidth of the multilayers, the double reflection provides extra spectral filtering at the laser wavelength. The wavefront is then split at the interferometer and brought to interference at the detector plane (CCD). The optical path difference between the arms of the interferometer can be varied displacing one of the bi-mirrors in the vertical direction as schematically shown in Fig. 1.

This produces a change in the visibility of the interference fringes with the different path differences that allows the construction of the visibility curve. A typical

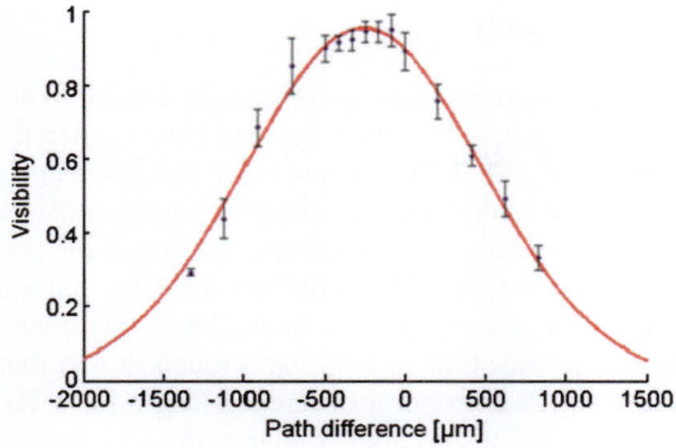

Fig. 2 Visibility curve obtained for a single capillary length. Experimental data with a calculated fit

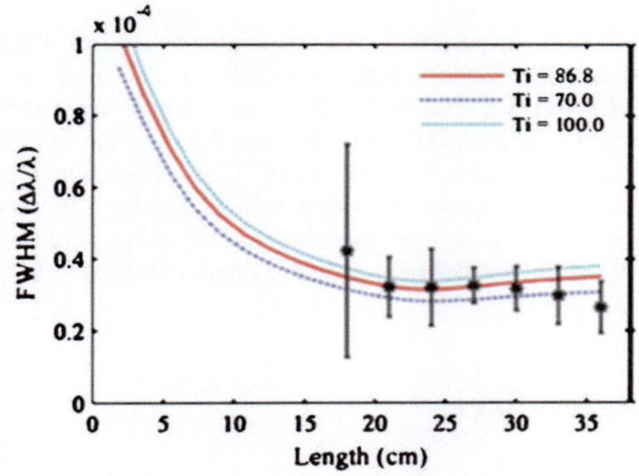

Fig. 3 Experimental results. Linewidth as a function of the amplifier length. Data set is fitted with numerical calculation

visibility curve is shown in Fig. 2. The coherence length is obtained from the optical path difference at which the fit drops to 1/e of its maximum in the Gaussian fit of the curve. From this measurement the linewidth is calculated. The results of the linewidth measurement are illustrated in Fig. 3.

The bandwidth decreases slowly as the plasma length increases maintaining an almost constant value for long capillaries, without any significant rebroadening for the longest lengths even though the laser was operated above saturation. Each data point in Fig. 3 corresponds to an average of 5 measurements and the error bars were selected to include all measurements. The lines show the computed linewidth assuming ion temperatures of 100, 87 and 70 eV from top to bottom, respectively. The model simulates the amplification of the laser line by solving the frequency-dependent intensity and population equations accounting for Doppler and collisional broadening, gain saturation, beam refraction, and collisional redistribution. The calculations predict a weak rebroadening of the line beyond $L = 24$ cm. However this effect is small compared to our experimental error bars, and no conclusive rebroadening can be derived from the experimental data.

3 Summary

In conclusion, we have measured the linewidth and temporal coherence of a capillary discharge soft-x-ray laser, and have studied its variation as a function of amplifier length. While the line profile in this discharge amplifier is clearly dominated by Doppler broadening, the relatively small collisional broadening component is still sufficiently large to homogenize the line to the point at which no rebroadening was measured to take place as the line intensity continued to increase beyond gain saturation. The narrow relative linewidth of this laser $\Delta\lambda/\lambda = 3.5 \times 10^{-5}$ corresponds to a coherence time of 2 ps that is much shorter than the pulse duration of 1.2–1.8 ns [1, 5, 6]. The temporal coherence length $L_c = 700\ \mu m$ measured for this capillary discharge SXRL is either similar or in some cases even larger than that of other collisional soft-x-ray lasers [7–9], which facilitates applications requiring high temporal coherence, such as interferometry [1] and large-area Talbot nanopatterning [10].

Acknowledgements We acknowledge fruitful discussion with A. Calisti (PIIM, Universite de Marseille, France) on ionic correlations. The support and expertise from D. Joyeux and S. de Rossi (LCF, Institut d'Optique, Palaiseau, France) are greatly appreciated. This work was supported by NSF Grant No. PHY-1004295, the NSF Center for Extreme Ultraviolet Science and Technology under NSF Grant No. EEC-0310717, DTRA under Contract No. HDTRA 1-10-1-0070, and by the AMOS program of the Office of Basic Energy Sciences, US Department of Energy. M.B. acknowledges support for staff members at Oak Ridge National Laboratory managed by UT-Battelle, LLC, for the US Department of Energy under Contract No. DE-AC05-00OR22725.

References

1. Urbanski, L., Marconi, M.C., Meng, L.M., Berrill, M., Guilbaud, O., Klisnick, A., Rocca, J.J.: Spectral linewidth of a Ne-like Ar capillary discharge soft-x-ray laser and its dependence on amplification beyond gain saturation. Phys. Rev. A, Am. Phys. Soc. **85**, 033837 (2012)
2. Liu, Y., Seminario, M., Tomasel, F., Chang, C., Rocca, J., Attwood, D.: Achievement of essentially full spatial coherence in a high-average-power soft-x-ray laser. Phys. Rev. A **63**(3), 1–5 (2001)
3. Koch, J.A., MacGowan, B.J., Da Silva, L.B., Matthews, D.L., Underwood, J.H., Batson, P.J., et al.: Experimental and theoretical investigation of neonlike selenium x-ray laser spectral linewidths and their variation with amplification. Phys. Rev. A **50**, 1877–1898 (1994)
4. Klisnick, A., Guilbaud, O., Ros, D., Cassou, K., Kazamias, S., Jamelot, G., et al.: Experimental study of the temporal coherence and spectral profile of the 13.9 nm transient X-ray laser. J. Quant. Spectrosc. Radiat. Transf. **99**(13), 370–380 (2006). Radiative Properties of Hot Dense Matter
5. Benware, B.R., Macchietto, C.D., Moreno, C.H., Rocca, J.J.: Demonstration of a high average power tabletop soft X-ray laser. Phys. Rev. Lett. **81**, 5804–5807 (1998)
6. Macchietto, C.D., Benware, B.R., Rocca, J.J.: Generation of millijoule-level soft x-ray laser pulses at a 4-Hz repetition rate in a highly saturated tabletop capillary discharge amplifier. Opt. Lett. **24**(16), 1115–1117 (1999)
7. Smith, R.F., Dunn, J., Hunter, J.R., Nilsen, J., Hubert, S., Jacquemot, S., et al.: Longitudinal coherence measurements of a transient collisional x-ray laser. Opt. Lett. **28**(22), 2261–2263 (2003)
8. Meng, L.M., Alessi, D., Guilbaud, O., Wang, Y., Berrill, M., Luther, B.M., et al.: Temporal coherence and spectral linewidth of an injection-seeded transient collisional soft x-ray laser. Opt. Express **19**(13), 12087–12092 (2011)

9. Meng, L., Klisnick, A., Kozlova, M., Bohacek, K., Krus, M., Prokupek, J., Urbanski, L., Marconi, M.C., Berrill, M., Rocca, J.J., Guilbaud, O., Tissandier, F., Sebban, S., Zeitoun, P., Calisti, A., Ferri, S., Mossé, C., Talin, B.: Temporal coherence and spectral linewidth of neon-like XUV lasers pumped in the quasi-steady state regime. In: Sebban, S., et al. (eds.) X-Ray Lasers 2012. Springer Proceedings in Physics, vol. 147. Springer, Cham (2013). Chapter 28 in this book
10. Isoyan, A., Jiang, F., Cheng, Y.C., Cerrina, F., Wachulak, P., Urbanski, L., et al.: Talbot lithography: self-imaging of complex structures. J. Vac. Sci. Technol. **27**, 2931–2937 (2009)

Chapter 42
Defect Tolerant Talbot Nanopatterning

L. Urbanski, M.C. Marconi, A. Isoyan, A. Stein, C.S. Menoni, and J.J. Rocca

Abstract We report on a defect tolerant extreme ultraviolet (EUV) nanopatterning method. This technique is capable of printing arbitrarily shaped features arranged into periodic arrays with sub-micron resolution. The method is based on the Talbot effect. Masks with different defect layouts were fabricated and tested producing defect free prints regardless of presence of large amount of defects covering up to 1 % of the area of the mask. Numerical simulation was conducted in very good agreement with the experimental data.

1 Introduction

The demand for effective methods for nanostructures fabrication is driven by an increased interest in the field of nanosciences and advocates the emergence of alternative lithographic techniques. Nano structures have been widely investigated in multiple applications such as photonics, plasmonics, microfluidics, biosensors etc. [1–11]. A versatile and robust method to fabricate such structures will contribute to the continuing advancement of this field in the future.

In this paper, we present a nanofabrication technique which has the potential to print features in a photoresist with high spatial resolution and to produce faithful replicas of the original mask even though a significant amount of defects are present in the mask. The method is based on the classical Talbot effect, where a diffraction grating produces replicas of itself under coherent illumination [12, 13]. The image

L. Urbanski · M.C. Marconi (✉) · C.S. Menoni · J.J. Rocca
NSF Engineering Research Center for Extreme Ultraviolet Science & Technology and Department of Electrical and Computer Engineering, Colorado State University, Fort Collins, CO, USA
e-mail: marconi@engr.colostate.edu

A. Isoyan
Synopsys Inc., Hilsboro, OR, USA

A. Stein
Center for Functional Nanomaterials, Brookhaven National Laboratory, Upton, NY, USA

S. Sebban et al. (eds.), *X-Ray Lasers 2012*, Springer Proceedings in Physics 147,
DOI 10.1007/978-3-319-00696-3_42,

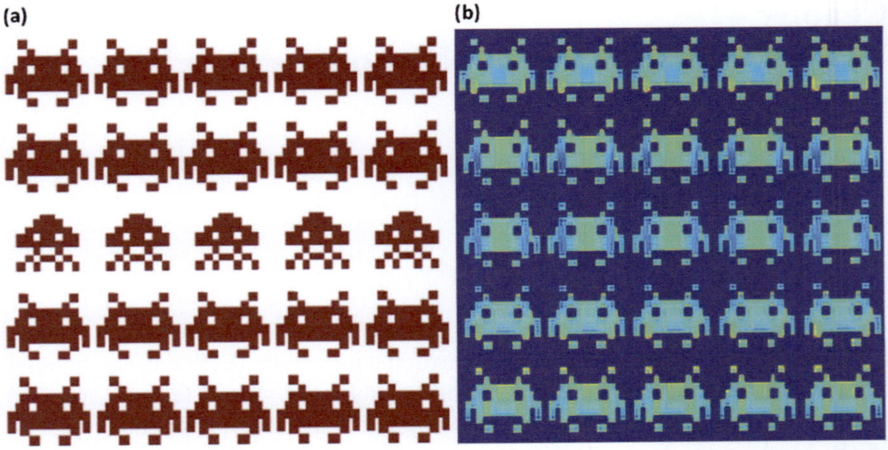

Fig. 1 Mask with 1 % defect (**a**) and corresponding reconstruction at the Talbot plane calculated numerically (**b**)

that replicates the diffraction grating emerges at a specific distance and its integer multiples from the master grating. These distances are called Talbot distances.

In an extension of this concept, the generalized Talbot effect (GTI), the grating is replaced by a periodic two dimensional array of arbitrarily designed motifs also referred to as the diffractive mask, or mask for brevity. Also the GTI renders replicas of the original mask at the Talbot distances, in a similar way as the classical Talbot effect. A more detailed description of the GTI can be found in [14].

The diffractive mask is composed of an array of periodically tiled cells of arbitrary design (native cells). For the purpose of this experiment, the native cell motif has been chosen to be a character form a 1970s video game, "Space Invaders". The defect was represented as a cell with a different motif, also corresponding to a different species of space invader. We will refer the defect as the impostor cell.

In order to characterize the extent of defect tolerance, we have designed masks with purposefully embedded defects. The masks were composed of 10,000 unit cells arranged in a square matrix where a certain number of impostor cells were included with different layouts. In one case a single impostor cell was introduced in the central cell of the mask. In the second case, the entire middle row of native cells was replaced with impostor cells. These two layouts correspond to a defect density of 0.01 % and 1 % respectively.

Figure 1(b) shows the numerical reconstruction of the mask with a row of impostor cells calculated at the first Talbot plane. The figure shows only the 5×5 central section of the 10,000 cells array. The reconstruction was performed using the Fresnel-Kirchhoff diffraction integral formalism to calculate the diffracted light intensity. No sign of defect is present in the reconstruction plane and the impostor cells are completely eliminated in the Talbot plane.

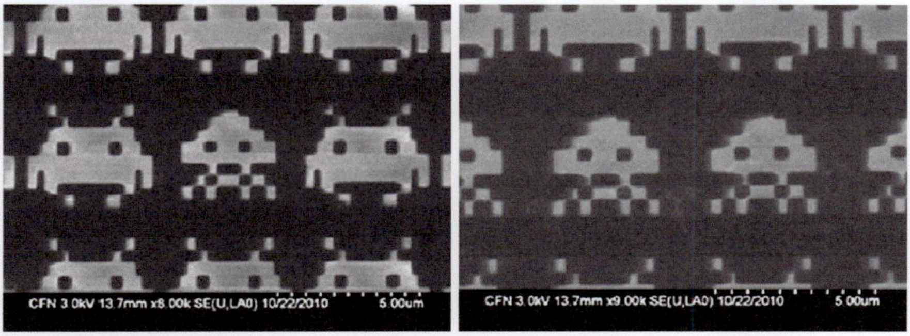

Fig. 2 The diffractive mask layout for two different defect concentrations. *Left*: 0.01 %, *right*: 1 %

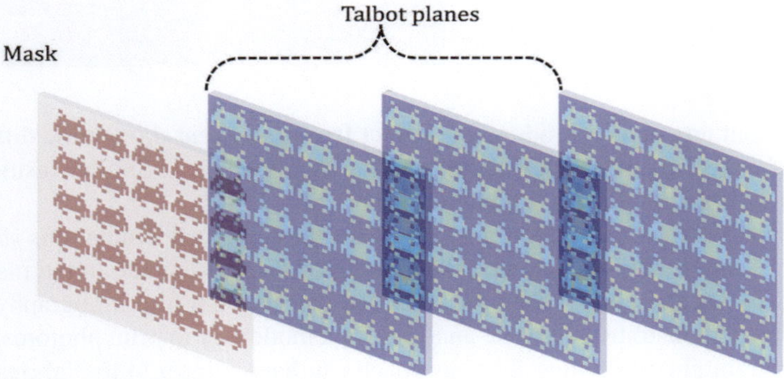

Fig. 3 Scheme of the experimental setup. In-line reconstruction of the mask

2 Experiment

In order to verify the numerical result experimentally, two masks were fabricated. To fabricate the masks with reasonable diffraction efficiency we used a thin (25 nm) silicon nitride membrane that provided an adequate transparent substrate on which the cells were defined in HSQ photoresist by electron beam lithography.

Figure 2 shows the SEM micrographs of the two masks. In the left hand side is pictured the central region of the mask with a single impostor cell. The right hand side of Fig. 2 is the micrograph of the second mask where the central row was replaced by the impostor motif. More details on the mask fabrication can be found in [14].

The schematic of the experimental setup is represented in Fig. 3. The mask with a defect was illuminated with coherent beam from a table top capillary discharge soft X-ray laser. At the first Talbot plane a resist-coated silicon wafer was placed in order to record the self-image. The exposed resist was then developed and scanned using an atomic force microscope (AFM). Figure 4 shows the AFM micrograph of the print obtained for the mask with a row of defects. The print recreates the original

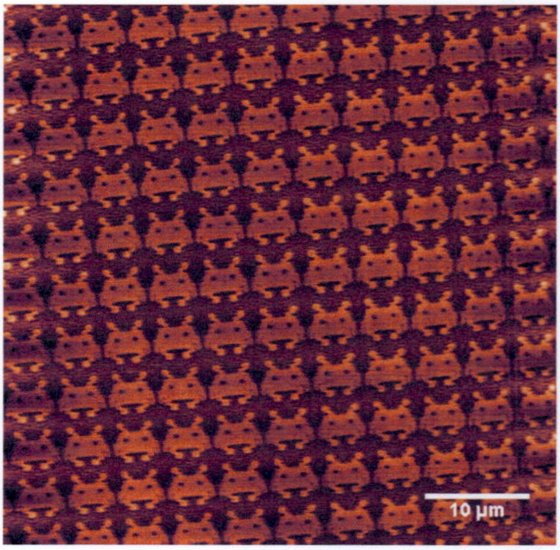

Fig. 4 Atomic force microscope scan of the resist patterned with a defective mask with 1 % defect

mask without any defect, producing a defect free print event though the diffractive mask had a severe defect equivalent to 1 % of the total area. Similar results were obtained with the mask with a single impostor cell.

This result is in very good correspondence with numerical simulations shown in Fig. 1. It is important to notice that the amount of defect included in the masks we tested is significantly larger than the tolerable defect count in a lithography mask. Our results demonstrate a versatile and robust methodology to print photoresist with defect tolerant characteristics. This approach can have impact in the fabrication of periodic structures like plasmonic surfaces, photonic crystal structures, calibration gratings, etc. and can contribute to the widespread of small scale nanofabrication.

Acknowledgements This work was supported by the National Science Foundation, award ECCS 0901806, the NSF ERC for Extreme Ultraviolet Science and Technology, award EEC 0310717. This research was carried out in part at the Center for Functional Nanomaterials, Brookhaven National Laboratory, which is supported by the U.S. Department of Energy, Office of Basic Energy Sciences, under Contract No. DE-AC02-98CH10886.

References

1. Kneipp, K., et al.: Single molecule detection using surface-enhanced Raman scattering (SERS). Phys. Rev. Lett. **78**(9), 1667–1670 (1997)
2. Liu, Z.W., et al.: Focusing surface plasmons with a plasmonic lens. Nano Lett. **5**(9), 1726–1729 (2005)
3. Liu, Z.W., Wei, Q.H., Zhang, X.: Surface plasmon interference nanolithography. Nano Lett. **5**(5), 957–961 (2005)
4. Luo, X.G., Ishihara, T.: Surface plasmon resonant interference nanolithography technique. Appl. Phys. Lett. **84**(23), 4780–4782 (2004)
5. Soukoulis, C.M., Wegener, M.: Optical metamaterials—more bulky and less lossy. Science **330**(6011), 1633–1634 (2010)

6. Vedantam, S., et al.: A plasmonic dimple lens for nanoscale focusing of light. Nano Lett. **9**(10), 3447–3452 (2009)
7. Brueck, S.R.J.: Optical and interferometric lithography—nanotechnology enablers. Proc. IEEE **93**(10), 1704–1721 (2005)
8. Chen, Y., et al.: Nanoimprint fabrication of micro-rings for magnetization reversal studies. Microelectron. Eng. **57–58**, 405–410 (2001)
9. Chou, S.Y., Krauss, P.R.: Imprint lithography with sub-10 nm feature size and high throughput. Microelectron. Eng. **35**(1–4), 237–240 (1997)
10. Himmelhaus, M., Takei, H.: Self-assembly of polystyrene nano particles into patterns of random-close-packed monolayers via chemically induced adsorption. Phys. Chem. Chem. Phys. **4**(3), 496–506 (2002)
11. Juillerat, F., et al.: Fabrication of large-area ordered arrays of nanoparticles on patterned substrates. Nanotechnology **16**(8), 1311–1316 (2005)
12. Talbot, W.H.F.: Facts relating to optical science. Philos. Mag. **9**, 403 (1836)
13. Rayleigh, Lord, F.: On copying of diffraction gratings, and on some phenomena connected therewith. Philos. Mag. (1881)
14. Isoyan, A., et al.: Talbot lithography: self-imaging of complex structures. J. Vac. Sci. Technol. **27**(6), 2931–2937 (2009)

Chapter 43
Imaging in Nanoscale Using Laser-Plasma Sources of Extreme Ultraviolet (EUV)

P.W. Wachulak, A. Bartnik, A. Baranowska-Korczyc, D. Pánek, P. Brůža, J. Kostecki, Ł. Węgrzyński, R. Jarocki, M. Szczurek, K. Fronc, D. Elbaum, and H. Fiedorowicz

Abstract New developments in nanoscience and nanotechnology require nanometer scale resolution imaging tools and techniques such as an extreme ultraviolet (EUV) and soft X-ray (SXR) microscopy, based on Fresnel zone plates. In this paper, we report on applications of a desk-top microscopy using a laser-plasma EUV source based on a gas-puff target for studies of morphology of thin silicon membranes coated with NaCl crystals and samples composed of ZnO nanofibers.

1 Introduction

Recent rapid developments of nanoscience and nanotechnology require nanometer scale resolution imaging tools and methods. One of the methods, extensively studied for the last few decades, is an extreme ultraviolet (EUV) and soft X-ray (SXR) microscopy, based on Fresnel zone plates (FZP) [1]. The introduction of compact sources of bright EUV and SXR radiation paved the way for the development of table-top microscopes that can render images of nanoscale objects with short exposures and spatial resolution approaching that of synchrotron-based microscopes [2–4].

In this paper, we report on applications of a desk-top microscopy using a laser-plasma EUV source, based on a gas-puff target, for thin-film studies of silicon membrane coated with NaCl crystals and imaging of ZnO nonofibers. The fact of recently increased interest in ZnO nanostructures in electronics is due to their morphology, availability, possibility of doping with other materials, low cost of fabrication and processing [5] and possibility to use in sensor applications [6]. Additionally, our

P.W. Wachulak (✉) · A. Bartnik · J. Kostecki · Ł. Węgrzyński · R. Jarocki · M. Szczurek · H. Fiedorowicz
Institute of Optoelectronics, Military University of Technology, Warsaw, Poland
e-mail: wachulak@gmail.com

A. Baranowska-Korczyc · K. Fronc · D. Elbaum
Institute of Physics, Polish Academy of Sciences, Warsaw, Poland

D. Pánek · P. Brůža
Czech Technical University, Prague, Czech Republic

S. Sebban et al. (eds.), *X-Ray Lasers 2012*, Springer Proceedings in Physics 147,
DOI 10.1007/978-3-319-00696-3_43,

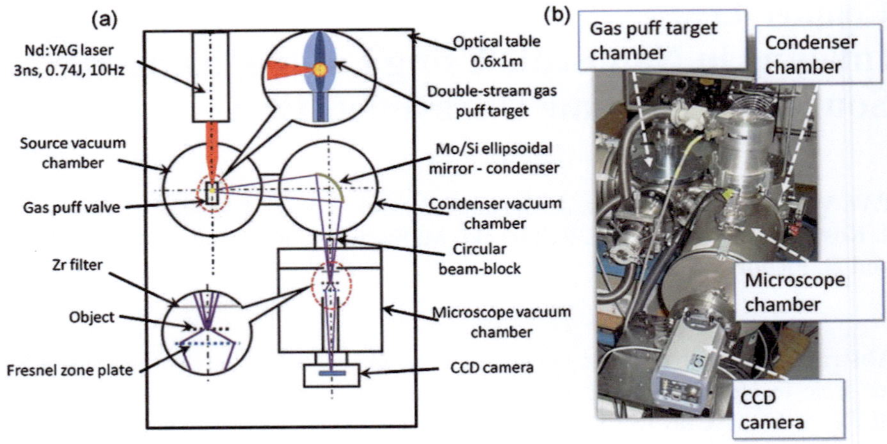

Fig. 1 An experimental arrangement of the EUV microscope (not to scale) using a laser-plasma EUV source (**a**) and its photograph in (**b**)

motivation is to test recently developed EUV microscopy setup on non-trivial objects. Utilization of the short wavelength EUV radiation allows to demonstrate the intrinsic advantage of this radiation for extraction of additional information about the investigated object, which cannot be obtained directly from optical micrographs and SEM images. Moreover, this microscope does not require additional sample preparation necessary for SEM microscopy.

2 Experimental Arrangement

The scheme of the experimental arrangement of the EUV microscope is shown in Fig. 1(a) and the photograph in (b). The source, emitting short wavelength radiation that is used for imaging, is a laser-plasma EUV source, presented in [7] and later modified for quasi-monochromatic emission in the 13–14 nm wavelength range, as described in [8]. The gaseous target does not produce any debris associated with solid targets and allow to change the emission bandwidth and the wavelength of emitted radiation simply by changing a gas. EUV radiation from the argon plasma was focused onto a sample and spectrally filtered by a Mo/Si multilayer ellipsoidal mirror.

We have imaged three different samples, to show advantages of EUV microscopy comparing to SEM and optical microscopy. Firstly we used NaCl crystallized on a 15 nm thick non-porous silicon membrane (SPI Supplies) with dimensions 0.07×1.5 mm^2. The substrate, a silicon wafer 3×3 mm^2 in size and 100 μm thick, contains two rectangular shape membrane windows. The NaCl crystals were prepared from phosphate buffered saline (PBS) solution (Lonza, Cat. No. BE17-516Q, contents: KH_2PO_4—144 mg/L, NaCl—9000 mg/L, Na_2HPO_4—795 mg/L). Five microliters of the solution were deposited onto the membrane. The water from the solution was evaporated at the temperature of 21 °C and atmospheric pressure, resulting in re-crystallization of NaCl. During the process a stress was introduced to the membrane causing random cracks of various sizes to appear in the NaCl crystal film and the silicon membrane. Second type of sample was composed of ZnO nanofibers. A 10 wt.% poly(vinyl alcohol) (PVA, MW = 72,000) solution was prepared from PVA powder and deionized water. Then 0.25 g of zinc acetate dihydrate ($Zn(CH_3COO)_2 \cdot 2H_2O$) was dissolved in 1.25 g solution of PVA as a precursor to obtain ZnO nanofibers. Solution, prepared in this way, a precursor, was placed in a syringe and then a constant flow rate of 0.1 ml/h through a syringe pump was established. For electrospinning a high electric potential of 10 kV was applied between the metal tip of a needle of the syringe and grounded collector. The fibers were formed in ambient temperature at 22 °C and humidity in the range from 30 % to 40 %. Finally, the process of calcination at 500 °C for 4 h in air was performed, producing high length/diameter aspect ratio ZnO fibers attached to gold mesh by van der Waals forces. The fibers, deposited in this way, can be seen in transmission mode in rectangular openings of the mesh. The third sample was similar to the second one: ZnO nanofibers were electrospun on top of a 50 nm thick silicon nitride membrane but without the calcination process.

The objects are imaged using a FZP objective (Zone Plates Ltd.) onto a EUV sensitive CCD camera iKon-M, DO934N-BN (Andor). Previously, different test objects (samples) were imaged using this microscope: Cu mesh with 5 μm bars and thickness of ~4 μm [3] and carbon foil, 70 nm thick, with 1.5 μm diameter holes [4], where the EUV images have been obtained with a half-pitch spatial resolution approaching ~50 nm (3.7λ) in a very compact setup (more technical information about the microscope can be found there). The required exposure for imaging of NaCl crystals, deposited on the silicon membrane, was 200 EUV pulses, while for the nanofiber samples −100 EUV pulses at 2 Hz repetition rate. The CCD camera was cooled down to −20 °C to decrease intrinsic noise during the image acquisition.

3 EUV Imaging Results

The objects were imaged using optical, SEM and EUV microscopes. Optical image of the NaCl crystals, deposited on top of a membrane, obtained with 100× (NA ~ 1) objective is shown in Fig. 2(a). Dashed box indicates the region imaged later with the EUV microscope. For the purpose of comparison, the same area of the sample was imaged with an SEM microscope (Tescan VEGA II SBU) and with the EUV microscope. SEM image was obtained after deposition of ~10 nm of Au to avoid charging during image acquisition at 30 kV at magnification of 4000× (sample to screen magnification). SEM image of the sample, ~54×54 μm^2 in size, is shown in Fig. 2(b) and corresponding EUV image—in Fig. 2(c). To show the same area, EUV image was obtained by stitching and partially overlapping 4 × 4 smaller EUV subimages. Typically, the magnification was 520× (FZP magnification). This results in 25.9×25.9 nm^2 pixel size and the field of view 26.4×26.4 μm^2.

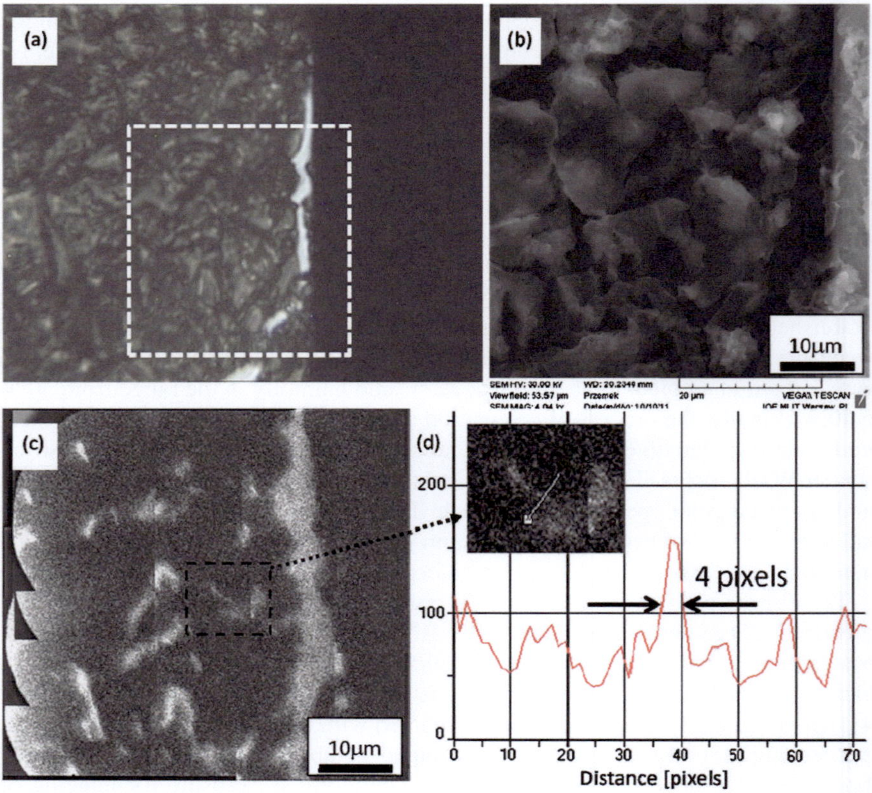

Fig. 2 Optical image of thin Si membrane with re-crystallized NaCl layer deposited on top, obtained with 100× (NA ~ 1) objective (**a**). *Dashed box* indicates the region imaged with the EUV and SEM microscopes. SEM image of the sample, ~54 × 54 μm^2 with re-crystallized NaCl (**b**) and corresponding EUV image (**c**) showing the same region of the sample. Intensity profiles of one of the smallest features in the EUV image showing visible features as small as 4 pixels (**d**)

The cracks in thin silicon membrane are usually quite large, observed with the optical microscope, in the range of one to tens of microns. Crystal film exhibits also a variable transmission to the visible light, but to extract any additional information another techniques need to be used. The SEM image shows NaCl crystal surface in detail. Its complex structure obscures, however, the information about the underlying membrane. In the SEM image, it is possible to see the surface features, such as cracks, but in some cases, it is not possible to determine their depth.

The EUV mosaic, on the other hand, does not show any surface features because the thick layer of crystallized salt has very low transmission at EUV wavelengths. However, the EUV image shows regions with low thickness, smaller than approx. ~100 nm, which are represented by the bright regions. These are very deep gaps in the NaCl layer and small cracks and holes in the supporting Si membrane. Intensity of these regions is in this case related to their transmission. Moreover, as can be seen

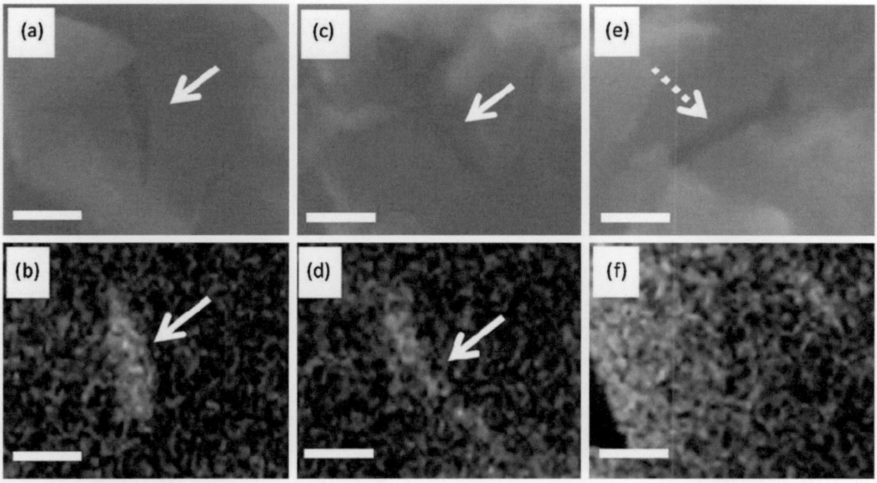

Fig. 3 Comparison between small sections of SEM images (**a**, **c**, **e**) and EUV images (**b**, **d**, **f**) showing different regions of the sample. Some cracks in the sample are deep all the way to the membrane and can be seen in both SEM and EUV images (marked with *white arrows*)—(**a–d**) and some of the features (marked with *dotted arrow*), are visible in the SEM image (**e**) but are not present in the corresponding EUV image (**f**). Scale bar is 2 μm

in Fig. 2(d), features with characteristic size as small as 4 pixels can be resolved. For the given magnification and pixel size of 25.9 nm square, the FWHM width of this small feature approaches 100 nm.

One can notice that there are some features, marked with white arrows in Fig. 3(a–d), that are visible in both SEM and EUV images, showing exactly the same region of the sample. There are also, however, some cracks in the crystal film, visible in the SEM image, such as one marked with dotted arrow in Fig. 3(e), but not present in the EUV image, as can be seen in Fig. 3(f).

Interestingly, some features, such as the cracks in Fig. 3(a, c, e), quite similar in the SEM images, show completely different depth in corresponding EUV images. Because of EUV images we can tell, that the ones, shown in Fig. 3(a, c), are undoubtedly deep all the way to the supporting membrane, while another one in Fig. 3(e) is only a near-surface feature. Thus, the EUV radiation enables additional, complementary information about the object to be obtained, which is based on an intrinsic property of EUV light, and that cannot be obtained directly using SEM microscopy or visible light microscopy.

Similarly, the second sample, composed of ZnO nanofibers, was imaged employing three, previously mentioned, microscopes. Optical micrograph of small region, $\sim 20 \times 20\ \mu m^2$ in size, of the mesh, containing the nanofibers, near one of the bars of the mesh, is shown in Fig. 4(a). The micrograph was obtained with the same objective. Again, for comparison, the same region was imaged using EUV microscope and SEM microscope (Quanta 3D FEG, from FEI). SEM image was obtained at 30 kV with magnification of $\sim 5000\times$ (sample to screen magnification). SEM im-

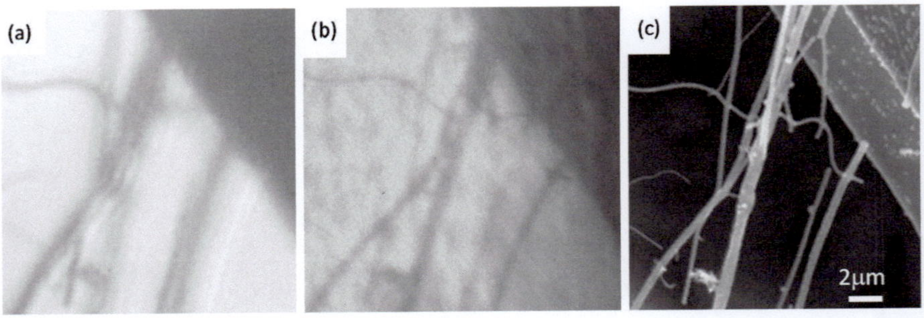

Fig. 4 Comparison between images of ZnO nanofibers, image size ~$20 \times 20\ \mu m^2$, obtained with different methods. Optical micrograph (**a**) with 100×, NA ~ 1 objective, EUV microscope image (**b**), 100 EUV pulses exposure, 690× magnification, NA ~ 0.138, (**c**) SEM micrograph at acceleration voltage of 30 kV, magnification ~5000× (sample to screen magnification)

Fig. 5 EUV microscope image showing ZnO nanofibers. Nanofibers as small as 100 nm are visible, especially in the contrast enhanced section

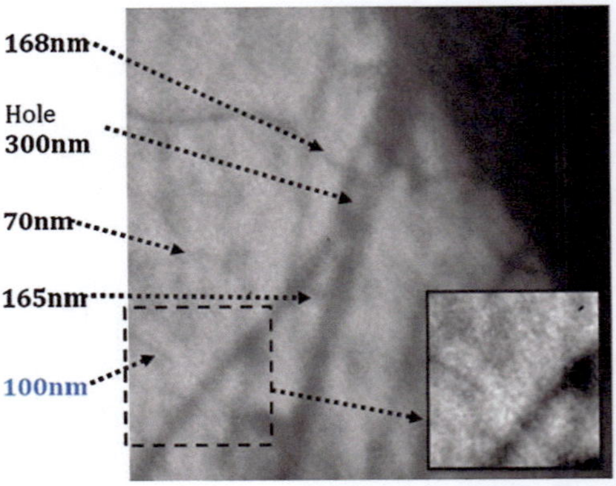

age of the sample, ~$20 \times 20\ \mu m^2$ in size, is shown in Fig. 4(c) and corresponding EUV image—in Fig. 4(b). For the EUV image the magnification was 690× (FZP magnification).

It is easy to notice, that EUV image of the nanofibers has much better spatial resolution than the diffraction limit of the optical microscope, theoretically equal to $0.61\lambda/NA \sim 300$ nm ($\lambda = 500$ nm, NA ~ 1).

Figure 5 shows some characteristic dimensions of the sample, where single, separated nanofibers, as small as 100 nm, are visible. Smaller nanofibers, ~70 nm in diameter, scatter much less light, their transmission is higher and they appear much wider due to the diffraction limit, $0.61\lambda/NA = 61$ nm ($\lambda = 13.8$ nm, NA ~ 0.138).

The last sample, composed of ZnO nanofibers on top of a silicon nitride membrane, imaged using EUV microscope can be seen in Fig. 6(a, b). Size of the EUV images is ~$20 \times 20\ \mu m^2$, 100 EUV pulses exposure, 690× magnification. White, solid arrows indicate cracks in 50 nm thick silicon nitride membrane, while dotted arrows show ZnO nanofibers. The fibers are larger here, usually between 380 and 570 nm in diameter, as measured from the EUV image.

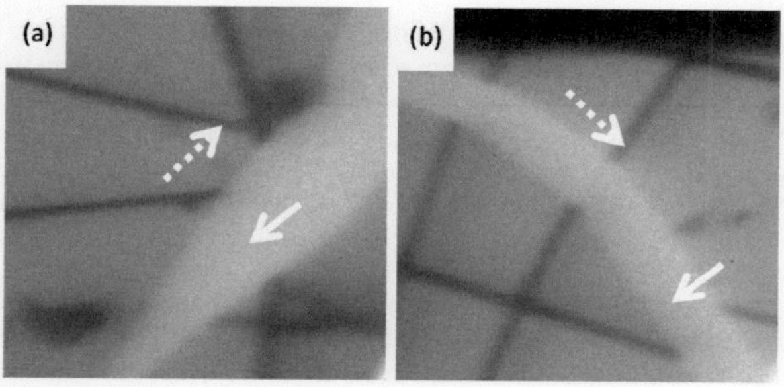

Fig. 6 Two examples of EUV images of ZnO nanofibers (no annealing) deposited on silicon nitride membrane (**a**, **b**), image size $\sim 20 \times 20\ \mu m^2$, 100 EUV pulses exposure, $690\times$ magnification, NA ~ 0.138. *Solid arrows* indicate cracks in 50 nm thick silicon nitride membrane, *dotted arrows* show ZnO nanofibers 380–570 nm in diameter

4 Conclusions

We have demonstrated imaging of nanostructures using a desk-top EUV transmission microscope based on laser-produced plasma from argon-based gas-puff target and Fresnel optics. Features with characteristic dimensions as small as 100 nm were resolved; moreover, we showed that EUV light allows straightforward extraction of meaningful information about sub-100 nm thin, dielectric films by a direct observation. No specific sample preparation is required, such as conductive coating for the SEM and it is still photon-based imaging, important in some applications, such as thin film studies of electron beam photoresists. The short attenuation length of EUV photons in almost any material can be utilized to perform imaging with enhanced optical contrast. Thus, EUV microscopy can be employed in thin layer studies providing information complementary to optical and SEM microscopy. We also presented results of imaging of ZnO nanofibers, showing improved spatial resolution and good optical contrast of this material at EUV wavelengths.

The work, presented in this paper, is a part of an effort to develop a novel compact high-resolution imaging system that may find applications in different areas of science and technology, such as material science, biomedicine, nanotechnology, and other fields. Our future plans also include development of a SXR microscope based on Ar gas puff target source [9], capable of imaging in the XUV "water window" spectral region.

Acknowledgements The research was supported by the Foundation for Polish Science, HOMING 2009 Programme, Grant number HOM/2009/14B, the Ministry of Science and Higher Education of Poland, LASERLAB-EUROPE III and GA AV CR KAN300100702 project. We thank L. Pina and T. Feigl for their help in the preparation of EUV optics.

References

1. Attwood, D.: Soft X-Rays and Extreme Ultraviolet Radiation. Cambridge University, Cambridge (1999)

2. Vaschenko, G., Brewer, C., Brizuela, F., Wang, Y., Larotonda, M.A., Luther, B.M., Marconi, M.C., Rocca, J.J., Menoni, C.S.: Sub-38 nm resolution tabletop microscopy with 13 nm wavelength laser light. Opt. Lett. **31**, 1214–1216 (2006)
3. Wachulak, P.W., Bartnik, A., Fiedorowicz, H.: Sub-70 nm resolution tabletop microscopy at 13.8 nm using a compact laser–plasma EUV source. Opt. Lett. **35**(14), 2337–2339 (2010)
4. Wachulak, P.W., Bartnik, A., Fiedorowicz, H., Kostecki, J.: A 50 nm spatial resolution EUV imaging–resolution dependence on object thickness and illumination bandwidth. Opt. Express **19**(10), 9541–9550 (2011)
5. Djurisic, A.B., Ng, A.M.C., Chen, X.Y.: ZnO nanostructures for optoelectronics: material properties and device applications. Prog. Quantum Electron. **34**, 191–259 (2010)
6. Stafiniak, A., Boratynski, B., Baranowska-Korczyc, A., Szyszka, A., Ramiaczek-Krasowska, M., Prazmowska, J., Fronc, K., Elbaum, D., Paszkiewicz, R., Tłaczała, M.: A novel electrospun ZnO nanofibers biosensor fabrication. Sens. Actuators B **160**, 413–1418 (2011)
7. Fiedorowicz, H., Bartnik, A., Jarocki, R., Kostecki, J., Krzywinski, J., Mikołajczyk, J., Rakowski, R., Szczurek, A., Szczurek, M.: Compact laser plasma EUV source based on a gas puff target for metrology applications. J. Alloys Compd. **401**, 99 (2005)
8. Wachulak, P.W., Bartnik, A., Fiedorowicz, H., Feigl, T., Jarocki, R., Kostecki, J., Rakowski, R., Rudawski, P., Sawicka, M., Szczurek, M., Szczurek, A., Zawadzki, Z.: A compact, quasi-monochromatic laser-plasma EUV source based on a double-stream gas-puff target at 13.8 nm wavelength. Appl. Phys. B **100**(3), 461–469 (2010)
9. Wachulak, P.W., Bartnik, A., Fiedorowicz, H., Rudawski, P., Jarocki, R., Kostecki, J., Szczurek, M.: "Water window" compact, table-top laser plasma soft X-ray sources based on a gas puff target. Nucl. Instrum. Methods Phys. Res., Sect. B, Beam Interact. Mater. Atoms **268**(10), 1692–1700 (2010)

Chapter 44
Experiments to Diagnose Plasma with a Soft X-Ray Laser Double-Frequency Grating Interferometry

C. Wang, H.H. An, Z.H. Fang, J.R. Sun, W. Wang, W.D. Zheng, X.M. Qiao, and S.J. Wang

Abstract We developed a kind of soft x-ray double frequency grating shearing interferometry to diagnose hot dense plasma electron density. The method is very suitable for the measure on high-Z material plasma near critical surface, because of its special advantages of suppression of plasma radiation noise and responding to density dispersion. Experiments on diagnosing a series of plasma, such as plasma of Au foil target, plasma inside a tube, collision plasma between different material, were carried out with this method. The experiments acquired good interference pattern images. The high quality images well proved the method was stable and practical.

1 Introduction

Experimental diagnosis on hot dense plasma near critical surface has important meaning to the research works of inertial control fusion (ICF) and plasma physics fields. There is special advantage of soft x-ray laser probe method, which has become an valid and practical technique. Many methods of measuring plasma have been developed with soft x-ray laser as a probe, such as interferometry [1–6], deflectometry [7, 8], shadow image [9] and Tomeson scatter [10, 11] in the past years. The interferometry is more advantage for understanding plasma developing process and comparing between theoretic simulations and experimental results.

The first demonstration of soft x-ray laser interferometry on measuring plasma was exhibited in 1995 by LLNL [1]. They used multilayered component as beamsplitter and made use of a tilt Mach-Zehnder interferometry. A few years later, the author's group also repeated the same experiments [5]. The well results of both experiments indicated the interferometry were feasible with advantages of clear optical road, big sight field and simple data processing. However there are some considerable points for real practice. The first one was surface figure problem of multilayered

C. Wang (✉) · H.H. An · Z.H. Fang · J.R. Sun · W. Wang · S.J. Wang
Shanghai Institute of Laser Plasma, Shanghai, China
e-mail: wangch@mail.shcnc.ac.cn

W.D. Zheng · X.M. Qiao
Institute of Applied Physics and Computational Mathematics, Beijing, China

S. Sebban et al. (eds.), *X-Ray Lasers 2012*, Springer Proceedings in Physics 147,
DOI 10.1007/978-3-319-00696-3_44,

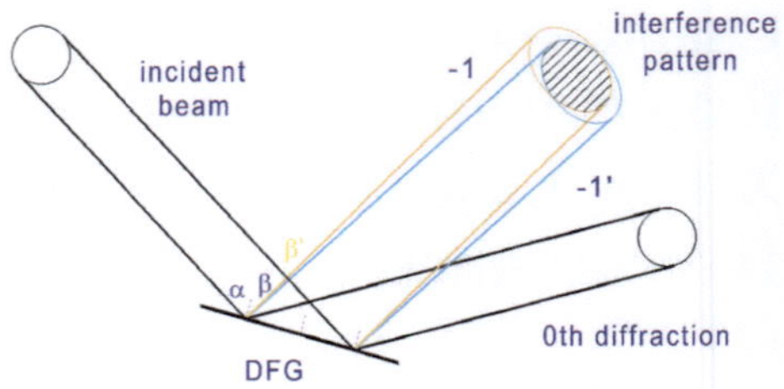

Fig. 1 Basic principle of a DFG interference

beam-splitter. The second one was very complicated regulation of optical road and the last one was hard request of soft x-ray laser probe source. A new technique, which used gratings as beam-splitter, was developed and well resolved the surface figure problem of multilayered beam-splitter in about 2000 [4]. But the others difficulties still existed. Besides of M-Z interferometry, some other interferometries, such as Lloyd's mirror [3] and Fresnel bi-mirrors, were tried to measure plasma, but these were less then M-Z method at both stripe quality and applied effect.

In recent years, we have developed a kind of method, so called soft x-ray laser double frequency grating (DFG) interferometry. The DFG interferometry is a common technique in visible light band [12, 13]. There is no report that this method was carried out in soft x-ray wave band. The method is good to avoid the difficulties of M-Z interferometry system, and the stripe quality was also very well.

2 Basic Principle of DFG Interference

The key element is a DFG that owns two kinds of different frequency periods, such as d and d'. When a beam of light injects a DFG, all the 0 order diffraction beams are at the same direction and the DFG is just like a plane mirror. But for other order diffraction, for example -1st order, each beam is divided to two beams with difference directions due to the different grating frequency periods. The diffraction angles β and β' accord to the formula:

$$\begin{cases} \sin\beta - \sin\alpha = -\lambda/d \\ \sin\beta' - \sin\alpha = -\lambda/d' \end{cases} \tag{1}$$

where α is incident angle, λ is the wavelength of light. There may be a large area where the two beams overlap and the interference may happen. The basic principle is shown in Fig. 1.

The two beams which participate an interference are in fact from a same beam of light by deviating a part of distances breadthwise each other, and the optical path difference comes from the subtraction value between different positions of the same beam. The optical path difference causes the deflexion of interference fringes, from which the electron density distribution can be calculated. The given result is not the real electrical density distribution but a subtraction difference one, labeled $\Delta n_e(z)$ if only taking one dimension condition as an example, where z is the vertical direction of interference fringe. Then the real electron density can be obtained by:

$$n_e(z) - n_e(z + z_\Delta) = \Delta n_e(z) \tag{2}$$

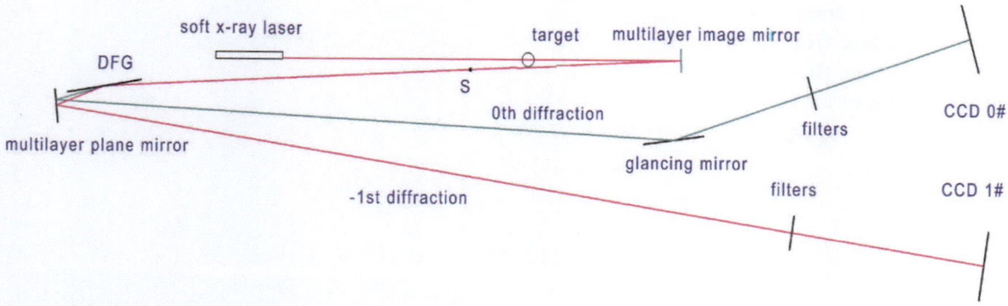

Fig. 2 The experimental setup of DFG interferometry

where z_{Δ} is the respective offset at z direction, and $n_e(z)$ means the electron density at the position of z.

3 Experimental Setup

The experiments were carried out with SG-II high power laser system of Chinese National Laboratory on High Power Laser and Physics. The experimental setup is schematically shown in Fig. 2. The system consisted of a soft x-ray laser source, an image multilayer mirror, a DFG and several plane multilayer mirrors.

A collisionally pumped nickel-like silver soft x-ray laser operating at 13.9 nm was used as the probe source. The x-ray laser was produced with double targets coupling technique by irradiating two 16 mm long Ag-coated glass targets with two beam from SG-II ($\lambda = 1.053$ μm, 70 ps, 70 J, 1 % prepulse ahead 3 ns) at an intensity of 6×10^{13} W cm^{-2}. The output energy of the x-ray laser is approach to several hundreds μJ. The divergence is approximately 3 mrad × 5 mrad and output pulse duration 30~50 ps. The short pulse and high brightness of the soft x-ray laser allowed us to obtain an interferogram in a single 50 ps exposure. The exposure time was so short that the effects of vibrations could be ignored. The probe source injected directly into plasma to be probed where was 500 mm from x-ray laser without collimation, so the field of view was about 1.5 mm × 2.5 mm and it is enough to measure all the interesting area of large scale plasma.

The plasma to be probed produced by another beam laser irradiating a target was imaged onto a CCD with a 10 mm radius of curvature multilayer mirror with the focus of about 270 mm. A series of plasma may be studied by changing different targets and drive conditions. We mainly payed attention to the 0 and −1st order diffraction, so two CCD were used to record the signals respectively. The CCD 0# recorded the 0 diffraction and the image was just like a shadow of plasma. The CCD 1# recorded the −1st diffraction that might show the interference patterns.

The DFG is the most critical element of this system. For our experiments we chose the frequency periods of 1000 and 1003 lines per mm. The incident angle was designed as about 84°. The diffraction angle difference $\beta - \beta'$ was about 0.012° and the fringe separation could be estimated to about 150 μm.

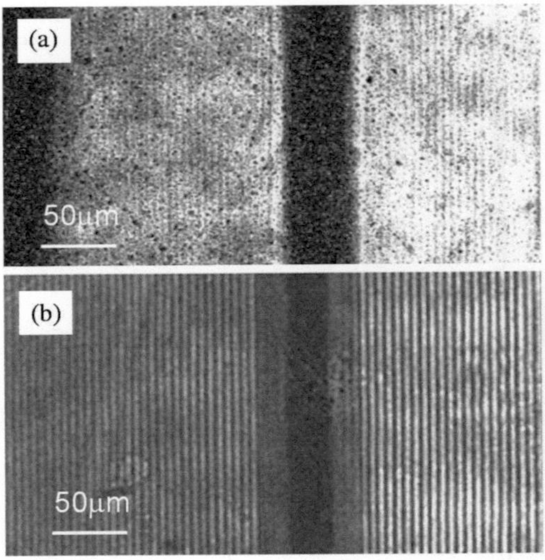

Fig. 3 Experimental static images. (**a**) came from the 0 order diffraction. (**b**) came from the −1st order diffraction

4 Experimental Results

Figure 3 shows a typical experimental static images. The static experiments means the object was only a target and no plasma was created. So the images were not disturbed by plasma. Figure 3(a) came from the 0 order diffraction signal, so it can be regarded as the static target shadow image. The shadow of target was clear, however some slight stripes can be seen, which could be caused by the effect of beat frequency of DFG. Figure 3(b) came from the -1 order diffraction signal, so it is a static DFG interference pattern. There were two clear target shadows overlapped each other. The interference stripes are clearly observed in Fig. 3(b), and the fringe visibility is about 0.5 ± 0.1.

Figure 4 shows a series of DFG interference experimental results with different plasmas. Figure 4(a) and (b) show the simple experimental setup and the result of collision plasma. The plasma was created by a drive laser irradiating a pair of vertical plane target with a lens array uniform focus of about 450 μm $\times$ 450 μm. The two targets were coated with 5 μm-thickness aurum, so the component of plasma was Au ions. The clear inference stripes and obvious stripe movement indicate abundant physical contents.

Figure 4(c) and (d) showed another experimental setup and the result of the plasma in a hohlraum. The hohlraum was indeed a tube cavity with a hole at side. The drive laser injected from the hole and heated the plasma in hohlraum, and a soft x-ray laser probe passed along the axis direction of hohlraum and diagnosed the plasma. (d) is the experimental image and the stripes in hohlraum was fairly clear.

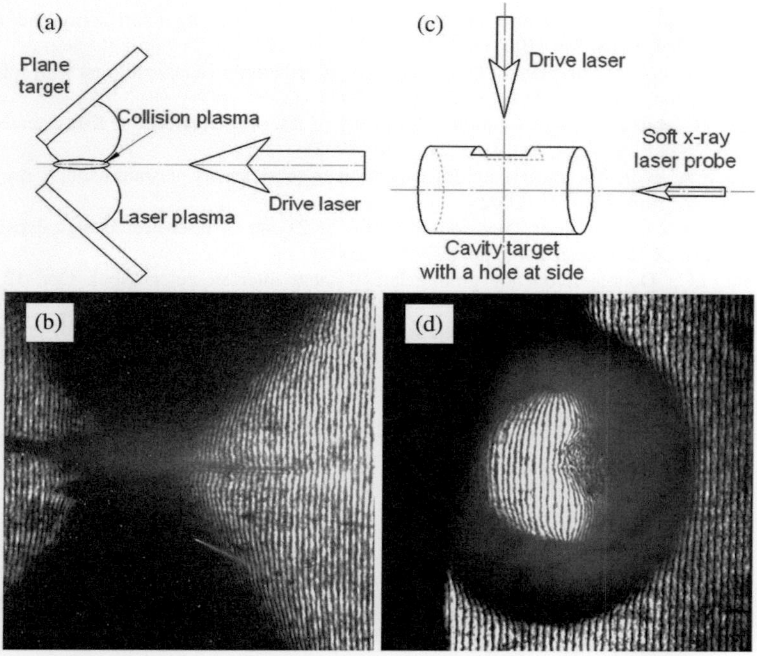

Fig. 4 Some DFG interference experimental results. (**a**) and (**b**) showed the setup and result about collision plasma. (**c**) and (**d**) showed the setup and result about the plasma in a hohlraum

5 Conclusion

The experiments show the method of soft x-ray laser DFG interferometry is a good tool to diagnose laser plasma near critical density, especially high-Z matter plasma. The experimental images of inference patterns are simply shown and the data processing are going now.

References

1. Da Silva, L.B., et al.: Electron density measurements of high density plasmas using soft X-ray laser interferometry. Phys. Rev. Lett. **74**, 3991 (1995)
2. Wan, A.S., et al.: Electron density measurement of a colliding plasma using soft-x-ray laser interferometry. Phys. Rev. E **55**, 6293 (1997)
3. Rocca, J.J., et al.: Soft-x-ray laser interferometry of a plasma with a tabletop laser and a Lloyd's mirror. Opt. Lett. **24**, 420 (1999)
4. Filevich, J., et al.: Dense plasma diagnostics with an amplitude-division soft-x-ray laser interferometer based on diffraction gratings. Opt. Lett. **25**, 356 (2000)
5. Wang, C., et al.: Experimental diagnoses of plasma electron density by interferometry using an x-ray laser as probe. Acta Phys. Sin. **54**, 202 (2005)
6. Purvis, M., et al.: Dynamics of converging laser-created plasmas in semicylindrical cavities studied using soft x-ray laser interferometry. Phys. Rev. E **76**, 046402 (2007)

7. Ress, D., et al.: Measurement of laser-plasma electron density with a soft X-ray laser deflectometer. Science **265**, 514 (1994)
8. Wang, C., et al.: Measurement of electron density distribution in a laser plasma with soft x-ray laser deflectometry. Acta Phys. Sin. **51**, 847 (2002)
9. Cauble, R., et al.: Micron-resolution radiography of laser-accelerated and X-ray heated foils with an X-ray laser. Phys. Rev. Lett. **74**, 3816 (1995)
10. Baidis, H.A., et al.: Thomson scattering diagnostic of solid density plasmas using X-ray lasers. Rev. Sci. Instrum. **73**, 4223 (2002)
11. Rus, B., et al.: X-ray laser Thomson scattering at 21 nm of laser-heated high-density foil plasmas. In: X-Ray Laser 2006, pp. 577–583 (2006)
12. Wyant, J.C.: Double frequency grating lateral shear interferometer. Appl. Opt. **12**, 2057 (1973)
13. Ming, H., et al.: Phase shifting in shearing interferometry with double frequency grating and semiconductor laser to measure temperature field. In: Proc. SPIE, vol. 2321, p. 246 (1994)

Chapter 45
Heavy-Ion Spectroscopy with X-Ray Lasers at GSI

B. Zielbauer, B. Ecker, P. Neumayer, K. Cassou, S. Daboussi, O. Guilbaud, S. Kazamias, D. Ros, T. Kuehl, U. Eisenbarth, S. Goette, D. Winters, V. Bagnoud, and T. Stoehlker

Abstract Different pumping schemes for soft X-ray lasers have been investigated at the PHELIX laser facility, including a double-target seeding approach at 18.9 nm. A technical feasibility study of using a Mo XRL beam of several μJ as an excitation source for heavy-ion spectroscopy in a storage ring has been carried out. XRL photon numbers and the beam transport under ultra-high vacuum conditions over almost 30 m are the major challenges.

1 Introduction

One of the strong motivations to develop XRL sources at GSI and FAIR was their applicability in the spectroscopy of highly charged heavy-ions in storage rings. For example, in Li-like systems, the measurement of fine structure transitions enables a direct experimental access to nuclear structure which allows the comparison to theoretical predictions [1]. Several such investigations have already been carried out using lasers in the visible wavelength range, but the application of XRLs with their short wavelength will allow a much deeper insight. And although the XRL wavelength cannot be tuned to perform scans, the effective wavelength exciting the stored ions can be changed very precisely by changing the ions' velocity and using the Doppler up-shift in counter-propagation.

B. Zielbauer (✉) · B. Ecker · T. Kuehl · V. Bagnoud · T. Stoehlker
Helmholtz Institute Jena, Helmholtzweg 4, 07743 Jena, Germany
e-mail: b.zielbauer@gsi.de

B. Zielbauer · T. Kuehl · U. Eisenbarth · S. Goette · D. Winters · V. Bagnoud · T. Stoehlker
GSI Helmholtz Center for Heavy-Ion Research, Planckstr. 1, 64291 Darmstadt, Germany

B. Ecker · T. Kuehl
Johannes Gutenberg University Mainz, Saarstr. 21, 55122 Mainz, Germany

P. Neumayer · T. Kuehl
ExtreMe Matter Institute EMMI, Planckstr. 1, 64291 Darmstadt, Germany

K. Cassou · S. Daboussi · O. Guilbaud · S. Kazamias · D. Ros
LASERIX-CLUPS/Laboratoire de Physique des Gaz et des Plasmas, 91400 Orsay, France

S. Sebban et al. (eds.), *X-Ray Lasers 2012*, Springer Proceedings in Physics 147,
DOI 10.1007/978-3-319-00696-3_45,

2 X-Ray Laser Development

Since the heavy-ions can be stored in the experimental storage ring (ESR) over, depending on the ion species, up to several tens of minutes, a high repetition rate pump laser system would be rather beneficial. Due to the high-energy (Nd:glass based) orientation of the PHELIX laser system at GSI, the high repetition rate developments are carried out off-site at the LASERIX facility [2], while first XRL spectroscopy experiments will be performed with the available PHELIX preamplifier with shots every 2 minutes. However, also here, an upgrade to 3 shots/min is under way.

The amount of XRL photons available for the excitation process proves to be the most critical factor, and calculations show that with reasonable efficiency assumptions on the XRL as well as the spectroscopy part, at least 1 μJ of XRL pulse energy is required in order to reach a fluorescence signal of more than a few counts/shot. Thus, the XRL development here is driven by optimization towards high and stable XRL photon numbers.

Our most recent approach to this, outlined in more detail in [3], is the amplification of a standard DGRIP Mo XRL in a comparably wide plasma column using a butterfly-shaped setup with two independent targets. Recent simulations with the ARWEN code [4] show that with this approach, it should be possible to overcome the transversal lasing in the amplifier plasma and achieve a high XRL output signal with good beam quality. The Mo XRL was chosen as a source because of the comparably low pump laser energy requirements, which gives a lot of margin with a 5 J/IR pulse pump laser, but which also allows for stable 10 Hz operation with a high yield of shots per target position [5]. Also, the Mo XRL wavelength of 18.895 nm has been measured [6] with a precision which is sufficient for the spectroscopy experiment where the uncertainty of the ion velocity is on the same order of magnitude.

3 Installation and Setup

Figure 1 shows a top view of the planned installation. The PHELIX preamplifier beam (60 mm diam., 10 J, 1053 nm) will be transported over more than 50 m from the PHELIX building to the ESR hall where it will be compressed to a few ps pulse duration in a dedicated double-pass compressor inside the ESR radiation shielding environment. Another beamline will transport the IR beam over about 5 m to the XRL target chamber with the butterfly setup. In order to minimize XRL beam losses, this chamber is situated as closely to the injection into the straight section of the ESR as possible and the XRL beam transport and collimation will be achieved with only one 90° off-axis parabola and one flat mirror, both XUV multilayer coated. After the flat (steering) mirror, the XRL beam has to propagate almost 15 m to the center of the straight section of the ESR which is the only convenient spot where the fluorescence light can be detected. The excited ions, propagating clockwise seen from top, emit this light in a cone in their forward direction, and since the decay is so fast, this happens mostly within the interaction region of about 1 m length

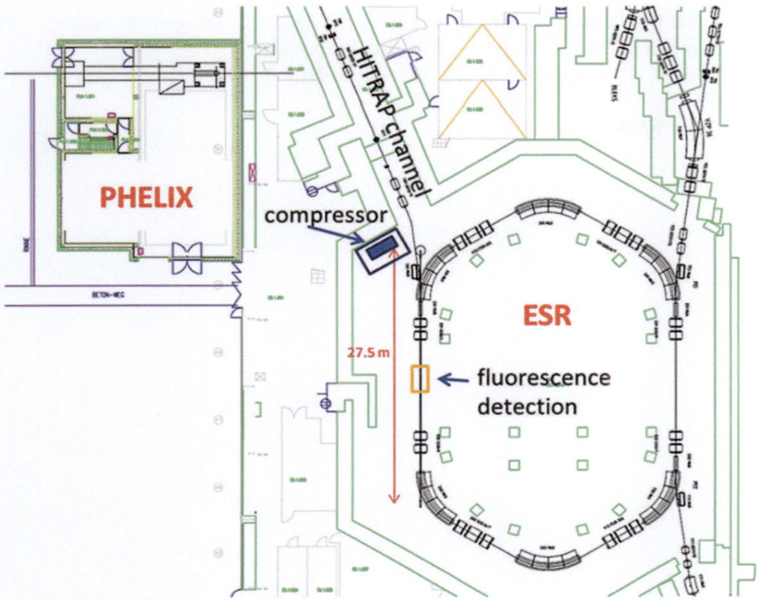

Fig. 1 Top view of the planned installation. The IR pump pulse is generated in the PHELIX building, transported to a compressor and then on into the XRL target chamber. The XRL beam is then propagating through the entire straight section of the ESR on the left side, exciting the ion bunch which is running clockwise

between the few-meter long ion bunch and the XRL beam. For alignment and on-shot diagnostics purposes, the XRL pulse is propagating on to the end of the straight section onto an in-vacuum XUV CCD camera.

A major challenge in this scenario comes from the fact that the ESR is operating at a vacuum level around 10^{-11} mbar which is not compatible with a typical XRL setup. Thus, the XRL beam needs to be coupled in and out of the ESR vacuum through a system of shutter valves, equipped with thin metal filters and differential pumping through small apertures. In addition, a set of rapid shutter valves equipped with glass windows allows for a pre-alignment of the beamline with a visible pilot laser as well as for maintaining the vacuum integrity in case of metal filter damage.

References

1. Otten, E.: Treatise on Heavy-Ion Sciences. Plenum, New York (1989)
2. Guilbaud, O., et al.: In: Sebban, S., et al. (eds.) X-Ray Lasers 2012. Springer Proceedings in Physics, vol. 147. Springer, Cham (2013). Chapter 18 in this book
3. Ecker, B., et al.: In: Sebban, S., et al. (eds.) X-Ray Lasers 2012. Springer Proceedings in Physics, vol. 147. Springer, Cham (2013). Chapter 16 in this book

4. Oliva, E., et al.: In: Sebban, S., et al. (eds.) X-Ray Lasers 2012. Springer Proceedings in Physics, vol. 147. Springer, Cham (2013). Chapter 33 in this book
5. Zielbauer, B., et al.: Appl. Phys. B **100**, 731–736 (2010)
6. Li, Y., et al.: Phys. Rev. A **58**, R2668 (1998)

Index

1–9
100 Hz repetition rate, 216
1D-radiative transfer code, 181, 182
8.85 nm, 215
8.85 nm wavelength, 216

A
Ablation, 148
Al/Mo/B4C multilayer mirror, 159
Amplification, 159
Amplification chain, 206, 207
Amplification of sub-picosecond duration, 154
Amplified line profile, 181–183
Amplified spontaneous emission, 203, 204
An atom in the external field, 7
Applications, 39
Argon laser, 231
Argon-based gas-puff target, 275
ARWEN code, 159
ASE, 203, 204, 206
ASE level, 89
Atomistic model, 122
Attenuation length, 275
Average-atom code, 194, 196, 197

B
B-integral, 33, 235–237
Barrier layer, 161, 163, 164
Baskakov-Popov-Papadakis, 106
Beam pointing, 129
Beam stability, 227
Biomedicine, 275
Brightness, 229
Brilliance, 3

C
Calcination, 271
Capillary discharge, 231, 232
Channel, 251
Chirped-pulse amplification, 35, 209
Classical molecular dynamics (MD) simulations, 50
Coherence volume, 129, 130
Coherent imaging of inclined objects, 19
Collector, 228
Collision plasma, 280
Collisional-radiative, 83
Collisionally pumped, 279
Complementary information, 273
Computational domain, 105
Condenser, 227, 228
Contrast, 132
Critical density, 154, 281
Crystal film, 272
Cut-off frequency (CF), 7, 9, 11, 12

D
Deflectometric technique, 155
Deflectometry, 155
Deflectometry technique, 159
Degeneracy factor, 130
Dense plasmas, 152, 154, 155, 159
Density gradient, 155, 156
DFG, 278
DGRIP, 89, 100, 101, 103, 284
Diffraction, 278
Diffraction imaging, 80
Diffraction limit, 274
Diffusion barrier layer, 161, 163
Discharge-pumped XUV laser, 231
Divergence, 227–230
Doppler broadening, 49
Double frequency grating, 278
Double pass, 152

S. Sebban et al. (eds.), *X-Ray Lasers 2012*, Springer Proceedings in Physics 147,
DOI 10.1007/978-3-319-00696-3,

Double targets coupling technique, 279
Double-stage, 97, 101

E
Eikonal, 171
Electron density, 278
Electrospinning, 271
Etendue, 229
EUV ablation, 144
EUV microscope, 270, 271
EUV microscopy, 270
EUV radiation, 270, 273
EUV transmission microscope, 275
Extreme ultraviolet (EUV), 263, 269
Extreme ultraviolet (XUV), 7, 8

F
Femtosecond laser, 63
Femtosecond laser ablation, 199
Elettra, 1, 2
Few-Hz repetition rate, 137
Flash imaging, 188
Fourier optics, 35
Fourier transform holography, 166
Fourth generation light source, 1
Free electron laser, 1, 39, 55
Fresnel bi-mirror interferometer, 258
Fresnel optics, 275
Fresnel zone plate, 165, 269
Fringe visibility, 258
Full coherence, 130
FZP magnification, 274
FZP objective, 271

G
Gain duration, 247
Gain media, 171
Gain-guiding, 41
Gain-length product at saturation, 41
Gain-narrowing effect, 128
Gas-puff target, 269
Generalized Talbot effect (GTI), 264
Generation of harmonics, 209
Grazing-incidence, 235
Grazing-incidence pump geometry (GRIP), 137
Grazing-incidence pumping (GRIP), 29, 102

H
Half-pitch spatial resolution, 271
High densities, 152
High electron density, 152
High gain harmonic generation (HGHG), 2
High gain single pass FELs, 2
High harmonic generation, 63, 67
High intensity optical lattice, 14
High order harmonics, 203, 204
High power OPA, 78
High-order harmonic generation (HHG), 7, 8
High-resolution imaging system, 275
Higher-order harmonics, 144, 146
HOH, 203–206
Hohlraum, 41
Hollow targets, 172
Homogeneous broadening, 50
Hydro-code, 171

I
Illumination, 227–229
Imaging, 269
Inhomogeneous broadening, 50
Injection-seeding, 97, 101, 103
Interference patterns, 279
Interferometer, 144, 147
Interferometry, 277
Intrinsic, 181
Intrinsic line profiles, 182
Intrinsic property of EUV light, 273
Irradiance, 235

L
Laboratory-scale sources, 41
Laser ablation dynamics, 117
Laser like pulses in the x-ray, 1
Laser wake field accelerator (LWFA), 55
Laser-plasma EUV source, 270
Laser-plasma sources, 269
Laser-produced plasma, 171, 275
Lensless imaging, 39
Light atom interaction theory, 7, 8
Line profile, 181
Line-focus, 236, 237
Line-focus generation, 235
Linewidth, 258
Linewidth measurement, 159
Low ablation threshold, 122

M
Magnification, 273, 274
Material science, 275
Maxwell-Bloch, 203, 204
Measuring plasma, 277
Microscopy, 185, 227, 230
Mo/Si mirror, 159
Multilayer, 161–163
Multilayer mirror, 161

N
NaCl crystals, 270
Nanofibers, 273
Nanometer scale resolution imaging, 269
Nanoscale resolution time resolved microscopy, 185
Nanostructures, 275
Nanotechnology, 275
Nd:YLF, 137
Ne-like Zn x-ray laser, 152, 156
Near-field, 229, 230
Ni-like, 29
Ni-like Ag laser $4d - 4p$ $(J = 0\text{–}1)$ line at 13.9 nm, 51
Nonlinear optics, 63
Non-perturbative theory, 8, 12
Normalized pulse integrated intensity, 130

O
OFI, 247
Optical contrast, 275
Optical-field-ionization, 244
Optical-parametric chirped-pulse amplification (OPCPA), 137
Optical path difference, 278
Optical transformation, 19

P
PALS centre, 151, 152
PALS facility, 159
PALS laser facility, 157
Parabolic wave equation, 19, 105
Partial coherence, 126
Partially coherent, 130
Particle correlation effects, 49
Plasma amplifier, 206
Plasma based seeded soft x-ray lasers, 204
Plasma channel, 249
Plasma guiding, 247
Plasma in a hohlraum, 280
Plasma physics, 277
Plasma self-emission, 156
Plasma waveguide, 247, 252
Plasma-based soft-x-ray lasers, 203
Pointing stability, 227–230
Polarization, 127, 143–145, 205, 206
Polarization of radiation emitted by the x-ray laser, 127
Pre-amplifier, 206
Pre-amplifier main-amplifier chain, 207
Prepulses, 89
Probability density, 130
Pulse-front tilt, 35

Q
QSS, 152, 175, 176
Quasi-steady state, 175

R
Rabi oscillations, 204
Radiation in the water window, 78
Radiation polarization, 126
Radiative shocks, 239, 240
Ray-tracing, 235–237
Recombination pumping, 252
Reflective optics, 235
Reflective x-ray microscopy, 201
Refraction, 171
Refractive optics, 235, 236
Repetitive laser, 231

S
Saturation rebroadening, 183
Seeded, 244
Seeding, 246, 283
Shadow of plasma, 279
Soft x-ray, 155, 185, 186
Soft x-ray holograms, 165
Soft x-ray laser, 29, 89, 121, 137, 143, 175, 239
Soft x-ray laser probe method, 277
Soft x-ray laser (SXRL) probe, 117, 118
Soft x-ray range, 2
Spatial coherence, 130, 245
Spatial resolution, 274, 275
Speckle, 126, 245
Speckle contrast, 132
Speckle effect, 126
Speckle structure, 126
Spectral broadening, 49
Spectral linewidth, 175–178
Spectral width, 176
Spectroscopy, 98, 100, 283, 284
Spontaneous emission, 205
SPring-8, 69, 71
Surface modifications, 122
SXRL, 175–179
SXRL ablation, 144, 147

T
Table-top soft x-ray laser, 216
Talbot distances, 264
Talbot effect, 263
Talbot pattern, 155
TCE, 152
Temperature effect on ASE, 46
Temporal coherence, 175–179, 246
TGRIP, 101

Thomson scattering, 193–195, 197
Time-dependent radiation transport, 83
Time-dependent Schrodinger equation (TDSE), 8
Transient collisional-excitation, 30, 228
Transmission mode, 271
Transparent boundary condition, 106
Traveling-wave, 31
Traveling-wave excitation, 35
Triple-pulse scheme, 30
Turning point, 172

U
Ultraviolet, 2
Undulator, 69, 74, 75

V
Van der Waals forces, 271
Velocity redistribution, 52
Visibility curve, 258
Visibility profile, 258

W
Wake, 203–206
Water window, 161
Waveguide, 250, 251
Waveguiding, 171
Wavelengths as short as 7.36 nm, 216

X
X-ray laser backlighting, 155
X-ray laser at 100 Hz repetition rate, 215
X-ray lasers, 155, 235
X-ray movie, 72–74
X-ray reflective microscope, 199
X-ray source, 159
X-ray techniques for real time observation of melting and solidification processes, 69
XRL, 153, 154, 156, 157, 159
XRL's amplifiers, 152
XUV laser, 231
XUV lasers pumped by collisional excitation of Ni-like ions, 50

Y
Young's interferometry, 126

Z
Zn x-ray laser, 151, 153, 154, 157, 159
ZnO nanofibers, 271

MIX
Papier aus verantwortungsvollen Quellen
Paper from responsible sources
FSC® C105338

If you have any concerns about our products,
you can contact us on
ProductSafety@springernature.com

In case Publisher is established outside the EU,
the EU authorized representative is:
Springer Nature Customer Service Center GmbH
Europaplatz 3, 69115 Heidelberg, Germany

Printed by Libri Plureos GmbH
in Hamburg, Germany

Career Skills for Surgeons

Charalambos Panayiotou Charalambous

Career Skills for Surgeons

Springer

Charalambos Panayiotou Charalambous
Department of Trauma and Orthopaedics
Blackpool Victoria Hospital
Blackpool
Lancashire
United Kingdom

ISBN 978-3-319-86176-0 ISBN 978-3-319-57490-5 (eBook)
DOI 10.1007/978-3-319-57490-5

Printed on acid-free paper

This Springer imprint is published by Springer Nature
The registered company is Springer International Publishing AG
The registered company address is: Gewerbestrasse 11, 6330 Cham, Switzerland

I dedicate this book to my parents and to all my special teachers and trainers

Foreword

This book contains a wealth of knowledge to better guide the aspiring surgeon through the many hoops of fire that they will encounter in their chosen career of surgery. It offers a guide to junior doctors in how to deal with the personal demands, difficulties and pressures of an evolving chosen surgical career path.

Written is an easy to understand, relaxed style, the reader is guided through a series of chapters dealing with all the major obstacles a junior trainee is likely to encounter in their career progression. Choosing surgery as a career ultimately commits and/or condemns a trainee to many years of hard study. Just concentrating on the basics of training preparing for progress assessments and ensuring log book numbers meet indicative targets is all very well, but the average trainee needs to be much more informed and savvy than this. This book will give trainees that extra edge by providing invaluable advice and insight into dealing with the complex challenges and uncertainties that trainees will encounter outside of their core surgical training practice. It will allow trainee surgeons to successfully steer through the many different traps and hurdles they may encounter.

Although primarily written for surgical trainees, one of the strengths of the book is its appeal to newly appointed or established consultants. Established consultants will find a wealth of material contained within its pages of great relevance to their own surgical practice. I particularly enjoyed reading the evidence-based chapter, and the all too easy to forget importance of departmental morbidity and mortality meetings and the correct spirit in which to conduct them.

This book contains information not easily found elsewhere. It has been written in an honest, straightforward manner, packed with pearls and gems of wisdom. It is not scared to deal with issues that are either controversial or not usually discussed for fear of political correctness. These are not easy subjects to write about.

I found this book fascinating to read. The book's content is precise, succinct and important. It is very cleverly written and draws on a wealth of personal experience and knowledge, gained through years of navigating through a career path that for many surgeons has never been particularly easy or straightforward.

The often quoted sentence "if only we knew then what we know now" could be more aptly rewritten as "if only this book had been available many years ago".

A word of congratulations to Professor Charalambous for writing this book. Hope you enjoy reading the book as much as I have done.

Paul A Banaszkiewicz, FRCS (Tr & Orth) MClinEd FAcadMEd FHEA
Queen Elizabeth Hospital and North East NHS Surgical Centre (NENSC)
Gateshead, UK

Northumbria University
Newcastle-upon-Tyne, UK

Preface

The primary objective of this book is to present skills that a surgeon may find helpful in day-to-day workplace practice and in career progression. The inspiration for this book came from personal experiences at various stages of my surgical career and through the observations of similar challenges that other surgeons faced.

The aim of this book is to present information in an easily read, succinct way and break down a vast subject into small, manageable sections. An attempt is made to provide the reader with knowledge and information, but also stimulate lateral thinking though repeated, often rhetoric, questioning.

I would like to thank Liz Pope, Associate Editor at Springer, and Vignesh Iyyadurai Suresh, Project Coordinator at Springer Nature, for their invaluable help and advice in the preparation and presentation of this book.

Charalambos Panayiotou Charalambous, BSc, MBChB, MSc, MD, FRCS (Orth)
Department of Trauma and Orthopaedics
Blackpool Victoria Hospital
Blackpool, UK

School of Medicine
University of Central Lancashire
Preston, UK

Contents

1 Introduction 1

2 Professionalism in Surgery 5
2.1 Offer Surgery for the Right Reasons 7
2.2 Maintain the Trust of Patients 9
2.3 Maintain the Trust of Other Staff 11
2.4 Respect All Staff 13
2.5 Respect Colleagues 14
2.6 Support Colleagues 15
2.7 Consider Emotions of Others 16
2.8 Listen and Act upon Concerns 17
2.9 Stick to the Core Principles 18
2.10 Go with the Flow 19
2.11 Swim Against the Tide 20
2.12 Act More Than a Technician 21
2.13 Stand Up for the Profession 22
2.14 Declare Acceptable Conflict 23
2.15 Stir Well Away from Unacceptable Conflict 25
2.16 Gather Experience 27
2.17 Seek Continuity of Care 28
2.18 Don't Make Speed Your Sole Aim 29
2.19 Understand Your Limitations 30
2.20 Walk Before You Can Fly 31
2.21 Seek and Give Feedback 32
2.22 Self-Reflect 33
2.23 Acknowledge Uncertainty 34
2.24 Aim for Accurate Communication 35
2.25 Follow Procedures 36
2.26 Welcome Innovation 37
2.27 Question Innovation 38
2.28 Choose Implants Wisely 40

2.29 Appreciate the Limitations of Technology 42
2.30 Pass on the Craft 43
References 44

3 Communication in Surgery 47
3.1 Basic Concepts in Communication 49
3.1.1 Introduce Self and Get to Know Others 49
3.1.2 Choose the Words and Tone 49
3.1.3 Body Language 50
3.1.4 Listen 50
3.1.5 Repeat in Different Ways 51
3.1.6 Do You Have Any Remaining Questions? 51
3.1.7 Check that Information Has Gone Across 52
3.1.8 Summarise 52
3.1.9 Pick the Cues 52
3.1.10 Empathy 53
3.1.11 Communicating Urgency 55
3.2 Shared Decision Making 56
3.3 Consent 57
3.3.1 Information to Be Communicated in Consenting 57
3.3.2 Who Obtains Consent 60
3.3.3 Who Can Provide Informed Consent 61
3.3.4 Timing of Obtaining Consent 63
References 64

4 Organising a Surgical Theatre List 67
4.1 The Patient's Journey to Surgery 69
4.2 Listing for Surgery 72
4.3 Putting Together a Theatre List 73
4.3.1 Ensuring Good Surgical Care 73
4.3.2 Surgical List Location 74
4.3.3 Order of the Surgical List 75
4.4 Improving Theatre List Utilisation 76
4.4.1 Prompt Start 76
4.4.2 Avoiding Early Finishes 77
4.4.3 Avoiding Interruptions 77
References 79

5 Technical Skills in Surgery 81
5.1 Planning for Surgery 83
5.2 Developing Surgical Skills 85
5.3 Get Accustomed to the Operating Theatre Environment 86
5.4 Developing Technical Skills 88
5.4.1 Model for Progression of Motor Skills Acquisition 88
5.4.2 Learning Curve 88
5.4.3 Deliberate Practise 89

5.4.4 Piecemeal Learning 90
5.4.5 Distributed Practise 91
5.4.6 Mental Practise. 91
5.4.7 Simulation 92
5.4.8 Feedback 93
5.4.9 Progress in the Levels of Participation. 93
5.4.10 Be an Active Assistant 94
5.5 Learning a Surgical Technique. 95
5.6 Getting Technically Fluent. 96
5.7 Limitations of Surgical Skills 97
References. 98

6 Situation Awareness in Surgery 101
6.1 Levels of Situation Awareness 103
6.2 Components of Situation Awareness 105
6.3 Situation Awareness Versus Decision Making. 106
6.4 Barriers to Situation Awareness 107
6.4.1 Failure of Situation Awareness in Aviation 108
6.4.2 Failure of Situation Awareness in Vehicle Driving 109
6.4.3 Failure of Situation Awareness in Surgery. 109
6.5 Improving Situation Awareness 111
6.5.1 Improve Information Gathering. 111
6.5.2 Aim for Manageable Workload 111
6.5.3 Improve Memory Utilisation 112
6.5.4 Minimise the Effect of Stress on SA 113
6.6 Team Situation Awareness 114
6.7 Recognising Loss of Situation Awareness 116
References. 118

7 Decision Making in Surgery. 121
7.1 The Spectrum of Surgical Decisions 123
7.2 Types of Decision Making. 124
7.3 Effect of Experience on the Choice of Decision Model 125
7.4 Switching Between Decision Models 126
7.5 Biases in Decision Making 127
7.6 Factors Influencing Surgical Decision Making 130
7.7 Decision Making in Offering Surgery 131
7.8 Emotions and Decision Making. 134
7.9 Improving Decision Making 136
7.10 Surgeons Are Allowed to Change Their Mind. 137
References. 138

8 Leadership in Surgery 141
8.1 Leader Versus Manager 143
8.2 Types of Leadership. 144
8.3 Leadership Matters. 145

8.4 Challenges of a Surgeon Leader . . . 147
8.5 Desirable Leadership Behaviours . . . 148
8.5.1 See the Big Picture . . . 149
8.5.2 Stick to the Core Values, Share Your Followers' Values . . . 150
8.5.3 Lead by Example . . . 150
8.5.4 Lead Change Wisely . . . 151
8.5.5 Welcome Alternative Voices . . . 154
8.5.6 Follow the Rules . . . 154
8.5.7 Consider How You Start . . . 155
8.5.8 Consider How You End . . . 155
8.6 Undesirable Leadership Behaviours . . . 157
8.7 Personality and Leadership . . . 159
8.8 Leadership Posts and Titles . . . 160
References . . . 161

9 Surgical Research . . . 165
9.1 Accepting the Need for Surgical Research . . . 167
9.2 Defining the Aim . . . 168
9.3 Defining the Study Groups . . . 170
9.4 Defining the Study Population . . . 171
9.5 Study Design . . . 172
9.6 Defining the Intervention . . . 176
9.7 Quality Assurance of Interventions . . . 178
9.8 Control Groups—Sham Surgery—The Surgical Placebo . . . 179
9.9 Sample Size Calculation . . . 181
9.10 Follow Up . . . 182
9.11 Surgical Waiting Lists and Research . . . 183
9.12 Outcomes . . . 184
9.13 Recruitment . . . 185
9.14 Improving Success of Surgical Clinical Research . . . 186
9.15 Putting Evidence into Practise . . . 189
9.16 Running a Journal Club . . . 191
References . . . 193

10 Surgical Audit . . . 197
10.1 Surgical Audit Cycle . . . 199
10.2 Surgical Audit Ladder . . . 201
10.2.1 Surgical Logbook . . . 201
10.2.2 Mortality and Morbidity Meetings . . . 202
10.2.3 Systematic Audit at Local Level . . . 203
10.2.4 Audit Databases . . . 204
10.3 Hawthorne Effect . . . 209
References . . . 210

11 Safety in Surgery . . . 213
11.1 The Safety Problem . . . 215
11.2 Contributors to Healthcare Safety . . . 216

11.3 Causes of Harm 217
11.4 Latent Failures in Aviation 219
11.5 Latent Failures in Healthcare 221
11.6 Isolated Versus Sequence of Events Leading to Harm 224
11.7 Strategies for Improving Safety 225
11.8 The Iceberg of Error 226
11.9 Individual's Actions to Improve Safety 227
11.9.1 Be Risk Aware 227
11.9.2 Avoid Taking Unnecessary Risks 228
11.9.3 Identify Error Prone Situations 229
11.9.4 Check and Check Again 230
11.9.5 Use Checklists Appropriately 231
11.9.6 Recognise Critical Steps 232
11.9.7 Aim for Effective Error Reduction Measures 233
11.9.8 Aim for Standardisation 234
11.9.9 Aim for Consistency 235
11.9.10 Recognise the Impact of Minor Failures 235
11.9.11 Learn from Near Misses 236
11.9.12 Use Incident Reporting Systems Appropriately 236
11.9.13 Lead in Safety by Example 237
11.9.14 Encourage Flat Hierarchy 237
11.9.15 Aim for a No-Blame Culture 238
11.9.16 Communicate for Safety 238
11.9.17 Avoid Disruptive Behaviours 241
11.9.18 Plan to Recover from Failures 242
References 243

12 When Things Go Wrong 247
12.1 Managing Adverse Events 249
12.2 Root Cause Analysis of an Adverse Event 250
12.3 Dealing with a Cluster of Recognised Complications 253
12.4 Stop Whilst You Investigate 256
12.5 Near Misses and No Harm Events 257
12.6 Learning from Adverse Events 258
12.7 Informing the Wider Surgical Community 259
12.8 Disclosure 260
12.8.1 Challenges of Disclosure 260
12.8.2 Delivering the Disclosure 261
12.8.3 Disclosure When Another Surgeon Is Involved 262
References 263

13 Surgical Interviews 265
13.1 Choosing a Surgical Specialty or Sub-Specialty 267
13.2 Surgical CV 269
13.3 Surgical Portfolio 271
13.4 Surgical Interview Questions 274

13.4.1 CV—Portfolio Questions. 274
13.4.2 Surgical Clinical Scenario Questions. 275
13.4.3 Leadership/Management Questions. 277
13.4.4 Awkward Scenarios 277
13.4.5 Communication Questions. 278
13.4.6 Political Issues Questions 279
References. 281

14 Emotions in Surgery 283
14.1 Types of Emotions 285
14.2 Emotional Intelligence. 286
14.3 Emotions and Stress. 287
14.4 Facing Complications 289
14.5 Dealing with Errors 292
14.6 Coping with the Stereotype 294
14.7 Struggles in the Workplace 296
14.8 Be Proud. 298
References. 299

Index. 301

Chapter 1
Introduction

The transition from medical school to being a junior doctor, and then a practicing surgeon is challenging. It requires knowledge, clinical skills and technical dexterity but also the ability to efficiently function in the surgical workplace and be a reliable team member. Being a surgeon is not just a profession but a way of life.

Medical school and postgraduate training often provide clinical and technical knowledge that are essential in formulating a clinical diagnosis, initiating medical treatment, and planning or carrying out invasive surgical procedures. However, surgery may be viewed not just as another medical specialty, but as an apprenticeship, that requires the acquisition, application, and constant refinement of lifelong skills. Surgery doesn't get taught simply through books, but has to be seen and experienced, over and over again.

This book presents some well-known career skills to help improve your day to day surgical performance, development and progression. The generic guidance presented in this book may be of use to aspiring surgeons, no matter their surgical subspecialty choice and no matter where they practice.

Professionalism is central in our day to day performance, guiding our interactions with patients, other doctors, or non-medical staff. The chapter on professionalism explores some of the behaviours that may help improve one's conduct.

Effective communication is necessary in transmitting and receiving information as well as developing and maintaining functional relationships. The first part of the communication chapter discusses basic concepts of communication and ways in which one may improve their communication skills. The second part focuses on how to communicate to patients surgical treatment options and the processes of shared decision making and informed consent.

The ability to organise a surgical theatre list and achieve efficient theatre utilisation is an inherent part of surgical practise. It involves planning the content and order of a surgical list, taking into account, amongst others, clinical priorities, safety factors, and available resources. Hence, a chapter is devoted discussing how to construct a theatre list, followed by ways in which theatre time utilisation may be enhanced.

C. Panayiotou Charalambous, *Career Skills for Surgeons*,
DOI 10.1007/978-3-319-57490-5_1

Acquiring, improving and maintaining technical surgical skills in the era of working hours restriction is a highly demanding task, hence training opportunities must be maximised and thoroughly utilised. The chapter on surgical technical skills discusses how to plan for a surgical procedure, along with ways of developing surgical motor skills.

Surgeons often practise under dynamic, constantly changing conditions. A clear understanding of the context in which one practises is essential in facilitating decision making and improving performance. Appreciating the process of situation awareness and recognising the factors that may assist or impair a surgeon from reaching that state are important skills to develop and are discussed in the subsequent chapter.

As surgeons we are often faced with a constant need of making decisions. Being aware of how we reach such decisions and recognising the potential limitations of our decision making are important skills to develop. The chapter on decision making discusses intuitive and analytical decision processes, and explores the various biases in making decisions. It also discusses some of the factors that may influence our decision to offer surgery, especially in the absence of strong scientific evidence.

Surgeons are often called upon to lead or manage, through formal or informal roles, and these are important skills in surgical practise. The chapter on leadership describes various leadership styles, discusses the challenges a surgeon leader may face, and presents both desirable and undesirable leadership behaviours for the surgeon leader. The need to distinguish between leading and holding leadership titles or posts is emphasised.

Surgical research may be considered as the process through which we can discover new information to improve our understanding and management of surgical conditions. The surgeon has a central role in identifying areas in which further research is needed, in carrying out surgical research and implementing its findings. The chapter on surgical research discusses the challenges faced in setting up and running clinical surgical studies as well as incorporating their findings in day to day practise. Advice is given on how to deal with such challenges, and also on how to contribute to the assessment of research as part of a Journal Club.

Audit is the process through which we can compare our surgical practise to a pre-defined ideal to help identify strengths and deficiencies, reflect, and further improve. The chapter on surgical audit describes the audit cycle and discusses the role of the surgical logbook, mortality and morbidity meetings, and national databases in assessing surgical outcomes.

The subsequent two chapters explore safety in surgery and discuss what to do when things go wrong. As surgeons we aim to improve our patients' condition but without causing harm. Understanding the factors that may lead to errors and harmful events, as well as the system processes and individuals' behaviours that may promote safety, are important skills to develop and are discussed. Nevertheless, despite all good intentions, it may be difficult, if not impossible, to achieve absolute safety, and on occasions adverse events will occur. Ways of dealing with an adverse event, exploring and understanding its root causes are presented. Particular reference is made to the process of evaluating clusters of uncommon, yet recognised, surgical complications.

The next chapter discusses how to prepare for postgraduate surgical interviews, along with potential questions and scenarios. Factors that may be considered in choosing a surgical specialty or sub-specialty are also given. Finally, the last chapter explores some aspects of a surgeon's world of emotions, with advice on how to deal with these.

As a Consultant Surgeon in Trauma and Orthopaedics who has done all my undergraduate and postgraduate training in the United Kingdom, the career skills presented in this book originate from personal experiences, the teaching and "wisdoms" of my senior trainers and from supervising and mentoring multiple junior surgeons.

Much of what is presented is commonly available knowledge, or is based on critical evaluation of the published literature, and every attempt has been made to acknowledge and reference its origins as warranted.

Some may not fully agree with what is presented, some may have opposite views, but that is understandable and acceptable. Nevertheless, I hope the reader will gain and benefit from what is said and incorporate some of the advice given in their surgical career.

Chapter 2
Professionalism in Surgery

Professionalism is the ability to practise and act in line with standards and values conferred by one's profession.

The Royal College of Physicians of London working party on Medical Professionalism defined professionalism as "a set of values, behaviours and relationships that underpins the trust the public has in doctors" [1]. Along similar lines there have been several definitions of professionalism in surgery, which nevertheless share a common theme—working in the best interests of patients and the wider society and public.

The Committee on Professionalism of the Canadian Association of General Surgeons refers to the essence of professionalism as "in return for professional autonomy, self-regulation and a recognition of their unique place in society, the public demands of physicians accountability, ethical standards and an altruistic manner of delivering care" [2].

The American College of Surgeons, in their Code of Professional Conduct refer to "individual and collective participation in ways that promote the good of both our patients and society" [3]. Geoffrey Davies, retired judge of the Supreme Court of Queensland, in delivering the Presidential Lecture at the 2010 Annual Scientific Conference of the Australian Orthopaedic Association, described as the main elements of professionalism a "duty, individually and collectively, to put the interests of patients (clients) ahead of those of your own, of your profession and of other bodies", a "wider duty to existing and future patients", "duties of public service" [4].

Professional behaviours may be rooted in one's own personal upbringing and development, general education, may be shaped by one's life experiences and character maturity. However, professionalism may also be taught or learned in daily practise through observation of our seniors, juniors, peers or colleagues, medical and non-medical professionals, through self-reflection, trial and error.

This chapter describes certain behaviours that may improve professional conduct in surgery, and may help strengthen your professional practise. It is not meant to be exhaustive or to replace any advice given by the medical professional body or other oversight institutions you may be accountable to.

C. Panayiotou Charalambous, *Career Skills for Surgeons*,
DOI 10.1007/978-3-319-57490-5_2

Well dear colleague…not sure how you got your-self into this trouble. I never had this complication…or any other complications

2.1 Offer Surgery for the Right Reasons

Do the right thing. It will gratify some people and astonish the rest.

Mark Twain [5]

As surgeons we are trained to operate, that's our main craft. But not one solution fits all, there is not a surgical solution for each disorder. Appreciating the role of surgery in the wider medical care is an important skill to develop. Just because we are surgeons it doesn't mean we have to operate, just because patients consult us it doesn't mean that they expect surgery or that we have to offer surgery.

Surgery may help to improve quality of life, extend or save life. Even when other management options are available, surgery may provide more speedy recovery, or enhance the effect of subsequently applied non-surgical interventions. But when faced with a particular condition, question as to whether surgery truly offers an advantage over other non-surgical treatments, any special gains, whether it changes the natural history of the condition. Could less invasive options have the same effect? Do the risks of an invasive intervention justify the gains? How does the surgical option fit with the patient's preferences or individual life circumstances?

One may consider a Ladder of Interventions, with the least invasive at the bottom, and the most invasive at the top (Fig. 2.1). In some situations it may be appropriate to climb up the ladder step by step, in other situations one may have to jump straight to the top.

Look at the whole patient, the whole situation, rather than simply concentrating on the bowel, the thyroid, the arteries, the bone. What may work in one context may not be applicable to another. Surgery offered in a chronic presentation may differ from what is done in an emergency or trauma. Availability of resources may also influence what you do, what may be routine practise in a developed country may be unheard of in less well developed societies.

Be pragmatic, don't exaggerate the benefits of surgery to promote a procedure, don't unnecessarily exaggerate the urgency to guide the way. Similarly don't dismiss the benefits of surgery when referring to procedures that you are not familiar or comfortable with.

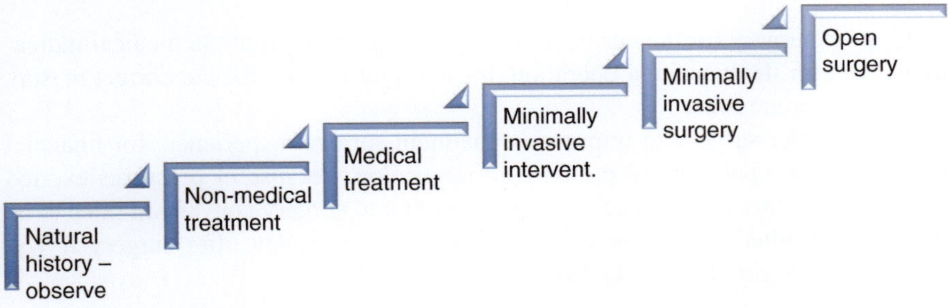

Fig. 2.1 Ladder of interventions

If we take our broken car to a garage we may want to know what the likely problem is and what any repairs may entail. We may like to know how significant the problem is, and whether we could leave it all alone. Would our safety be compromised if we kept driving with the flashing dashboard lamp, or could the car break down? We would expect to be given accurate and honest information, explained the various options in a truthful way. How would we feel if we are told that we have to make an immediate decision, leave the car in for extensive and costly servicing, as the car is about to fall apart, that the garage accepts no responsibility if we drove away, when we know that the dashboard lamp has been flashing for the last 5 years? And what would we think if a garage down the road sorts out the problem simply by changing the malfunctioned lamp?

Several studies have suggested a relationship between surgeon to population ratio and the number of surgical procedures performed, as well as between surgical fees reduction and the number of surgical interventions carried out. It has also been shown that cases treated in a fee-for-service system were significantly more likely to undergo invasive procedures as compared to those treated in salary-based settings. Previous authors have postulated that this was the result of a "provider induced demand", that is surgeons creating demand by changing their indications for financial or other incentives [6–10].

Geoffrey Davies, retired judge of the Supreme Court of Queensland, talks of the "scalpel happy surgeon". He postulates that reduction in surgeon fees, increased rates of legal action for negligence rates with consequent high damage costs, as well as the increased materialism encountered in modern societies, may be some of the factors that could lead to some surgeons offering surgery [4].

An article in the Times of India [11] warned patients to be cautious of surgeons offering surgery:

> *"Doctors are increasingly becoming scalpel-happy. If your doctor asks you to undergo a cataract surgery, hysterectomy, gall bladder or tonsils removal, or an operation of the lower-back, please take a second opinion, patients are forced to undergo unnecessary surgeries".*

The same article [11] reported that an Orthopaedic surgeon explained the reason behind the increasing surgery rates:

> *"We have a quota to meet every month. Many of us see patients as a potential candidate on our operating table".*

Propose surgery for the correct clinical reasons, for appropriate medical indications. Aim for the beneficial operation, for the right patient, for the correct reason, at the appropriate time.

Do not offer surgery to improve or maintain surgical experience, for financial gains, to gain reputation, to promote an innovation, because of pressures exerted upon you by other professionals, for your referral to surgery conversion rates to be in line with other practitioners, because someone else may offer surgery if you don't. Offer surgery for the right reasons.

2.2 Maintain the Trust of Patients

The best way to find out if you can trust somebody is to trust them.

Ernest Hemingway [5]

Maintain trust. Trust is defined in the Oxford Dictionary as the "firm belief in the reliability, truth, or ability of someone or something" [12].

Maintaining trust is vital in your professional relationships with patients, colleagues and other co-workers. It gives others the faith and confidence that what you say is what you mean, that your words are honest and truthful.

Trust of patients on surgeons forms the cornerstone of their relationship. The relationship between surgeon and patient is described as being fiduciary in nature. In fiduciary relationships one person is in a position of vulnerability and relies on another person for help, advice or protection. The fiduciary is required, expected and trusted to act at all times solely for the benefit and in the best interests of the one who entrusts [13–16]. The patient is considered as vulnerable at their time of need and turns to the surgeon who is in a stronger, more privileged position, for advice and support. Hence, the surgeon has the duty to maintain that trust, acting in the patient's best interests and doing no harm.

Beneficence and non-maleficence are actions in line with a fiduciary relationship. Beneficence refers to actions done for the benefits of others, the surgeon acting for the benefit of patients. Non-maleficence refers to doing no harm, the surgeon acting to prevent harm to patients [13–16] (Fig. 2.2).

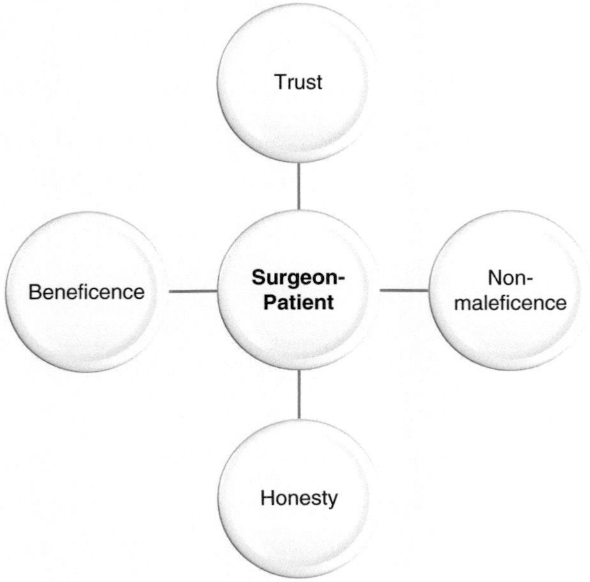

Fig. 2.2 Attributes central to surgeon-patient relationship. Adapted from Chamberlain et al. [13]

Giving an honest estimation of surgical risks and benefits, avoiding exaggerating or playing down the urgency of the need for surgery, acknowledging uncertainty, disclosing adverse events, and disclosing financial or other personal conflicts may help promote trust.

Loss of trust in the medical profession, may mean that patients shop around, take multiple opinions, compare them, weigh them up, trying to figure out who is trustworthy and who is not.

David Berger in an article in the British Medical Journal [17] on the administration of unnecessary treatments and tests, summarises the detrimental effect that loss of trust can have, based on his experiences as a volunteer physician in a small Himalayan hospital:

> *"A common complaint I heard from poor and middle class people is that they don't trust their doctors. They don't trust them to be competent or to be honest, and they live in fear of having to consult them, which results in high levels of doctor shopping".*

As Richard Lunniss (a father whose son had heart surgery at Bristol), told the Bristol enquiry (an enquiry into failures that led to unnecessary deaths following paediatric heart surgery at Bristol hospitals) [18]:

> *"You cannot trust people if you do not think they are being honest, even if they are being nice. Once you think that they might not say the thing as it is, then you can never believe quite — there is no working relationship from that point on".*

2.3 Maintain the Trust of Other Staff

Trust, but verify.

Ronald Reagan [5]

Trust is essential in your relationships with colleagues, co-workers and other staff. Professional trust is essential for a team to function, to thrive, to progress.

Professional trust is not about disclosing your private affairs, your personal life, what you did on your holidays or on your last night out in town. Professional trust refers to the ability of someone feeling that what you say can be believed, that you correctly portray the situation, that your words are true and accurate, that you have no secret agendas, that you don't say one thing but mean another.

Consider your actions, do they maintain trust, do they make you trustworthy? You may consider:

- Should you be reassuring a colleague that you will finish the outstanding ward tasks, simply to leave them untouched for when they return the next day?
- Should you be telling your seniors in the ward round that the blood tests are ok, when you actually forgot to check them on that day?
- Should you be exaggerating your surgical experience so your trainer lets you do more operating, when your real skills will be easily become apparent?
- Should be rubbishing a colleague who asks your advice about a case, or questioning their competency on unfounded grounds?
- If a co-worker confides to you their medical problems and the need to take time off for sickness, should you be sharing their misfortunes with others in the coffee room?
- If a colleague is worried about their complications, the high infection rate they have recently observed, should you be dismissive, denying that you have ever faced any, when the reality is very different?
- If in a leadership position should you be talking about engagement, when whatever your followers say is falling on deaf ears and all you are doing is simply informing them of your decisions?
- Should you be championing safety, but then ignore concerns about high infection rates, for fear of productivity and financial losses?
- Should you be championing teaching and training, but in theatre you find excuses to ensure you do the surgery rather than your resident?
- Should you be telling the anaesthetist that you estimate the next operation will take an hour, when you know that it will likely take all afternoon?
- Should you be telling your leaders that you fully endorse their actions, but undermine them by encouraging others to refuse their directives?
- Should you be trying to limit a colleague's practise simply to enhance yours?
- Should you be using pretexts of efficiency or quality to limit the practice of others?
- Should you be changing the rules of tenure simply to get rid of someone who disagrees with you in the committee?

- Should you be (in the name of quality) refusing to use certain equipment in one hospital, when you are prepared to use the same equipment in another hospital where such refusal may lead to personal financial loss?
- Should you calling an urgent meeting and promising that important decisions will be made, when you know that most of those you invite won't be able to make it at such short notice?
- Should you be making alliances to achieve majority decisions, when you know that individuals in the minority group may be disadvantaged by such decisions?
- Should you be turning up for an interview, and accept the offer, saying this is your "dream job", simply to turn it down a week later, when another post turns up?

Maintain trust. Trust lost may be difficult to regain.

2.4 Respect All Staff

Few people can see genius in someone who has offended them.

Robertson Davies [5]

As a surgeon you often have a central role in what goes on. You may be the one, or may be part of the surgical team, tying off the bleeding aneurysm, drilling the burr holes, ex-fixing the smashed pelvis, replacing the broken hip, suturing the prosthetic heart valve, inserting the transplanted kidney, taking out the gangrenous appendix, bypassing the obstructing coronaries. You may be the one operating the arthroscope, the laparoscope, the robot.

But a whole group of non-surgeons is usually essential for you to do these vital acts. For you to operate someone else may have to prepare the patient medically, give them details for their surgery attendance, ensure that necessary surgical instruments and equipment are sterilised, ready and available, that the patient gets into their gown and is brought to theatre on time. For you to carry out your high skilled operating your assistant may need to keep up the pace, the scrub nurse may need to know and anticipate what instrument to hand over next.

The administrators, the receptionists, the porters, the ward and theatre nurses, the scrub staff, the theatre runner, the recovery staff, and many of those others around you, they too have a vital role to play. Treat them with the respect they deserve. Be courteous, be polite. Respect their opinions, value their input. They may not be surgeons, but are professionals in their own right. Treat them as part of the team, make them feel at ease.

How you deal with other staff, how you exactly behave is a personal choice, and may be shaped by one's character, personality, values, previous experiences, and expectations. But try not to leave your humanity at the hospital gate, not to leave your interpersonal skills at the theatre suite door. Respect all staff, surgeons and non-surgeons alike.

2.5 Respect Colleagues

Colleagues should take care of each other, have fun, celebrate success, learn by failure, look for reasons to praise not to criticize, communicate freely and respect each other.

Richard Branson [5]

Respect your colleagues, respect your peers. Treat them in ways that you would like to be treated, like you would like them to treat you. Evaluate your actions, consider what effect they may have on others, their training, their clinical service, their work pattern.

You may consider:

- You may be the favourite of the head of training, you may be family friends or even ski together. Should you be trying to use this to ensure you get allocated the best training posts, when you know that would disadvantage your colleague trainees?
- Should you be jumping into the theatre lists of your fellow trainees, to improve your own training, when you know they would prefer to be on a one to one with their trainer?
- Should you be trying to squeeze a colleague out of their theatre sessions simply because it suits you better operating on that day?
- Should you be cherry picking patients from colleagues' waiting lists so that you can do "overtime", when you know that your colleague is able and keen to accommodate those in their normal working week lists?
- Should you feel threatened by the talent of colleagues or embrace them for it? Should you criticise and undermine them or be supportive to them? A talented colleague may help further improve the service you provide, rather than limit or reduce your practice. A skilled colleague may be someone you can rely upon to look after your service when you are not around, when on holiday or other leave.
- Respect your surgical colleagues with years of experience. You may be the fresh blood, the one with up to date knowledge, lots of energy and enthusiasm, they may have slowed down, become not so easily impressed and more cynical. Don't suddenly aim to push them out of the way, out of sight. You may have lots to learn from their years of practise, their words of wisdom. What you may be championing as novelty they may have repeatedly seen tried and failed.

Respecting your colleagues, being considerate to others could encourage long-term positive interpersonal relationships, which may be much more valuable than other short term gains.

In a competitive world, in a sometimes perceived competitive profession, one may advance and better own-self without obstructing the pathway of others.

2.6 Support Colleagues

Tis not enough to help the feeble up, but to support them after.

William Shakespeare [5]

Support your colleagues, your peers, juniors or seniors.

A colleague may be having difficult times at home, their husband, wife or mother may be ill. Their own health may be failing, the work pressures may be overwhelming. They may be facing a complex surgical case, having a spell of bad complications, they may have to answer to claims or complaints. They may turn to you for advice, for support, a listening ear, a comforting voice. You may be busy, have your own troubles, your own targets and deadlines to meet. Support your colleagues, devote time and stand by them at times of need.

You may be aware of a colleague's passion, of their interest in a surgical condition or technique, in leading a service, in teaching, in training, in researching, of their need for having Wednesdays off for child care. Should you be taking advantage of their soft spot? Should you be trying to wind them up? If appropriate and if you can support others in their goals, facilitate their ambitions, help them meet their needs.

A junior may just be starting training and have limited clinical or surgical experience, not much knowledge, all may seem strange and difficult to comprehend. They may have already sat the exam twice, but haven't been able to pass that hurdle. Support your juniors when they are starting, help them develop until they can stand on their own feet. The relation between junior and senior, trainee and trainer is often one of dependency; the junior or trainee seeks to learn, be taught, relies on the superior's help to develop and progress. It is not a sin to lack experience, knowledge, surgical skills. You may feel that you are carrying them on your shoulders, but did not some of us have, at some point, to be helped by others, saved by someone else?

Support those in need, rather than simply standing by the strong and powerful.

2.7 Consider Emotions of Others

Work dehumanises people.

Pope Francis [5]

We may find ourselves concentrating on the physical aspects of a disease, on the tasks or assignments we have to complete, the reorganisation in services we have to oversee, the targets we have to deliver.

However, even in highly demanding times, one may also consider the effect our actions, inactions, or reactions may have upon the emotional world of those around us, our patients, our colleagues, co-workers, our juniors or followers. This may help to remind us that we are dealing with human beings, individuals with emotions and feelings, with personal and private lives, rather than simply dealing with problems we have to solve, buildings blocks with which we may construct our ambitions, walls that we need to demolish to get our own way.

You may consider:

- Should you simply concentrate on the physical aspects of the disease you are treating, concentrate only on the technical complexities of the surgical intervention you are proposing or should you also be paying attention to the worries, concerns, and anxieties of your patients? There may be fear that one cannot take all the burden of others, a belief that one has to stay detached. But is it not that "all round care" encompasses both physical and emotional wellbeing? Can you fully ignore one for the other?
- Should you be looking at your followers as names or grades, that you can slot in your newly proposed rota, or should you be also considering what effect your reorganisation may have on their emotional world, as it may interfere with their training priorities, career development, ambitions or home circumstances? Should you be considering what emotional effect the proposal of changes may be having on them, as it creates a state of uncertainty, a state of limbo, a constant concern about the future? Could the "keep them on their toes" motto have the opposite effect, leading to less engagement, less enthusiasm, less strive, less productivity?
- Should you be looking at your colleague as someone you have to over shine, to get ahead of, to get an advantage over, in order to impress your seniors or your employers? How would your colleague or junior feel if
  - You ask them to stay quiet in the ward round so you do all the talking?
  - Collect all the information for you to stand up and do the presentation?
  - Complete the pre-op work up but being sent to the ward when the time comes for them to operate?

Could one not also shine by supporting others, helping them progress, evolve and develop?

Consider emotions, consider the humane aspects of your professional interactions.

2.8 Listen and Act upon Concerns

Chief Brody: This is a Great White, Larry, a big one. And any shark expert in the world'll tell ya it's a killer........And there's no limit to what he's gonna do. I mean, we've already had three incidents...You open the beaches on the 4th of July, it's like ringing the dinner bell.... we have to close the beaches!

Mayor Larry Vaughn: ....Look, we depend on the summer people here for our very lives.... And if you close those beaches, we're finished.....I don't think either one of you are familiar with our problems.....Tomorrow is the 4th of July, and we are going to open for business. It's going to be our best summer in years.....with you or without you, those beaches will be open for this weekend.

Adapted from Jaws [19]

You may be asked to act upon concerns raised by your team, your juniors, your peers, the nursing staff, the administrative staff.

They may be truly concerned about how things are done, about the inadequate specialist medical cover in the ward, the stretching of resources, the delays in typing clinic letters, the rise in wound infections of the unit, the mixing of clean surgery patients with those admitted with chest infections, the unsafe handovers, the spread of your patients round the hospital, the poor training experience that an attachment provides.

Treat concerns with the seriousness they deserve, give ample time and thought, don't assume that this is simply another "moaning exercise", just another part of day to day complaining.

Look at the concern in a systematic way:

- What is the concern?
- Do the data justify the concern?
- Is safety compromised?
- Should we pause until data is gathered?
- What is the solution and how can I facilitate the solution?
- Whom do I need to involve to implement the solution?
- Can an interim solution be found?

Some decisions may be really challenging to make, the pressure may be high from all sites:

- Your surgical wound infection rate may have gone sky high, but cases are waiting, personal, institutional, or national targets are at stake, personal and institutional financial pressures may be mounting, how can you stop the flow?
- Your juniors may be reporting that they get no meaningful surgical training, that their logbooks are failing, that they simply keep holding retractors, but the rest of the team keeps insisting that you do the cases yourself to finish on-time, your cases may be pooled and never make it to your theatre, lack of beds means cases keep getting cancelled on the day.

Listen and act upon concerns, even when they come from those you never hear praises from, even when they come from those who constantly complain.

2.9 Stick to the Core Principles

Change your opinions, keep to your principles; change your leaves, keep intact your roots.
Victor Hugo [5]

Stick to the core principles of surgery, of providing safe high quality care, whilst wisely using resources, contributing to the advancement and progress of the profession, passing on knowledge and skills.

Aim to improve the overall experience of care you provide but avoid being distracted from your main goals and aims. There may be a need to improve your surgical skills, improve your surgical experience to advance and progress in your career, to gain reputation and status. There may be system pressures, targets to meet, cutbacks in resources. There may be competition with other providers, a need to attract referrals for your practise to survive. But even at difficult times, at times of turmoil, core principles remain the same.

When getting in a plane, one's priority is to get to their destination, and do so safely. Choosing an airline which has a user friendly website, easy to check in apps, shiny couches in its waiting lounge, comfortable seats with extra leg space, one which sends out regular newsletters, or allows three pieces of hand luggage on board, may be very welcoming, but would we exchange safety for any of these spoils? Even if we are in a rush and there are delays we may be forgiving once we safely land. And when at 36,000 ft, in bad turbulence, or facing strong head-on winds, would we really appreciate our low cost ticket, if the airline we chose was advising its pilots to keep any extra fuel on board to the bare minimum [20]?

Should you be sacrificing quality and safety for improving surgical experience, for financial gains, for pleasing an employer or a political master?

2.10 Go with the Flow

I go with the flow. Whatever music you play for me, I'll dance.

Gael Garcia Bernal [5]

In the workplace there are often different ways of doing things, different opinions in dealing with clinical or non-clinical issues, different approaches or different solutions. There may be various ways of investigating a particular clinical problem, various management options for a clinical presentation, various ways of setting up the order of the theatre list, various manufacturers from which to purchase surgical equipment, various ways of allocating resources or staff.

You may have strong views, be strongly opinionated, you may want things done your own way. Argue your case, try and find a preferable or acceptable midway solution.

But it is unlikely that anyone can win all battles, many situations are not even worthy battlefields. If it doesn't matter, if it won't affect the quality and safety of the care you provide, if it is not limiting your practise, if it is not detrimental on the service provision, training or patient care does it worth the confrontation, does it worth the conflict?

Develop the skill of recognising your priorities, the skill of knowing when to stand and fight and when to let go, when to go with the flow. It is not a failure not getting our way every time. If we fight for all, even those that don't really matter, how can we be really listened to when it comes to the vital issues?

Consider going with the flow if it doesn't matter, if it is an acceptable alternative to what you would really prefer, if what will be gained does not justify the conflict. It is professional to accept that for some things you have to give way and not demand that all goes your way. Go with the flow, but not for personal gains, not to attract gratitude, financial incentives, for fear of consequences, for keeping a peace of mind.

2.11 Swim Against the Tide

It's OK to not be fine.

Frasier Crane [5]

There may be times when you need to stand up and be counted. Just because many around you are going in one direction does not mean that has to be your way too. There are instances when groups of individuals, organisations, institutions, professions, or even whole societies go down an unwanted route, and may take time and even lone voices to bring them back to a better way.

Some of you may be practising in healthcare systems driven by targets put in place to mean good, but are consistently (consciously or unconsciously) misused. In the National Health Service (NHS) of the United Kingdom (UK), multiple targets were introduced, such as the 18 week target from referral to elective surgery and the four hour target for either admission or discharge from the Emergency Department [21, 22]. These aimed to reduce unnecessary delays in both elective and urgent treatment. However, as a surgeon, you may be aware that patients who breach their wait for surgery target are put on the site, with those waiting less brought in to be operated upon. You may see patients unnecessary admitted to avoid breaching their Emergency Department target stay, when they could be kept there a bit longer and be discharged once their tests are back [23, 24]. If you see the wrong, call it out.

Going against the tide can be tiring, wearing you out, unnerving, may feel threatening to your well-being, employment status, progression or career. It may feel like a constant struggle, like no one is listening, you may be made to feel as the one at fault, as the one to blame, as the argumentative, the one obstructing progress. Discuss your worries with those close to you, your professional support body, your seniors, your mentors. Evaluate your actions, question your reasoning, but if you still feel that you have to go against the tide then do so. If it matters stand up and speak out. Sometimes all it takes is for some to speak up.

2.12 Act More Than a Technician

The universe is not indifferent to our existence – it depends on it.

Stephen Hawking [5]

You may really enjoy the technical part of surgery, being scrubbed in theatre, cutting up or stitching, fixing the broken, putting them back together. But being a surgeon may not just be about the time you spend in theatre, the technical aspect of surgery, but also about the input you make to all else that goes around you.

Should you not be contributing to the shape of your society's healthcare, the design of a new operative theatre block, the development of perioperative pathways and protocols, the construction of you theatre list? Should you act as a mere technician that turns up, is given a theatre list and operates? You may enjoy being a technician, others may prefer you just being a technician,

Should you be a bystander when political decisions are made about which surgical services will remain and which will be centralised, when commissioning decisions are made about cost effective or not cost effective procedures, when ward management decisions are made as to whether your hip replacement case will be nursed next to a patient with a thigh abscess or in a ring fenced clean surgery ward, decisions about the procurement of cheaper but also poorer quality sutures, waiting list decisions as to who goes on next week's theatre list and who waits longer, who is classified as urgent and who as routine, decisions about shortening the length of training, the production of generalists rather than specialists.

Surgical care is not just about the patient and the surgeon. Healthcare has complex dynamics, may be subjected to political influences, subjected to organisational and non-medical professionals' motives and priorities. Systems that aim to facilitate the patient-surgeon relationship may not be functioning, may be consciously or unconsciously counterproductive. Your priorities as a surgeon, and what you perceive as your patients' priorities, may be different form the priorities of your colleagues or co-workers.

2.13 Stand Up for the Profession

I learned a long time ago the wisest thing I can do is be on my own side, be an advocate for myself and others like me.

Maya Angelou [5]

Healthcare is changing. There may be more political interference, the public may be, and rightly so, becoming more demanding. There is constant flow of information arguing for rapid, immediate actions.

The surgical profession may be challenged by attempts to shorten training to produce more generalists than specialists, by targets for productivity and delivery that threaten quality standards, by the lack of enough juniors to support senior surgeons, public and private commissioners limiting surgical funding, by moves to increase shift and weekend working with further disturbance of work-life balance.

Be an advocate for the profession [25], put forwards its case, argue and fight. You are in a good position to know what's needed for a safe, high quality surgical care. Whether it is by speaking out to raise concerns, promoting the good work delivered by the profession, changing misconceptions, advising commissioners and politicians, engaging in reducing waste and improving efficiency, acting as a follower or a leader. Stand up not only for issues that affect you in person, but also those that may affect your colleagues, your peers, your juniors. Vocalise your views, write emails or letters, complete surveys, demonstrate, support or vote for those that will stand up for the profession, in your department, your institution, locally, nationally or internationally. Act as part of a group, part of a professional body, or even act alone.

Be an advocate for the profession, not only in words, but also through your day to day actions, your personal conduct, your behaviour and relations with colleagues and other staff, by maintaining above all safe patient centred care. Be an advocate by avoiding behaviours that can question your integrity or motives as an individual, and which, through you, can also question the wider profession.

2.14 Declare Acceptable Conflict

If I, taking care of everyone's interests, also take care of my own, you can't talk about a conflict of interest.

Silvio Berlusconi [5]

Clinicians may depend on industry to develop surgical equipment, devices, procedures or technologies that can make surgical care simpler, safer, more reproducible. Industry may want to develop products that are user friendly, endurable, reliable, well sold and profitable. Industry and surgeons may have some diverse but also some common goals.

Surgeons may work with industry to design, develop, or assess new technologies, solely or as part of an extended group. They may be on the lecture circuit or have training roles, teaching other surgeons to use devices or surgical equipment. They may have a research role, evaluating the outcomes of an implant, in a research post funded partly or wholly by its manufacturer. Such roles may be demanding, time consuming, and reimbursement may be given, monetary or otherwise; royalties, licensing fees, hospitality, travel expenses, consulting fees.

Such relations with industry raise the possibility of conflict, if the surgeon is then making decisions or recommendations to institutions, patients or colleagues, as to which implant or device to purchase or use. Such relations may also raise the possibility of conflict if the surgeon presents research findings concerning implants or devices, obtained in a post funded by the manufacturer of those devices.

There may be concerns that physicians could be biased by industry consciously or subconsciously, and may even fail to recognise this vulnerability. Physicians may feel that the practice of others but not their own can be vulnerable to industry influences [26].

Relationships between doctors and the industry do exist. In 2007, a survey of about 3000 physicians in the Unites States of America (USA), showed that 94% of those responding had some sort of relationship with industry; 83% received food in the workplace and 28% received payments for services such as consulting or participating in research [27]. In the USA, the Physician Payments Sunshine Act was passed into law in 2010, in an aim to encourage transparency on financial interactions between producers and consumers (physicians, hospitals). It requires all drug, medical device, and biologics companies to report any transfers of value to physicians and teaching hospitals, on an annual basis. Companies began recording data in 2013 and the first report was released in September of 2014. First data reported on 4.4 million individual payments, worth approximately 3.5 billion dollars, made to 546,000 physicians at 1306 teaching hospitals [28, 29].

Cooperating or working with industry does not mean that such relationships have to cloud one's judgement, guide their actions, influence their preferences or choices. One may well be able to separate the two. There is evidence that patients approve of financial relationships between physicians and industry that support innovation and research, and trust their surgeons in making patient-centred choices [30].

However, declaring such relationships and any potential conflict when it exists, can help clear the air, minimise the doubts as to the exact surgeons' reasoning, thoughts or motives. Okike et al. [31] analysed the self-disclosure statements of authors of a scientific presentation and committee members at the 2008 Annual Meeting of the American Academy of Orthopaedic Surgeons, to determine whether payments received from the manufacturers of hip and knee prostheses were disclosed. The overall rate of disclosure was about 70%. In a subsequent survey exploring the reasons for non-disclosure, factors cited by surgeons included a lack of relation between the presented topic with the payment received, and misunderstanding of the disclosure requirements of the scientific meeting.

Follow the rules of your institution, the rules of the scientific meeting or journal at which you may be presenting your research, the guidelines of your professional or regulatory body, the law of the land. Declare acceptable conflict, get it out first before others do.

2.15 Stir Well Away from Unacceptable Conflict

The end may justify the means as long as there is something that justifies the end.

Leon Trotsky [5]

Don't let financial or other interests erode your trustworthiness in referring on, in accepting referrals, in recommending devices or implants. Stir away from inappropriate or illegal practises, from unacceptable conflict.

Transparency International, a non-governmental organization set up to combat corruption, defines corruption as 'the abuse of entrusted power for private gain' [32]. In a 2006 publication [33], it refers to corruption in the healthcare as the

> *"bribery of regulators and medical professionals, manipulation of information on drug trials, the diversion of medicines and supplies, corruption in procurement, and overbilling of insurance companies",*

and stresses that

> *"society frequently entrusts private actors in health care with important public roles. When hospital administrators, insurers, physicians or pharmaceutical company executives dishonestly enrich themselves, they are not formally abusing a public office, but they are abusing entrusted power and stealing precious resources needed to improve health".*

Transparency International refers to facilitation payments as a

> *"small bribe, also called a 'facilitating', 'speed' or 'grease' payment; made to secure or expedite the performance of a routine or necessary action to which the payer has legal or other entitlement"*

and describes that in healthcare

> *"medical staff can charge unofficial fees to attend to patients. They may demand bribes for medication which should be free. Or they may let patients who bribe them queue-jump"* [32].

It is a type of corruption not hidden behind doors, but one that patients may face day in day out.

Inappropriate or illegal practises unfortunately do exist round the world. David Berger, in his article in the British Medical Journal [17], described his observations of such practices:

> *"Investigations attract a 10-15% kickback to the referring doctor. One day, the marketing executive for this clinic had turned up at the hospital with an envelope full of cash—the commission for investigations ordered in the past few months".*

Helena Smith, the Guardian's correspondent in Greece, in a 2014 article [34] describes the practice of "fakelakia", or bribe little envelopes, well known of in Greece:

> *"This summer, a lift at Athens's largest public hospital broke down. When a repair technician arrived he couldn't believe his eyes: knee-deep at the bottom of the shaft were hundreds of envelopes, the vessels for bribes to doctors who then dispensed with the telltale fakelakia".*

Interestingly a Transparency International 2012 National Survey on Corruption in Greece [35] suggests that even little envelopes could not stay immune to the country's financial troubles:

> *"The average bribe to a public service cost €1228 in 2012. However, the bribes paid to public services have also been affected from austerity, as there is substantial reduction in their average amount from 2011, when the average bribe cost €1399. A bribe to get"... .."surgery in a hospital can cost between €100 and €30.000".*

Charles Piper, a Healthcare fraud investigator in the USA, author of the Healthcare Fraud Investigation Guidebook, describes common encounters in falsifying healthcare insurance claims, including billing for services never delivered, billing for non-covered services, misrepresenting dates, locations, or providers of a service [36].

Stir away from such practices for financial or other gains, even if asked by patients to help them inappropriately claim, to alter the details of the person you treated, the type of procedure, the cost, the date. Stir well away from inappropriate or illegal practises, from unacceptable conflict, from actions that can impair your individual integrity, but can also undermine the wider profession, question its honesty or challenge its trust. Stir away from such practises, even if practising in unhealthy systems, even if doing the right makes you the exception.

2.16 Gather Experience

We are all apprentices in a craft where no one ever becomes a master.

Ernest Hemingway [5]

A surgeon's professional life is one of constant evaluation and learning, a constant strive to learn new skills, maintain skills, refine skills learnt, keep up with the progress. Aim to actively seek the acquisition of experience.

Clinical experience may refer to history taking, examinations skills, formulation of a working diagnosis, consideration of differentials, treatment strategy, technical skills, communication skills, the whole spectrum of surgical clinical professional life.

Experience requires devotion of time, requires one to be present, rather than simply reading or thinking about an event. However, experience is not a passive process, a mere presence in a situation, but an active participation along with a reflection upon what occurred. In this way, as summarised by Kolb's cycle of experiential learning [37], one may re-adjust their actions to help improve performance if a similar situation is encountered in the future (Fig. 2.3).

Aim to gain knowledge, but also the ability to acquire principles and concepts that could be used to solve newly encountered problems, deal with situations you haven't previously faced.

Learn from all, whether senior or junior; do not be put off by their titles, grades, or professional status. Surgery is a craft, the master of which may be improved through repetition. One may not have progressed in professional status or in career pathways, but may have mastered that craft, may be really skilled in what they do, and you may have lots to gain by joining them. Your junior may have acquired concepts from other trainers that you haven't really considered, you may still listen and learn.

Learn from experts in their fields; some may be skilled in a specific surgical procedure, some may be skilled researchers, excellent communicators, some may have clarity in their thought process and in formulating of clinical diagnoses. Aim to learn and gain from all.

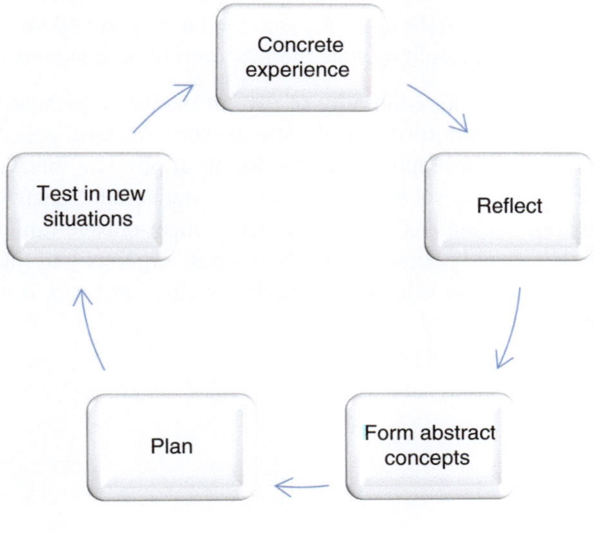

Fig. 2.3 Kolb's cycle of learning. Adapted from Kolb [37]

2.17 Seek Continuity of Care

80% of health apps are abandoned within two weeks.

Marco Della Torre [38]

Aim for continuity of care, aim for following a patient through their treatment journey, whether that journey involves surgical or non-surgical treatment. Continuity of care may facilitate inter-personal relationships between patients and surgeons, and be of direct benefit to both.

Patients may prefer to be seen by a familiar face, someone who knows their story, someone who is likely to catch up from where they last left. They may feel more reassured that they are seeing someone that knows their case first hand, someone who was physically present at their surgery, rather than someone who transmits messages relying on records and notes.

Surgeons may similarly feel confident that they are dealing with a familiar case, more certain of the information they are relaying, more able to assess their patient's progress since their last encounter. Surgical learning may also be facilitated by directly observing changes that occur with time; how a disorder deteriorates or improves, the milestones through which a patient recovers post-surgery, how the bone unites radiologically, how a huge open wound heals with vacuum dressings. Such first-hand observations may give you the confidence to anticipate what will likely happen in the future, much greater confidence than simply reading about it in books or hearing seniors say so.

Continuity of care is even more important in those cases where difficult decisions had to be made:

- Was your diagnosis of acute appendicitis correct or did it turn out to be a non-inflamed appendix?
- Was it truly a pseudo-obstruction or did it turn out to be a mechanical block?
- Were you right to accept the not so well perfused bowel anastomosis or should you have opted for a stoma?
- Did the mal-united fracture remodel like you had counselled the child's parents?
- Were you right to fix the fractured hip or should you have replaced it?
- Were you right to plate the fractured tibia or should you have nailed it?

Maintaining continuity of care in modern healthcare delivery is often an uphill struggle. High turnover of patients, shift working patterns, healthcare systems that place greater emphasis on the timing of physical interventions rather than the context in which such interventions are delivered, transfer of patients between teams without clinicians' input in order to meet elusive (and sometimes unjustified) targets, are only some of the challenges surgeons may face. Despite the difficulties, appreciate the value of continuity of care, and seek it as much as possible in your practise.

2.18 Don't Make Speed Your Sole Aim

Time me, gentlemen, time me.

Robert Liston [39]

Dr. Robert Liston was a London based surgeon, practising in the eighteenth century. He was considered one of the finest of his time, a really fast pair of hands, a very quick surgeon. Legend has it that once he was moving so fast that he even took off the fingers of his surgical assistant [40, 41]. His speed may sound careless but he was carrying out leg amputations in the pre-anaesthetic era, when speed was essential to minimise patients' pain and agony.

But even now, in the era of sound anaesthetic techniques, and effective pain control measures, minimising surgical time may confer a clinical advantage, reduce the risk of complications such as surgical wound infections or venous thromboembolism [42–45].

Improve your practises and techniques, cut unnecessary steps, to help minimise your operating time. However, do not compromise on quality simply in pursue of speed.

The world is moving fast. Healthcare time is expensive. Pressures may be mounting for fast turnovers. You may hear of surgeons referred to as "fast" or "slow", according to their operating times. You may be aware of "fast" surgeons being more popular amongst co-workers, as the list can finish early, more popular amongst employers as more cases can be squeezed on the list.

There may be times in theatre when things don't work out, operations don't go according to plan, your assistants may sound tired of the prolonged standing and retracting, the scrub staff may want to plan their lunch, the theatre coordinator may be eager to know if you will have sufficient time for the next case, times when you may feel that you have had enough. Avoid losing your nerve, rushing the job and compromising quality. Devote the time that's necessary, devote the time that's needed.

And even in clinics you may hear of surgeons that tend to "finish early" or of others who religiously "overrun". But consider the patients who took time to come and see you, got time off work, took the bus or drove from the other end of town, struggled to park, and have being waiting all this time in the waiting room to consult you, in search of information, solutions or help to their problems. You may be running late, the clinic may be overrunning, but don't look at them as just another appointment to get it done and over, devote the time needed. One may forget and forgive the wait if it was worthy waiting for.

You are more likely to be judged on the outcomes and quality of your work by those who matter the most, rather than the time it takes you to get there. Be fast, but don't make speed your sole aim.

2.19 Understand Your Limitations

Refusing to ask for help when you need it is refusing someone the chance to be helpful.
Ric Ocasek [5]

Understand your limitations in terms of clinical skills, surgical skills, diagnostic ability, your ability to communicate, to persuade.

Understand your own personal limitations:

- You may have taken an adequate history, performed a thorough examination, carried out several investigations, but still you are unclear as to the working diagnosis. A second opinion, may help see or hear what you may not be able to do. Seek further opinion, even if you know it is highly likely the further opinion will be in line with yours.
- You may be a skilful surgeon, a specialist in an area, but a patient presents with a severe form of a condition and you feel this will be a challenging, demanding case. If needed, ask for another surgeon to join you in the surgical theatre for help and support. It is not weakness to scrub for surgery with another senior surgeon, no matter how senior one is; instead, it may be considered a sign of strength to recognise own limitations and put egos on the side in the name of quality and safety.
- You may be practising in an advanced healthcare system, but faced with a condition which is rare in your country but much more common in another country, where surgeons may be treating in a week what you may see in a year. If needed, refer patients on to those who may be able to do a better job.

Appreciate the limitations of your team. You may have just visited a centre of excellence in joint replacement surgery and impressed by their high turnover, their ability to perform six hip replacements in a day. Aspire to improve your team's performance, to reduce theatre time loss, enhance patient flow. But recognise what may be easily achieved by other teams may not directly apply to yours. You may face high turnover of scrub staff, inexperienced staff that are still learning, processes and regulations by which you have to abbey.

Recognise the limitations of your institution or the healthcare system you are working in. Your institution may be having difficulties with bed availability, theatre time availability, organisational structures, too many cooks spoiling the broth. You may not be able to immediately influence what is going on. Work hard, engage, argue to change things, but you may appreciate that change takes time, that there is so much that can be achieved. If you feel that your patients may be served better by being referred or directed to another provider then consider that option.

You may be under pressure by patients, relatives, colleagues, nursing staff, clinicians from other specialties, or management, to make a quick decision, to admit or discharge, operate or mobilise. If you feel that what you are being asked is beyond your own limits say so, seek advice and help. Don't stretch your limits simply due to external pressures. One may be initially praised for quick, decisive actions, simply to be later condemned for acting out of their safe zone.

Knowing your limitations doesn't mean to be defensive, indecisive, refusing to take anything risky on, but acting within your safe, high quality boundaries. Seek help as you need it. You are likely to be judged by the outcome and not by whom you listen to or whom you consult in the process of getting there.

2.20 Walk Before You Can Fly

He who would learn to fly one day must first learn to stand and walk and run and climb and dance; one cannot fly into flying.

Friedrich Nietzsche [5]

As a surgeon you may find yourself in new situations. As a junior you may be moving from hospital to hospital, from one team to another. But even if you have completed your training and you are independent practising you may find yourself operating for the first time in a new hospital, in a new theatre, with anaesthetic or scrub staff you haven't worked with before. Others may not have come across the ways you do things, and may even be surprised with what you are doing. You may be asked "why are you doing things this way", rather than "do you really know what you are doing?".

All eyes may be on you, trying to work out what you are made of, whether you can really rise to the challenge. Your senior may be keen to see if you can really fix unaided the nasty hip fracture (as you have already claimed in your log book), the waiting list coordinator may be really keen to get that long awaited cases operated. But should you be rushing to take them on, or should you go at your own pace?

You may have the skills to take really complex cases on, but should you be rushing into doing those if they can safely wait? Performance in the operating theatre may be influenced by many factors, not just knowledge, technical abilities, decision making, and endurance, but also by our relations with the rest of the team, the confidence they have in us, the trust we put upon them.

If you end up struggling to reduce a fracture in an environment you have not worked in before, can the rest be assured it is the fracture complexity rather than you to blame? If you keep asking for the fracture reduction rod but you are told there isn't one, can you be assured that's the case rather than it being shelved somewhere difficult to find?

Should a singer going to an audition pick a really hard song to sing, one that would really challenge their vocals, or should take it easy, and pick one that can allow their voice to be heard, one that can allow their voice to shine? Should an athlete turning up for international games high jump with the bar set at the 2.4 m barrier from the beginning, or start from lower heights, get used to the track, and then attempt the big jump?

Walk before you can fly. A pilot doesn't suddenly get in front of the controls of an Airbus 380, and it is unlikely that one can go straight from medical school to carrying complex neurosurgical, cardiac or other surgical procedures.

Understanding your limits is vital when working in a new environment, a new team, under a new senior. Is it better to stay within safe limits and gain trust or take unnecessary risks and end up losing face? First impressions may matter, and acting within limits may help an initial good impression to be made.

2.21 Seek and Give Feedback

It is impossible to begin to learn that which one thinks one already knows.

Epictetus [5]

Feedback needs to be exchanged between team members, between colleagues, seniors and juniors, trainers and trainees, amongst surgeons, between surgeons and other medical or non-medical specialists.

Seek and give feedback. Learn and educate; for what went well, for what could have gone better. Establish well intended, constructive feedback channels between you and the professionals with whom you routinely collaborate. Exchange feedback soon after an event, whilst memories are still sharp and facts are still clear. Use feedback not for criticising or getting at others, but for promoting your own and their professional development.

A surgeon may exchange feedback from a variety of sources as summarised in Fig. 2.4.

- You may receive feedback from a radiologist about the cruciate ligament you clinically suspected had ruptured; you may inform the radiologist of your arthroscopic findings of a meniscus reported on Magnetic Resonance Imaging as having a large tear.
- You may receive feedback from the pathologist about the margins clearance of a breast lesion you considered fully excised; you may inform the pathologist of your surgical findings based on an initial aspirate that was reported as possibly malignant.
- You may receive feedback from a physiotherapist about the effects of your prolonged postoperative immobilisation regime on wrist stiffness; you may inform the physiotherapist of how well their early rehabilitation techniques speeded recovery and return to play.
- You may receive feedback from your anaesthetist about the effect of your intra-articular local aesthetic infiltration; you may give feedback to the anaesthetist about the effectiveness of the regional block they employed.

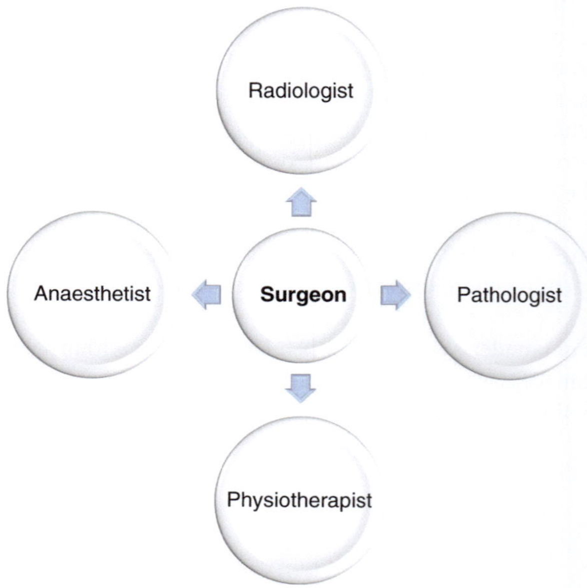

Fig. 2.4 Surgeon exchanges feedback with collaborating professionals

2.22 Self-Reflect

If I speak of myself in different ways, that is because I look at myself in different ways.
Michel de Montaigne [46]

Self-reflection refers to one looking at own self and evaluating own actions, beliefs, or motives [47]. It may allow one to evaluate whether they communicated effectively, evaluate their clinical decision making, diagnosis, investigations and treatment choice, prescription patterns, technical actions.

Self-reflection is increasingly recognised as an important component of personal professional development for the training surgeon, and a requirement of surgical portfolios [48]. However, self-reflection shouldn't be looked at as just another tick box, another entry in your portfolio, but a way of practise, a way of constant evaluation and development.

Self-reflection doesn't mean that you did things wrong, it just asks could I have done better? Self-reflection does not imply weakness, but strength, as it is an indication of a constant strive for improvement. It takes maturity, self-consciousness and self-confidence to recognise that our actions may have been different, to learn and adjust from previous experience.

You may consider:

- You may have picked an argument with a casualty doctor trying to refer an abdominal pain case to you. It might have been an inappropriate referral but could you have handled it better, rather than ending up shouting "NO" on the phone?
- Could you have explained better to the ward sister that you will be there as soon as you finish clerking the new admissions in casualty? Should you have raised your voice? Should you have slammed the phone down?
- You may have just had a consultation that you thought went well, but as soon as the patient goes home they ring to complain. Could you have done anything else, could you have communicated better?
- A patient may still be asking you the same questions even after you explained all—why is your message not getting across? Should you reconsider how you explain things?
- You may have just spoken to a patient about a complication they developed, but it was a struggle to vocalise that an error occurred. How could you be more upfront next time?
- You may have been unhappy about an anaesthetist's decision to cancel your case, but would your interpersonal relations been served better if you spoke to them directly and argued your case, rather than complaining to their director the next day?
- You may have opened up a shoulder fracture with the intention to fix it but you find that it cannot be put back together; should you have anticipated this and made sure that the replacement equipment was available?
- You may have tried to remove an appendix laparoscopically but it was just not happening due to multiple adhesions; should you have converted to an open procedure earlier?
- You may have encountered a high rate of post-operative surgical complications in your practise. Is it just bad luck or could you have been doing something differently?

Self-reflect, learn and get better.

2.23 Acknowledge Uncertainty

There is no such uncertainty as a sure thing.

Robert Burns [5]

As surgeons we may have strong dogmatic views as to how things should be done, demand our own ways—which prep to use, which venous thromboprophylaxis to instate, which fractures to fix and which ones to treat in plaster, which surgical procedure to choose out of the many, how many screws to use and in what form and shape, whether to glue, stitch or clip the skin, which postoperative analgesia to prescribe. Strong views often based on previous practise and experiences, our training and teaching, on trial and error; strong views occasionally based on high quality evidence and research.

We may discuss, argue, or even give elaborate explanations as to why our approach is the preferable one, the right one.

However, in many aspects of your work you may be faced with uncertainty. You may consider:

- Recognising uncertainty in clinical diagnosis may allow one to reach a plan of action and initiate treatment, rather than unnecessarily waiting until all pieces of information are gathered and a definite conclusion reached.
- Recognising uncertainty of outcomes may allow you to communicate realistic expectations to patients as part of shared decision making and the consent process, rather than painting an over-optimistic or over-pessimistic picture.
- Recognising uncertainty as to which treatment method is best, may allow one to appreciate that what other colleagues follow may be an accepted alternative option.
- Recognising uncertainty in the effectiveness of treatment methods we employ may allow one to set up research trials or recruit patients for such trials.

Acknowledge uncertainty and keep an open mind.

2.24 Aim for Accurate Communication

Accuracy is the twin brother of honesty.

Nathaniel Hawthorne [5]

Be truthful, honest and accurate in communication with colleagues, co-workers, or other professionals. Avoid over exaggerating, or playing things down. You may consider:

- "How long will surgery take?" The anaesthetist needs to know to adjust the muscle relaxants, use a laryngeal mask or intubate, whether to administer more anaesthetic. Enquiring about how long you will take, how long more to go, is not a criticism of your speed. Simply stating "it won't take long" or "not too long to go" will not decrease your operating time. Take into account how long to close and dress the wound, come off the operating table.
- "How much blood has been lost?" The anaesthetist needs to know to explain the observed hypotension, to guide the need for haemoglobin testing, fluid resuscitation or blood transfusion. How much is in the wound, in the swabs, on the floor? Simply stating "just a bit of oozing, not much at all" will not reduce the volume of actual blood loss.
- "Is your patient unwell enough to justify jumping the queue on the shared emergency list?" Is your case really unwell or can it wait? Enquiring about the urgency of your case, may not be an attempt to avoid operating or postponing your case, but of prioritising amongst several cases that are waiting.
- "How painful will the procedure be?" The anaesthetist needs to know to adjust the intra- and post-operative analgesia, consider regional blocks, continuous or regular painkillers. Enquiring about pain is not a criticism of your surgical technique or soft tissue handling. Simply stating "not very painful" or "some local anaesthetic will be ok" will not reduce the post-operative pain.

Nurok et al. [49] surveyed surgeons and anaesthetists in the USA enquiring how often they intentionally mis-reported information in the peri-operative setting. Seven percent of the responding surgeons admitted to such intentional mis-representation at least once a month, with two percent of the responding surgeons reporting that patients came to harm due to them representing the truth. Justifications for such mis-representation included that the other party would not understand, or that they would demand unreasonable actions. Almost 70% of anaesthetists were concerned that surgeons mis-reported information to them at least once a month.

Aim for accurate and honest communication.

2.25 Follow Procedures

If I'd observed all the rules, I'd never have got anywhere.

Marilyn Monroe [5]

There may be established procedures in your working environment, existing policies, rules, protocols, pathways. There may be procedures about ordering surgical equipment for theatre cases, how to trial new theatre kit, procedures about on call allocation, job planning, overtime pay, referrals to other providers, admissions from the community.

Such procedures may have been set up with or without your knowledge or contribution, they may seem out of touch or out of date, bureaucratic, slowing progress, unnecessarily labour consuming or expensive. Question the procedures, fight to change them, make your views heard, but try to follow these whilst in place.

- You may be overwhelmed with admissions; should you suddenly send an ultimatum saying that you no longer accept referrals from the Accident and Emergency Department?
- You may feel that your juniors should be more ward based rather than being in theatre; do you just inform them one morning as to where they are now rostered?
- You may feel that your juniors are not delivering the work they get paid for, that their on-calls are not busy enough, should you simply write to them and inform them of your decision to cut their pay?
- You may be the rota master and you are not happy of the rolling rota that is already in place. You feel that all should be on call on one of the upcoming festive days. Do you simply inform them of the new on call allocation?
- You may have just been to a course and you really liked the bowel stapler exhibited. Do you just tell the manufacturer's rep to bring it along to your next theatre list to try?
- You may need a new junior in the department, do you just ask for your friend's cousin to be added to the payroll?

McDonald et al. [50] evaluated the views of doctors and nurses to guidelines and rules in the operating department of a teaching hospital in the North of England. Unlike nurses, who viewed adherence to guidelines and written procedures as essential for professionalism, doctors seemed to reject written rules and be more inclined to follow what they considered to be the unwritten "norms" of the profession.

For a team to function, an institution to thrive, procedures are to be followed, rules to be obeyed. Any change you may want to bring, sound or noble, well-motivated and worthy, may struggle to make it through if procedures and rules are not abided.

2.26 Welcome Innovation

The arrogance of success is to think that what you did yesterday will be sufficient for tomorrow.

William Pollard [5]

We live in a fast changing world, a revolution of innovations. Healthcare and surgery cannot stay immune from such change. We are constantly faced with new ways of doing things, new practices, new techniques, implants, devices. We are faced with innovations aimed at improving outcomes, improving safety, making operative procedures more reliable and reproducible, making surgical instruments and devices more user friendly, allowing treatments to be accessible to the wider community, to the wider world. Innovations that aim to streamline flow, reduce hospital stay, improve efficiency, or reduce costs.

You may face ward rounds with electronic tablets rather than paper lists, electronic radiographs rather than printed films, laparoscopic rather than extensive open surgical techniques, mini rather than lengthy incisions, custom made rather than off the shelf implants, 3D printed rather than factory manufactured splints, electronic rather than hand written records, robotic arms rather than human hands.

Welcome innovation than can improve care, innovation that can enhance the quality of service you provide. Avoid simply staring in the past, refusing to look forwards to the future.

As a surgeon you may have a key role to play in identifying the need for such innovations, in their design, testing, use and dissemination. Try and be the one who can shape innovation, the one who can set the priorities, show the direction. Innovations are likely to happen one way or another, try and be an active participant rather than a bystander.

2.27 Question Innovation

> *It would be a terrific innovation if you could get your mind to stretch a little further than the next wisecrack.*
>
> Katharine Hepburn [5]

Innovations may come about for noble reasons, to improve, to better. Nevertheless, despite their good intentions, question innovations, question their effectiveness, their benefits, their costs, their added new risks. Don't just take them for granted. Something new may not always be so much better, or even just better.

There may be moves to minimally invasive surgery, laparoscopic, arthroscopic, endoscopic, computer assisted, robotic. What is the gain they confer, do they justify any extra costs or time that such procedures may entail? There have been moves to smaller incisions, tiny cuts, but are multiple incisions so much smaller than a single one when you add them all up? Question innovations:

- What is their real aim, their real benefits?
- Do they have improved outcomes, do their improved outcomes matter? If their main outcome is equivalent to existent technologies, could they be better in secondary outcomes?
- Do they get rid of old recognised complications? Do they bring new complications?
- What follow ups do they have? Do we really know what their long term outcomes are?
- Do the described outcomes originate from the innovators or the wider community?
- Is the promised better performance based on simulation, laboratory data or real life outcomes?
- Upon what numbers are the reported outcomes and complications based? Could we just not be fully aware yet of their potential complications?
- What were the characteristics of the population upon which the innovations were tested? Are those results applicable to your patients, to the wider community?

The IDEAL framework [51] is used to describe the stages of innovation in surgical procedures or interventions and gives guidance as to research design types that are appropriate for assessing each stage (Fig. 2.5).

In the idea stage a concept is proven, by describing a case-report to demonstrate that it is possible. In the development stage the number of cases is increased with the aim to demonstrate the safety and success or failures of the procedure. This is further expanded in the exploration phase with prospective studies that have even more cases. In the assessment phase one aims to compare the innovation to another current acceptable intervention through randomised controlled trials. Once the benefits and advantages of the innovation are established and it becomes part of regular practise, long-term surveillance studies (through large registries or databases) may look for rare events or long term outcomes. You may critically examine your practise and question how many of the innovations you encounter have gone through these stages.

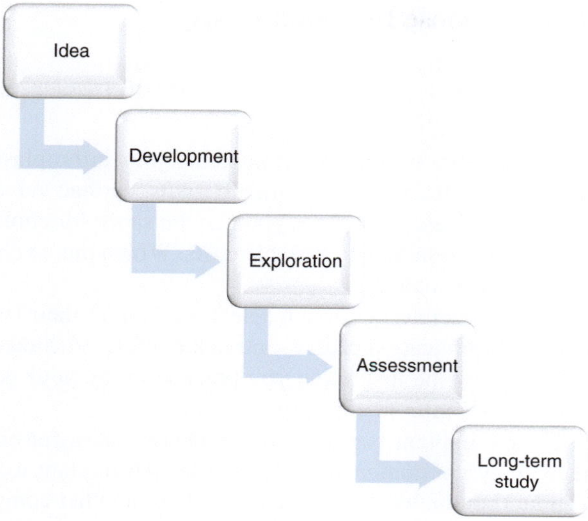

Fig. 2.5 IDEAL pathway to innovation

Listen to the promoters of innovations, to the manufacturers and their sellers, read the promotion leaflets, but also examine the wider literature, speak to colleagues who have used and tried the new proposals. What difficulties did they face? What complications have they seen? How does the new really compare with the old?

If you are satisfied with what you use, if what you use has been tried and tested, has proved the test of time, should you be changing for a promised yet unproven future? New innovations may solve old problems but may also introduce new ones, new risks, new troubles.

If you are flying to the other side of the world would you like to be on a plane design that has done the miles, flown through storms and turbulence, by experienced and not so experienced crew, one that has a safe track record, or would you choose a plane that has just come out of test flights but promises bigger leg space and an uninterrupted smooth flight? The wing cracks that the giant Airbus A380s initially faced [52], and the easily inflammable lithium batteries that the Boeing's Dreamliners initially battled [53], may help focus minds. The choice is personal but at least consider what you are getting into.

Nieuwenhuijse et al. [54] carried out a systematic review to assess the safety and effectiveness of five recently introduced innovations in joint replacement (hip replacement ceramic-on-ceramic bearings, modular femoral necks, un-cemented monoblock cups, high flexion and gender specific knee replacements). They examined published research studies and national joint registry data, but could not find high quality evidence supporting the use of these innovations. In contrast, the authors reported that the available evidence suggested that existing hip and knee replacement implants may be safer than the newly introduced innovations.

You may decide to err on the cautious site, and wait for long term results before you make the transition. But, even if you make the jump early on, do so in a controlled way, explaining the limitations of the new innovation to your patients, closely monitoring your results and the results of your peers.

2.28 Choose Implants Wisely

If we choose, we can live in a world of comforting illusion.

Noam Chomsky [5]

As surgeons we may insert implants, joint arthroplasties, tendon substitutes, hernia mesh grafts, vascular synthetic grafts, cardiac valves, breast implants, fracture fixation devices. Sometimes you may be spoilt for choice—multiple implant types, multiple manufacturers, variable costs. Which one to choose, which one is better, it can be a real challenge.

Be considerate in implant selection, look at their specifications, long term outcomes, complication profile, and make informed choices. Even if your options are restricted by the institution you practise in, by your partners or colleagues, be an active contributor to any choice made.

A new implant based on a previously successful one, made by the same company, may not confer the same results. An implant made by one company that is similar (but not identical) to an implant of another company, may not give the same outcomes. Changes in material composition, geometry, surface finish, shape and size, may exert substantial effects on an implant's behaviour. Small design or manufacturing changes may have profound effects on durability or longevity.

Hip replacement surgery is a commonly performed procedure, which gives reliably good results. There are multiple hip replacement implants available, some of which have been in use for more than 20 years and have been shown to have successful outcomes. Sir John Charnley is considered one of the fathers of hip replacement surgery, with his hip replacement design showing long lasting successful results with more than 90% survival rate at 10 years, and more than 70% survival at 20 years following implantation [55, 56]. In the early 1990s the company 3M Health Care Ltd. introduced a new hip implant, the Capital hip design. The Capital hip was described as a low cost copy of the Charnley hip replacement implant. However, there were minor differences between the two, with regards to surface roughness and geometry. These differences were small, but critical enough to lead to early stem loosening, 16–26% at just over two years, a catastrophic high early failure rate. The capital hip was withdrawn from the market in 1997 but before its withdrawal, almost 5000 such prostheses were implanted in multiple centres throughout Britain [57].

High failure rates have also been observed more recently with metal on metal hip replacement implants. Traditional hip replacement prostheses involve a metal ball articulating with a polyethylene cup. In an attempt to reduce wear and increase implant longevity, metal on metal implants were introduced, which involved a metal ball articulating with a metallic cup. Mr. Derek McMinn, an Orthopaedic surgeon, introduced the Birmingham hip metal on metal replacement system, which has been shown to confer good long term results, with an estimated ten year revision rate of 4.5% [58]. Another company, De-Puy Orthopaedics introduced its own version of metal on metal replacement, the ASR, which although similar, had critical geometrical differences. The ASR led to a high failure rate, 5.6% to 31% at four to seven year follow up, and was withdrawn in 2010 [59–61].

Similar troubles have been seen for breast silicone implants made by Poly Implant Prothèse, a French company. Inferior silicone was used, leading to higher than expected rupture rates, causing anxiety and fear in thousands of patients who had these implants [62, 63].

Be cautious in implant selection. Choose implants based on their record, their long term outcomes and results. A proactive or persistent manufacturer's salesperson, aggressive marketing, low cost, or use by folks or rivals are not appropriate selection criteria.

2.29 Appreciate the Limitations of Technology

It's supposed to be automatic, but actually you have to push this button.
John Brunner, Stand on Zanzibar [5]

Recent years have seen a technological explosion with high resolution cameras for minimally invasive surgery, computer assisted or computer guided surgery, or even robotic surgery. Technologies that aim amongst others to improve visualisation during surgery, allowing one to get to spaces that haven't been scoped before, fine tune the traditional eye-balling of jigs and cutting blocks to facilitate implant positioning and alignment, or to improve hand steadiness and dexterity. Embrace technology but appreciate its limitations.

Appreciate that in some cases the output of electronic devices may depend on your input. If your entry is incorrect, the complex calculations and instructions that the device provides may also come out wrongly. A computer may be able to fly the plane, but relies on the correct destination coordinates being keyed in. Too much reliance on technology may lead to unwanted results. Faulty instruments, complex situations, over reliance on default values, may lead you in the wrong direction.

In computer assisted knee surgery an electronic stylus that can record 3D position in space may be used by the surgeon to record the spatial position of bony landmarks around the knee, as well as lower down the ankle. These landmarks are then used to calculate the optimal alignment axes of the leg and guide as to the direction and orientation of bone cuts, in preparing the bones for prosthesis implantation [64]. But if the initial stylus marking of landmarks is not accurate, then subsequent calculations and directions may be wrong [65].

Get familiar with the devices you employ, know the technology you use. You may find yourself in the surgical theatre, with staff that are inexperienced, who haven't joined you before. They may not be familiar with the equipment used, and their supervisor may be out on a lunch break. Know how to set up the arthroscopy pump, the laparoscopy camera, the diathermy, the tourniquet, the shaver or the drill. Be able to recognise an equipment failure, know how to deal with it, how to take over, or adjust course if technology were to let you down. What does the beep from the diathermy mean? Why are the bright lights on the laparoscopy stack flashing?

Pilots recognise the limitations of technology and are prepared to take over, to fly the plane if the autopilot were to disengage. Shouldn't be that surgeons appreciate the limitations of technology and train to adjust if technology let you down?

2.30 Pass on the Craft

If I don't have wisdom, I can teach you only ignorance.

Leo Buscaglia [5]

Aim to pass on knowledge, clinical and surgical skills. Train your juniors, your seniors, your colleagues, you peers. For a profession to survive, skills must be passed on.

You may not have a formal trainer's role, you may feel that you are not a skilled lecturer or teacher, but you may still try in every opportunity, no matter how short that is, to pass on skills. You may not feel that you have something innovative to say, that your students may have heard it all before, but repetition reinforces knowledge, may confirm that what is known is correct, that it is main stream practise.

Even if you don't communicate knowledge, how to handle the scalpel or use the robot, one may pass on skills of decision making, critical thinking, questioning mind, coping under pressure, the skills of acceptable professional conduct.

In a fast demanding world of pressurised healthcare care systems one may be absorbed in productivity, financial or other targets. Co-workers may simply want you to finish the clinic early, avoid overruns in theatre. Teaching or training activities may not be formally recognised, or co-workers may even complain that time spent teaching your juniors is under-used time.

Pass on the craft. One may achieve much more by training others who then go on and successfully practise, rather than simply through their own personal surgical practise.

References

1. Doctors in society: Medical professionalism in a changing world. Royal College of Physicians; 2005. https://shop.rcplondon.ac.uk. Accessed 25 Mar 2017.
2. Christian F, Pitt DF, Bond J, Davison P, Gomes A, Committee on Professionalism, Canadian Association of General Surgeons, Bond J. Professionalism–connecting the past and the present and a blueprint for the Canadian Association of General Surgeons. Can J Surg. 2008;51(2):88–91.
3. Code of Professional Conduct. The American College of Surgeons. ACS Task Force on Professionalism; 2003. https://afmc.ca/pdf/AmerCollSurg.pdf. Accessed 25 Mar 2017.
4. Davies G. Professionalism of surgeons: a collective responsibility. ANZ J Surg. 2011;81(4):219–26.
5. Famous Quotes at BrainyQuote. https://www.brainyquote.com/quotes/quotes. Accessed 25 Mar 2017.
6. Cromwell J, Mitchell JB. Physician-induced demand for surgery. J Health Econ. 1986;5(4):293–313.
7. Fuchs VR. The supply of surgeons and the demand for operations. J Hum Resour. 1978;13(Suppl):35–56.
8. Christensen S. Volume responses to exogenous changes in Medicare's payment policies. Health Serv Res. 1992;27(1):65–79.
9. Nguyen NX, Derrick FW. Physician behavioural response to a Medicare price reduction. Health Serv Res. 1997;32(3):283–98.
10. Nguyen LL, Smith AD, Scully RE, Jiang W, Learn PA, Lipsitz SR, Weissman JS, Helmchen LA, Koehlmoos T, Hoburg A, Kimsey LG. Provider-induced demand in the treatment of carotid artery stenosis: variation in treatment decisions between private sector fee-for-service vs salary-based military physicians. JAMA Surg. 2017; doi:10.1001/jamasurg.2017.0077.
11. Ekatha AJ. Doctors do avoidable surgeries to meet corporate targets. The Times of India; 2015. http://timesofindia.indiatimes.com/city/chennai. Accessed 25 Mar 2017.
12. English Oxford Lining Dictionaries https://en.oxforddictionaries.com/definition/trust. Accessed 25 Mar 2017.
13. Chamberlain CJ, Koniaris LG, AW W, Pawlik TM. Disclosure of "nonharmful" medical errors and other events: duty to disclose. Arch Surg. 2012;147(3):282–6.
14. Institute of Medicine (US). In: Gray BH, editor. The new health care for profit: doctors and hospitals in a competitive environment. Washington: National Academies Press (US); 1983.
15. Thurman AE. Institutional responses to medical mistakes: ethical and legal perspectives. Kennedy Inst Ethics J. 2001;11(2):147–56.
16. Pellegrino ED, Thomasma DC. For the patient's good the restoration of beneficence in health care. New York: Oxford University Press; 1995.
17. Berger D. Corruption ruins the doctor-patient relationship in India. BMJ. 2014;348:g3169. doi:10.1136/bmj.g3169.
18. Learning from Bristol. The Report of the Public Inquiry into children's heart surgery at the Bristol Royal Infirmary 1984–1995. Kennedy I, 2001. http://webarchive.nationalarchives.gov.uk/20090811143745/http:/www.bristol-inquiry.org.uk. Accessed 1 Apr 2017.
19. Jaws Quotes. http://www.moviequotedb.com/movies/jaws/quote. Accessed 31 Mar 2017.
20. Wright O. Safety warning as budget airlines such as Ryanair cut fuel levels for flights. The independent. 2013; http://www.independent.co.uk/travel/news-and-advice/exclusive-safety-warning-as-budget-airlines-such-as-ryanair-cut-fuel-levels-for-flights-8749046.html. Accessed 1 Apr 2017.
21. The NHS Improvement Plan. Putting People at the Heart of Public Services. Department of Health 2004. http://1nj5ms2lli5hdggbe3mm7ms5.wpengine.netdna-cdn.com/files/2010/03/pnsuk3.pdf. Accessed 1 Apr 2017.
22. Department of Health Reforming emergency care. London: HMSO; 2001.

23. White C. Health secretary announces drive to clear 12 month waiting list backlog. BMJ. 2014;349:g5011. doi:10.1136/bmj.g5011.
24. Gubb J. Have targets done more harm than good in the English NHS? Yes. BMJ. 2009;338:a3130. doi:10.1136/bmj.a3130.
25. Lee AK, Sethi MK. Do Politics Matter to an Orthopaedic Surgeon? (They Should!). J Orthop Trauma. 2015;29(Suppl 11):S3–4.
26. Chren MM. Interactions between physicians and drug company representatives. Am J Med. 1999;107(2):182–3.
27. Campbell EG, Gruen RL, Mountford J, Miller LG, Cleary PD, Blumenthal D. A national survey of physician-industry relationships. N Engl J Med. 2007;356(17):1742–50.
28. Carpenter D, Joffe SA. Unique researcher identifier for the physician payments sunshine act. JAMA. 2011;305(19):2007–8. doi:10.1001/jama.2011.651.
29. Parisi TJ, Ferre IM, Rubash HE. The basics of the sunshine act: how it pertains to the practicing orthopaedic surgeon. J Am Acad Orthop Surg. 2015;23(8):455–67.
30. Camp MW, Gross AE, McKneally MF. Patient views on financial relationships between surgeons and surgical device manufacturers. Can J Surg. 2015;58(5):323–9.
31. Okike K, Kocher MS, Wei EX, Mehlman CT, Bhandari M. Accuracy of conflict-of-interest disclosures reported by physicians. N Engl J Med. 2009;361(15):1466–74.
32. Transparency international. http://www.transparency.org/what-is-corruption/. Accessed 31 Mar 2017.
33. Global Corruption Report. Transparency International. London: Pluto Press; 2006.
34. Smith H. Corruption still alive and well in post-bailout Greece. The Guardian; 2014. https://www.theguardian.com/world/2014/dec/03/greece-corruption-alive-and-well. Accessed 31 Mar 2017.
35. Less and cheaper "fakelaki" in Greece: The key results of 2012 National Survey on Corruption in Greece. Transparency International Greece. http://www.transparency.org/news/pressrelease/20130307_less_and_cheaper_fakelaki_in_greece_the_key_results_of_2012_nation. Accessed 1 Apr 2017.
36. Piper CE. Healthcare Fraud Investigation Guidebook: CRC Press; 2016.
37. Kolb D. Experiential learning: experience as the source of learning and development. Englewood Cliffs: Prentice Hall; 1984.
38. Munro D. Top Ten Healthcare Quotes For 2013. https://www.forbes.com/sites/danmunro/2013/12/22/top-ten-healthcare-quotes-for-2013/#6ab487fa2096. Accessed 1 Apr 2017.
39. Jones AJ, Nesbit RR, Holsten SB. Time me, gentlemen! The bravado and bravery of Robert Liston. American College of Surgeons. https://www.facs.org/~/media/files/archives/05_liston.ashx. Accessed 1 Apr 2017.
40. Magee R. Robert Liston:Surgeon extraordinary. Aust N Z J Surg. 1999;69(12):878–81.
41. Gordon R. Great medical disasters. Looe: House of Stratus; 2012.
42. Reames BN, Bacal D, Krell RW, Birkmeyer JD, Birkmeyer NJ, Finks JF. Influence of median surgeon operative duration on adverse outcomes in bariatric surgery. Surg Obes Relat Dis. 2015;11(1):207–13.
43. Leong G, Wilson J, Charlett A. Duration of operation as a risk factor for surgical site infection: comparison of English and US data. J Hosp Infect. 2006;63:255–62.
44. Tan TW, Kalish JA, Hamburg NM, et al. Shorter duration of femoral-popliteal bypass is associated with decreased surgical site infection and shorter hospital length of stay. J Am Coll Surg. 2012;215:512–8.
45. Chan MM, Hamza N, Ammori BJ. Duration of surgery independently influences risk of venous thromboembolism after laparoscopic bariatric surgery. Surg Obes Relat Dis. 2013;9:88–93.
46. Popular Quotes. Goodreads. http://www.goodreads.com/quotes. Accessed 1 Apr 2017.
47. McGlinn EP, Chung KC. A pause for reflection: incorporating reflection into surgical training. Ann Plast Surg. 2014;73(2):117–20.
48. Webb TP, Merkley TR. An evaluation of the success of a surgical resident learning portfolio. J Surg Educ. 2012;69(1):1–7.

49. Nurok M, Lee YY, Ma Y, Kirwan A, Wynia M, Segal S. Are surgeons and anesthesiologists lying to each other or gaming the system? A national random sample survey about "truth-telling practices" in the perioperative setting in the United States. Patient Saf Surg. 2015;9:34. doi:10.1186/s13037-015-0080-7. eCollection 2015.
50. McDonald R, Waring J, Harrison S, Walshe K, Boaden R. Rules and guidelines in clinical practice: a qualitative study in operating theatres of doctors' and nurses' views. Qual Saf Health Care. 2005;14(4):290–4.
51. Barkun JS, Aronson JK, Feldman LS, Maddern GJ, Strasberg SM, Balliol Collaboration, Altman DG, Barkun JS, Blazeby JM, Boutron IC, Campbell WB, Clavien PA, Cook JA, Ergina PL, Flum DR, Glasziou P, Marshall JC, McCulloch P, Nicholl J, Reeves BC, Seiler CM, Meakins JL, Ashby D, Black N, Bunker J, Burton M, Campbell M, Chalkidou K, Chalmers I, de Leval M, Deeks J, Grant A, Gray M, Greenhalgh R, Jenicek M, Kehoe S, Lilford R, Littlejohns P, Loke Y, Madhock R, McPherson K, Rothwell P, Summerskill B, Taggart D, Tekkis P, Thompson M, Treasure T, Trohler U, Vandenbroucke J. Evaluation and stages of surgical innovations. Lancet. 2009;374(9695):1089–96. doi:10.1016/S0140-6736(09)61083-7.
52. Milmo D. All A380 superjumbos to be checked for wing cracks. The guardian, 2012. https://www.theguardian.com/business/2012/feb/08/a380-superjumbo-checks-wing-cracks. Accessed 1 Apr 2017.
53. Dreamliner: Boeing 787 planes grounded on safety fears". BBC News. January 17, 2013. http://www.bbc.co.uk/news/business-21054089. Accessed 1 Apr 2017.
54. Nieuwenhuijse MJ, Nelissen RG, Schoones JW, Sedrakyan A. Appraisal of evidence base for introduction of new implants in hip and knee replacement: a systematic review of five widely used device technologies. BMJ. 2014;349:g5133. doi:10.1136/bmj.g5133.
55. Joshi AB, Porter ML, Trail IA, Hunt LP, Murphy JC, Hardinge K. Long-term results of Charnley low-friction arthroplasty in young patients. J Bone Joint Surg Br. 1993;75(4):616–23.
56. Halley DK, Glassman AH. Twenty- to twenty-six-year radiographic review in patients 50 years of age or younger with cemented Charnley low-friction arthroplasty. J Arthroplast. 2003;18(7 Suppl 1):79–85.
57. Muirhead-Allwood SK. Lessons of a hip failure. BMJ. 1998;316(7132):644.
58. Treacy RB, McBryde CW, Shears E, Pynsent PB. Birmingham hip resurfacing: a minimum follow-up of ten years. J Bone Joint Surg Br. 2011;93(1):27–33.
59. Seppänen M, Karvonen M, Virolainen P, Remes V, Pulkkinen P, Eskelinen A, Liukas A, Mäkelä KT. Poor 10-year survivorship of hip resurfacing arthroplasty. Acta Orthop. 2016;87(6):554–9.
60. Laaksonen I, Donahue GS, Madanat R, Makela KT, Malchau H. Outcomes of the recalled articular surface replacement metal-on-metal hip implant system: a systematic review. J Arthroplast. 2017;32(1):341–6.
61. Shemesh S, Kosashvili Y, Heller S, Sidon E, Yaari L, Cohen N, Velkes S. Hip arthroplasty with the articular surface replacement (ASR) system: survivorship analysis and functional outcomes. Eur J Orthop Surg Traumatol. 2014;24(6):925–30.
62. Aktouf A, Auquit-Auckbur I, Coquerel-Beghin D, Delpierre V, Milliez PY. Breast augmentation by Poly Implant Prothèses silicone implants: retrospective study about 99 patients. Rupture analysis and management. Ann Chir Plast Esthet. 2012;57(6):558–66.
63. Dieterich M, Stubert J, Stachs A, Radke A, Reimer T, Gerber B. Ruptured poly-implant protheses breast implant after aesthetic breast augmentation: diagnosis, case management, and histologic evaluation. Aesthet Plast Surg. 2013;37(1):91–4.
64. Krackow KA, Bayers-Thering M, Phillips MJ, Bayers-Thering M, Mihalko WMA. new technique for determining proper mechanical axis alignment during total knee arthroplasty: progress toward computer-assisted TKA. Orthopedics. 1999;22(7):698–702.
65. Kim Y, Lee BH, Mekuria K, Cho H, Park S, Wang JH, Lee D. Registration accuracy enhancement of a surgical navigation system for anterior cruciate ligament reconstruction: a phantom and cadaveric study. Knee. 2017;24(2):329–39.

Chapter 3
Communication in Surgery

Communication is defined as the "exchanging of information by speaking, writing, or using some other medium" [1].

However, communication is not simply the exchange of information, the sole passage or receipt of facts and knowledge. Communication also forms the cornerstone of interpersonal relationships [2], the ways in which we interact with our patients, colleagues, juniors and peers.

Communication is a two way process whereby we speak and talk but also listen and hear. Communication may be verbal or written but also expressed through body language, facial expressions, attitudes and gestures. Communication may be emotional, transmit or carry emotions.

Understanding the importance of communication in our daily surgical practise and recognising the need to communicate effectively are important skills to develop. The first part of this chapter discusses some basic concepts of communication and ways in which one may improve their communication skills. The second part focuses on the process of communicating to patients surgical treatment options, shared decision making, and obtaining consent for surgical interventions.

C. Panayiotou Charalambous, *Career Skills for Surgeons*,
DOI 10.1007/978-3-319-57490-5_3

I would greatly value your suggestions. Please speak up.

3.1 Basic Concepts in Communication

In considering communication you may consider the following basic concepts that may help critically question and possibly improve your communication skills. These may be applied in your interactions both with patients but also co-workers.

3.1.1 Introduce Self and Get to Know Others

Lesson one, introduce yourself to everyone when you walk into a room. Don't act like you're too bougie to say, 'Hello.'

Estelle [3]

State who you are, what your role is, why you are there. Ask who you are dealing with, who they are, how they happen to be there.

Should you be spending the whole day in the theatre without knowing the name of the anaesthetist putting your cases to sleep, the name of your junior holding the pelvic retractor, the name of the radiographer trying to give you a sharper image intensifier picture? They may be agency rather than regular staff, they may have been dragged in because your regular co-workers are on leave, they may have just started a new post, or you may be operating in a new hospital and haven't met them before. You may end up spending years as co-workers or just a few hours together. Should you not know who they are, should they just assume as to who you are?

3.1.2 Choose the Words and Tone

People may hear your words, but they feel your attitude.

John C. Maxwell [3]

Choose the words wisely, based on the situation, based on those you are communicating with. Choose the words that will allow you to clearly transfer the message across, ones that can be easily understood by the other party, words that will help strengthen relationships and build bridges rather than construct walls.

- Consider the complexity level of your speech; will others understand what you are trying to say?
- Avoid jumping into saying just for the sake of telling; will what I say upset the other party? Can I be more courteous in what I say?

Use simple language. Avoid technical terms, medical, surgical or managerial jargon, abbreviations, terms that even technocrats may have difficulty to grasp; they may sound clever but may not mean much to your listeners [4]. Tailor your language to your audience. Start with the basics and then build up. The aim is not to impress with your knowledge but to pass a message.

It may also be important as to how we say things rather than simply what we say. Ambady et al. [5] looked at the relationship between the voice tone of general and orthopaedic surgeons whilst conducting clinic consultations and related that to their malpractice history. Surgeons were taped whilst speaking to their patients and voice tone only recordings were then rated for several parameters including dominance and concern/anxiety. They reported that surgeons rated as more dominant and less concerned/anxious on their voice tone were more likely to have had malpractice claims against them rather than those rated as less dominant and more concerned. The authors of that study suggested that voice tone dominance (characterised by loud, fast, clear, unaccented speech) and lack of anxiety in the tone may indicate to patients lack of empathy and indifference by the surgeon, hence leading to a higher chance of getting sued.

3.1.3 Body Language

What I really like doing is storytelling, finding the body language that is necessary for the story.

Patricia Birch [3]

Use body language [6–9] appropriately.

- Maintain eye contact, avoid distractions such as speaking whilst simultaneously typing on the computer, writing notes, reviewing radiographs or scans.
- Face the patient when talking—turn your chair, adjust your seat.
- Be at the same level as the patient—sit down, on the bed or chair, it doesn't take long to stand up again.
- Use appropriate body cues—head nodding, direct body orientation, maintaining eye contact, leaning forwards.
- Avoid distancing behaviours—absence of smiling, avoiding eye contact, crossing arms.
- Be close to the patient—avoid simply standing at the other end of the consultation room, looking distant and detached, looking as if you are about to open the door and get out.
- Make it clear that you attempt to minimise distractions—close the door, draw the curtain. Give the impression that any time spent is truly devoted to the individual, is not a rushed consultation.

3.1.4 Listen

Two monologues do not make a dialogue.

Jeff Daly [10]

Communication is a two way process and it's not just about talking but also about listening.

Listen to all. Listen to those who are well spoken, who are articulate with high pitched voice. But listen equally to those who are not great masters of the spoken word, who may stumble, who may be soft spoken. Tell them that you are not dismissing anything, make it clear you are listening to their worries or concerns.

Listening is not just about receiving information, or acknowledging the receipt of such information, but also about acting based on that information, rather than just putting it on the side. What may sound trivial to you may be vital to your patients or your co-workers. Listening is more than mere hearing.

Many of us like our coffee in a specific way, with no sugar, skimmed milk, no cream, with an extra shot, half full cup, not too hot or on the ice. And even though the menu just says "coffee" we still expect the barista to listen to our small but so vital special requests. And how does one feel if our requests are dismissed or told "it's not possible", when we have had our "custom made coffee" over and over again? How does one feel if the barista looks at us, hears what we say, nods, and still makes us a "coffee" rather than our "individualised coffee"? How do we feel if the barista sounds abrupt or dismissive? Do we enjoy the rest of our coffee shop visit, or do we regret stepping in there in the first place?

If we expect to be heard at the coffee shop, should one not expect to be listened to and their individuals circumstances accommodated at their clinic visit, at their surgical consultation? Should one not expected to be listened to by their leaders, seniors, colleagues, juniors, or followers?

3.1.5 Repeat in Different Ways

Repeat in order to get a message across. Repeating in the same or different words, may help re-inforce what you are trying to say.

And if you are asked a question in an unclear way, you may unpick the question and repeat it back so you all understand the question correctly, rather than simply jumping straight into an answer.

3.1.6 Do You Have Any Remaining Questions?

It's one of the last chances not to leave anything unanswered. It shows that you are not there just to talk but also listen, it is an opportunity to clarify any unresolved issues. It may also show that you are not rushing to get out of the consultation room, to go and do something else. And if after you walk out of that door, you still feel that the consultation didn't go as it should, that you may have been abrupt, cutting the conversation short, that your patient may have felt rushed, nothing stops you from going back, and asking once more

- "Is there anything else you would like to know, anything else to be discussed?"

3.1.7 Check that Information Has Gone Across

We often pass information but are not really sure if it has gone across, if it has been understood and registered.

Checking that information has gone across, can help ensure that everyone is on board, all share a common understanding. In doing so you may consider:

- Warn the other person that you will ask them to repeat the information they are about to receive.
- Ask the receiver "What is your understanding of what I have just said?".

3.1.8 Summarise

- You may explain to your patient the pros and cons of surgery, the risks and benefits, the multiple treatment options.
- You may give a long history about the patient's presenting complaint and examination findings to your senior.
- You may provide your junior with a detailed analysis of your thought process in coming up with your management plan.
- You may have written a long letter to a referring physician to explain the need for the patient's medical condition to be optimised prior to surgery.

There is so much attention one can pay to what we say, so much information one can process, understand and easily recall. Develop the ability to try and summarise what you have said, to give a succinct snapshot of what has just been heard. Once you go through all the information, try and summarise as to what it all comes down to.

3.1.9 Pick the Cues

The most important thing in communication is hearing what isn't said.

Peter Drucker [3]

Sometimes others don't come out with what really concerns them, don't vocalise their worries, and simply try to give verbal or no-verbal cues. They may leave what they want to say behind, with the hope that you will somehow work it out, that you will be sensing their anxiety.

- A patient may be asking you about the usual length of stay following a knee replacement, but may be worried of going home early as they won't have anyone to look after them.

- A patient may be asking you about the type of anaesthesia they will be having, but may be really worried about getting sick like last time they had a general anaesthetic.
- A junior may be showing you a fracture on a radiograph asking you what to do, but is really worried about the possibility of this being a secondary given the history of prostate cancer.
- A junior may be enquiring your management decision trying to put across that there may be facts that you have overlooked.
- You may be introducing a new pattern of ward based working for juniors. You explained to all, you feel it's all well laid out, but someone keeps enquiring about what it will all mean. Are they simply being awkward or should you be re-thinking your proposals?
- Someone may read by mistake the wrong hospital number whilst checking a patient for surgery. Was it a slip of the tongue or have they just identified an error?

It may be unusual that someone directly says that "I'm really worried" or "I'm really anxious", but it may be said by the tone of their voice, their facial expression, the hesitation in their speech, the pause in their talk, their deep thought. They may keep revisiting a subject, asking an unusual question or a question out of context, repeating a question, looking unsure, avoiding eye contact, sitting with arms crossed, keep nodding, covering their mouth, rubbing their neck, lowering their tone, slowing down, not saying many words or staying silent. These are just some cues that may make you think, is there anything else going on, anything else I should know? An open question along the lines of:

- "Why are you saying that?"
- "What are you worried about?"

may help them open their heart and express their worries.

Sometimes, the cues are not so obvious for us to recognise. An ending question may help gather important left out information:

- "Is there anything else you are concerned about?"

3.1.10 Empathy

No one cares how much you know, until they know how much you care

Theodore Roosevelt [3]

In our out of work interactions with friends, relatives or acquaintances, we often acknowledge the suffering of others, their concerns, the difficult times they are going through. We may do so because we feel that recognising such difficulties is part of human courtesy and may help strengthen our inter-personal relationships. We may also understand that even though we cannot take over their worries, a comforting word and an acknowledgment of the troubles others experience can give hope in times of despair.

Similarly, we often acknowledge the enthusiasm, joy and happiness our friends feel when things are going well, when getting a new job, moving into a new house, passing a difficult exam, getting married, having a new-born baby. We may feel that acknowledging success may be just as helpful as sharing troubles.

In dealing with patients we may be taught as part of our training, in medical school and afterwards, to distance ourselves, not to get too close, not to get involved emotionally with other people's suffering, to help avoid psychological overloading, excessive stress and burnout. But one may still express empathy whilst not taking someone else's weight on their shoulders. Steinhausen et al. [11] showed that physician's empathy was the strongest predictor of a better subjective evaluation of treatment outcomes by patients at 6 weeks and 12 months following general surgical trauma. Empathy has been also associated with adherence to treatment and overall satisfaction [12, 13]. However, Levinson et al. [14], in a systematic review of studies evaluating surgeon-patient communication, reported that even though surgeons were thorough and devoted ample time in providing information to help patients decide about treatments, they did not perform well in exploring the concerns or emotions of their patients or in expressing empathy.

Although patients seek consultation about their own health problems, in their conversation with doctors they may mention other events happening in their life, positive or negative. One may decide to ignore these if not relevant to what the consultation is about, or listen and respond appropriately. You may acknowledge the pain of a patient losing a close relative, the financial hardships due to their recent job redundancy, or you may congratulate them on mentioning that they just got back from their honeymoon, that their daughter has just started university.

Similarly, in our communication with colleagues and co-workers, we may consider their emotions and feelings, rather than simply focusing on emotionless exchange of work related information.

- We may acknowledge the stress, worry and pressure a colleague faces due to an unexpected complaint or medico legal suit, rather than just sticking to the technical advice as to how to construct a reply letter or which lawyer to consult.
- We may acknowledge how hard it must be for a colleague returning to work from a personal bereavement, rather than just advising them on ways to reduce their follow up clinic appointments.
- We may acknowledge how difficult it must be for our junior to keep studying for the upcoming fellowship exams whilst caring for an unwell parent, rather than simply stating possible viva questions they may be asked.

We may sometimes hear of the "nothing personal, it's just business" [15] but one may consider that humanity and professionalism are not mutually exclusive.

3.1.11 Communicating Urgency

Lightning shall be slow to my hasting.

E.R. Eddison [16]

Communicating urgency or how critical a situation is, may be vital when other members of the team do not have all the knowledge, training or skills to see the full picture.

Rather than simply stating the need for urgency, explaining the reasoning for such urgency may help. Explain the dangers from not acting quickly, and the gains from acting promptly:

- You may have a patient with a nasty elbow fracture with a pulseless arm, or an acute appendicitis case who is becoming septic and you need to take them to theatre for immediate surgery. You ring the emergency theatre bleep holder to book the patient and ask them to send for the case. Explain why the case has to take priority, why it has to be dealt with straight away, the grave consequences if you don't get them to theatre on time.
- You may need to send a referral letter to the oncology team based in a different hospital, for an urgent case review. You ask the secretary to send the letter through. Explain why the patient has to be seen promptly, to allow the secretary to fax the referral, ring through and confirm an early appointment date, rather than just putting the letter in the routine post.
- You may be asking the on-call physiotherapist to see your post-laparotomy case for chest exercises. Explain the need to be seen over the weekend rather than waiting for Monday, as the patient has signs of atelectasis and it is only a short window before developing a full blown chest infection.

3.2 Shared Decision Making

My job is to make images and leave the decision-making and conclusion-drawing to other people.

Laurie Anderson [3]

In clinical practise decisions may have to be taken with regards to investigations or treatment. Reaching such decisions requires close communication between patient and surgeon.

Shared decision making involves a bi-directional exchange of information between surgeon and patient, to help reach a mutually accepted decision with regards to investigations or treatment, rather than simply the surgeon disclosing information and recommending a line of action.

Shared decision making, refers to approaching each case in an individualised way, as well as communicating and considering those unique factors that come into play. It involves exploring the preferences of the patient, and their special circumstances that may influence the intervention or treatment choice [17, 18].

Shared decision making may facilitate communication, increasing patients' knowledge of risks and benefits, allowing more accurate risk understanding, recognition of any uncertainty that may exist, helping patients reach a decision, and leading to greater acceptance of the decision. Decision sharing may involve questions and answers which can take place over multiple meetings. It starts at the first encounter in clinic and up the point where the patient has the investigation or intervention. It is often said "never talk a patient into surgery" [19] as those who may have hesitations about proceeding but who are eventually persuaded, may look back with regret, more likely to be unsatisfied or have poorer outcomes [20, 21]. Shared decision making involves reaching a mutual decision with which both parties are comfortable and accept.

An integral component of shared decision making is the consent process, and this is discussed next. Risks and benefits which may be acceptable to one patient may not be acceptable to another and this must be carefully considered.

3.3 Consent

Consent is defined as "permission for something to happen" [1], a prior agreement as to what will occur. Informed consent in surgery is considered as permission to proceed with a surgical procedure, given in knowledge of the possible consequences. Obtaining informed consent is an integral part of a surgeon's practice and one that combines communication as well as documentation skills.

Even though the exact rules and regulations around consenting may vary between institutions, regulatory bodies, states or countries, the principles of obtaining informed consent are similar and the ability to obtain such consent is an important skill to develop.

Like in other forms of communication, use plain language, easily understood by the patient, avoid medical jargon and any abbreviations. "Translate" to the patient any medical terms you use or write down; a laparotomy may be described as a "cut though the tummy to open up and have a look at the internal organs" or arthroscopic surgery may be described as "keyhole surgery".

Informed consent may be documented using a pre-designed form of your institution. It is important to adequately fill those forms, and any parts of the form which are not applicable to the specific case marked as such, rather than simply left blank, to avoid confusion. If adequate space is not provided on the form to document all that has been explained, additional information may be recorded on institution headed paper which can then be attached to the consent form and filed in the patient's notes. Inadequate space on the form is not a reason for not documenting communicated information. In completing the consent form it may be better to use capital letters, and write clearly so there is no doubt as to what is documented.

In addition to the above, there is a need to consider what information needs to be communicated to the patient, who can obtain informed consent, who can provide informed consent, and the timing of obtaining consent in relation to surgery. These considerations are discussed next.

3.3.1 Information to Be Communicated in Consenting

The following information may be communicated to the patient as part of informed consent.

<u>The surgical procedure to be performed</u>: state the exact procedure to be performed (the one that will definitely be carried out) and any procedures that may be performed (depending on findings).

Be specific in describing the procedure: e.g. if one describes a "total knee replacement" but the patella is not routinely replaced then explain that to the patient. Document that it will be a "knee replacement +/− patellar replacement" to avoid someone having a "total knee replacement" only to find out later that their patella was not replaced.

If one is not certain of the exact procedure to be performed, as that may depend on the intra-operative findings, this needs to be communicated and documented appropriately: e.g. you may state "arthroscopic evaluation of the knee + proceed depending on findings" or "exploratory laparotomy of the abdomen + proceed depending on findings". Alternatively, you may state the procedure options: e.g. a patient may be consented to have a uni-compartmental knee replacement but if at the time of surgery they were found to have extensive arthritis then that will be converted to a total knee replacement.

Describe the mode of anaesthesia, and any anaesthetic risks. Often the anaesthetist explains the exact anaesthetic measures to be employed, as well as the specific risks associated with anaesthesia. However, the surgeon can still confirm that this explanation has taken place and help clarify any remaining queries. The patient should have a clear understanding as to whether the procedure will be carried out under a local anaesthetic, under a regional block, under spinal or epidural anaesthesia or under a general anaesthetic (GA). The possibility of having to revert from non-general anaesthesia to GA intra-operatively also needs to be discussed. Patients may have strong preferences for one type of anaesthesia over the other, and this should be taken into consideration. If the patient is refusing to accept the possibility of GA you may come to the shared decision that you could attempt to perform the surgery under regional or local anaesthesia (if feasible) but the procedure may have to be abandoned if adequate pain relief could not be obtained.

Potential benefits of the procedure: state why the procedure is performed and what the potential gains are. This may be amongst others to help diagnose a disorder, to improve pain, to improve function, to improve mobility, to prolong survival, to save limb or life. Explain to the patient the potential benefits in a balanced way, without exaggerating the benefits or playing down any potential gains.

Risks of surgery: state what could go wrong, what complications could arise. Explaining risks is vital, as it allows the patient to appreciate what they are getting into, and further helps in making an informed decision. You may be more forgiving for those plane bumps or shakes if the pilot warned you of some expected turbulence ahead (even though deeply inside you hoped they wouldn't be encountered) rather than if you were promised clear skies and a smooth flight simply to be thrown off your seat halfway through.

You may divide complications into:

- Intra-operative, early post-operative, and long term complications.
- Generic complications (such as infection, failure of surgery to improve symptoms, neurovascular injury, bleeding, wound healing problems, haematoma formation, tender scar, prominent scar) versus procedure specific complications.
- Complications associated with the underlying condition itself versus those directly related to surgery e.g. avascular necrosis is a recognised complication of intra-capsular hip fractures, even when these fractures are perfectly fixed, as this complication is related to the vascular disruption which occurs at the time of the bone fracturing rather than being due to the surgical fixation per se.

It may be useful to have a list of the various risks to be explained, to avoid relying on memory, as the day you forget to mention the risk may be the day that a complication arises.

Give adequate information about the risks of surgery, but also explain the probability of such events occurring. Explain the individualised risk that may exist in some cases, such as the increased likelihood of diabetics developing infection or wound healing problems, or a history of keloid scars predisposing to further prominent scars.

There is debate amongst the surgical world as to what potential complications should be explained to the patient; one may consider that in addition to explaining common complications, any rare but important complications should be described. In the UK, this follows a case law judgement (Chester vs. Afshar), whereby the patient was not informed with regards the risk of cauda equina during spinal surgery (a very rare complication) and was left partially paralysed [22]. It was concluded that failure to warn the patient of the risk related to surgery, no matter how rare the risk is, could deny a patient the ability of making a fully informed decision.

Explain what the potential complications mean, what may happen if those occur, what their impact could be.

- Infection following knee replacement may be perceived as a "superficial wound infection that settles with few days of antibiotics", but may also mean the need of revision surgery, whereby implants have to be removed, a spacer implanted and further implants inserted at a later stage, a long protracted and painful course.
- Instability following hip replacement may be perceived as "a bit wobbly" which is different from the hip coming out of joint and needing further revision surgery.
- Arterial damage may sound like "a bit of bleeding", but in severe vascular injury it could lead to someone losing a leg.

It may help to remind patients that you are not trying to scare them by explaining these risks, but it is part of your professional obligation, and that you tend to explain such risks to all patients having a particular procedure.

In explaining the risks of surgery one may consider the benefits of the surgical procedure versus the possible consequences if a complication arose. You may be questioned as to whether it is worthwhile having surgery and taking the risks. What the author uses is that:

- "It is almost equivalent to considering whether to cross the road; if we need to cross the road then we do so with the understanding that there is a risk of being run over by a car, a tiny but recognised risk that accompanies this frequent activity of daily life. If we do not need to cross the road, then, of course, we would not unnecessarily take the risk of getting run over".

You may be questioned whether the procedure you are describing is a "big" or "small" operation, "minor" or "major" surgery. You may consider that any surgery is "big" or "major" enough for the individual having the procedure. Describing a procedure as a "routine" versus "infrequently performed" or "short" versus "lengthy", a "low risk" versus "high risk" may be more appropriate.

Post- operative course: State what will happen post surgery if all goes as planned, the expected course of improvement and rate of recovery, the need for gradual adjustment and rehabilitation, the possible need for splintage, length of hospital stay, weight bearing status, return to work, return to drive. State as to whether any follow up procedures may be needed such as removal of metalwork inserted to fix an ankle fracture, reversal of an abdominal de-functioning stoma, as well as the timing and the factors that may guide to performing those. Explain the possibility of needing an intra-operative or post-operative blood transfusion and discuss alternatives in those who object, such as Jenovah's witnesses.

Operating surgeon: State who will be doing the procedure, especially if trainees will be involved. The patient may have strong preferences as to who will be the surgeon. You may be working in an institution where trainee surgeons may perform surgical procedures under the direct supervision of their seniors. Explain that to the patient prior to the procedure and respect their wishes. If applicable to the procedure in question, it may be useful to explain to patients that there is evidence that a trainee doing the operation under direct supervision is not associated with substantially more complications or greater length of surgery [23–25].

If you are not the person directly in charge of the patient's care (you may be a trainee or a more junior level staff) and you are to perform a procedure in the absence of direct supervision by the surgeon in charge, explain that to the patient. If applicable, you may explain to the patient that you are confident in performing that procedure, that you have performed that procedure multiple times previously, even though you are not "the boss". You may explain the possible delay that may occur if they want to have surgery by a specific surgeon, that they may have to be cancelled on that day, or postponed. Nevertheless, it would be unwise to pressurise in order to convince a patient otherwise, once they have decided that they want the particular procedure to be performed by a specific surgeon.

The same may also apply to more senior surgeons, if the patient is enquiring about whether you are highly experienced in a particular procedure as they are looking for a super-specialist. It is better to be open and transparent early on about your experience with the procedure in question, rather than having to explain in retrospect after a complication occurs. Being questioned about your level of experience should be understandable and acceptable; after all, isn't it that when booking a flight some of us would like to know how experienced the pilot will be, how may flying hours they have clocked in that type of aircraft, how many storms they have flown through?

3.3.2 Who Obtains Consent

In an ideal practise, if you are performing the procedure it is better to obtain the consent yourself. Nevertheless, in a busy surgical practice you may delegate the task, but ensure that your delegate is able to obtain adequate consent.

If you are going to delegate consent to another professional, you may still explain yourself the benefits and risks at the time of the initial clinic consultation and go through the consent again just prior to surgery, to ensure that all relevant information has been communicated and thoroughly explained.

If you will not be performing the procedure and you are only obtaining consent make that clear to the patient. If you are not able to answer all the patient's queries, accept your limitations and offer to find and provide that information.

Follow the rules of your institution, your regulatory body, the healthcare system you are practising in. In the UK, the General Medical Council (GMC) advises that informed consent should be obtained by someone who is appropriately trained and qualified, someone who has adequate knowledge of the proposed procedure, its potential benefits and risks, to provide an adequate explanation as to what the procedure entails [26].

3.3.3 Who Can Provide Informed Consent

Consent to surgery may be given voluntarily by an individual who is adequately informed, and who has the capacity to do so.

What constitutes capacity to give Consent, may vary in different countries, but in the UK the Mental Capacity Act of 2005 [27] states:

"a person lacks capacity in relation to a matter if at the material time he is unable to make a decision for himself in relation to the matter because of an impairment of, or a disturbance in the functioning of, the mind or brain".

The act [27] also describes that a person is considered as not able to decide for themselves if unable to:

1. Understand the information needed in deciding.
2. Retain the relevant information.
3. Use the relevant information in making a decision.
4. Communicate their decision by any means.

Mental illness does not necessarily preclude from being able to make an informed decision. Similarly, young age does not preclude from providing informed consent if one has the capacity to do so.

If an impairment of capacity exists and one is unable to make informed decisions other arrangements may need to be made. It is important to understand the regulations of the regulatory body and legal framework in which you practise. In cases where there is lack of capacity the GMC, in the UK, advises that in your considerations you should have above all the patient's care, treat them as individuals, respect their dignity, support them in getting involved as much as possible and as much as they wish in decisions about their care, and not discriminate against them [26].

The individual may have previously expressed what would have liked to happen in such circumstances, and stated such preferences in an advanced directive, or relatives and friends may be aware of such wishes. Discuss with the individual who has legal authority to decide on the patient's behalf, and try and reach a consensus about any proposed treatment. In many cases consensus will be reached but you should be aware of the escalation pathway in your healthcare system if there were to be disagreement. In cases where agreement as to the proposed action cannot be reached, one may consider involving an independent advocate, more experienced colleagues, or mitigation services. Legal advice and court rulings may have to be sought and obtained.

In emergency situations whereby the patient cannot consent, and their exact wishes cannot be promptly obtained, one may act in what they consider to be the best interests of the patient, without obtaining consent. Again be aware of the legal framework in which you are practising but in the UK the GMC states [26]:

> *"when an emergency arises in a clinical setting and it is not possible to find out a patient's wishes, you can treat them without their consent, provided the treatment is immediately necessary to save their life or to prevent a serious deterioration of their condition".*

Other issues to consider include the legal age of consent, how is consent obtained in people younger than this legal age but who have the maturity and capacity to consent, who gives consent on behalf of young people who lack the capacity to do so, and refusal of a child's parents to treatment.

In the UK young people are presumed to have capacity to give consent at 16 years of age [28, 29]. Nevertheless, children younger than 16 may have the capacity to give consent and they should be assessed on an individual basis. A young person under 16 who is considered capable of consenting, is referred to as being "Gillick competent". This is based on the Gillick vs. West Norfolk and Wisbeck Area Health Authority (1986) case whereby Mrs. Gillick challenged the relevant health authority about the practise of doctors to provide contraceptive advice and treatment to girls under 16 without parental consent. The House of Lords concluded that a doctor could give contraceptive advice and treatment to a child younger than 16 without the parents' consent or knowledge, if this is in the best interests of the girl, the girl can understand the advice and cannot be persuaded to tell her parents or to allow her doctor to tell them [30]. Young people can be encouraged to involve their parents in making important decisions but one must follow their decision if they have the capacity to decide themselves. It is of note that although the Gillick case was specifically about contraceptive advice and treatment, the judges also spoke more generally in reference to competence of children in deciding medical treatment [31]. The situation whereby a child is able to consent but refuses treatment is particularly challenging, and seeking legal guidance is advised.

When a young person is not competent to provide informed consent, follow the regulations of your healthcare system as to who needs to be involved in such a decision. In the UK when a child under the age of 16 lacks the capacity to consent, consent can be given on their behalf by a person who has parental responsibility or if there is disagreement with the proposed intervention, legal advice and court input are sought [26].

Even when a person has capacity to give informed consent ensure that they fully understand the information you are providing. If there is a language barrier, a translator can facilitate communication. If your accent can not be easily understood ask someone else to help, a nurse, other professional or relative, someone who is able to explain in a clearer way.

3.3.4 Timing of Obtaining Consent

Discuss any benefits and risks of a procedure with the patient well in advance to surgery, to allow adequate time to consider those in detail before proceeding. It may be better to consent the patient several days before the operation, allowing them time to take away the information, consider it and if necessary change their mind [26, 32]. This may not be possible in emergency surgery but is usually feasible in elective procedures.

Avoid leaving the consent process to just a few minutes prior to surgery, just before the patient is about to put on the surgical gown. If you are to leave the completion of the consent form for the day of surgery, give the patient verbal or written information about the risks and benefits well in advance, so that they can consider it with a cool mind. In that way, signing the consent form on the day of surgery has a more confirmatory role.

If you have obtained consent prior to the day of the surgery, then on the day of surgery run through the consent form again, to ensure the patient is still agreeable to what has been previously said and that nothing has changed.

References

1. English Oxford living dictionaries. https://en.oxforddictionaries.com/definition/communication. Accessed 1 Apr 2017.
2. Guerrero L, Andersen PA, Afifi W. Close encounters: communication in relationships. 4th ed. Thousand Oaks: SAGE Publications Inc.; 2013.
3. Famous Quotes at BrainyQuote. https://www.brainyquote.com/quotes/quotes. Accessed 25 Mar 2017.
4. Farrington C. Lost in translation: the impact of medical jargon on patient-centred care. The Guardian. 2014. https://www.theguardian.com/healthcare-network/2014/jul/01/impact-medical-jargon-patient-centred-care. Accessed 1 Apr 2017.
5. Ambady N, Laplante D, Nguyen T, Rosenthal R, Chaumeton N, Levinson W. Surgeons' tone of voice: a clue to malpractice history. Surgery. 2002;132(1):5–9.
6. Marcinowicz L, Konstantynowicz J, Godlewski C. Patients' perceptions of GP non-verbal communication: a qualitative study. Br J Gen Pract. 2010;60(571):83–7.
7. Byrne PS, Heath CC. Practitioners' use of non-verbal behaviour in real consultations. J R Coll Gen Pract. 1980;30(215):327–31.
8. Sawada NO, Mendes IA, Galvão CM, Trevizan MA. The importance of nonverbal communication during the preanesthesia period. Clin Nurs Res. 1992;1(2):207–13.
9. Montague E, Chen P, Xu J, Chewning B, Barrett B. Nonverbal interpersonal interactions in clinical encounters and patient perceptions of empathy. J Participat Med. 2013;5:e33.
10. AZ Quotes. http://www.azquotes.com/quote/667458. Accessed 1 Apr 2017.
11. Steinhausen S, Ommen O, Antoine SL, Koehler T, Pfaff H, Neugebauer E. Short- and long-term subjective medical treatment outcome of trauma surgery patients: the importance of physician empathy. Patient Prefer Adherence. 2014;8:1239–53.
12. Flickinger TE, Saha S, Roter D, Korthuis PT, Sharp V, Cohn J, Eggly S, Moore RD, Beach MC. Clinician empathy is associated with differences in patient-clinician communication behaviors and higher medication self-efficacy in HIV care. Patient Educ Couns. 2016;99(2):220–6.
13. Di Blasi Z, Harkness E, Ernst E, Georgiou A, Kleijnen J. Influence of context effects on health outcomes: a systematic review. Lancet. 2001;357(9258):757–62.
14. Levinson W, Hudak P, Tricco AC. A systematic review of surgeon-patient communication: strengths and opportunities for improvement. Patient Educ Couns. 2013 Oct;93(1):3–17.
15. Popular Quotes. Goodreads. http://www.goodreads.com/quotes. Accessed 1 Apr 2017.
16. Boss EF, Mehta N, Nagarajan N, Links A, Benke JR, Berger Z, Espinel A, Meier J, Lipstein EA. Shared decision making and choice for elective surgical care: a systematic review. Otolaryngol Head Neck Surg. 2016;154(3):405–20.
17. Weckbach S, Kocak T, Reichel H, Lattig F. A survey on patients' knowledge and expectations during informed consent for spinal surgery: can we improve the shared decision-making process? Patient Saf Surg. 2016;10:15.
18. Dizon DS, Politi MC, Back AL. The power of words: discussing decision making and prognosis. Am Soc Clin Oncol Educ Book. 2013:442–6. doi:10.1200/EdBook_AM.2013.33.442.
19. Page A. Safety in surgery: the role of shared decision-making. Patient Saf Surg. 2015;9:24.
20. King JS, Moulton BW. Rethinking informed consent: the case for shared medical decision-making. Am J Law Med. 2006;32:429–501.
21. Roter D. The patient-physician relationship and its implications for malpractice litigation. J Health Care Law Policy. 2006;9:304–14.
22. Judgments - Chester (Respondent) v. Afshar (Appellant). House of Lords. https://www.publications.parliament.uk/pa/ld200304/ldjudgmt/jd041014/cheste-1.htm. Accessed 1 Apr 2017.
23. Wakasugi M, Tsujimura N, Nakahara Y, Matsumoto T, Takemoto H, Takachi K, Nishioka K, Oshima S. Single-incision laparoscopically assisted appendectomy performed by residents is safe and feasible: a single institution, retrospective case series. Ann Med Surg (Lond). 2017;15:43–6.

24. Dull MB, Gier CP, Carroll JT, Hutchison DD, Hobbs DJ, Gawel JC. Resident impact on operative duration for elective general surgical procedures. Am J Surg. 2017;213(3):–456, 459. doi:10.1016/j.amjsurg.2016.10.031.
25. Ross SW, Oommen B, Kim M, Walters AL, Green JM, Heniford BT, Augenstein VA. A little slower, but just as good: postgraduate year resident versus attending outcomes in laparoscopic ventral hernia repair. Surg Endosc. 2014;28(11):3092–100.
26. Consent: patients and doctors making decisions together. GMC 2008. http://www.gmc-uk.org/Consent___English_1015.pdf_48903482.pdf. Accessed 1 Apr 2017.
27. Mental Capacity Act 2005. http://www.legislation.gov.uk/ukpga/2005/9/pdfs/ukpga_20050009_en.pdf. Accessed 1 Apr 2017.
28. Reference guide to consent for examination or treatment. Department of Health 2009. https://www.gov.uk/government/uploads/system/uploads/attachment_data/file/138296/dh_103653__1_.pdf. Accessed 1 Apr 2017.
29. Family Law Reform Act 1969. http://www.legislation.gov.uk/ukpga/1969/46. Accessed 1 Apr 2017.
30. Gillick v West Norfolk & Wisbeck Area Health Authority [1986] AC 112 House of Lords. http://www.bailii.org/uk/cases/UKHL/1985/7.html. Accessed 1 Apr 2017.
31. Hayhoe B. Decision making in children and young people: Gillick competent? InnovAiT. 2008;1(11):764–70.
32. Anderson OA, Wearne IM. Informed consent for elective surgery—what is best practice? J R Soc Med. 2007;100(2):97–100.

Chapter 4
Organising a Surgical Theatre List

Putting together a theatre list is a task surgeons are called upon to complete on a regular basis. Organising an elective, semi-elective or urgent list, for inpatient or day surgery attendants, requires organisation skills as well as the ability to prioritise. It involves consideration of what goes on the list and in which order, taking into account, amongst others, clinical priorities, safety factors, and available resources.

This chapter discusses some of the factors one may consider in successfully organising a theatre list and in improving surgical theatre list efficiency, to facilitate patient flow. The patient's journey from deciding to operate to reaching the surgical theatre is initially discussed, along with the parameters that need to be communicated in listing a case for surgery. Considerations in constructing a theatre list are then presented, followed by ways in which theatre time utilisation may be enhanced.

C. Panayiotou Charalambous, *Career Skills for Surgeons*,
DOI 10.1007/978-3-319-57490-5_4

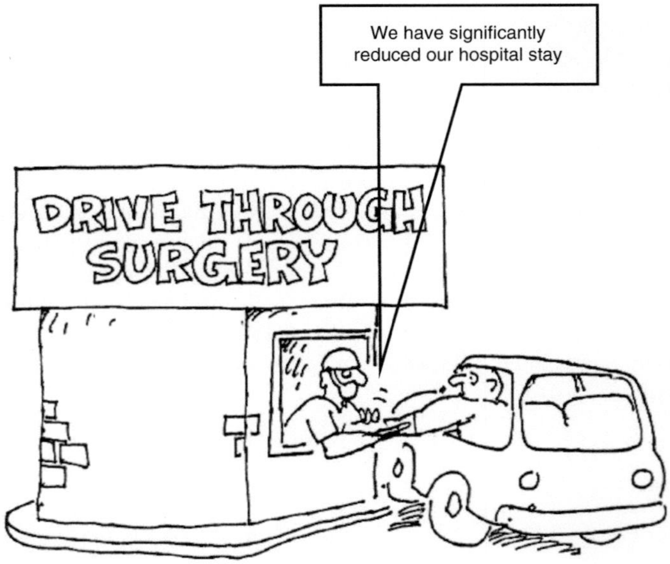
We have significantly reduced our hospital stay
DRIVE THROUGH SURGERY

4.1 The Patient's Journey to Surgery

The patient's journey for elective surgery may be divided into various steps as shown in Fig. 4.1.

Patients are increasingly admitted for elective surgery on the same day of surgery—this aims to minimise hospital length of stay, minimise hospital bed usage, reduce associated costs, and reduce hospital infections such as acquired pneumonias.

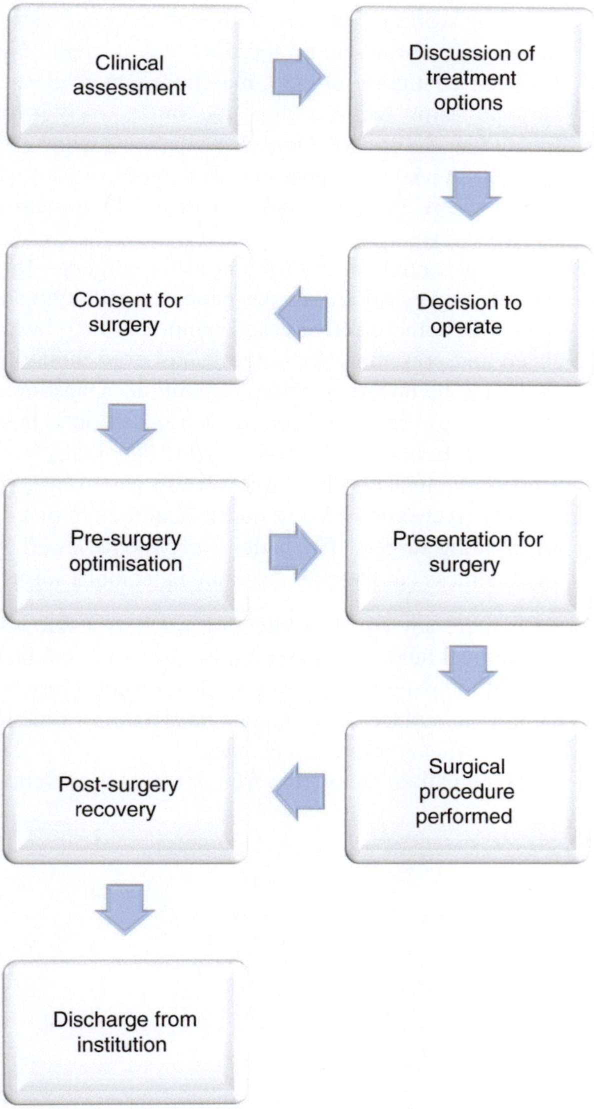

Fig. 4.1 Patient's path to elective surgery

As part of good surgical care the following may be performed along this pathway to surgery:

1. Maintain continuity of care from clinical assessment to post- operative management, with active involvement of the same surgeon or surgical team. Surgeon and patient establish a relation, a shared understanding of treatment, that can see them through the surgery journey ahead. Continuity of care may help strengthen such a bond, ensuring consistent communication and exchange of information [1, 2]. There may be pressures in some healthcare systems in meeting targets, need for pooling cases to match demands, patients passed between teams. Evolving working patterns, with working hour restrictions, may mean that none of us can be available all the time, day or night. Nevertheless, recognising the need for continuity of care, may help you achieve effective handover between surgical teams, whilst at the same time actively involving and updating patients in any transfer of care. Don't leave patients wondering as to who is looking after them, as to who will perform their surgery, or confused by being told one plan of action when initially listed and then told something completely different by someone else.
2. Carry out a final check of the need for surgery—by questioning patients on the day of surgery, to ensure the patient is still symptomatic, and still keen to go ahead. In some elective cases symptoms may have improved between getting listed and presenting for surgery, and even though patients turn up on the day, they may be unsure as what to really do. Questioning about current symptoms may help guide them through such uncertainty, helping some to avoid surgery and some to reaffirm that this is what they really want to do.
3. Carry out a final check of fitness for surgery—by assessing patients on the day of surgery to ensure there are no medical factors that could necessitate postponing or avoiding surgery. The patients may be reviewed by the anaesthetist for anaesthetic fitness but there are also others factors a surgeon may have to consider:

   - Is there any focus of infection, such as a skin boil, close to or distant to the surgical field, that may prohibit the safe implantation of a prosthesis?
   - Has the patient suffered a further, recent injury to the knee that you are planning to replace, and you may need to allow time for the acute inflammation to settle down before going ahead?
   - Is the patient recovering from a recent deep venous thrombosis?

4. Consent patients well before the day of surgery, to allow them sufficient time to consider all pros and cons. Even if not able to go through the full consent documentation and get them to sign the form, describe the risks and inform. Would one be really able to make an informed choice if simply told of the risks just before or just after changing into their theatre gown in the surgical admission suite?
5. If practicalities allow, see the patient in the recovery suite, post-surgery, to ensure there are no immediate post-operative complications:
   - What are their overall observations?
   - Are they coming round from the anaesthetic?
   - Is the limb neurovascularly intact?
   - Is there any staining of the dressing to suggest oozing from the wound?
   - How much is in the drains?
6. See the patient in the recovery suite to ensure the immediate post-operative plan has been followed:
   - Are the prophylactic antibiotics prescribed?
   - Are the thrombo-prophylaxis stockings on?
   - Is deltaparin prescribed in the correct dose?
   - Is the elevation sling applied?
7. See the patient in the post-operative ward, prior to discharge, when they are more alert, to explain the surgical findings, what was done and any further plans. Information exchanged with the patient in the recovery suite, whilst the patient is coming round, may not be registered or recalled, hence better to see them later in the ward or clinic to explain what was found, what was done, and what happens from then on.

4.2 Listing for Surgery

A list is only as strong as its weakest link.

Donald Knuth [3]

Listing for surgery refers to documentation of the need to proceed with surgery for a case, and the plan as to what will be done. Depending on the healthcare system, or institution you are working in, this aims to inform the waiting list coordinator, the theatre staff, the pre-operative assessment staff, the anaesthetic team. Clear, thorough documentation is essential in encouraging safety, avoiding confusion and interruptions to the patient's journey to surgery. The following describes some the information that you may need to provide in filing such a listing:

1. Patient details.
2. Dates of non-availability for surgery.
3. Ensure that the correct contact details are up to date.
4. Procedure to be performed.
5. Side of the procedure.
6. Surgeon in charge of patient's care.
7. Estimated time needed for procedure (taking into account anaesthetic time, positioning time, time for other activities such as awaiting the results of an intra-operative biopsy).
8. Specific anaesthetist to be present—such as one who has the skills of administering a regional nerve block, managing a difficult airway.
9. Proposed type of anaesthetic: general, regional or local.
10. Urgency of time to surgery—state the required time frame for surgery rather than simply stating urgent.
11. Surgical equipment or instruments that will be needed.
12. Special surgical equipment or instruments which will need to be ordered, their provider, and whether the company representative will be required to attend.
13. Whether the patient is a day case or an inpatient.
14. Date and sign entry.

4.3 Putting Together a Theatre List

My to-do list is so long that it doesn't have an end; it has an event horizon.

Craig Bruce [3]

Organising a theatre list forms a big part of a surgeon's working life. Acquiring the ability to organise the theatre list to allow the provision of high quality surgical care, but also efficient and cost effective utilisation of theatre time, is an importance skill to develop. It requires surgical knowledge, understanding of the theatre environment in which such lists are performed, organisational abilities, leadership as well as team working.

The theatre list refers to a specified theatre space, time and location used for performing a particular surgical task. This may be to accommodate emergency or elective patients, to accommodate patients operated upon as day cases, hospital in stays or a combination of both. In planning a theatre operating list the aims are to:

1. Ensure the list will provide high quality care, both anaesthetic and surgical.
2. Allocate patients to the appropriate list.
3. Allow efficient and cost effective utilisation of theatre time, to allow maximum care provision and avoid wastage of resources.

4.3.1 Ensuring Good Surgical Care

The first essentials, of course, is to know what you want.

Robert Collier [3]

In planning a list, one must consider the availability of essential clinical skills, (anaesthetic, surgical, nursing, pre and post- operative care). Carrying out surgery involves having a surgeon along with necessary assistants who are able and well skilled to carry out the surgical procedure, an anaesthetist along with anaesthetic assistants who are able to provide the necessary anaesthesia and analgesia, theatre scrub staff who are familiar with and able to support the procedure, ward or admission suite staff who look after the patient pre-operatively and prepare the patient for theatre, recovery staff who provide care in the early post-operative period, as well as ward and nursing staff who look after the patient up to safe discharge.

Surgeons' increasing subspecialisation may mean that as the depth of their skills increases the breadth of their skills diminishes. In a similar way, non- surgical medical staff may be subspecialised and one grade does not fit all. It may be that only a particular anaesthetist can provide specialised intubation techniques for the difficult airway, or administer specific regional blocks to avoid a general anaesthetic or aid post-operative analgesia. A scrub nurse who has trained in arthroscopic shoulder surgery may not be familiar with various jigs and implants to allow scrubbing for a knee replacement procedure. A cardio-thoracic recovery nurse, highly competent in looking after chest drains and monitoring arterial pressures, may feel unsure as to how to manage an abduction wedge following hip replacement. Along similar lines, a low grade assistant may suffice in soft tissue retraction during open hernia repair, but a more experienced, skilful assistant may be necessary in open oesophagectomy. A junior assistant may be competent at retracting the open shoulder, but may still be struggling to hold steady the camera to allow suturing in arthroscopic shoulder surgery.

Hospital ward staff numbers may fluctuate with less staffing at nights, weekends or bank-holidays. Hence, the ability of ward staff to closely monitor patients returning from surgery may be better early on in the day and less late at night. Similar considerations may apply to medical staff—if there is leakage from the bowel anastomosis will the shift breast surgeon be able to return the case to theatre or will a more specialist colorectal surgeon be needed? If a complex hip replacement is carried out in a patient who has substantial medical comorbidities on a Friday afternoon, will there be sufficient numbers of adequately skilled medical staff to look after the patient over the weekend?

4.3.2 Surgical List Location

Mr. Breton didn't know about location, location, location.

Daniel Pinkwater [3]

Consider whether the proposed procedure will be performed as elective inpatient surgery or day surgery. Day surgery refers to admission and discharge on the same day. Improvements in anaesthetic techniques as well as advances in minimal invasive surgical procedures mean that an increasing number of operations can be performed on a day case basis rather than patients having to stay in hospital. Cost of hospital beds, hospital related post-operative complications (such as acquired pneumonias) have led to a drive to encourage the utilisation of day surgery. Guidelines by the British Association of Day Surgery [4] suggest that cases meeting the following criteria could be considered for day surgery care:

- Patient consenting to day surgery.
- An adult can escort the patient home and look after them for the first 24 h.
- Medically fit to undergo the procedure.
- Surgery not associated with a significant risk of complications that require medical input (such as haemorrhage).
- Post-surgery symptoms (such as pain, sickness) should be controllable with oral agents or local anaesthetics.
- Patient able to restart oral intake within a few hours post-surgery.
- Patient should be able to at least partially mobilise.

Hence, it is important to identify those patients that potentially could be done as day surgery, and place them on the appropriate list.

In cases where the theatre list is mixed including both inpatients and day surgery cases then the order of the list may influence whether a patient will be able to be discharged safely home on the same day. A patient operated last on the list may not have sufficient time for safe recovery to allow a same day discharge, hence another factor to consider in the list order.

4.3.3 Order of the Surgical List

Like it or not, the world evolves, priorities change and so do you.

Marilu Henner [3]

The main priority in guiding the order of a surgical list, should be clinical urgency or other clinical factors to ensure safe, high quality care. Emergency, life-saving or limb-saving procedures take priority over more semi-urgent cases. Assuming these factors are equal then other parameters may be taken into consideration:

- Diabetics have traditionally been operated first on the list, to avoid extensive fasting and consequent blood glucose derangement. However, it is increasingly recognised that such a strict approach may not be essential [4], but where possible this is accommodated.
- Children are preferably done first on the list, again to avoid unnecessary periods of fasting. If a child cannot be accommodated early in the morning, then a plan may be made for later on in the day, at a predetermined time, and the child fasted in reference to that. Hence, if a child cannot be operated first on the morning list, may be better operated first on the afternoon list and fasted for that.
- Latex allergy—patients who have such allergy, should be done first (after theatre cleansing) on the list, with the necessary precautions taken to provide a latex free environment [5].
- Infected cases should be done last on the list, rather than prior to procedures involving insertion of implants, provided that they can wait. If they are to be done prior to clean procedures, adequate time should be given for thorough theatre cleansing prior to the next case.
- Complex cases:
  - Medically complex cases are better done early in the day to ensure adequate anaesthetic support in the post-operative period.
  - Complex surgical cases preferably done early in the day when the staff are at their best, rather than towards the end of the list when some of us may be getting tired and our concentration may decrease [6, 7].

4.4 Improving Theatre List Utilisation

I'll take fifty percent efficiency to get one hundred percent loyalty.

Samuel Goldwyn [3]

Theatre time is valuable both with regards to opportunity for delivering surgical care but also actual monetary costs. It is estimated that in the UK an hour of operating theatre suite utilisation costs about £1200 [8]. Hence theatre time wastage should be kept to the minimum, but too often this is not happening.

Sultan and the current author [9] evaluated prospectively theatre time utilisation in a district teaching hospital in the UK. They reported that only 54% of theatre time was utilised for operating, whilst 21% was utilised for anaesthetic time, positioning and draping. In addition, 25% of operating time was not utilised due to delays in starting the list and turnover time. At the same institution Singhal et al. [10] reported a 10% same day cancellations in elective orthopaedic cases for patient reasons (such as medical fitness, symptoms improving).

Similar results were reported on the other side of the world, at the Ipswich General Hospital in Australia [11], where it was reported that only about 50% of theatre time was utilised for actual operating, with 16% of time wasted due to late starts, early finishes or as changeover time. Of 3014 elective procedures, 14.3% were cancelled on the day of surgery due to lack of surgical fitness or due to lack of operative time.

In principle, theatre time wastage may be minimised by:

1. Prompt start.
2. On time finish.
3. Avoiding in between interruptions.

4.4.1 *Prompt Start*

This may be ensured by several factors such as:

1. Having as first case on the list a patient who is medically fit and worked up for surgery.
2. Ensuring the admission suite prepares the patient for theatre on time.
3. Prompt arrival of anaesthetic and surgical staff to theatre.
4. Carrying out the consent process and pre-operative medical assessment prior to the day of surgery.
5. Avoid putting first on the list cases for which specialised equipment must be checked on the day of surgery.
6. Ensuring bed availability for inpatient cases, to avoid delays in starting the list.

4.4.2 Avoiding Early Finishes

This may be achieved by:

1. Avoiding same day cancellations occurring due to various reasons including:
   - Patients not turning up for surgery.
   - Patients found medically unfit on the day of surgery.
   - Clinical symptoms having improved or settled.

   Contacting patients, in the week prior to surgery, to ensure they are still symptomatic, to check that they do not have any new onset medical symptoms that would make them unfit for surgery, and confirm they are aware of the arrangements for presentation for surgery, may help reduce the rate of same day cancellations. The above may be achieved through a phone consultation [10], or by brining patients to a pre-operative clinic [12, 13].

2. Having an adequate number of theatre trays and equipment, along with adequate back up, to avoid relying on same day turn around in sterilisation of equipment.
3. Allocating the right number of cases to the time available. Regular evaluation of theatre lists may help identify ones that repeatedly finish early, or have long break interruptions with the aim of adding more cases.
4. Having some flexibility in allowing overrunning of the lists (on the unusual situations that is necessary), to avoid same day cancellations. This can be a challenging arrangement as different staff teams have to be agreeable and on-board.
5. Avoiding compartmentalisation—if one is very rigid into what procedures can be done on a list, then those lists may be under-filled if those cases are not available.
6. Allowing a mixture of long and short cases—even though it may be difficult to squeeze a hip replacement at the end of the list, it may be possible to add a couple of carpal tunnel decompressions; even though it may be difficult to add another laparotomy it may be possible to add a hernia repair.

4.4.3 Avoiding Interruptions

Several factors may account for interruptions including:

- Availability of theatre scrub staff—may be helped by staggering of lunches.
- Availability of anaesthetic assistants—may improve by having one anaesthetist anaesthetise a patient whilst another is recovering the previous case.
- Availability of surgical staff—may be helped by having a surgeon carrying out the full day list or a surgeon staying on until another one arrives to take over. However, there may not be much scope in ensuring that a full day surgeon is available, if other staffing levels do not allow an uninterrupted running of the theatre list.

- Order of the list that requires changing equipment or moving around equipment—may be helped by planning the list to avoid having to change from arthroscopic to open procedures and then back to arthroscopic. May be also helped by avoiding having to swap from a left to a right shoulder and then back to the left.
- Delays in patients arriving to theatre from the ward—may be helped by anticipating the finish of the current case, and sending for the next patient on time, or warning the ward that you will be sending shortly to ensure the patient is ready.
- Delays in patients transferred from the recovery suite to the ward—may be helped by adequate staffing of recovery, and by utilising anaesthetic and analgesia techniques that allow early recovery. In addition, it may be helped by identifying patients who may need high dependency or intensive care and pre-arranging beds in those settings prior to surgery, to avoid having to stay in recovery awaiting for a bed on the day. On occasions, if recovery is blocked then theatres may have to pause.
- Lack of hospital beds—in some institutions or healthcare systems there may be fluctuations in the availability of in-patient beds (due to winter or other pressures) which may lead to same day cancellations. This may be helped by having adequate number of beds, ring-fencing beds (reserved or protected for elective surgery), ensuring that there is a clear process in confirming the availability of beds on the day to allow surgery to proceed promptly, having day case patients on standby who can be brought in at short notice if overnight beds do not become available.
- Nevertheless, the planning of operating theatre lists is not an exact science, and may be influenced by the surgical and overall theatre team, as well as the operative environment and institution in which such surgery is carried out. What applies to one institution, one surgeon, one surgical specialty, one group of patients, may not apply to another. Comparing surgeons and institutions with regards to crude number of cases per theatre list, may be an over-simplified approach, but one should look into and learn from better practices.

References

1. Fischer JE. On the uniqueness of surgery. Am J Surg. 2005;189(3):259–63.
2. Kjaergaard J, Sørensen LT, Bigaard JV, Jørgensen T. The effect of patient-specialist-continuity in a surgical department. Evaluation of seven years of experiences. Ugeskr Laeger. 2000;162(16):2324–9.
3. Famous Quotes at BrainyQuote. https://www.brainyquote.com/quotes/quotes. Accessed 25 Mar 2017.
4. Day Case and Short Stay Surgery. The Association of Anaesthetists of Great Britain & Ireland and The British Association of Day Surgery. 2011. https://www.aagbi.org/sites/default/files/Day%20Case%20for%20web.pdf. Accessed 1 Apr 2017.
5. Latex Allergy Management (Guidelines). July 2014. http://www.aana.com/resources2/professionalpractice/Pages/Latex-Allergy-Protocol.aspx. Accessed 1 Apr 2017.
6. Hänecke K, Tiedemann S, Nachreiner F, Grzech-Sukalo H. Accident risk as a function of hour at work and time of day as determined from accident data and exposure models for the German working population. Scand J Work Environ Health. 1998;24(Suppl. 3):43–8.
7. Teng TY, Khor SN, Kailasam M, Cheah WK, Lau CC. Morning colonoscopies are associated with improved adenoma detection rates. Surg Endosc. 2016;30(5):1796–803.
8. The productive operating theatre. NHS. Institute for innovation and improvement 2009. http://harmfreecare.org/wp-content/files_mf/Improving-quality-and-efficiency-in-the-operating-theatre.pdf. Accessed 31 Mar 2017.
9. Sultan J, Charalambous CP. Theatre time utilisation in elective orthopaedic surgery. J Perioper Pract. 2012;22(8):262–5.
10. Singhal R, Warburton T, Charalambous CP. Reducing same day cancellations due to patient related factors in elective orthopaedic surgery: experience of a centre in the UK. J Perioper Pract. 2014;24(4):70–4.
11. Keller A, Ashrafi A, Ali A. Causes of elective surgery cancellation and theatre throughput efficiency in an Australian urology unit. F1000Res. 2014;3:197. doi:10.12688/f1000research.4824.1.
12. Emanuel A, Macpherseon R. The anaesthetic pre-admission clinic is effective in minimising surgical cancellation rates. Anaesth Intensive Care. 2013;41(1):90–4.
13. McKendrick DR, Cumming GP, Lee AJ. A 5-year observational study of cancellations in the operating room: does the introduction of preoperative preparation have an impact? Saudi J Anaesth. 2014;8(Suppl 1):S8–S14. doi:10.4103/1658-354X.144053.

Chapter 5
Technical Skills in Surgery

Much of surgery is operating, a technical craft. Much of what we do is with our hands, whether it's using the scalpel, the scissors, the forceps, the needle, the drill or saw, the endoscope, the arthroscope or laparoscope, the robot or microscope. Technical dexterity is an important skill to develop. At the same time the ability to plan for surgery, and the ability to be movement and time efficient, are important skills to acquire.

This chapter discusses the considerations in planning for a surgical procedure, the challenges faced in surgical training, ways of maximising surgical learning opportunities, ways of improving and maintaining surgical motor skills, learning a surgical technique, and improving surgical efficiency.

C. Panayiotou Charalambous, *Career Skills for Surgeons*,
DOI 10.1007/978-3-319-57490-5_5

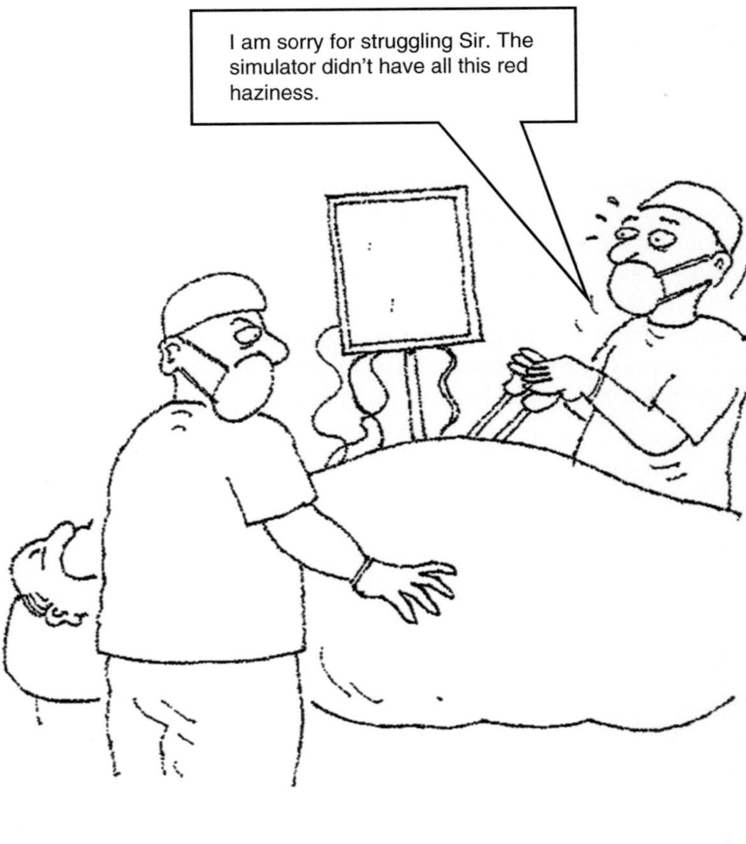
I am sorry for struggling Sir. The simulator didn't have all this red haziness.

5.1 Planning for Surgery

Adventure is just bad planning.

Roald Amundsen [1]

In planning for a surgical procedure, consider the actual surgical steps to be performed, but also all the factors needed to allow successful completion of those surgical steps. Some of these are described below:

General considerations

- Type of anaesthesia
- Analgesia
- Intra-operative monitoring (vital signs)
- Intravenous access
- Antibiotic prophylaxis
- Temperature control
- Glycaemic control
- Thrombo-prophylaxis
- Urinary catheterisation

Personnel choice

- Surgeon
- Surgical assistants
- Scrub staff
- Theatre runners
- Anaesthetist
- Recovery staff
- Equipment support staff

Theatre choice

- Airflow system—for clean surgery
- Spacious to accommodate personnel and equipment

Equipment choice

- Prepping, draping, scrubbing, gowning
- Theatre table to allow positioning, facilitate intra-operative imaging, support patient's weight
- Surgical instruments for performing procedure
- Equipment for performing procedure
- Equipment for supporting patient during procedure
- Equipment for haemostasis
- Sutures
- Implants
- Grafts

Patient positioning

- To allow access
- To minimise bleeding

Equipment positioning

- Position theatre table so it is within the clean area, and to allow access to anaesthetic machine, image intensifier or other equipment
- From the ceiling—theatre lights to allow illumination of the surgical field taking into account where the surgeon and assistants stand so no obstruction
- Surgeon's eye level—camera screens, image intensifier screens to allow easy visualisation
- Surgeon's elbow level—diathermy and suction. Where will the tray with instruments be in relation to the surgeon so you don't need to keep turning round to get instruments from scrub nurse?
- On the floor—diathermy pedals, arthroscopic shaver pedals, to allow easy foot access
- Image intensifier, fluid pressure pump, laparoscopic stack

Staff positioning

- Surgeon
- Assistants
- Scrub
- Anaesthetic

Prepping and draping

- Extensive prepping to allow good access and appropriate draping
- Draping to allow access to incision area taking into account the possibility of extending the incision

Incision

- To allow adequate visualisation and access to perform the procedure
- One that minimises damage, less risk to surrounding structures
- Can be extended

Procedure

- Steps
- Sequence
- Beginning
- End

Wound closure

- Sutures
- Drainage
- Dressing
- Splinting

Post-operative care

- In recovery
- In ward
- Post-hospital discharge

5.2 Developing Surgical Skills

The most complicated skill is to be simple.

Dejan Stojanovic [2]

Developing surgical technical skills is a challenging task. Restrictions in weekly working hours and increased utilisation of shift patterns, may lead to reduction in surgical exposure with fewer training opportunities. But even when one manages to make it to theatre, learning can still be difficult.

Learning in the surgical theatre often has to be achieved under time constraints, having to keep watching the clock; not only due to the need to balance service provision with delivery of training but also to potentially minimise the time patients are kept under anaesthetic. The time left to surgical training may be additionally limited by having multiple teams in theatre, which may have different hierarchies, time constraints, working patterns, responsibilities. The surgical team often has to conform to such patterns, rather than setting the time frames.

Hence, trainers may be under triple pressure; pressurised by their institutions to accommodate more cases on their list, under pressure by co-workers for lists to finish on-time and not to overrun, and by their trainees for devoting time in quality training (Fig. 5.1).

Increased medico-legal challenges and public reporting of surgeon-specific outcomes, may make trainers anxious and nervous of letting trainees operate due to fear of complications and the consequences that may entail. There is sometimes a tendency for "the more you can do the more to be allowed to do". But then we all have to start somewhere.

Learning may also have to be achieved whilst under high levels of stress, when dealing with familiar or unexpected conditions, whilst faced with anticipated or unanticipated events. Each case is different, has unique characteristics, which may deviate from the textbook norm, and one has to learn how to adjust for those. The ability to deal with intra-operative complications rather than simply with cases where all goes smoothly, means there is need to encounter these, to get involved in complex cases.

Hence, learning and developing surgical skills is a demanding task. A trainee needs to maximise all opportunities, to make the most of this challenging task. In improving your surgical skills and increasing your operating opportunities you may consider some of the approaches described next.

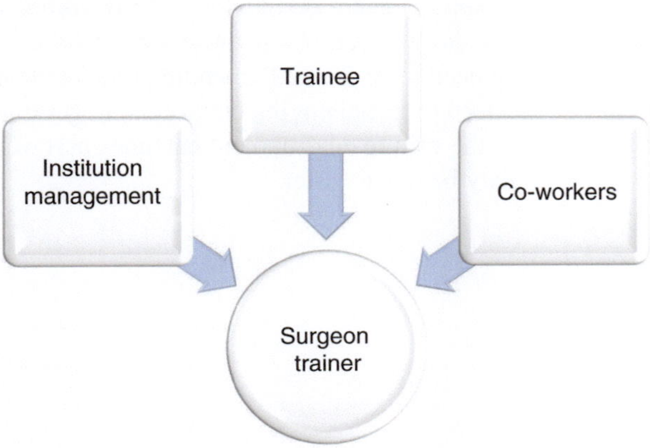

Fig. 5.1 Surgical theatre pressures that may influence delivery of surgical training

5.3 Get Accustomed to the Operating Theatre Environment

You are affected by the surroundings, the mood of people, by confidence. I am no different.
Michael Owen [1]

Get used to the surgical theatre environment. Spend time in the surgical theatre, even if you are not operating or assisting. No theatre time is a waste of time. Being in theatre may increase your familiarity with the environment, give you confidence of how to act and behave in such an environment, how to cooperate with other teams and theatre participants.

Find your bearings. Get to know where the patient is brought and checked in, where the anaesthetic is administered, where recovery is located, where is the theatre manager's office, the stores' room, the coffee room. Get to know where you scrub, where the gowns and gloves are shelved, where the theatre table side supports are stored, where the arthroscopic stack or radio-protection gowns are kept.

Get used to common surgical theatre activities, learn how you dress, prep, glove and gown, what you change into if you want to go out of the theatre suite, how you transfer the patient to and from the operating table, what the buttons on the table's remote control mean, how close to the surgical field you can get, where you can stand, what's sterile and what you can touch, how you position your arms whilst standing and waiting, how you move the heavy theatre lights around.

Get used to the roles of all your co-workers, the anaesthetist, the scrub nurse, the theatre runner, the radiographer, the perfusionist, the operating department practitioner, the theatre coordinator, the theatre sister. What do they do, what are their responsibilities, whom do you need to ask to help solve a specific problem?

When learning to swim you may stick to one pool, swim by the side, just in case, and then gradually move away towards the centre of the water. But even when you can confidently go up and down that pool, swim multiple laps, you may still feel out of your depth and need time to settle in if you find yourself in new, un-accustomed waters.

Likewise, initially each surgical theatre may look strange, but as you become familiar with the environment, theatres will look similar whether in the same block, one hospital or in different institutions.

Every moment in theatre is valuable, even the times when you don't scrub, when you scrub but don't touch, even those long lists of "lumps and bumps", those abscesses that need to be drained. Spending time in theatre can give you confidence, help you get used to a challenging environment, get used to how theatre staff act or behave. Once the theatre environment becomes part of you, carrying out the procedure per se may be much easier.

In improving your exposure to the operating theatre you may consider:

- Join additional lists taking place in your department. See which lists may need an extra pair of hands and offer to join.
- Ask your seniors or colleagues to let you know if any unusual cases come during out of hours, you may be living on site and able to attend to those at short notice.
- Make the most of attachments where you can get to spend ample time in theatre and get to operate. In an ideal situation each training post should offer similar operating opportunities, but this may not be the case. Hospitals vary in number of cases they can put on a list due to organisational factors, seasonal or other external pressures (with winter cancellations at short notice), there may be a run of complex cases, trainers may differ in experience, confidence, willingness to let you do things. In some attachments you may find yourself doing most of the operating, in some mostly holding the retractor. But even if you end up simply assisting, you can still gain lots from being there. You may see how things should be done, or how they shouldn't be done, what to avoid.

5.4 Developing Technical Skills

Technique is noticed most markedly in the case of those who have not mastered it.
Leon Trotsky [1]

In developing your surgical motor skills you may consider various models and parameters previously described as able to facilitate motor skills acquisition. These are presented next.

5.4.1 Model for Progression of Motor Skills Acquisition

Fitts and Posner [3] described the three stage model of motor skill acquisition (Fig. 5.2).

In the cognitive stage the learner tries to understand the task, has to think what they are doing, analyse the actions, how to position or move the hands; the task is carried out in separate steps, going back and forth, often in an irregular pattern. In the integrative stage the learner understands the task and the movements of carrying it. Hand motions are more fluent and faster but the learner still has to think about the task.

In the autonomous stage the task is carried out with little if any thinking, fast and effectively, in an automatic mode.

This model may allow you to appreciate the gradual process of acquiring motor skills, and hence view your performance and progress in the context of this process.

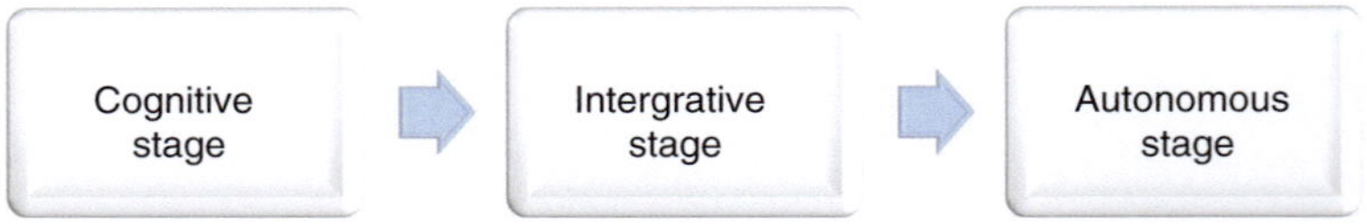

Fig. 5.2 Three stage model of motor skill acquisition. Based on Fitts and Posner [3]

5.4.2 Learning Curve

Learning is not attained by chance, it must be sought for with ardor and diligence.

Abigail Adams [1]

Motor skills or tasks have a learning curve that describes the relationship between performance and exposure [4–12]. As one does more and more, performance improves, initially rapidly, but then more slowly. Eventually, a plateau may be reached where it is more about staying good rather than getting better.

Appreciating the learning curve may allow one understand the difficulties faced when beginning to learn a new skill, but also seeing that there can be light at the end of the tunnel, that one may be able to reach their peak. It may also remind the need for support, time and resources (assistants or supervisors) whilst at the ascending part of the curve.

Recognising that each part of the procedure, each hand movement, use of each surgical instrument, or use of each surgical kit has a learning curve may help one appreciate the need for trying to find and settle into a routine. It may allow one to be more critical as to the need to keep trialing new instruments or techniques.

And even when reaching the plateau phase of the learning curve, one may need to keep practising skills to maintain performance. Several studies have correlated surgeon procedure-specific experience (as described by procedure volume—the number of specific procedures performed annually) to improved outcomes in laparoscopic cholecystectomy [11, 12], colorectal surgery [13], thyroid [14] and cardiac surgery [15, 16].

Ascending the learning curve does not simply refer to getting better at operations where all goes well and smoothly, but also facing and dealing with complications. Flying through smooth conditions is different to dealing with bumpy turbulence, dumping fuel for an emergency landing or averting a landing at the last minute.

5.4.3 Deliberate Practise

Every artist was first an amateur.

Ralph Waldo Emerson [1]

Acquisition of motor skills and ascend of the learning curve are not just about exposure to an action or task. Improvement of surgical skills requires seeing more but also seeing wisely. Experience is achieved through a high volume of exposure, but also by each exposure challenging the learner, and hence acting as a learning event. The aim is not simply to do more, but also to do better.

Anders Ericsson, a Swedish Professor of Psychology at Florida State University, describes that the difference between experts and normal practitioners is the constant deliberate effort of the former to improve their performance in a particular task. Ericsson reports that performance is not so much influenced by mere repetition of an action but by deliberating trying to improve performance in that action, and trying to go beyond one's comfort zone in carrying out a particular task. This may be achieved by breaking down the task in its components, concentrating on one component at a time, putting more emphasis on components that have not been fully mastered [17, 18].

Deliberate practise also involves:

- Coaching—with a trainer setting goals of performance, providing guidance and support.
- Immediate feedback—on performance to build on strengths and address deficiencies.

Acquisition of motor skills may be more closely related to time spent in deliberate practise rather than any practise. During surgical training you may focus on specific parts of a surgical procedure, concentrating on the steps you need to get better at, such as specific hand movements in which you need to improve. As you get competent in carrying out a procedure in straightforward cases, without much mental effort, you may seek to progress to more complex cases, get better and better.

5.4.4 Piecemeal Learning

I have never met a man so ignorant that I couldn't learn something from him.

Galileo Galilei [1]

Learn is steps, bit by bit. In learning a surgical procedure one may wonder whether doing just part of the procedure is of value, or whether only those cases where the whole procedure is performed (skin to skin) are of benefit. In the surgical environment a trainee and trainer may face the pressure challenges described earlier in this chapter, and the ability to allow a trainee to go through the whole of the procedure may be limited.

There is no convincing evidence that carrying out wholly tasks is better than doing parts of the procedure, and doing parts of the procedure may help in the same way as doing the procedure from start to end [19, 20]. Early on, in learning a surgical procedure, one may aim to learn doing each individual step [21]. As skills develop and confidence grows, doing the full procedure, from start to end, is preferable.

Break the operation into multiple steps. Do parts of the operation at a time, with the rest performed by your senior. As you get better at one step, you may swap that step with your senior and get to carry out another step, before eventually being able to carry out the whole procedure. You may initially do one step at a time, and then multiple steps before proceeding to the whole procedure.

- In training to do laparotomies for bowel resection, you may start by opening or closing the abdomen and once you master that you may move into dissecting and resecting the bowel, and then onto fashioning a stoma.
- In doing a hip replacement you may master the approach, before learning how to prepare the femur or the acetabular components.

Discuss at the beginning of the list as to which cases or which parts of those cases would be appropriate for training. Identify which steps you need to get better at and concentrate on those.

5.4.5 Distributed Practise

We hear and apprehend only what we already half know.

Henry David Thoreau [2]

Distributed practise refers to having periods of practise separated by periods of rest, whereas massed practise refers to having periods of continuous practise with little if any rest in between. Distributed practise may allow more effective acquisition of surgical skills than massed practise [22–24]. This may be related to activation of certain parts of the brain that aid consolidation of learning and to the mental rehearsal as well as retrieval from memory required in each episode of practise [24–26].

Moulton et al. [24] compared the two in a randomised trial whereby 38 surgical residents were assigned to distributed (weekly) or massed practise arrangements in learning microvascular anastomosis. Even though both groups improved their performance by practising, the distributed practise group scored much better in the retention of skills gained.

Hence, you may aim to have repeated exposures in a particular task rather than trying to get it all in at once; any exposure to a surgical procedure may facilitate learning.

5.4.6 Mental Practise

Practice doesn't make perfect. Practice reduces the imperfection.

Toba Beta [2]

Mental practice refers to mentally rehearsing a task prior to carrying it out physically [27–29]. It aims to help learn a motor skill or a surgical procedure without physically carrying it out.

Mental practise is not just about thinking of the steps of the procedure, but carrying out the procedure mentally, equivalent to cognitive simulation. It is thought that mental practise may activate motor networks in the brain that overlap with those activated during physical execution of motor activities which can lead to motor skill acquisition [30, 31].

Arora et al. [32] examined whether mental practise can enhance performance in laparoscopic surgery. They randomised 18 novice surgeons to performing 30 min of mental practise or watching an online lecture prior to carrying out laparoscopic cholecystectomies on a virtual simulator. Although there was no difference in baseline technical ability between the two groups, learning curves were significantly better in the mental practise group. Mental practise has also been shown to aid medical students develop basic surgical skills [27, 33] and crico-thyroidomy [34].

Hence, you may facilitate your surgical learning by mentally rehearsing certain actions or whole surgical procedures; you may rehearse how to position yourself, how to move your hands, even imagine the tactile feedback you may experience.

5.4.7 Simulation

For the things we have to learn before we can do them, we learn by doing them.
Aristotle [2]

Learning means doing. Motor skills are thought to have a learning curve. A surgical procedure has an overall learning curve but each of its individual steps has to be mastered and has its own unique learning curve.

Simulation offers the ability to practise motor skills without time pressures and without fear of causing harm to patients. Simulation may help ascend the first part of the learning curve, grasp the basics of the skill, upon which one can further build in the surgical theatre. Simulation may also allow practise in handling rare events, unusual intra-operative complications; like pilots who may train to deal with a stall or an engine failure well before they ever encounter one.

Simulation may be in the form of:

- Mannequins, synthetic models or body parts (such as bones)
- Animal models
- Human cadavers
- Virtual reality simulators
- Theatre environment

Simulators may vary from simple and low cost, to advanced, high cost, almost real life simulations. One may develop basic surgical skills on synthetic or animal models; knot tying, suturing, drilling, sawing, inserting bone screws, fixing fractures with plates or nails, undertaking arthroscopic or laparoscopic tasks are only some of the skills that can be learned and developed. Cadaveric training allows near life exposure, better grasp of anatomy and the handling of soft tissues, clearer appreciation of the whole 3D orientation. Virtual simulators use the last edge of technology, to match a real life surgical procedure, incorporating visual, auditory, and tactile cues; one may see the effect of their hands moving on the screen but also feel resistance to using instruments or handling tissues, getting to know how much force to exert for a specific task. Virtual simulators may give a running commentary on how the operator is doing, or a summary performance feedback after task completion. And once individual motor skills or procedures are mastered, simulation may help prepare for the real surgical experience, in enacting the whole surgical theatre environment; one may simulate the whole surgical theatre team in expected and unexpected scenarios.

Several studies have examined the role of simulation demonstrating their benefit in acquiring surgical skills, but also the transferability of such skills to the real life surgical theatre [35–40].

Make the most of any simulation you can get your hands on. Maximise your learning opportunities out of the surgical theatre. Simulation may not substitute the real theatre experience or real surgical procedure, but they may help you get faster there.

5.4.8 Feedback

A man only learns in two ways, one by reading, and the other by association with smarter people.

Will Rogers [1]

Seek feedback to help you reflect and further develop your surgical skills. Feedback is not just about what to do but also how to do it. Feedback may be concurrent, given whilst the task is carried out, or terminal, given at the end of a task. Feedback may be part of a computer or simulator response or given by an "expert" such as your senior or supervisor directly observing you during surgery. Porte et al. [41] demonstrated, in a randomised trial, that verbal feedback from an expert is more effective than a computer based motion efficiency feedback (self-accessed) in the retention of suturing and knot tying skills amongst medical students.

Self-reflect, but also aim to carry out tasks under the direct observation of someone who can comment on how you are doing. Even when you feel competent in carrying out a procedure, an "expert's" observation and words may help to further refine or fine tune your skills. Previous studies have highlighted the role of both concurrent and terminal feedback [42, 43]. Xeroulis et al. [42] compared concurrent and summary expert feedback in a randomised trial of medical students performing suturing and knot-tying skills. They demonstrated that both methods were effective for the initial instruction of these skills to novice students, but only summary feedback improved retention of superior skills. Walsh et al. [43] assessed novice endoscopists in performing simulated colonoscopy tasks. Both the concurrent and terminal feedback groups performed similarly at post-practise and in retention of skills, but the terminal feedback group performed significantly better on transfer of skills.

In the early stages one may benefit from getting feedback with each step as they go along (addressing the individual hand movements or other actions). In later stages, when proficiency develops, terminal feedback, may be of greater use. Depending on the stage of your surgical development you may ask your senior to give you feedback as you go along, or at the end of the procedure, go through what went well and what could have improved.

5.4.9 Progress in the Levels of Participation

I am still learning.

Michelangelo [1]

As one improves their surgical skills and gets more competent, may progress to operating under less and less supervision (Fig. 5.3). Operating with a senior unscrubbed or in the absence of a senior, are important steps in gaining technical confidence, and at some point you will have to make that jump, if working towards independent practise.

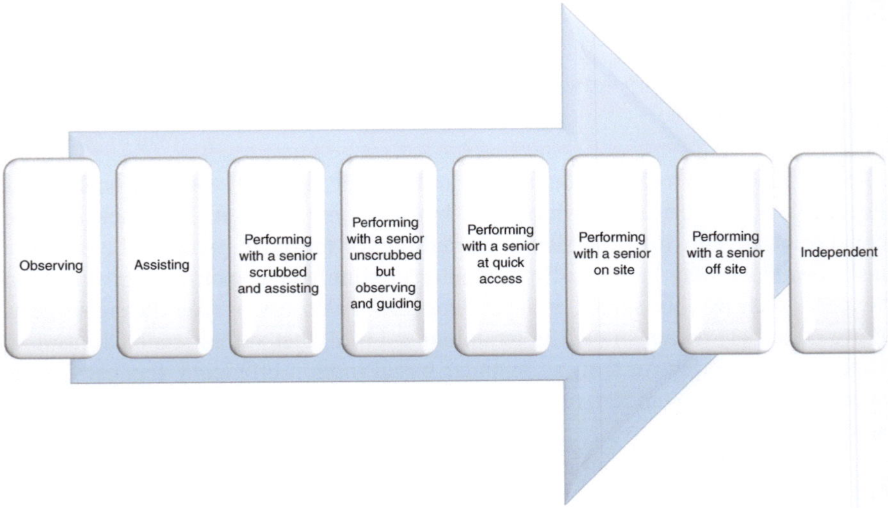

Fig. 5.3 Progress of surgeon's participation in surgery

Finding yourself with no senior in theatre for the first time can be a taunting experience. Follow the regulations of the healthcare system you are practising in, as some training programs may not allow unsupervised operating until fully qualified for independent practise. If you have to make the step only once you are fully qualified, then consider asking a colleague to support you in those initial stages.

5.4.10 *Be an Active Assistant*

Learning without thought is labor lost; thought without learning is perilous.

Confucius [1]

Sometimes we may have to assist and assist, clock the hours, before we get the driving seat. But learning is not just about doing it yourself, you can also learn by being an engaged assistant, by following the procedure through, anticipating the next step, questioning what was done and how else you could have approached it.

Discuss with your trainer the procedure and its steps, confirm that you understand what they are doing, that you are keeping up with their pace, show you recognise the anatomy and the dissection planes. Offer to retract before being asked, or, once you know what is expected, carry out the task without having to be asked. Show that you have thought about the procedure, you studied, you checked that the patient and theatres are ready. Demonstrate to your senior that you know the basic but essential steps of the procedure, how to position, drape and prep. This may give more confidence to your trainer about your abilities, help them take you over to the next step.

5.5 Learning a Surgical Technique

The beautiful thing about learning is nobody can take it away from you.

BB King [1]

Learning a new surgical technique and getting better at it, is often a challenging and demanding task. Once a technique is learnt it may still not be fully mastered, and the skills may need to be refined. LOAD [44] describes the essential steps in learning a new procedure and stands for:

L-earn about the procedure.

O-bserve the procedure.

A-ssist in the procedure.

D-o the procedure.

You may read up the procedure, hear about it in lectures or from peers, study the relevant anatomy. You may watch the animation or video, observe or assist in the simulator or real life, before undertaking the procedure with or without supervision, in full or in small progressive steps. But this is not a one way process, and you can go back and read more once you have attempted it, again observe and assist others to fine tune and further develop your technique (Fig. 5.4).

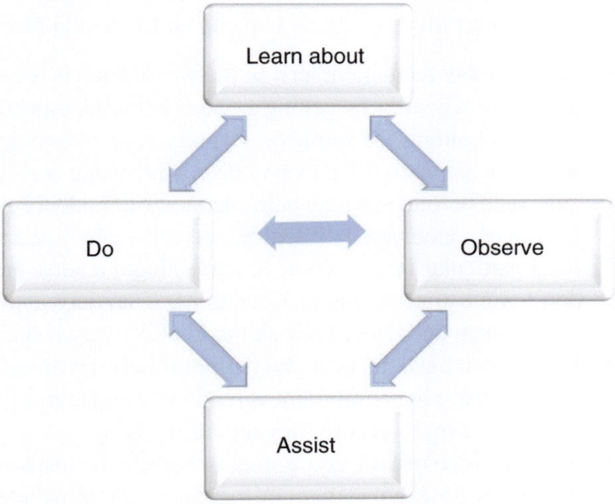

Fig. 5.4 Procedural learning cycle—LOAD. Adapted from Charalambous CP [44]

5.6 Getting Technically Fluent

The noblest pleasure is the joy of understanding.

Leonardo da Vinci [1]

In increasing your efficiency you may improve your manual dexterity, speed of movements, knowledge of what the steps involve and their exact sequence. At the same time you may also consider:

- Plan what you will do, and how. Perform a mental rehearsal. Run the operation through your mind before you set off.
- Have a logical sequence. So that one step leads to the next, avoid having to go back and forth.
- Eliminate unnecessary steps, unnecessary movements.
- Keep it simple.
- Know when to stop, know what you can accept.
- Know when to change course, if a plan does not work use another, if one approach fails use an alternative. Be able to adapt your actions in the context of each patient and each procedure.
- Work with your assistant, so they can facilitate your actions. Explain what you are asking them to achieve, for them to try and help you out. Delegate tasks:
  - "Look after the nerve whilst I put the plate on".
  - "Show me a good view with the camera whilst I tie the knots".
  - "Retract this muscle so you can show me the bleeding vessel".
- Engage other team members, so the whole team is focused on the task. Engage your assistants, who may be getting anxious or bored, especially if it is felt that all they are doing is holding the retractor. Explain your reasoning as they may be wondering what you are up to if the technique you are using is slightly different from what they have seen before; this may help encourage confidence and hence active participation. Inform of encountered difficulties, describe why you are taking longer than expected for a particular step (such as in harvesting a tendon graft which feels tight and you don't want to prematurely lacerate it). If anxiety and boredom set in, the level of engagement and hence their ability to effectively contribute may wear out.
- Ensure you have the necessary support before you start–staff, instruments, equipment. Ensure the equipment is ready to use. Have a list of instruments and equipment you use, to ensure they are all ready.
- Know how to use the equipment or handle the instruments. Do not assume others will know how to use them or that they will know what to do if they malfunction. Don't wait until the equipment fails to find out that the personnel who normally bail you out are on holidays.
- Check the experience of the scrub nurse and your assistant; not just their general theatre experience but also their experience in carrying out the required tasks for the operation concerned. Know what you can expect from them.
- Have pictures or photos showing how you aim to position the patient, and how you expect the image intensifier to come in, to facilitate communication to the rest of the team as to what you are trying to achieve.

5.7 Limitations of Surgical Skills

It is not that I'm so smart. But I stay with the questions much longer.

Albert Einstein [2]

As surgeons we often place much emphasis on the need of improving our surgical skills, getting better and better at what we technically do. However, being good with our hands may be only one of the parameters necessary in enhancing our surgical practise.

Regenbogen [45]examined the characteristics of technical errors in malpractice claims of four liability insurers. They found that most technical errors occurred in routine operations carried out by experienced surgeons, rather than in advanced procedures or in procedures carried by less experienced or in-training staff.

Hence, addressing technical experience could address only a minority of deficient surgical care. As surgeons we may also need to consider other parameters such as situation awareness and decision making which can affect our technical performance, and these are addressed in subsequent chapters.

References

1. Famous Quotes at Brainy Quote. https://www.brainyquote.com/quotes/quotes. Accessed 25 Mar 2017.
2. Popular Quotes. Goodreads. http://www.goodreads.com/quotes. Accessed 31 Mar 2017.
3. Fitts PM, Posner MI. Human performance. Belmont, CA: Brooks/Cole Pub. Co; 1967.
4. Arora KS, Khan N, Abboudi H, Khan MS, Dasgupta P, Ahmed K. Learning curves for cardiothoracic and vascular surgical procedures–a systematic review. Postgrad Med. 2015;127(2):202–14.
5. Arai T, Romano M, Lefèvre T, Hovasse T, Bouvier E, Morice MC, Farge A, Garot P, Chevalier B. Impact of procedural volume on outcome optimization in transaortic transcatheter aortic valve implantation. Int J Cardiol. 2016;223:292–6.
6. Hartford JM, Bellino MJ. The learning curve for the direct anterior approach for total hip arthroplasty: a single surgeon's first 500 cases. Hip Int. 2017;17:0. doi:10.5301/hipint.5000488.
7. Al-Harazi A, Goel R, Tan CT, Cheah WK, Lomanto D. Laparoscopic ventral hernia repair: defining the learning curve. Surg Laparosc Endosc Percutan Tech. 2014;24(6):475–7.
8. Yoshizawa J, Ashizuka S, Kuwashima N, Kurobe M, Tanaka K, Ohashi S, Hiramatsu T, Baba Y, Kanamori D, Kaji S, Ohki T. Laparoscopic percutaneous extraperitoneal closure for inguinal hernia: learning curve for attending surgeons and residents. Pediatr Surg Int. 2013;29(12):1281–5.
9. Hohmann E, Bryant A, Tetsworth K. Tunnel positioning in anterior cruciate ligament reconstruction: how long is the learning curve? Knee Surg Sports Traumatol Arthrosc. 2010;18(11):1576–82. doi:10.1007/s00167-010-1183-4.
10. Yount KW, Yarboro LT, Narahari AK, Ghanta RK, Tribble CG, Kron IL, Kern JA, Ailawadi G. outcomes of trainees performing coronary artery bypass grafting: does resident experience matter? Ann Thorac Surg. 2017;103(3):975–81. doi:10.1016/j.athoracsur.2016.10.016.
11. Sakpal SV, Bindra SS, Chamberlain RS. Laparoscopic cholecystectomy conversion rates two decades later. JSLS. 2010;14(4):476–83.
12. Csikesz NG, Singla A, Murphy MM, Tseng JF, Shah SA. Surgeon volume metrics in laparoscopic cholecystectomy. Dig Dis Sci. 2010;55(8):2398–405.
13. Yeo HL, Abelson JS, Mao J, O'Mahoney PR, Milsom JW, Sedrakyan A. Surgeon annual and cumulative volumes predict early postoperative outcomes after rectal cancer resection. Ann Surg. 2017;265(1):151–7.
14. Adam MA, Thomas S, Youngwirth L, Hyslop T, Reed SD, Scheri RP, Roman SA, Sosa JA. Is there a minimum number of thyroidectomies a surgeon should perform to optimize patient outcomes? Ann Surg. 2017;265(2):402–7.
15. Karamlou T, Jacobs ML, Pasquali S, He X, Hill K, O'Brien S, McMullan DM, Jacobs JP. Surgeon and center volume influence on outcomes after arterial switch operation: analysis of the STS congenital heart surgery database. Ann Thorac Surg. 2014;98(3):904–11.
16. Weiner MM, Hofer I, Lin HM, Castillo JG, Adams DH, Fischer GW. Relationship among surgical volume, repair quality, and perioperative outcomes for repair of mitral insufficiency in a mitral valve reference center. J Thorac Cardiovasc Surg. 2014;148(5):2021–6.
17. Ericsson AK, Krampe RT, Tesch-Romer C. The role of deliberate practice in the acquisition of expert performance. Psychol Rev. 1993;100(3):363–406.
18. Ericsson AK. Deliberate practice and the acquisition and maintenance of expert performance in medicine and related domains. Acad Med. 2004;79:S70–81.
19. Brydges R, Carnahan H, Backstein D, Dubrowski A. Application of motor learning principles to complex surgical tasks: searching for the optimal practice schedule. J Mot Behav. 2007;39(1):40–8.
20. Willaert W, Aggarwal R, Harvey K, Cochennec F, Nestel D, Darzi A, et al. Efficient implementation of patient-specific simulated rehearsal for the carotid artery stenting procedure: part-task rehearsal. Eur J Vasc Endovasc Surg. 2011;42(2):158–66.

21. Wickens CD, Hutchins S, Carolan T, Cumming J. Effectiveness of part-task training and increasing-difficulty training strategies: a meta-analysis approach. Hum Factors. 2013; 55(2):461–70.
22. Lee TD, Genovese ED. Distribution of practice in motor skill acquisition: learning and performance effects reconsidered. Res Q Exerc Sport. 1988;59:277–87.
23. Donovan JJ, Radosevich DJ. A meta-analytic review of the distribution of practice effect: now you see it, now you don't. J Appl Psychol. 1999;84:795–805.
24. Moulton CA, Dubrowski A, Macrae H, Graham B, Grober E, Reznick R. Teaching surgical skills: what kind of practice makes perfect?: a randomized, controlled trial. Ann Surg. 2006;244(3):400–9.
25. Luft AR, Buitrago MM. Stages of motor skill learning. Mol Neurobiol. 2005;32:205–16.
26. Whitley JD. Effects of practice distribution on learning a fine motor task. Res Q. 1970; 41:576–83.
27. Sanders CW, Sadoski M, van Walsum K, Bramson R, Wiprud R, Fossum TW. Learning basic surgical skills with mental imagery: using the simulation centre in the mind. Med Educ. 2008;42(6):607–12.
28. Cocks M, Moulton CA, Luu S, Cil T. What surgeons can learn from athletes: mental practice in sports and surgery. J Surg Educ. 2014;71(2):262–9.
29. Malouin F, Richards CL. Mental practice for relearning locomotor skills. Phys Ther. 2010;90(2):240–51.
30. Malouin F, Richards CL, Jackson PL, Dumas F, Doyon J. Brain activations during motor imagery of locomotor-related tasks: a PET study. Hum Brain Mapp. 2003;19:47–62.
31. Bakker M, De Lange FP, Helmich RC, et al. Cerebral correlates of motor imagery of normal and precision gait. Neuroimage. 2008;41:998–1010.
32. Arora S, Aggarwal R, Sirimanna P, Moran A, Grantcharov T, Kneebone R, Sevdalis N, Darzi A. Mental practice enhances surgical technical skills: a randomized controlled study. Ann Surg. 2011;253(2):265–70.
33. Sanders CW, Sadoski M, Bramson R, Wiprud R, Van Walsum K. Comparing the effects of physical practice and mental imagery rehearsal on learning basic surgical skills by medical students. Am J Obstet Gynecol. 2004;191(5):1811–4.
34. Bathalon S, Dorion D, Darveau S, Martin M. Cognitive skills analysis, kinesiology, and mental imagery in the acquisition of surgical skills. J Otolaryngol. 2005;34(5):328–32.
35. Thomas GW, Johns BD, Marsh JL, Anderson DD. A review of the role of simulation in developing and assessing orthopaedic surgical skills. Iowa Orthop J. 2014;34:181–9.
36. Tan SS, Sarker SK. Simulation in surgery: a review. Scott Med J. 2011;56(2):104–9.
37. Weller JM, Cumin D, Civil ID, Torrie J, Garden A, MacCormick AD, Gurusinghe N, Boyd MJ, Frampton C, Cokorilo M, Tranvik M, Carlsson L, Lee T, Ng WL, Crossan M, Merry AF. Improved scores for observed teamwork in the clinical environment following a multidisciplinary operating room simulation intervention. N Z Med J. 2016;129(1439):59–67.
38. Franzeck FM, Rosenthal R, Muller MK, Nocito A, Wittich F, Maurus C, Dindo D, Clavien PA, Hahnloser D. Prospective randomized controlled trial of simulator-based versus traditional in-surgery laparoscopic camera navigation training. Surg Endosc. 2012;26(1):235–41. doi:10.1007/s00464-011-1860-5.
39. Waterman BR, Martin KD, Cameron KL, Owens BD, Belmont PJ Jr. Simulation training improves surgical proficiency and safety during diagnostic shoulder arthroscopy performed by residents. Orthopedics. 2016;39(3):e479–85.
40. Dawe SR, Pena GN, Windsor JA, Broeders JA, Cregan PC, Hewett PJ, Maddern GJ. Systematic review of skills transfer after surgical simulation-based training. Br J Surg. 2014;101(9):1063–76.
41. Porte MC, Xeroulis G, Reznick RK, Dubrowski A. Verbal feedback from an expert is more effective than self-accessed feedback about motion efficiency in learning new surgical skills. Am J Surg. 2007;193(1):105–10.

42. Xeroulis GJ, Park J, Moulton CA, Reznick RK, Leblanc V, Dubrowski A. Teaching suturing and knot-tying skills to medical students: a randomized controlled study comparing computer-based video instruction and (concurrent and summary) expert feedback. Surgery. 2007;141(4):442–9.
43. Walsh CM, Ling SC, Wang CS, Carnahan H. Concurrent versus terminal feedback: it may be better to wait. Acad Med. 2009;84(10 Suppl):S54–7.
44. Charalambous CP. Procedural skills. In career skills for doctors. Switzerland: Springer international publishing; 2015. p. 109–18.
45. Regenbogen SE, Greenberg CC, Studdert DM, Lipsitz SR, Zinner MJ, Gawande AA. Patterns of technical error among surgical malpractice claims: an analysis of strategies to prevent injury to surgical patients. Ann Surg. 2007;246(5):705–11.

Chapter 6
Situation Awareness in Surgery

Surgeons practise under constantly changing conditions that require fast and accurate decisions, correct and precise actions. In achieving appropriate decision making, and desirable performance, it is essential to initially understand the context in which one operates.

Situation awareness may be described as the ability to read the situation, the ability to see the bigger picture.

Understanding the process of situation awareness and recognising the factors that may assist or impair reaching that state are important skills to develop. This chapter describes the levels and components of situation awareness, looks at how situational awareness may fail, and considers measures for improving situation awareness. Ways of identifying lapses in situation awareness are also presented.

C. Panayiotou Charalambous, *Career Skills for Surgeons*,
DOI 10.1007/978-3-319-57490-5_6

I keep pressing the button but is not doing much. Put it to maximum.
DIATHERMY

6.1 Levels of Situation Awareness

Anyone who isn't confused really doesn't understand the situation.

Edward R. Murrow [1]

Dr. Mica Endsley, an engineer and former Chief Scientist of the United States Air Force, published extensively in the 1990s on situation awareness (SA) [2–7]. SA was described as a "state of knowledge that individuals operating in demanding and dynamic situations should aim to achieve".

Endsley defined SA in the form of three levels [2–7] (Fig. 6.1):

- Level 1—Perception of elements in the environment
- Level 2—Comprehension of the situation
- Level 3—Projection of future status

Level 1—Perception refers to seeing or noticing all the information/data relevant to the task in hand. Such information may be perceived by visual, auditory, tactile, or even olfactory means. One may be presented only with the relevant information, but often one may have to sieve information to obtain the relevant parts.

Level 2—Comprehension refers to combining all relevant information, to help understand the overall situation; putting together all available pieces of the jigsaw to see the overall picture. This requires having predetermined mental models or patterns of potential situations, and selecting the one appropriate for the information received; like having a library of jigsaw puzzle pictures and based on the pieces gathered, to work out which puzzle picture it is likely to be.

Level 3—Projection refers to considering the various pieces of information gathered, along with the understanding of what they all together mean, to help predict what may happen in the future. It requires thorough understanding of the various situations that may be encountered, and knowledge as to where these situations could lead.

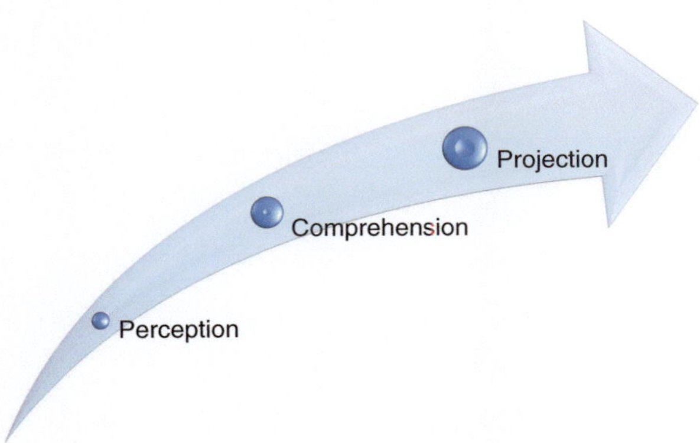

Fig. 6.1 Levels of situations awareness. Based on Endsley [2]

Hence, SA refers to gathering information, and based on the information gathered recognising as to what you are dealing with; once you recognise what you are dealing with, predicting what will likely happen next (in the immediate or near future) based on knowledge about such situations and experience of previous similar encounters.

The concept of SA refers mainly to dynamic situations [2], such as those encountered in the surgical theatre, whereby the pieces of data keep changing or some data only become apparent with the passage of time, requiring constant re-evaluation and understanding of the new picture. Hence, SA is often not acquired instantaneously, but over a period of time as more and more pieces of the jigsaw are added. The dynamics of the situation may also mean that time pressures exist for establishing full awareness, to allow prompt decision making and initiation of action [2–7]. If the required actions are delayed then such responses may cease to be relevant to the current situation.

6.2 Components of Situation Awareness

Harwood et al. [8] described four domains of situation awareness:

1. Spatial awareness (where)—knowledge of one's location in space and spatial relationships between components
2. Temporal awareness (when)—knowledge of dynamics, sequence and evolution of events, time availability
3. Responsibility awareness (who)—knowledge of who is in charge for what (including operator vs. automated system)
4. Identity awareness (what)—knowledge of the presence of threats and operating system variables

In performing surgery you may need to maintain:

- Spatial awareness—in choosing the correct dissecting planes, identifying the correct anatomical structures, retracting carefully without stretching the underlying nerves, drilling without lacerating the vessel under the bone.
- Temporal awareness—in knowing the time left for the task, how more to go before the cement sets, how more to go before the tourniquet has to be deflated.
- Responsibility awareness—how are the roles distributed, who is looking after the nerve, who is retracting, does the surgeon set the fluid pressure settings or are they automatically adjusted by the fluid management pump?
- Identity awareness—recognising the threats, in knowing that the appendix is gangrenous and it may burst whilst dissected, the hip fracture configuration makes it unstable and may spin whilst inserting the screw, that the nerve may be stretched if extensively retracted. It may also involve recognising the system variables, the diathermy readings, the significance of the laparoscopy console displays, the configuration of the screw head holes on the fixation locking plates.

6.3 Situation Awareness Versus Decision Making

You can't do anything sensible until you know what the situation is that you're in.

Julian Assange [1]

SA can influence decision making and performance, but is different from either of those [2]. Once you work out what you are dealing with, and what is likely to happen (that is once you become aware of the situation) you then go on to decide what to do.

If your understanding of what you are dealing with is correct, and your prediction as to where it could lead you is right, then you may make the appropriate decision for the situation encountered. However, it is also possible for your SA to be correct and still make the wrong decision, if the decision making process is impaired.

Alternatively if your understanding of what you are dealing with is wrong, or your prediction as to where it could lead you is incorrect, then you may make the right decision for the situation you think you are dealing with, but this would be the wrong decision for the real situation you are actually facing (but which you misread). Even the best decision makers may make the wrong decisions if faced with an incomplete, inaccurate, or wrong SA. Hence, achieving correct SA is the first step in appropriate decision making and desirable performance (Fig. 6.2).

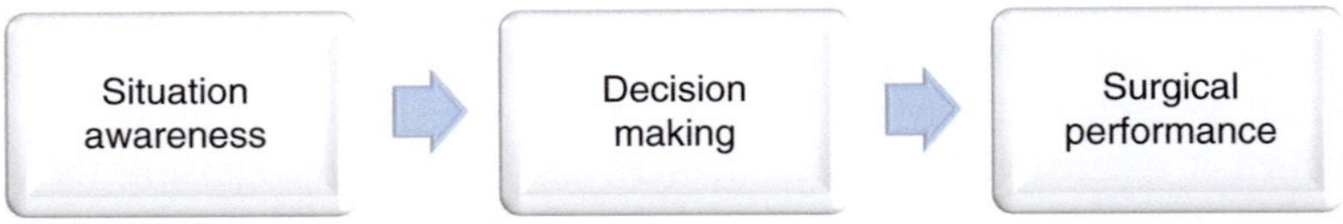

Fig. 6.2 Relation of situation awareness to decision making and performance

6.4 Barriers to Situation Awareness

I am very short-sighted, and if I don't like a situation I take my glasses off.

Jenny Éclair [1]

Failure to acquire the correct SA may occur at one or multiple levels, due to various factors (Fig. 6.3) [7].

In level 1 the necessary information may simply not be available (it doesn't exist, is not captured or is not presented), you may fail to see it (you are not paying attention, the monitor is not user friendly, you are distracted by other stuff, you are fixated on another task, you are bombarded with tons of information and you cannot see it all), you may see it wrongly (the measurement units are not stated and your assumption is wrong, displays are unclear), or see it but forget about it (you are stressed, physically or emotionally exhausted).

In level 2 you may not be able to combine all pieces of information together into a pattern, you may not be aware of the correct pattern, you may pick the correct pattern but not fully understand it, or you may pick an incorrect pattern; you may lack knowledge or experience to allow you to develop a correct pattern, you may have lots to do and something has to give, or you may prioritise another task.

In level 3 you may not have recognised the true situation, you may lack knowledge as to where things are heading or have a wrong understanding as to where things are going; you may lack experience or knowledge, or faced with too many tasks.

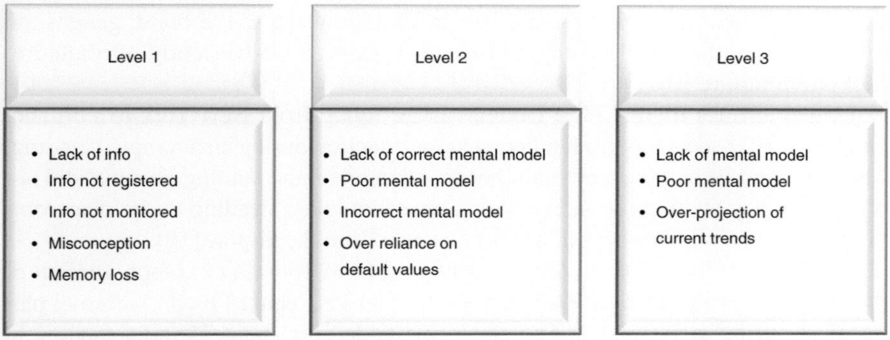

Fig. 6.3 Potential causes of failure of situation awareness at each level. Based on Endsley [2]

6.4.1 Failure of Situation Awareness in Aviation

It's not the situation... It's your reaction to the situation.

Robert Conklin [2]

Situation awareness has been extensively studied in aviation, from where we may draw examples as how things could go wrong.

Planes need fuel to fly, and a pilot must ensure they have enough fuel for the journey ahead. Fuel requirements depend on the weight to be transported, the distance to be travelled, the front or rear winds, parameters which may be changing during a journey. The pilot is required to be aware as to what the fuel situation is and predict as to whether it will suffice. One would think that having adequate fuel for a flight duration is a given fact, and that running out of fuel would be a non-encountered event. Yet, there are multiple reports of flights running out of fuel, some of which with devastating consequences. Failure of SA may have contributed to some of these events as described below.

Air Canada Flight 143, a Boeing 767 flying from Montreal to Edmonton, ran out of fuel whilst at about 35,000 ft half way through its journey. The experienced captain was able to glide and safely land the plane at Gimli, a closed air-force base whose decommissioned runway was used as a motor sports race track. The subsequent investigation found that the plane had been refuelled with about a quarter the amount of fuel necessary, due to wrong fuel calculations (a result of a mix up between the newly introduced metric system and the previously used imperial unit system meaning that fuel was calculated in pounds rather than kilograms). In this situation, valuable information about the actual fuel on board was not available due to malfunctioning of the cockpit fuel gauges. Unaware of the true fuel on board, and not having live information due to the blank gauges, an initial alarm indicating low engine fuel pressure was inadvertently attributed to fuel pump failure [9].

United airlines flight 173, a Douglas DC8, flying from New York to Portland, Oregon in 1978, developed a landing gear malfunction during an attempted landing. The crew, and its experienced pilot, became fixated on the landing gear and did not effectively pay attention to information about fuel levels, leading to the plane running out of fuel and crashing at a short distance from the airport [10].

Air Transat Flight 236, an Airbus 330 flying from Toronto to Lisbon, run out of fuel whilst over the Atlantic Ocean, due to a fuel leak caused by an incorrect part installed during maintenance. The plane was glided for about 75 miles landing in Azores. However, the situation was initially misread, and although initial cockpit warnings about a lower than expected fuel quantity were recognised, they did not lead to initiation of the "FUEL LEAK procedure, until later in the flight". The crew "continued to believe that the low quantity indications were caused by some type of computer error" and continued with this mental model up to and even beyond one of the engines flaming out [11].

6.4.2 *Failure of Situation Awareness in Vehicle Driving*

It is not only pilots that have to be aware of the fuel situation. Whenever in our car we consider the fuel in the tank and the distance we have to travel, to predict whether we need to refuel or not.

We take into account as to whether we are on the M6, one of the major motorways in England with frequent petrol stations on the way, or on the Route 23 in Chile's Atacama desert. Similarly, we may take into account as to whether we are driving our own car with its predictable fuel consumption and reliable fuel indicator, or a rental car with a shaky fuel gauze and uncertain fuel economy.

But on occasions we may get it wrong. The fuel indicator may be broken giving no reading. We may not look at it, or we may look at it but forget about its low reading after getting pre-occupied with the possible cause of the traffic jam we have just hit. We may notice the fuel indicator going down rapidly but attribute this to the aged engine, rather than recognising that the fuel pipe has burst. The low fuel warning light may have just come on but we may consider there is still a large reserve. Hence, we may wrongly predict that the fuel will last till our destination.

Driving is also a dynamic situation with events changing fast. We may plan to refuel at the next station, but it may have shut down, or we may simply pass it whilst daydreaming with no near exit on the motorway ahead.

6.4.3 *Failure of Situation Awareness in Surgery*

In surgery, one must keep re-evaluating the patient's intraoperative haemodynamic status and haemoglobin level. But how could you mis-read the low blood pressure occurring whilst plating a femoral shaft fracture, not considering the possibility of excessive bleeding and ending up with low haemoglobin and haemodynamic collapse?

Level 1—you may lack information of the low pre-op haemoglobin level because it wasn't checked, you did not look at the results, the measurement was wrong. You may lack information on the current haemoglobin level because it's not monitored regularly during surgery, or is checked on a blood gas sample which gives an inaccurate reading. You may be fixated on trying to reduce the fracture (which is very unstable and keeps slipping) and not notice the constant dribbling of blood from the bone ends, your assistant may be struggling to retract to help you see better and forgets to mention the red pool on the floor, the scrub nurse may be looking for a bigger bone holder to help your reduction and puts the used swabs on the site without considering how damp they are, the anaesthetist may be separated from the surgical field by drapes, concentrated on the new anaesthetic machine monitor.

Level 2—The blood pressure may start dropping but the anaesthetist is junior and puts this down to a spinal effect. The anaesthetist mentions the drop in blood pressure but your assistant feels it's the blood pressure tablets the patient took this morning starting to kick in, whilst you feel that it's probably small fat emboli due to moving the fracture about, in trying to achieve reduction.

Level 3—Hence you all predict that the situation will soon stabilise.

….and you decide to administer intravenous fluids and observe the response, leading to further haemodilution and further drop in haemoglobin.

In literature there are multiple reports of surgery going dramatically wrong, with the surgeon operating upon or removing the wrong structure. There are published reports of harvesting the median nerve rather than the palmaris longus tendon, excising the scaphoid rather than the trapezium in treating thumb arthritis, ligating and dividing the hepatic ducts rather than the cystic duct in laparoscopic cholecystectomy [12–15]. One may question how it is possible to go so wrong, and the answer is probably multifactorial. Although it would be difficult to know the exact reasons for the reported situations, one may postulate that loss of situation awareness could lead to similar harmful events.

Dekker and Hugh [15] explored the factors leading to bile duct injury during laparoscopic surgery, encountered even amongst experienced surgeons. They reported that duct mis-identification is the cause of most cases of such injury; they postulate that construction of a wrong mental image that convinces the surgeon that the common bile duct or the hepatic duct is the cystic duct has a major role to play. They suggest that visual perception is not necessarily a replica of what is truly seen, but a mental model based partly on what one expects to see. Wrong perception can be perpetuated, whereby new information or cues suggesting that one is dealing with the wrong structure are dismissed. The authors state that.

> *"once surgeons have committed them-selves to a particular course of action they will build an explanation that justifies that action and that also guides further action. This explanation tends to persist and gets transformed into an assumption".*

You may assess a patient presenting with a hip injury after a fall. They complain of hip pain, have limited painful hip movements, you feel you are encountering a neck of femur fracture. You look at the hip radiograph, you identify a possible hair-crack line and, although there is a similar artefact line on the opposite hip, you consider this to be an un-displaced neck of femur fracture, the more you look at it the more you get convinced, it fits the story. Based on your working mental model, you decide to proceed with screw fixation of the fracture, only for the patient to continue with post-operative pain. You initially attribute the ongoing pain to recovering from a fracture, but when eventually you arrange an MRI scan you find no femoral neck fracture but identify an acetabular fracture, not evident on the initial plain radiographs.

6.5 Improving Situation Awareness

Let us train our minds to desire what the situation demands.

Lucius Annaeus Seneca [1]

Given the ways in which SA could fail, we may look at measures to try to improve our ability of being situation aware during surgery. These are described next.

6.5.1 Improve Information Gathering

- Gain familiarity with your operating environment so you know where to find relevant information.
- Aim for user friendly information sources such as monitors, consoles, displays.
- Arrange your working environment to facilitate access to information—
  - Placing monitors or image intensifier screens so they can be easily seen.
  - Raising alarm volume of devices so they are easily heard in the noisy theatre environment.
- Distribute attention, so that you keep looking at the whole picture, avoiding narrowing all your attention to one task and ignoring others.
  - You may develop a system where by you concentrate at the task at hand, but at regular intervals you consciously aim to scan all other parameters—haemodynamics, blood loss, tourniquet time.
  - Share the role of attention; if all are staring at the fracture ends who is looking at the rest of the patient? Who is looking after the nerve whilst you are freshening the bone ends?

6.5.2 Aim for Manageable Workload

Workload may impair the ability of SA both during routine but also unexpected events [2, 16].

Consider a situation whereby a pilot has to deal with the auto-pilot dis-engaging in a bad bout of turbulence, and has to keep scanning the cockpit displays for information to help understand what is going on, remembering what the indicators say, co-ordinating the activities of the co-pilot, communicating with the air traffic control for a possible altitude climb, informing the cabin crew to fasten the seat belts, deciding what measures to take, whilst at the same time keep flying the plane.

Similarly, consider a situation whereby a surgeon faces a rush of blood whilst trying to fix a pelvic fracture, and has to minimise the bleeding whilst directing the junior assistant to retract and use the suction effectively, communicating with the anaesthetist about the blood loss and need of transfusion, scanning the diathermy displays to see why it is not working, directing the theatre coordinator to contact the vascular surgeons, recalling the proximity of the sacral vessels to the fracture on the pre-operative CT scan, whilst trying to develop a mental model as to where the bleeding is coming from.

The level of workload may be reduced by:

- Reducing systems' complexity—user-friendly equipment, eliminating unnecessary steps of surgical procedures.
- Pre-planning—pre-operative planning, minimising the steps that will need careful thoughtful consideration.
- Matching staff to workload—having more experienced assistants or more assistants in complex cases.
- Distributing workload—assigning specific roles to specific team members, such as asking an assistant to keep an eye on the tourniquet time, the scrub nurse to inform you of the cement setting clock during arthroplasty surgery.

6.5.3 Improve Memory Utilisation

A great memory does not make a mind, any more than a dictionary is a piece of literature.

John Henry Newman [1]

Appreciate the influence of memory on SA, and hence aim to maximise memory utilisation for effective SA. Human memory can be divided into:

- Working memory (short memory)—utilised in the conscious awareness and processing of information
- Long term memory—the storage of information and skills gained over the years. In relation to SA, long term memory may store mental models or patterns along with their projections, which are then identified and retrieved from long term memory and transferred to working memory [2]

Working memory has limited storage capacity and when it gets full an individual may not be able to process new incoming information. Hence, short term memory space should not be wasted but used effectively to facilitate SA by:

- Focussing on the essential inputs.
- Cutting down on background noise (turning the music off).
- Refusing interruptions (reassuring your assistant that you will answer their constant, knowledge gathering, questions at the end of surgery).
- Gaining experience so that more and more tasks become familiar and automatic (thus occupying a smaller amount of your working memory).

St-Martin et al. [17] describe the need to be able to slow down during a surgical procedure and move from an automatic mode to a more focused, effortful mode. The authors describe that slowing down may be pre-planned in anticipation of challenging steps of the procedure or in dealing with a challenging case. Alternatively, slowing down may be in response to the encounter of an unanticipated event. Slowing down may involve:

- Stopping—to re-assess or inform the team of any critical steps, to ensure that all is ready for the critical step.
- Removing distractions.
- Focusing intensely—withdraw from conversations, stop teaching.

With regards to long term memory you may aim through knowledge and experience to gain a large bank of mental models or patterns relevant to your practise, which can then be retrieved as required. Zheng et al. [16] evaluated the role of mental workload on performing secondary tasks (such as monitoring activities in the surgical theatre) other than the main action surgeons worked on. Surgeons were asked to perform as many laparoscopic sutures as possible within 6 min, whilst an adjacent monitor displayed a random sequence of true and false visual signals. Surgeons were asked to identify the true visual signals whilst they kept suturing. The rate of suture completion and correct identification of true visual signals was higher amongst experienced surgeons as compared to novices. They postulated that practise and experience leads to automaticity which in turn reduces the mental workload needed for suturing, hence freeing mental resources to attend to a secondary task.

6.5.4 *Minimise the Effect of Stress on SA*

I'm picking and choosing in terms of the stress factor. If it's not fun, I'm not going to do it.

Anita Baker [1]

Situations which individuals perceive as stressful may reduce SA by narrowing attention to few central sources of information (cognitive tunnel vision), and also by reducing the capacity of working memory [2, 18, 19]. The effects of stress on SA may be reduced by:

- Simulating for difficult situations—anticipating what may go wrong and pre-enacting your behaviours may help minimise stress when such conditions are encountered
- Recognising the potential adverse effects of stress on SA—to develop and pre-plan coping mechanisms to help maintain SA when faced with stressful circumstances (such as actively scanning for all information rather than fixating on one task)
- Minimise stressors—if you find the actions of a junior assistant (who keeps putting their hands in your way) very unsettling, the actions of a scrub nurse (that starts shouting every time you ask for an instrument) very disruptive, or the action of a radiographer (who keeps moving the image intensifier in a purposeless way every time you ask them to screen) very un-easing, should you be starting a complex case with them?
- Develop techniques to improve resilience to stress, and help maintain composure and self-control when unexpected situations arise. Appreciate that loss of either is more likely to make matters worse rather than help a difficult situation.

6.6 Team Situation Awareness

I get nervous watching teammates. I get nervous for them. Late in the game, pressure situation, I'm nervous for them.

Derek Jeter [1]

A special situation of SA is team SA [2]. Complex dynamic situations, such as surgical operating theatres, involve the close interaction between various team members, surgeons, anaesthetists, scrub nurses, perfusionists, radiographers. Such team members have different responsibilities, parallel roles and individual sub-goals, but are also working towards a shared task, a shared overall goal, the team's goal.

Team SA is essential for the performance of the team as a whole, and has been described as the "degree to which every team member possesses the SA required for his or her responsibilities" [2]. This refers to each team member having the necessary SA for achieving their own sub-goal. By meeting their own sub-goals team members contribute to the team's overall goal [2, 20].

However, the sub-goals of some team members may partly overlap, hence part of their SA may also overlap (shared) to facilitate these common requirements. It should be noted that this is sharing SA, rather than giving it completely away, as that could compromise their own sub-goal, making them the weak link in the chain (Fig. 6.4).

The surgeon and anaesthetist are part of the same team and are working towards a common overall goal, but it is essential that some data or information are received by both in order to allow them to perform their individual sub-goals; just the

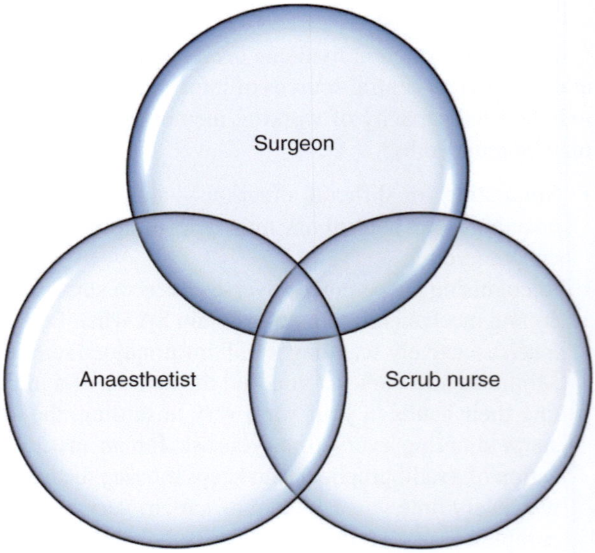

Fig. 6.4 Situation awareness sharing by the surgical team members—part overlaps between team members, part is limited to individual team members. Adapted from Endsley [2]

surgeon or the anaesthetist knowing may not be enough. Both may need to be aware of the case's previous medical history of cerebrovascular disease which may limit the ability to drop the intra-operative blood pressure to achieve a bloodless field, or both may need to be aware that the gall bladder looks gangrenous, and the laparoscopic attempt may need to be converted to an open laparotomy half way through.

Hence, team coordination involves SA sharing which may be achieved by exchange of information between team members, or by team members receiving information from a common source (such as displays accessible to all). The team needs to recognise what information needs to be shared, and how this can be achieved.

Information may be shared by explicit communications during pre-operative team briefs or intra-operatively, but other less direct means of communication are also possible. Gillespie et al. [20] looked at the effects of various aspects of communication on shared SA between surgical theatre team members. They reported that strategies that enhanced shared SA included:

- "Self-talk"—the surgeon talking though critical steps of the operation, telling the rest of the team what is doing, and what care needs to be taken.
- Closed loop communication—repeating the message received, to confirm that it has been understood or that the required action has been performed.
- "Overhearing conversations"—between other team members in close proximity, to help one stay in the loop, and coordinate their actions according to the need of the situation; such as the scrub nurse overhearing the surgeon and assistant discussing how complex the case is, or the surgeon overhearing discussions between scrub nurses about staff shortages that may influence the smooth running of the list.

If you closely pay attention next time you are in theatre you may notice the scrub nurse trying to hear what you are telling your assistant or senior surgeon, trying to work out how the operation is going, and then communicating that to the fellow scrub staff. Alternatively, you may notice your-self "thinking aloud", trying to let others know as to what you are doing.

Sharing of communication can directly impact patient outcomes. Mazzocco et al. [21] examined the relationship between certain team behaviours and patient outcomes, and reported a greater risk of complications or death when encountering less information sharing intra-operatively and at handovers. Along similar lines, Siu et al. [22], in an observational study, examined the behaviours associated with intra-operative incidents in general and vascular surgery at the Royal Infirmary of Edinburgh. Poor SA, poor communication and poor teamwork across the whole surgical theatre team were closely related to such intra-operative events.

Aim for functional communication, functional SA sharing.

6.7 Recognising Loss of Situation Awareness

Exploring how you could make a bad situation worse can sometimes tell you what not to do.
Harvey Mackay [1]

The ability to pick cues that your SA may be incorrect, is an important skill to develop. You may consider that your impression of the big picture is wrong if:

- Getting conflicting messages—
  - "The bone must be very weak to have broken so easily but it feels really hard to drill through".
  - "The diathermy is set to maximum, the suction is working, but it is not coagulating the vessel".
- Pieces of information do not fit to a recognisable pattern—
  - "Does not look like a tendon—can't follow it through to a muscle".
  - "The nerve is passing vertically rather than crossing horizontally –is it the main trunk or an unusually large branch?".
- You feel confused as to what is happening—
  - "Does not make sense, it looks like bursa rather than tendon but it merges with the muscle".
  - "Not sure what dissecting plane I am in".
- All teams members seem to be concentrating on one task.

If the received information does not fit with your expectations, consider the possibility that a different situation may exist to the one that you assumed and try to:

- Re-evaluate the information received—what else could it be?
  - "Is it possible this is the medial knee ligament rather than the hamstring tendon?"
  - "Could the drop in haemoglobin be due to taking blood from the cannulated limb with fluids going in?"
- Look for more information with an open mind—
  - "Let's screen with a radiograph to ensure this is the correct fracture level".
- Avoid trying to make the information fit the model, concentrating only on what fits with your assumption, and ignoring or disputing the rest. Recognise that once an interpretation of the situation is made and a mental model is established there may be a tendency to look for and preferentially select information that supports that mental model (confirmation bias) [23, 24]. Go back to the basics and reconsider [2, 7].

- Review the existing model and adjust it according to what may be the current true situation. Go back to the beginning, start again from a definitely accurate stage, and re-evaluate what you are facing:
  - "The hip hemi-arthroplasty is not reducing; but we did a reduction trial, the stem is still at the correct height, no swabs or soft tissue in the acetabulum, could it be that the spinal is wearing and need some muscle relaxation?".
- Adopt a new mental model.
  - "This is not a gangrenous appendix, we are dealing with a perforated caecum".
  - "This is not an isolated fracture of the capitellum, we are dealing with a fracture involving the whole of the distal humeral articulation".
- Change the initial goals to allow for the new situation
  - "The bone is too weak to fix the fracture, lets replace the hip".
  - "The anastomosis is poorly perfused, let's convert to a stoma".
  - "The wound is too tight to close, let's leave it open and try again in 48 h".

References

1. Famous Quotes at BrainyQuote. https://www.brainyquote.com/quotes/quotes. Accessed 31 Mar 2017.
2. Endsley MR. Towards a theory of situation awareness in dynamic systems. Hum Factors. 1995;37(1):32–64.
3. Endsley MR. Measurement of situation awareness in dynamic systems. Hum Factors. 1995;37(1):65–84.
4. Endsley MR, Kaber DB. Level of automation effects on performance, situation awareness and workload in a dynamic control task. Ergonomics. 1999;42(3):462–92.
5. Endsley MR. Automation and situation awareness. In: Parasuraman R, Mouloua M, editors. Automation and human performance: theory and applications. Mahwah, NJ: Lawrence Eribaum; 1996. p. 163–81.
6. Endsley MR. Situation awareness in aviation systems. In: Garland DJ, Wise JA, Hopkins VD, editors. Handbook of aviation human factors. Mahwah, NJ: Lawrence Erlbaum Associates; 1999.
7. Jones DG, Endsley MR. Sources of situation awareness errors in aviation. Aviat Space Environ Med. 1996 Jun;67(6):507–12.
8. Harwood K, Barnett B, Wickens CD. Situational awareness: a conceptual and methodological framework. In: McIntire FE, editor. Proceedings of the 11th biennial psychology in the department of defense symposium. Colorado Springs, CO: U.S. Air Force Academy; 1988. p. 23–7.
9. Final report of the Board of Inquiry investigating the circumstances of an accident involving the Air Canada Boeing 767 aircraft C-GAUN that effected an emergency landing at Gimli, Manitoba on the 23rd day of July, 1983/Commissioner, George H. Lockwood. Canadian Government publishing centre. 1985. http://data2.collectionscanada.gc.ca/e/e444/e011083519.pdf. Accessed 6 Apr 2017.
10. United Airlines Inc., McDonnell-Douglas, DC-8-61, Portland, Oregon, December 1978. National transportation safety Board, Report, 1979. http://lessonslearned.faa.gov/UAL173/AAR79-07.pdf. Accessed 1 Apr 2017.
11. Accident Investigation Final Report. Aviation Accidents Prevention and Investigation Department. Government of Portugal MINISTÉRIO DAS OBRAS PÚBLICAS, TRANSPORTES E COMUNICAÇÕES. 2004. https://www.fss.aero/accident-reports/dvdfiles/PT/2001-08-24-PT.pdf. Accessed 1 Apr 2017.
12. Marcuzzi A, Lando M, Sartini S, Petit A. Scaphoidectomy (instead of trapezectomy) in the treatment of rhizoarthrosis. Report of a case. Chir Main. 2012;31(1):30–3.
13. Choo J, Wilhelmi BJ, Kasdan ML. Iatrogenic injury to the median nerve during palmaris longus harvest: an overview of safe harvesting techniques. Hand (N Y). 2017;12(1):NP6–9.
14. Weber RV, Mackinnon SE. Median nerve mistaken for palmaris longus tendon: restoration of function with sensory nerve transfers. Hand (N Y). 2007;2(1):1–4.
15. Dekker SW, Hugh TB. Laparoscopic bile duct injury: understanding the psychology and heuristics of the error. ANZ J Surg. 2008;78(12):1109–14.
16. Zheng B, Cassera MA, Martinec DV, Spaun GO, Swanström LL. Measuring mental workload during the performance of advanced laparoscopic tasks. Surg Endosc. 2010;24(1):45–50.
17. St-Martin L, Patel P, Gallinger J, Moulton CA. Teaching the slowing-down moments of operative judgment. Surg Clin North Am. 2012;92(1):125–35.
18. Gillis JS. Effects of life stress and dysphoria on complex judgments. Psychol Rep. 1993;72:1355–63.
19. Kontogiannis T, Kossiavelou Z. Stress and team performance: principles and challenges for intelligent decision aids'. Saf Sci. 1999;33(3):103–28.
20. Gillespie BM, Gwinner K, Fairweather N, Chaboyer W. Building shared situational awareness in surgery through distributed dialog. J Multidiscip Healthc. 2013;6:109–18.
21. Mazzocco K, Petitti DB, Fong KT, Bonacum D, Brookey J, Graham S, Lasky RE, Sexton JB, Thomas EJ. Surgical team behaviours and patient outcomes. Am J Surg. 2009;197(5):678–85.

22. Siu J, Maran N, Paterson-Brown S. Observation of behavioural markers of non-technical skills in the operating room and their relationship to intra-operative incidents. Surgeon. 2016;14(3):119–28.
23. Croskerry P, Singhal G, Mamede S. Cognitive debiasing 1: origins of bias and theory of debiasing. BMJ Qual Saf. 2013;22(Suppl 2):ii58–64.
24. Pines JM. Profiles in patient safety: confirmation bias in emergency medicine. Acad Emerg Med. 2006;13(1):90–4.

Chapter 7
Decision Making in Surgery

As surgeons we are often faced with a constant need for making decisions; decisions in the surgical theatre, the clinic, the ward, in consultations, in meetings. We are faced with situations where fast decisions, immediate decisive actions are demanded, but also situations where all facts must be carefully considered so that balanced, well measured decisions are made. Decisions can influence our performance, our patients' outcomes, our relationships with colleagues, our professional standing.

Being aware of the process of our decision making, and understanding the biases and other factors that may influence our decisions are important skills to develop. They may help us achieve greater mental influence on our decision making, and enhance our ability to self-reflect, improve, learn from successes and errors.

This chapter discusses intuitive and analytical decision making, and explores the various decision biases that a surgeon may face. It also discusses some of the factors that may influence our decision to offer surgery, and considerations that may influence our decision making in the absence of strong scientific evidence. The ability of a surgeon to alter their decision once that is taken, is also discussed.

C. Panayiotou Charalambous, *Career Skills for Surgeons*,
DOI 10.1007/978-3-319-57490-5_7

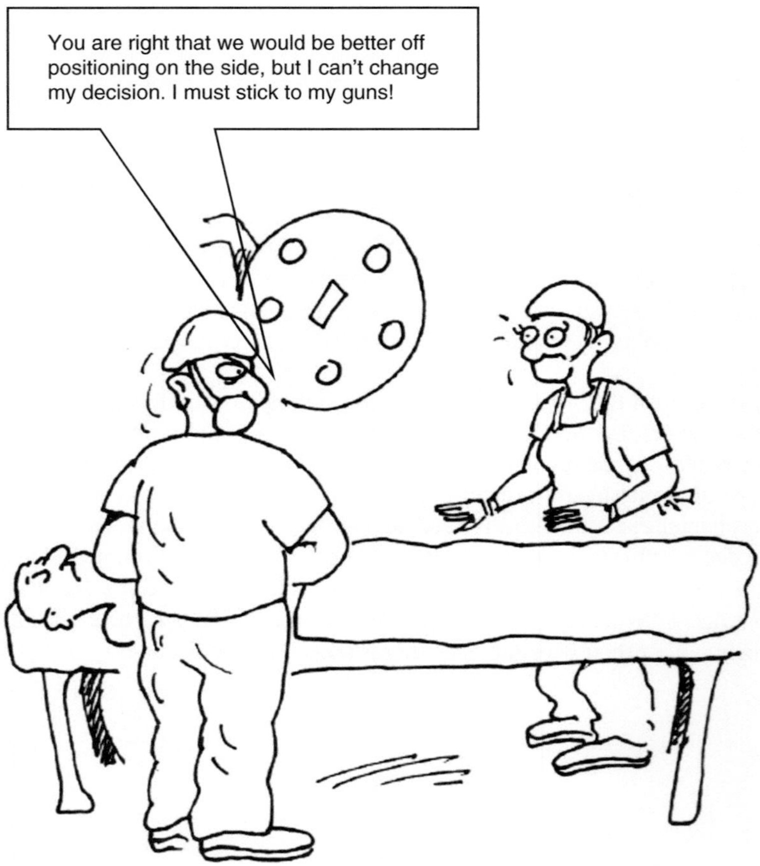
You are right that we would be better off positioning on the side, but I can't change my decision. I must stick to my guns!

7.1 The Spectrum of Surgical Decisions

Sometimes you make the right decision, sometimes you make the decision right.

Phillip C McGraw [1]

As a surgeon you may face multiple decisions in a normal working day. You may face multiple clinical decisions in formulating a working diagnosis, choosing the appropriate investigations, coming up with a management plan, whether and when to operate, how to position the patient on the operating table, selecting a surgical approach to get adequate exposure, the anatomical plane in which to dissect, whether to accept the less than perfect achieved fracture reduction, when to change course and convert the laparoscopic cholecystectomy to an open one, how to close the surgical wound, what dressing to apply, what restrictions to initiate post-surgery, when to discharge home, when to review in clinic.

At the same time you may have to decide about the choice and allocation of resources, how to organise the working environment and allocate available time. You may have to decide what assistance you need in theatre, what advice to give to a colleague asking for help, whom to consult for a further opinion.

You often have to make individual decisions, sometimes collective decisions in collaboration with colleagues, patients, relatives or other staff. You may have to decide on the spot, within tight deadlines, or you may have ample time for consideration.

Part of surgical training is to develop appropriate decision making, alongside skilful surgical techniques as on occasions decision making may be more important than the sleekest of surgical skills.

7.2 Types of Decision Making

I think there's a difference between a gamble and a calculated risk.

Edmund H. North [2]

Several decision making models have been described [3, 4] including:

- Intuitive—the problem is identified and the solution is chosen based on recognition of a pattern in which the problem fits. The solution may be based on procedures, or previous experience (of actions taken and the resultant outcomes).
- Analytical—the problem is identified, different options are compared, alternative solutions are weighted up, to carefully select the preferable one, to make a rational decision.
- Rule based—the problem is identified and rules that apply to the situation are recalled or looked up. This type of decision making requires awareness of the rules and also knowledge as to how to apply them.
- Creative—when faced with a rare problem, one comes up with a novel solution; make it up as you go along.

Decision making may be looked at as a broad spectrum process, ranging from the conscious and analytical based approach at one end, to the subconscious, intuitive, automatic approach at the other end (Fig. 7.1). Hence, at one end decision making is slow, thoughtful, requiring large mental concentration whereas at the other end it is fast and conscious-less, requiring minimal concentration. Somewhere in between these extremes is the rule based decision making.

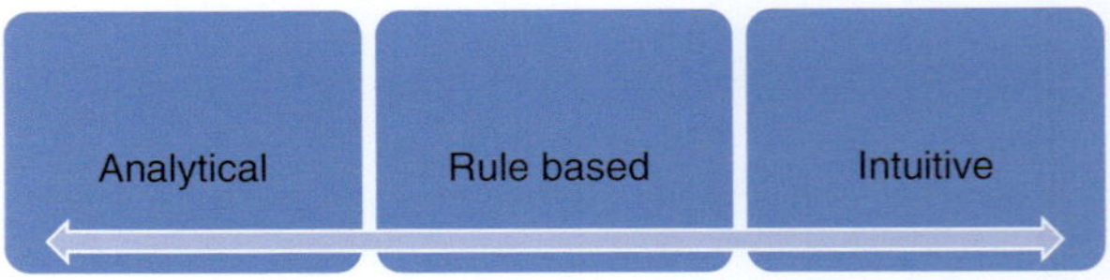

Fig. 7.1 Spectrum of decision making

7.3 Effect of Experience on the Choice of Decision Model

Good decisions come from experience. Experience comes from making bad decisions.

Mark Twain [5]

A problem that initially requires analytical decision making, may be solved by intuition as one develops experience and knowledge. Intuition may be considered pattern recognition, which doesn't consider rules or analyse deeply. Based on knowledge and previous experience, important cues are picked from the problem or situation, from the clinical history or examination findings to reach a decision.

Intuition occurs fast, at a subconscious level, but knowledge and experience are needed to allow this instant recognition. Hence, at the beginning of learning and training a novice surgeon may mainly utilise analytical processes, looking at each component in detail, analysing each in turn, and trying to determine how they relate and how they fit together.

However, as more cases are encountered, and one develops greater knowledge and expertise, and as the "mental database" of previously seen clinical presentations expands, there is increased utilisation of intuition, without a need to break down each problem into its sub-components. In this way, a repeated presentation of the same problem may help one to convert from analytical to intuitive decision making for that particular problem [4].

- At the beginning of learning, in dealing with a patient with a hip fracture, a surgeon may need to consider in detail the fracture pattern, previous medical history, pre-fracture mobility, and various treatment options in deciding which treatment to choose. In contrast, with more experience, one may be able to quickly decide how to treat the patient, by examining the radiograph and hearing the history summary.
- At the beginning of learning, in dealing with acute abdominal pain, a surgeon may consider in detail the nature, location, radiation, duration of pain, associated symptoms, previous medical history, haematological and other investigations to come up with a list of possible diagnoses. In contrast, with increasing experience one may be able to come up with an "end of bed" spot diagnosis utilising much less clinical information.

7.4 Switching Between Decision Models

If you do not change direction, you may end up where you are heading.

Gautama Buddha [1]

Under certain circumstances one may need to convert from intuitive to analytical decision making, and the ability to recognise the circumstances where this may be essential is an important skill to develop.

A component of intuition is to see when a situation does not fit with the expected, in order to allow one to consciously pause and think, that is to convert to more analytical decision making (Fig. 7.2). If what you see does not match what you expect, or if substantial pieces do not fit into the jigsaw you just put together, then you may need to reconsider. If what you see does not really fit into a pre-stored mental model or pattern, decision making may need to be slowed down, for a careful analysis to be performed [6].

- You may have just assessed a patient with a hip fracture, reviewed the radiographs, and be convinced this is just another case on the list for a Dynamic Hip Screw (DHS) fixation. But then, as you are writing your clerking notes, you recall the patient telling you of feeling pain in that hip for a few weeks prior to the fall; you suddenly stop and reconsider:
  - Could this be a pathological fracture?
  - Anything in previous medical history to suggest that?
  - Do we need to investigate further prior to surgery?

Hence, analytical decision making may override intuitive thinking, and allow change of course.

It may also be possible for one to allow intuitive decision making override the analytical process; one may not be able to hold back their "gut feeling" despite the fact that thorough analysis of the facts points to a different direction [6].

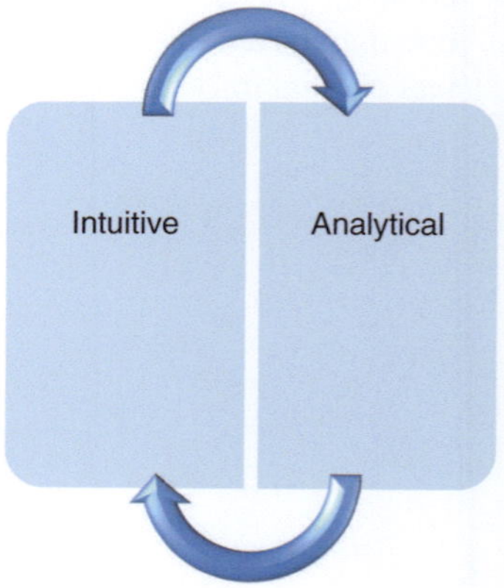

Fig. 7.2 Analytical and intuitive decision making may override each other

7.5 Biases in Decision Making

Decision making is an art only until the person understands the science.

Pearl Zhu [1]

In a similar way to how we evaluate our technical skills, one can evaluate their decisions and the process of reaching those decisions. Such evaluation includes appreciating that our decision making may not always be rational but on occasions may be subjected to biases. Such biases may be more likely to occur when we are in the intuitive mode, when decisions are made in the absence of detailed analysis.

Heuristics are mental shortcuts, or rules of thumbs, which allow fast decision making and avoid the need of deep, analytical, time consuming consideration when a problem is encountered. Heuristics are useful because they allow us to manage complex problems quickly, without devoting much time and "brain energy". However, heuristics may be more prone to biases and error [7–16]. Several decision making biases have been described [7–16] and you may recognise some of the following in surgical practise:

Status quo bias—choosing options that maintain the status quo:

- "Why convert to using oral anti-coagulants in thrombo-prophylaxis when deltaparin works pretty well?"

Aggregate bias—considering that information gathered from a group of patients may not apply to an individual. Surgeons may find reasons to justify that a specific patient is "atypical" hence it is appropriate to deviate from recognised guidelines:

- "It is a nasty open tibial fracture; better give antibiotics for a week rather than the recommended three doses".
- "The patient is mobilising outside her house independently, but is quite elderly. Let's perform a hemi-arthroplasty for her hip fracture rather than the total hip replacement that national guidelines would advice".

Commission bias—choosing action over inaction:

- "We can't leave it. We have to fix the fracture. How can we possibly leave it alone?"

Omission bias—choosing inaction over action to avoid taking responsibility for an adverse outcome.

- "Better see them back again in clinic in 3 months to make sure all is ok. Give them another follow up rather than discharging".

Ego bias—a tendency to overestimate the outcomes of own patients:

- "Not sure what they are talking about. My results are so much better".

Anchoring—the tendency to stick to specific features encountered early in the presentation of an illness, committing to a diagnosis, and resisting subsequent adjustment:

- "Long history of similar back pain, gone worse last few days…unlikely the scan will show up much".

Alternatively, a label may have been attached to a patient by another healthcare worker, by the General Practitioner or the triage nurse at the Accident and Emergency department, and this label continues to stay with that patient without much consideration or challenge. One may find it difficult to consider alternatives, to consider that the assigned label could be wrong:

- "Another referral of PR bleeding from haemorrhoids. Not sure why they keep referring them through".
- "Another case of hand numbness due to carpal tunnel syndrome, let's see if keen to have surgery".

Confirmation bias—findings or observations that appear to confirm a working hypothesis are given more weight, with less attention paid to evidence that could challenge these beliefs;

- "The knee aspirate is heavily blood stained…it looks like haematoma rather than infection…the raised CRP could be due to other causes…let's hold off antibiotics".

Search satisfying—one stops searching for any further abnormalities having been satisfied from finding the first; an orthopaedic surgeon having identified a cervical spine fracture may fail to look for and find any fractures in the rest of the spine or other associated injuries.

Availability bias—the prevalence of a disease is over estimated if it comes easily to mind. If one has recently encountered a disorder, read about it, been to a relevant lecture or course, then that disorder may be the first to be considered in the differential diagnosis next time a patient presenting with similar symptoms and signs is encountered, even though the diagnosis in question is unusual or rare;

- "Acute abdominal pain in a 30 year old; could be another hypercalcaemia and pituitary adenoma".

The opposite may also apply whereby one fails to consider a diagnosis simply because they have not come across it for a long time; one may consider that the 80 year old with back pain is another presentation of acute mechanical pain or cauda equina, but fail to consider the possibility of a ruptured aneurysm, as they have not encountered one before.

Hindsight bias—one looks back at an event and overestimates what was known at the time, thus falsely concluding that the event could have been predicted and appropriately managed. However, this may not be the case, as the context and others factors that could have influenced decision making or performance are not fully taken into account. This could lead to false reassurance that if a similar situation was encountered in the future, it would be appropriately managed. Thus one may fail to truly self-reflect and learn from a past failure.

- "How could we have delayed the diagnosis of appendicitis? It was obvious… there was iliac fossa tenderness and temperature".

But when considering decision making, with the benefit of hindsight, one must take into account the context in which such decisions were made. This may involve considering additional, conflicting information which was available, other cases dealt with at the same time, limitation of resources, the physical or mental exhaustion of any healthcare practitioners involved. This may allow a better evaluation of decision making and hence help development and improvement.

Outcome bias—tendency to judge the quality of a decision based on its outcome:

- "Maybe we should have left the fracture alone...it would have been acceptable to do so, and it would have avoided the infection".

Framing bias—when faced with decisions involving risks, one may consider potential gains and losses. However, in considering losses there may be a tendency to prefer a loss which has a lower chance of occurring but when it occurs it is more damaging, rather than a loss which will definitely happen but its effects are not as devastating;

- "The patient is diabetic, if we reduce and fix the ankle fracture it could get infected and in severe infections one could even lose the leg, but if we leave it alone it will heal in that bad position and cause long term pain and limping".

Recognising the existence of biases in decision making may help construct de-biasing defences, to help minimise their adverse effects.

7.6 Factors Influencing Surgical Decision Making

Ethical decisions ensure that everyone's best interests are protected. When in doubt, don't.
Harvey Mackay [2]

In surgical decision making, one would aim to make rational decisions, that above all aim to benefit the patient concerned. But even when it comes to conscious decisions, a surgeon may be faced with multiple, sometimes conflicting factors, that may guide their decision making process. Leung et al. [17] refer to three types of factors that may influence a surgeon's decision making:

1. Avowed factors—what is in the patient's best interests.
2. Un-avowed factors—that would benefit the healthcare system in general but not essentially the particular patient. These may be necessary to allow surgeons to address multiple priorities and commitments such as teaching, research, relations with colleagues. These may be include decisions such as:
   (a) when to allow a trainee to operate (despite knowing the operation will take longer).
   (b) when to invite a patient to participate in a surgical trial (despite having a strong preference for a particular treatment option).
   (c) when to rush through an operation to avoid cancelling the next case.
   (d) when to change a colleague's surgical management plan (knowing that could question their authority and impair relationships).
3. Dis-avowed factors—that would benefit the surgeon but not the patient such as monetary incentives or the need to maintain personal reputation. These may include deciding to perform surgical procedures which have limited clinical effectiveness because they pay well or because other surgeons do.

Recognising and accepting the decision making conflicts one may encounter, could aid self-reflection and help us appropriately adjust our decision making.

7.7 Decision Making in Offering Surgery

In any moment of decision, the best thing you can do is the right thing, the next best thing is the wrong thing, and the worst thing you can do is nothing.

Theodore Roosevelt [2]

The decision as to whether or not to recommend surgery is one that a surgeon regularly has to make. A challenge faced by surgeons is how to make reliable decisions in the absence of robust supporting evidence. In particular, how do you decide to offer surgery when it is not certain whether it will help, how do you decide between different treatment options?

Hageman et al. [18] surveyed 337 surgeons to determine how they make decisions when evidence is inconclusive. Firstly, they looked at decision making when there is no conclusive evidence as to whether treatment is better than allowing the disease to take its natural course. The most popular responses of surgeons were:

- "Works in my hands"
- "Familiarity with the treatment"
- "What my mentor taught me"

Secondly, they asked surgeons as to how they decide between two surgical treatments, when the evidence shows no difference between the two, and the most important factors reported were:

- "Fewer complications"
- "Quicker recovery"
- "Burns fewer bridges"
- "Works in my hands"
- "Familiarity with the procedure"

Some of the other factors reported by some respondents were "what others are doing", "patients are requesting the procedure", "highest reimbursement", "shorter procedure", and "cheapest/most resourceful".

Teunis et al. [19], presented 15 upper limb injuries (which could be treated surgically or non-surgically) to 223 surgeons and asked them whether they would recommend surgery. Participating surgeons also completed a test, designed to assess personality traits. The authors found that a higher score on the personality test (suggestive of innovation, creativity, and acceptance of higher risk levels) was related to higher rates of preference for surgery.

Janssen et al. [20] asked 254 orthopaedic, plastic and trauma surgeons whether they would recommend surgical or non-surgical treatment in each of 21 fictional cases presenting with upper limb disorders. Half of the surgeons were randomised to choose a treatment for themselves had they had the disorder in question, and half were randomised to choose a treatment for a patient of their age and gender presenting with that disorder. Surgeons were more likely to recommend surgery for patients rather than choosing surgery for themselves, but were more confident in deciding for own-selves than for their patients.

Recent events may also influence our decision making, a concept known as the recency effect. What happened recently to one patient may alter our decision process, guide us to one line of action, even when rational thinking points us in a different direction. Simianu et al. [21] evaluated the recency effect amongst colorectal surgeons. They evaluated the use of preventative leak testing in bowel anastomosis, prior to and after encountering operations with anastomotic leaks. They found that 36% of 25 surgeons showed a recency effect, increasing their leak testing, after encountering leaks where the anastomosis had not been tested.

Along similar lines Dale et al. [22] examined whether the decision making of vascular surgeons could be influenced by a negative previous experience. Surgeons were exposed either to an expanding balloon that eventually burst (considered as a bad outcome) or to a demonstration of an expanding balloon that didn't go on to burst (considered a good outcome). They were then evaluated on a simulator as to when they would opt out from an expanding abdominal aortic aneurysm situation and decide to operate. Despite evidence based guidelines as to when such aneurysms should be operated upon, the group that had been subjected to the bursting balloon were more likely to opt out and decide to intervene early. In accord with this Dale et al. [23] evaluated on a simulator, the decision of surgeons to operate on an abdominal aortic aneurysm. They showed that surgeons tended to proceed with surgery much earlier than evidence based guidelines would recommend, but in addition they demonstrated that previous exposure to a death during watchful waiting led to surgeons operating significantly sooner on a subsequent patient.

Szatmary et al. [24] investigated the factors affecting the decision of surgeons to operate in emergency surgery situations, through semi-structured interviews and consideration of clinical vignettes. Surgeons' decisions to operate were mainly influenced by perceived likely outcomes, deciding to operate when they perceived the outcome with surgery was better than without. Interestingly, surgeons with less than 5 years of experience were less certain of potential outcomes and were also more likely to decide to operate than more experienced surgeons.

In line with this finding (that suggests limited clinical confidence may lead to a greater tendency to operate) are the results of a national survey of American surgeons carried out by Wilson et al. [25].These authors evaluated the factors associated with surgeons' decision to operate in clinical scenarios where there was no strong indication for surgical or non-surgical treatment, and decision was upon the surgeon's discretion. They found that surgeons were less likely than their colleagues to recommend surgery for conditions that fell within their subspecialisation area, suggesting they were more confident in managing non-surgically conditions they had more experience with. Older surgeons were also less likely to offer surgery compared to younger ones.

Along similar lines Sacks et al. [26] presented to a national sample of 767 surgeons four clinical vignettes of cases of appendicitis, bowel obstruction, mesenteric ischaemia and gastro-interstinal bleed and asked them to judge the risks and benefits of surgery versus non-surgery and also state as to whether they would operate. They reported a huge variation in risk/benefit assessments of surgeons as well as in their decision to operate. Surgeons were more likely to operate as their perceptions of the benefits of surgery increased, and the risks of non-surgical treatment increased. However, differences in risk/benefit perceptions amongst surgeons accounted for only 39% of the variations reported in the decisions to operate. It is also of interest that in an associated study [27] the authors demonstrated that exposure of surgeons to a risk calculator improved the assessment of risks but did not influence the likelihood of recommending surgery.

Rutkow et al. [28] evaluated the reliability of surgical decision making in offering surgery for seven elective procedures (breast surgery, varicose vein, cholecystectomy, hysterectomy, tonsillectomy and adenoidectomy, cataract extraction, prostatectomy). They presented clinical scenarios of patients seeking consultation and determined the agreement in offering surgery, amongst surgeons in Maryland and Columbia in the USA. They reported a wide variation amongst surgeons in their decision to operate, with 12 of the 19 test cases examined showing less than 80% agreement. In a follow up study [29] surgeons were again asked (2 years later) as to whether they would offer surgery in those scenarios, and their responses were compared to those given initially. A large intra-observer variation was noted suggesting that even a surgeon's own judgement may differ with time.

As surgeons we seem to make important treatment decisions based on personal preferences, previous experiences, surgical knowledge and training, and may also be influenced by our personality characteristics. Appreciating the factors that may guide your decision making could help recognise the uncertainty that often exists in what you do, and the limitations of the decisions you make. Hence, you can be more forgiving to those who make different decisions to yours, and more forgiving to own-self when faced with the consequences of the decisions you once made.

7.8 Emotions and Decision Making

The essential difference between emotion and reason is that emotion leads to action while reason leads to conclusions.

Donald Calne [30]

Decision making and clinical judgement may be susceptible to emotional influences. Several factors may influence a doctor's emotional state such as patient characteristics, environmental conditions, physical and psychological fitness of the doctor themselves [11].

Doctors may develop emotional biases towards their patients, which may guide their treatment approach. Such biases include:

- Countertransference—the situation where past feelings of a doctor define the present relationship with a patient. Certain characteristics of a patient such as appearance or behaviour may remind the doctor of similar patients encountered in the past, creating a biased response and consequently biased decision making [11]. James Groves, of the Psychiatry Service in the Massachusetts General Hospital, wrote in the 1970s of patient characteristics that produced negative emotions to the treating doctor—the "dependent clingers, entitled demanders, manipulative help-rejecters and self-destructive deniers" [31].
- Fundamental attribution bias—the situation where an event is blamed on the personal qualities of a patient, rather than on the context in which the event occurs. This may lead to patients being judged negatively and disliked, minimising or overlooking their complaints leading to wrong decisions and management errors [11]:
  - An intravenous drug abuser may present with a nasty groin abscess and some abdominal pain, but one may consider it to be another superficial abscess due to groin injection with dirty needles rather than considering the possibility of a more complex iliopsoas collection tracking from above.

Our decision making may also be influenced by the potential emotions we may later experience due to those decisions (Fig. 7.3).

Anticipated regret—the situation whereby in making decisions we take into consideration our own emotional reactions that may occur in the future as a result of those decisions. This may lead to deciding to go with the option which may lead to the least anticipated unpleasant emotion or regret [32].

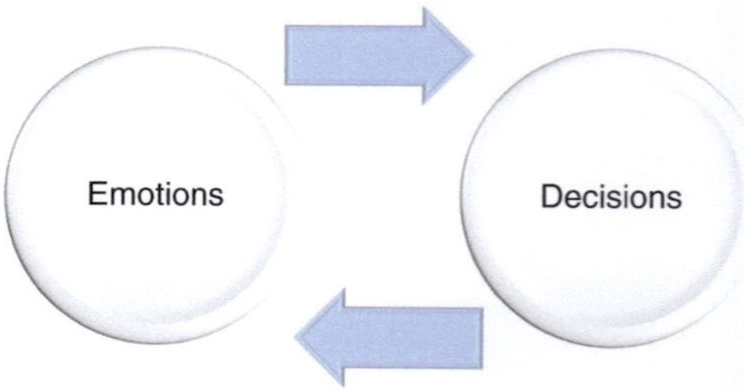

Fig. 7.3 Bi-directional relationship between emotions and decision making

Sorum et al. [32] asked primary care physicians in France and the USA how much regret they would experience if they found advanced prostate cancer in patients for whom they had refused to order Prostate Specific Antigen(PSA) tests several years earlier. The authors showed a positive correlation between the expressed levels of regret and the likelihood of ordering a PSA in a subsequent group of hypothetical patients presenting for routine physical examination.

Furthermore, being upset or preoccupied with other matters may influence our decision making process. The workplace, with its pressures and stressors, may affect your emotions, and hence potentially your decision making. Others, who interact with you, may not be aware of the impact of their actions on your emotional state, or may be aware but decide to ignore it:

- The rota master may keep asking you to cover the on call shift at short notice, ignoring that you have just finished three days on call in a row and not being aware that you already have social plans in place.
- The department manager may be asking you to go at short notice to do ward cover without considering that it would mean missing another day from theatre, another lost training opportunity.
- Your senior may suddenly decide to question why you have applied to go to a course, without considering that has been long agreed as part of your appraisal process.
- Your executive may inform you that "change is inevitable" making you wondering and anxious as to what that change may entail.

It is believed that stress can affect decision making but its exact influences are unclear [33–35]. Both improved performance and performance degradation have been associated with increased stress, and may be influenced by the stress levels. The effect of stress may be influenced by how the individual perceives stress, hence stress, and potentially its effects on decision making, may be minimised in emergency situations through simulation and preparation.

It is important to develop the skill of appreciating the effects that emotions and emotion related biases can have on decision making and develop strategies to deal with these. Not all of us are affected by events in the same way. Some may be able to shrug off troubles and move on, some may be less able to do so. You are likely to know better than others how it affects you, what to ignore or what to avoid. You may consider:

- Avoid any factor that can impair your emotional state prior to critical decision making (but then that may not be always possible). If you know that reading your work emails is likely to make you upset, create constant thoughts or worries, should you be reading these in the morning just before the commencement of a busy day clinic or a day of challenging operating?
- Acknowledge such emotions and aim for decision making that is objective and rational. It does not mean trying to suppress your feelings, but acknowledging those feelings. The advice is not to block your emotional reactions (which could lead to anxiety or even clinical burnout) but to recognise these reactions and take them into account in your decision making [11]. It's almost like being on an interview panel, interviewing a candidate that you know well and like as a colleague. One may acknowledge the affection, but still make an objective and impartial decision.
- If you feel that emotions are impairing you decisions, consider taking time off, seek help and support.

7.9 Improving Decision Making

An executive is a man who can make quick decisions and is sometimes right.

Elbert Hubbard [36]

In improving decision making you may consider:

- Meta-cognition refers to the ability to reflect on the decision process and adjust appropriately [37–42]. Develop the ability to recognise situations where your decision process may be impaired or biased, and plan strategies to deal with those:
  - You may be aware of your tendency to just concentrate on the fractured tibia and end up missing any other injuries—you may deliberately do initially a full system evaluation and leave the tibia last.
  - You may recognise that when faced with patients reporting symptoms out of proportion, you tend to question the validity of such symptoms, and hence endangering dismissing those symptoms—you may deliberately "give the benefit of doubt" to ensure that your usual investigation pathway is followed.
- Aim to continuously self-reflect on your decisions, like you do for technical skills. Keep enquiring whether the correct decision has been taken, and, if not, how could the decision making process be improved next time?
- Lack of insight, poor self –assessment, over confidence, distractions, competing interests, "surgical image" pressures, may impair the decision making process, hence examine own-self for these and deal with them if found.
- Develop the ability to slow down, focus concentration at critical times, at particularly difficult steps in a surgical procedure, when encountering unexpected complication [43].
- Identify attitudes that may adversely influence your decision making and address those. Kadzielski et al. [44] examined the rates of hazardous attitudes (including macho, anti-authority, impulsivity, resignation, self-confidence) amongst orthopaedic surgeons at the Massachusetts General Hospital. They reported that 30% of those surveyed had at least one elevated hazardous attitude level. In addition, a macho attitude correlated with an increased number of re-admissions and re-operations.
- Develop the ability to reflect in action. Reflection in action enables a surgeon to continuously monitor the progress of a task (such as a surgical procedure) [45]. This dynamic decision making model consists of:
  - Assessing the situation
  - Taking the appropriate action
  - Re-evaluating the results of your decision
- Ensure you have the necessary knowledge to facilitate your decision making. Clarke [46] evaluated the role of decision making skills and medical knowledge in clinical judgement of surgical residents. They presented surgical management problems to surgical residents, the solution of which required textbook knowledge as well as logical decision making. They reported that correct decisions correlated positively with more accurate knowledge of textbook information. Hence, the authors suggested that decision analysis can supplement but not replace gathering of scientific knowledge.

7.10 Surgeons Are Allowed to Change Their Mind

Reconsider, v. To seek a justification for a decision already made.

Ambrose Bierce [47]

Surgeons are often perceived as decisive, confident practitioners who cannot tolerate uncertainty. As surgeons we may be taught to make quick, definite decisions, justify our case, stick to our guns. We may be given the perception that re-thinking or re-evaluating are signs of weakness, hesitancy, dithering, vacillating. We may feel that revision or reversal of decisions could impair our ability to lead or could diminish the trust and confidence others put upon us.

But as a surgeon you may often practise in complex, dynamic situations, with constant change of information, emergence of new facts. As in other settings, out of surgical life, you may give an immediate intuitive decision, but then momentarily stop and re-think, consider a bit more the facts, be more analytical, and come to a different conclusion.

If you feel that your revised decision is the correct one, be prepared to change your mind. Is it more important to save face, match a surgeon's image, meet the perceived expectations of others, or is it more important to take, communicate and act upon a better decision?

If you have reconsidered and changed your mind let others know:

- "I am changing my mind..."
- "I am sorry, following further consideration…."
- "You are right, it would be better to do that…."
- "What I have just said is not right…."
- "Forget what I have just said, lets…"
- "No lets do this instead…."

If chancellors make dramatic public U-turns shortly after announcing the annual budget [48], if prime-ministers [49], presidents [50] and even footballers [51] change their mind, is it so difficult for a surgeon to reconsider?

References

1. Popular Quotes. Goodreads. http://www.goodreads.com/quotes. Accessed 31 Mar 2017.
2. Famous Quotes at BrainyQuote. https://www.brainyquote.com/quotes/quotes. Accessed 31 Mar 2017.
3. Marcum JA. An integrated model of clinical reasoning: dual-process theory of cognition and metacognition. J Eval Clin Pract. 2012;18(5):954–61.
4. Flin R, Youngson G, Yule S. How do surgeons make intraoperative decisions? Qual Saf Health Care. 2007;16(3):235–9.
5. AZ Quotes. http://www.azquotes.com/quote/667458. Accessed 1 Apr 2017.
6. Moulton CA, Regehr G, Mylopoulos M, MacRae HM. Slowing down when you should: a new model of expert judgment. Acad Med. 2007;82(10 Suppl):S109–16.
7. Croskerry P, Singhal G, Mamede S. Cognitive debiasing 1: origins of bias and theory of debiasing. BMJ Qual Saf. 2013;22(Suppl 2):ii58–64.
8. Croskerry P, Singhal G, Mamede S. Cognitive debiasing 2: impediments to and strategies for change. BMJ Qual Saf. 2013;22(Suppl 2):ii65–72.
9. Croskerry P. The cognitive imperative: thinking about how we think. Acad Emerg Med. 2000;7(11):1223–31.
10. Croskerry P. From mindless to mindful practice--cognitive bias and clinical decision making. N Engl J Med. 2013;368(26):2445–8.
11. Croskerry P, Abbass A, Wu AW. Emotional influences in patient safety. J Patient Saf. 2010;6(4):199–205.
12. Pines JM. Profiles in patient safety: confirmation bias in emergency medicine. Acad Emerg Med. 2006;13(1):90–4.
13. Croskerry P. Context is everything or how could I have been that stupid? Healthc Q. 2009;12 Spec No Patient:e171–6.
14. Croskerry P. To err is human--and let's not forget it. CMAJ. 2010;182(5):524.
15. Sinclair D, Croskerry P. Patient safety and diagnostic error: tips for your next shift. Can Fam Physician. 2010;56(1):28–30.
16. Campbell SG, Croskerry P, Bond WF. Profiles in patient safety: a "perfect storm" in the emergency department. Acad Emerg Med. 2007;14(8):743–9.
17. Leung A, Luu S, Regehr G, Murnaghan ML, Gallinger S, Moulton CA. "First, do no harm": balancing competing priorities in surgical practice. Acad Med. 2012;87(10):1368–74.
18. Hageman MG, Guitton TG, Ring D. Science of variation group. How surgeons make decisions when the evidence is inconclusive. J Hand Surg Am. 2013;38(6):1202–8.
19. Teunis T, Janssen SJ, Guitton TG, Vranceanu AM, Goos B, Ring D. Surgeon personality is associated with recommendation for operative treatment. Hand (N Y). 2015;10(4):779–84.
20. Janssen SJ, Teunis T, Guitton TG, Ring D. Science of variation group. Do surgeons treat their patients like they would treat themselves? Clin Orthop Relat Res. 2015;473(11):3564–72.
21. Simianu VV, Basu A, Alfonso-Cristancho R, Thirlby RC, Flaxman AD, Flum DR. Assessing surgeon behavior change after anastomotic leak in colorectal surgery. J Surg Res. 2016;205(2):378–83. doi:10.1016/j.jss.2016.06.075.
22. Dale W, Hemmerich J, Ghini EA, Schwarze ML. Can induced anxiety from a negative earlier experience influence vascular surgeons' statistical decision-making? A randomized field experiment with an abdominal aortic aneurysm analog. J Am Coll Surg. 2006;203(5):642–52.
23. Dale W, Hemmerich J, Moliski E, Schwarze ML, Tung A. Effect of specialty and recent experience on perioperative decision-making for abdominal aortic aneurysm repair. J Am Geriatr Soc. 2012;60(10):1889–94.
24. Szatmary P, Arora S, Sevdalis N. To operate or not to operate? A multi-method analysis of decision-making in emergency surgery. Am J Surg. 2010;200(2):298–304. doi:10.1016/j.amjsurg.2009.10.020.

25. Wilson NP, Wilson FP, Neuman M, Epstein A, Bell R, Armstrong K, Murayama K. Determinants of surgical decision making: a national survey. Am J Surg. 2013;206(6):970–7. doi:10.1016/j.amjsurg.2013.08.018. discussion 977-8
26. Sacks GD, Dawes AJ, Ettner SL, Brook RH, Fox CR, Maggard-Gibbons M, Ko CY, Russell MM. Surgeon perception of risk and benefit in the decision to operate. Ann Surg. 2016;264(6):896–903.
27. Sacks GD, Dawes AJ, Ettner SL, Brook RH, Fox CR, Russell MM, Ko CY, Maggard-Gibbons M. Impact of a risk calculator on risk perception and surgical decision making: a randomized trial. Ann Surg. 2016;264(6):889–95.
28. Rutkow IM, Gittelsohn AM, Zuidema GD. Surgical decision making. The reliability of clinical judgment. Ann Surg. 1979;190(3):409–19.
29. Rutkow IM. Surgical decision making. The reproducibility of clinical judgement. Arch Surg. 1982;117(3):337–40.
30. Emotional Quotes Related to Decision Making. http://www.decision-making-solutions.com/emotional_quotes.html. Accessed 1 Apr 2017.
31. Groves JE. Taking care of the hateful patient. N Engl J Med. 1978;298(16):883–7.
32. Sorum PC, Shim J, Chasseigne G, Bonnin-Scaon S, Cogneau J, Mullet E. Why do primary care physicians in the United States and France order prostate-specific antigen tests for asymptomatic patients? Med Decis Making. 2003;23(4):301–13.
33. Gillis JS. Effects of life stress and dysphoria on complex judgments. Psychol Rep. 1993;72:1355–63.
34. Kontogiannis T, Kossiavelou Z. Stress and team performance: principles and challenges for intelligent decision aids. Saf Sci. 1999;33(3):103–28.
35. Poulton EC. Arousing environmental stresses can improve performance, whatever people say. Aviat Space Environ Med. 1976;47:1193–204.
36. Management Quotes About Decision Making. http://www.decision-making-solutions.com/management_quotes.html. Accessed 1 Apr 2017.
37. Metcalfe J, Shimamura AP. Preface. In: Metcalfe J, Shimamura AP, editors. Metacoginition: knowing about knowing. Cambridge: MIT Press; 1994. p. xi–xii.
38. Uemura M, Tomikawa M, Nagao Y, Yamashita N, Kumashiro R, Tsutsumi N, Ohuchida K, Ieiri S, Ohdaira T, Hashizume M. Significance of metacognitive skills in laparoscopic surgery assessed by essential task simulation. Minim Invasive Ther Allied Technol. 2014;23(3):165–72.
39. Wokke ME, Cleeremans A, Ridderinkhof KR. Sure I'm sure: prefrontal oscillations support metacognitive monitoring of decision making. J Neurosci. 2017;37(4):781–9.
40. Trimble M, Hamilton P. The thinking doctor: clinical decision making in contemporary medicine. Clin Med (Lond). 2016;16(4):343–6.
41. Chew KS, Durning SJ, van Merriënboer JJ. Teaching metacognition in clinical decision-making using a novel mnemonic checklist: an exploratory study. Singapore Med J. 2016;57(12):694–700.
42. Grimaldi P, Lau H, Basso MA. There are things that we know that we know, and there are things that we do not know we do not know: confidence in decision-making. Neurosci Biobehav Rev. 2015;55:88–97.
43. St-Martin L, Patel P, Gallinger J, Moulton CA. Teaching the slowing-down moments of operative judgment. Surg Clin North Am. 2012;92(1):125–35.
44. Kadzielski J, McCormick F, Herndon JH, Rubash H, Ring D. Surgeons' attitudes are associated with reoperation and readmission rates. Clin Orthop Relat Res. 2015;473(5):1544–51.
45. Schön D. The reflective practitioner: how professionals think in action. London: Temple Smith; 1983.
46. Clarke JR. The role of decision skills and medical knowledge in the clinical judgment of surgical residents. Surgery. 1982;92(2):153–8.
47. The quote garden. http://www.quotegarden.com/decisions.html. Accessed 1 Apr 2017.
48. Horton H, McCann K. Philip Hammond's NICs Budget U-turn: 7 embarrassing times politicians have had to wheel back pledges after backlash. 2017. http://www.telegraph.co.uk/

news/2017/03/15/philip-hammonds-nics-budget-u-turn-7-embarrassing-times-politicians/. Accessed 1 Apr 2017.

49. U-turn if you want to: 7 policies the UK government has scrapped. Financial times. 2016. https://www.ft.com/content/d476b856-16bb-11e6-9d98-00386a18e39d. Accessed 1 Apr 2017.
50. Pathan M, McCrum K. Donald Trump's U-turns since winning election are leading supporters to question what they actually voted for. Mirror. 2016. http://www.mirror.co.uk/news/world-news/donald-trumps-u-turns-winning-9317371. Accessed 1 Apr 2017.
51. Bird S. Five of the best football U-turns following Willian's dramatic move towards Chelsea. Mirror. 2013. http://www.mirror.co.uk/sport/football/news/willian-transfer-five-u-turns-following-2215163. Accessed 1 Apr 2017.

Chapter 8
Leadership in Surgery

Surgeons are often called upon to lead or manage amongst others processes, activities, people, teams, departments or institutions, through formal or informal roles. Successful leading requires a clear understanding of the task at hand, a clear vision of what one aims to achieve, but also the ability to establish and maintain effective inter-personal relations with followers and the wider team.

This chapter examines the similarities and differences between leading and managing, describes various leadership styles, and discusses the importance of leadership along with the challenges a surgeon leader may face. Both desirable and undesirable leadership behaviours are then presented, whilst finally emphasising the distinction between leading and holding leadership titles or posts.

C. Panayiotou Charalambous, *Career Skills for Surgeons*,
DOI 10.1007/978-3-319-57490-5_8

As an all-inclusive leader, I would welcome your views as how to manage this emergency

8.1 Leader Versus Manager

Management is efficiency in climbing the ladder of success; leadership determines whether the ladder is leaning against the right wall.

Stephen Covey [1]

A surgeon may have to lead and manage their team and service, their department, institution or organisation. The terms "leader" and "manager" are often used loosely and interchangeably. Leadership and managerial roles do overlap and one does not preclude the other. Both may share common values, and the need of certain skills to achieve their objectives.

Nevertheless, despite being close, the two roles may differ in their inner core. A leader may be considered as the one who sets the vision, the manager the one who makes the vision reality. A leader identifies the priorities, eliminates the unnecessary and highlights the essential. A leader sets the direction, questions the current path and convinces others of the need to change course. A leader sets the way, a manager ensures the way is followed and the goal is reached. The leader sets the target, the manager hits the target. A manager ensures the wheel keeps rolling whilst a leader questions the need for a wheel.

A leader is the driver who chooses the destination, or the one that alters a destination that has already been pre-set. A manager is a co-driver who organises the journey, reads the map, sets the exact route, allocates the re-fuel stops, and ensures the engine and tyres are up to scratch to get to the end.

The leader is like a pilot, the manager the autopilot. The autopilot will manage most of the flight, make the turns, adjust the speed, control the thrust and throttle to counteract turbulence, to keep cruising smoothly. But no matter how efficient the autopilot is, the pilot is needed to dial in the destination, lead the taking off, and often fine tune the final descent to landing. When trouble hits hard, when in the middle of a storm, the pitot tubes get ice blocked, and the autopilot fails, the pilot once again needs to step in, take control, work out what's wrong and stir the plane back on course.

In the era of target driven healthcare, a leader may question the preference of waiting time targets over quality based assessments. A leader may recognise that waiting time targets are often counterproductive, not necessarily achieving what they mean to, collect supporting evidence, and argue to the stakeholders the need for course change. A manager faced with such targets may strive to make those system adjustments to ensure that target waiting times are not exceeded. A leader questions the task, a manager considers how to cope with the assigned task.

8.2 Types of Leadership

When the best leader's work is done the people say, 'We did it ourselves.'

Lao Tzu [2]

Several types of leadership have been described [3, 4] but three well known ones are:

1. Authoritative—dictates as what to be done.
2. Democratic—consults on decisions, aims to reach consensus, majorities decide but minorities are supported and respected.
3. Laisse-faire—delegates the tasks to followers who are given the freedom and flexibility in how to bring them about.

Leadership styles may also be described as transactional or transformational [5–7];

1. Transactional leader—focused on performing the task at hand, through assignments of roles, actions and responsibilities. Engages in transaction with follower to reward success or penalise failure.
2. Transformational leader—does not only focus on carrying out a task but also on the well-being, and development of the followers and the whole team. Focuses on the team rather than simply on own-self, and is driven by a sense of collective mission. Values and supports individuals, and encourages their personal and professional development and advancement beyond what is the minimum required for carrying out a task. Readily shares information with the rest of the team.

Different leadership styles have their pros and cons, and the one to choose may be influenced by your position, personality, your followers, the task at hand. Leadership styles however do not have to be all or none, and one may move from one to another, according to the circumstances faced. One may be able to move between enabling and more forceful leadership styles but without overdoing either [8]:

- Reaching a decision about long term departmental staffing may require a different leader's approach to that required in the allocation of tasks amongst a resuscitation team awaiting the trauma ambulance.
- Leading the team in deciding the order of an elective theatre list may require a different approach to resuscitating a crashed patient with an acutely ruptured aortic aneurysm.

You may aim for an all-inclusive, empowering, leadership style, where diverse views are heard and respected, where team skills are cultivated and talents are nourished. But there may be occasions where you have to be assertive, direct as to what has to happen, as the urgency of the moment dictates.

8.3 Leadership Matters

You manage things; you lead people.

Rear Admiral Grace Murray Hopper [9]

Leadership is vital at multiple levels, as it can affect [9]:

- The individual follower—promote commitment, encourage loyalty, enhance performance, inspire one to go that extra mile.
- The team—promote communication, organisation, coordination, conflict resolution, goal setting, role allocation, sharing of aspirations and values.
- The institution—set strategy, define direction.

Effective leadership may improve the wellbeing of individuals, and by doing so improve the performance and efficiency of the whole team. Spinelli [10] looked at the effectiveness of transformational, transactional, and laissez-faire leadership amongst hospital managers, and reported that a transformational style was associated with individuals being more satisfied with their leaders, and more willing to exert extra effort in achieving their goals.

Poor leadership may demoralise followers, creating anxiety and misery, with some organisational surveys reporting that for up to 75% of workers the most stressful aspect of their job is their immediate boss [11, 12]. Rasch et al. [13], in their description of ineffective leadership, reported the "failure to consider human needs" as having the worst impact on staff morale.

In assessing effectiveness of leadership at a team, departmental, or institutional level, one may consider its effects on productivity, control of finances, service quality but also the effects of leadership on staff retention and turnover, sickness levels, staff morale, staff engagement and on the sense of belonging and being valued.

In surgery, an effective leadership may not only get things done, improve efficiency and performance, but also facilitate safe care. Hu et al. [14] evaluated the effects of surgeon's intra-operative leadership style on the behaviour of the operating theatre team. They video-recorded five surgeons performing complex operations, who were then scored on their leadership style. All surgeons scored similarly on transactional leadership but varied widely on their transformational style. A transformational leadership was characterised by more supportive and less poor behaviours towards the team. Such a leadership style encouraged cooperation and information-sharing within the team and empowered team members to speak up constructively, to suggest improvements or warn of unsafe situations.

Decreased information sharing during surgery has been shown to be related to increased risk of complications or death. Leach et al. [15, 16] evaluated the factors and conditions that influence performance in the operating theatre, through observations of surgical procedures and interviews of members of the operating theatre team. They concluded that a surgeon's leadership can help "set the tone" in the surgical theatre, with good leaders proposing an environment where all individuals truly cooperate rather than simply carrying out their bit of the task, an environment where team members can work together and respond to any deficiencies of the system.

Kissane-Lee et al. [17] surveyed 40 surgical residents and interns on leadership styles in the surgical theatre. They described four styles:

- Authoritative—decides and informs firmly.
- Explanatory—decides and explains decisions.
- Consultive—consults in making decisions.
- Delegative—makes decisions by majority.

Residents and interns preferred mainly explanatory and consultive leadership styles but they mainly encountered authoritative leadership styles amongst attending surgeons.

You may find yourself in an operating theatre where everyone has to stand attention in front of the "master", follow clear instructions, where no dissent can be tolerated. Alternatively, you may happen to be in a theatre where tasks are delegated, roles are established, tasks are effectively performed, but at the same time hierarchies are put aside, where all team members are comfortable in speaking out on an equal level, speaking up or questioning what may not be right. A good leader may encourage such an atmosphere. Where would you choose to be?

8.4 Challenges of a Surgeon Leader

The function of leadership is to produce more leaders, not more followers.

Ralph Nader [1]

Leading or managing a surgical team can be a challenging task, due to the dynamics of the working environment, the demands of the tasks to be performed, but also due to your own or your followers' attributes. Appreciating these may help you examine own-self and see others in a different light, may enable you to be ambitious but also pragmatic and realistic.

When it comes to your own attributes you may consider:

- As a surgeon you may be used to giving instructions and orders, in dealing with complex emergency clinical cases, where fast, often immediate decisions must be made. Consultation, negotiation, constructive listening, reaching out for midway solutions may not apply to such circumstances, and hence may not easily come to mind in other leadership tasks.
- You may be used to the certainty of the impact of your actions; if you stitch a vessel the bleeding will stop, if you twist the screwdriver the screw will go in, but your words may not have the same predictable effect on people. You may be used to an action having a positive reaction but in dealing with human beings the reactions may be unpredictable.
- Your expectations may be high, demanding attention to detail, but in doing so you may end up setting too high standards. You may be highly gifted, very knowledgeable, technically skilful, a high striver, you may be looking others to match you, but others may not be able, or really wishing, to achieve that.

You may be leading a group of surgeons, who them-selves can have challenging qualities:

- They may be highly gifted, highly qualified, high striving, ambitious individuals, even more than you are.
- They may be highly opinionated and be prepared to express their opinions even when not asked for one.
- They may be resistant to change, not because they want to sabotage your plans, but because of the inherent belief that "if something works in my hands why change it?" an approach encountered in surgical technical decisions.
- They may be perfectionists who expect the same standards in their leaders or managers, and may be able to identify and pick on any signs of uncertainty or sloppiness you may have.
- They may not be forgetting or forgiving as they don't do in technical surgery. They may notice that you keep changing your mind, making up plans on the back of an envelope, making promises that you don't keep, engage only in what interests you and not in what interests the team, implementing changes without fully thinking things through.
- They may be expecting more from you, after all you are one of them, you went through similar hardships, you should know better.

However, despite their challenges, your followers may well be like you, likely to share the same goals and core values, share the aim to provide some of the best care to patients. Try and utilise their strengths, rather than feel threatened by them.

8.5 Desirable Leadership Behaviours

Before you try to revolutionise my business, I'd like to know that you actually know my business.

George Clooney as Ryan in the Up in the Air [18]

Desirable behaviours of successful leaders have been extensively described by several authors and these share some common characteristics.

The Domain Model of Performance organises personal competencies into four categories [19]:

1. Intra-personal skills—being self-aware.
2. Inter-personal skills—social skills, relationships.
3. Work skills—plan, organise, resource use.
4. Leadership skills—build and lead a team.

According to this model, skills in each category build on those developed in the previous one, so that maintaining positive relationships depends on appropriate self-control, and effective use of resources depends on interpersonal skills with other co-workers. In this model, the ability to lead depends upon effective intrapersonal, interpersonal and work skills.

Dr. Thomas Gordon, a well-known Psychologist and a recognised figure in the teaching of communication and conflict resolution, described that an effective leader must be both a productivity specialist (meeting organisation needs) but also a human relations specialist (who is people centred, meeting their followers' needs) [20]. A leader must aim to achieve productivity and meet the goals of the organisation but also achieve followers' satisfaction and engagement.

Lombardo et al. [21] reported that successful leaders perform well in eight areas:

- Directing and motivating followers
- Handling complexity
- Honour
- Drive for excellence
- Organisational knowledge
- Composure
- Sensitivity
- Staffing

Several studies reported as to what constitutes successful leadership behaviours in surgery. The Surgical Leadership Inventory is a description of surgeons' intraoperative leadership behaviours that are expected to improve patient safety and team performance. It was derived from the literature on surgical leadership, from observations of surgeons at work, and interviews of theatre teams [22]. It has eight components:

1. Effective communication
2. Coping with pressure
3. Maintaining standards
4. Decision making
5. Managing resources

6. Directing
7. Training
8. Supporting others

Hjortdahl et al. [23] interviewed members of trauma teams in Norway with regards their views of the necessary qualities of a team surgical leader. The responses suggested that the ideal leader should be:

- Trustworthy
- Able to transmit confidence and calmness
- Clear communicator
- Decisive
- Technically competent, good in what they do

Patel et al. [24] reviewed published literature to determine the essential qualities for surgical leaders. They suggested that a surgical leader should be:

- Professional
- Technically competent—competent surgeons
- Motivated
- Innovative
- Team builder
- Communicator
- Decision maker
- Emotionally competent
- Resilient
- Teacher
- Business aware

In a study of new technology adoption Edmondson [25] found that effective surgical leaders supported change, encouraged speaking up, and acted to reduce any power differentials between surgeons and other team members.

In addition to the above desirable behaviours previously described for successful leaders, several other characteristics may also be considered and these are described next.

8.5.1 See the Big Picture

You have to get a picture not a photograph.

Doreen Spooner [26]

An important aspect of leadership and management is not to lose sight of the big picture, not to get absorbed too much in the fine details, in the small components of systems. It is like a Monet's painting, you have to stand away to appreciate it. If you stand too close, all you see are coarse detached brushings, random blobs of paint. And the larger the painting the further back you may have to stand.

It is not however uncommon to fall for the "don't sacrifice the forest for the tree". It's a fine balance between not losing sight of the big picture, the big project, the big change, and ignoring the small but important components, the processes, the individual staff.

See the big picture but bear in mind that the forest is made of individual trees. And unless each of those is looked after and taken care of the forest may not exist.

Don't sacrifice the trees for the forest. A system may consist of people, equipment, rules, or procedures. Each component has its role, and may not be able to fully replace the other. A team is made of individual members, with their individual skills and talents. Don't sacrifice the loyalty, the enthusiasm, the morality, the contribution that each individual brings, in pursuit of an imaginary concept, novel or not.

Instead aim to make the most of each individual, to achieve the main goal. Respect each individual, look after them, as they may help you keep the task on track. Success often relies on cooperation and engagement of people across the system, and people often going that extra mile that cannot be easily quantified or measured. Don't give up on processes, regulations, checks and structures simply because they stand in the way of a big achievement. Aim to work within such frameworks, but the same time improve these to make them more efficient or streamlined.

8.5.2 Stick to the Core Values, Share Your Followers' Values

It's strange how over time what is crazy gets normalised.

Obama [1]

As a leader you may be under pressure, but don't belittle the core values of the profession to justify processes, resource cutting, or change.

- Should you be telling your followers that research or audit activities they undertake are of no benefit to the organisation, when such activities promote an academic environment and encourage high quality practise?
- Should you be telling your followers that the ward rounds they do, or the time they spend doing administration work are not essential and should cut down, simply in an attempt to justify a reduction in working hours?
- Should you be doubting their requests for urgent investigations, or listing patients for urgent surgery, without properly evaluating their clinical reasoning and judgement, simply because such requests are interfering with you meeting imposed time to surgery targets?

8.5.3 Lead by Example

Not the cry, but the flight of a wild duck, leads the flock to fly and follow.

Chinese Proverb [27]

Demand from others what you are prepared to give, motivate by inspiration. Let your vision be shown by your actions not your words. Inspire through actions rather than big words. Set up standards but lead by example:

- Should you be telling others that they are not hard working when you are the last to arrive and first to leave from clinics?
- Should you be stressing the importance of surgical infections but keep walking in the surgical theatre without a mask, sneezing in front of open surgical trays?
- Should you be championing checklists, but keep scrubbing in the side room whilst the time out checklist is taking place?

Respect cannot be enforced, but simply gained, through actions and not promises. Be straight, honest, aiming for consensus, having a clear vision, learn of the problems first hand, rather than relying on messengers who may twist the facts to suit their own agenda.

8.5.4 Lead Change Wisely

I kept the same suit for six years and the same dialogue. They just changed the title of the picture and the leading lady.

Robert Mitchum [1]

As a leader you may have to bring about change. Changing the on all rota, the team structure, the role allocation, the operating schedules. Encourage change, yet control the flow of change.

Surgical training sometimes teaches radical approaches, definitive treatments, complete resections, a single rather than multiple operations. But bringing about real life change may be a bit more challenging than that.

There may be times when the current practise is so rotten, when the status quo can simply not go on, when there is a need to start afresh, but that is likely to be the exception rather than the rule. Erasing the past and starting all over again may not be necessary in many cases, and may not be well received if attempted. Instead small changes, fine tunings, minor adjustments may achieve the wishful outcome, the same result. Small changes may not sound as big achievements of a great leader, but then one ought to look at the outcome rather than the difficulty of getting there. Legend has it that Babylon was not build in a day; small changes can have a huge impact, potentially greater than hitting the reset button which may not achieve much other than setting many into disarray.

We sometimes associate leaders with difficult, painstaking decisions. But change does not have to mean confrontation or conflict, as in many areas a middle solution may be found, a win-win situation for all may be reached.

We sometimes associate change with an emotional turmoil. The reaction of individuals to major change has been likened to the five stages of the grieving process of terminally ill patients learning their diagnosis, described by Elizabeth Kubler-Ross, an American psychiatrist in the 1960s. These five stages are shock and denial, anger, bargaining, depression, and eventually acceptance [28]. But one doesn't have to be made to grieve, before they can come to terms with the encountered change. Change can be welcomed, can be sought for, can be fulfilling. Many of your followers may thirst for change, but it may simply be that their concept of change, or their route to change are different from yours.

Sometimes one sees great plans for change not taking off or collapsing half way through. Share the vision with your followers and let them work it out, or join them in working it out. What you may not see because you spend all day in theatre, your juniors in the ward may have already mapped it out.

Engaging, listening, and appropriately acting, may help to maintain trust, build bridges, enable the desirable outcome. The audience that needs to be targeted may be wide and not just be limited to one professional group, not just surgeons but theatre or nursing staff too.

Part of change may be the introduction of a protocol, a new procedure, another paper form you are asking others to fill in; introduction of venous thromboembolism assessments for hospitalised patients, and surgical theatre safety checklists, are two some of the processes that surgical leaders have been asked to implement in UK hospitals in recent years. How does one go about achieving this? Healy et al. [29] evaluated the implementation strategies used by hospital leaders in Australian public hospitals in achieving compliance with a "safe surgery" checklist. Their strategies fell into "support" or "sanction" measures depicted by two corresponding pyramids (Figs. 8.1 and 8.2).

Recognise that one implementation approach may not suffice. We often refer to stick or carrot but then there may be so many carrots to go round. But if one could get away with carrot why revert to the stick? Avoid rushing into imposing sanctions, which may achieve a short term effect but not necessarily a long lasting improved

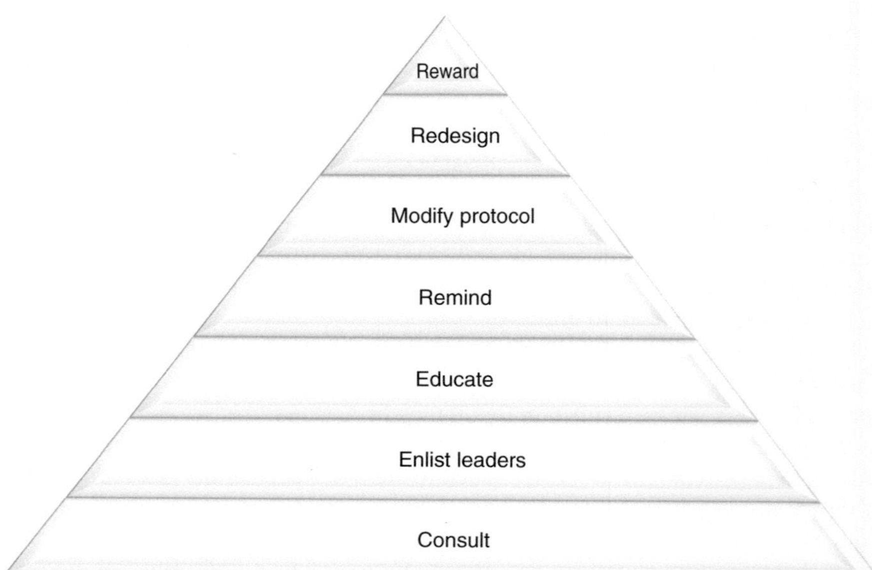

Fig. 8.1 Support pyramid in implementing change; modified from Healy et al. [29]

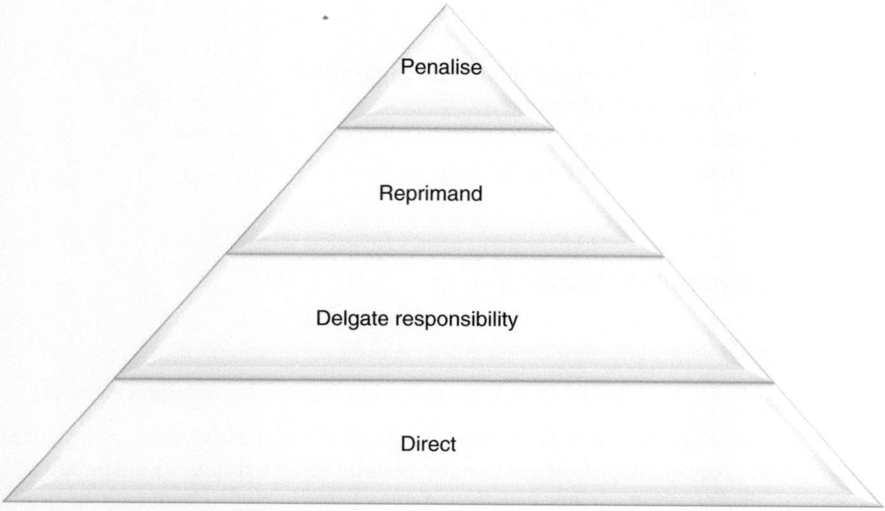

Fig. 8.2 Sanction pyramid in implementing change; modified from Healy et al. [29]

performance. Work with those at the frontline of change, as they have to be engaged and buy into it. Listen, and consider modifying your plans according to what they say. In introducing a new protocol or checklist you may examine how that can fit in, how can it be the least disruptive to the current routine, how could an increase in workload be minimised? Establish an understanding of the need for change, an acceptance for the need for change rather than rushing to instruct or direct.

Bergs et al. [30] carried out a systematic review of studies evaluating barriers and facilitators to safety check-list implementation. They reported that the introduction of such lists required not only changes in the workflow of healthcare workers but also changes in the overall perception of these workers of safety, and changes in their perceptions of the specific checklist to be introduced (such as perceptions of importance, evidence base, time consumption, legal implications).

And if things are not working, if there is no compliance with the introduced change, avoid simply blaming the whole team and disengaging all, but instead aim for personal feedback along with support. Lau et al. [31] demonstrated that individualised performance feedback to surgical residents, along with coaching, significantly increased the rate of appropriate venous thromboembolism prophylaxis prescription for hospital admitted patients.

In bringing about change you may consider the following:

1. Explain why there is a need for change.
2. Identify the task, what change aims to achieve.
3. Define the new order, once change comes about.
4. Define how change will be achieved.
5. Break the task of change into individual components.
6. Define the time frame for changing.

7. Involve the various stakeholders, consult wisely.
8. Plan for alternatives.
9. Perform a risk assessment of things not working out.
10. Be flexible and adaptable along the way.
11. Collect data to assess progress.
12. Communicate progress along the way.

8.5.5 Welcome Alternative Voices

If everyone is thinking alike then somebody isn't thinking.

George S Patton [1]

You may face opposition to your well thought plans, to the upcoming reorganisation that you propose, to your ways that are bound to make things so much better.

Accept dissent, accept being questioned, being challenged. There may be many ways to solve a problem. And for those that may be in constant dissent, who may question all you do, listen to those too, they may be on occasion right.

In the movie "World of the Wars" [32] a Mossad agent explains the "tenth man rule". According to that, any major challenge Israel faces is put in front of a ten man council. If nine agree then the tenth has to, as a rule, oppose it and investigate the possibility of all others not getting it right.

Fact or fiction, reality or legend, the "tenth man rule" reminds of the need to welcome alternative voices. They may be on occasion right.

8.5.6 Follow the Rules

In a courtroom it is not the loudest voice that prevails, it's the Constitution.

B Ferguson [33]

Follow the rules. Just because you are the leader it doesn't mean you ignore policies already in place or that you set up new ones without consultation or due process.

- You may feel that employees should be working in theatre until the list finishes, even though they are contracted only nine to five.
- You may feel that trainees must attend the wards every day, even though they are allowed a day off for regional teaching.
- You may have noticed that trainees do not turn up to teaching. Should you be contemplating using a fingerprint recognition system? Where will the fingerprints be stored? What about those refusing?

What may be acceptable to you may not be acceptable to others. Follow the rules and regulations, rather than jumping to impulsive decisions. If not sure what the rules are, then find out, seek clarification from the experts in the area – such as the human resources department for employment issues, the post-graduate department for training issues, the legal department for controversial medico-legal problems. Resist the temptation of making up rules on the spot, making up regulations that suit the moment.

8.5.7 Consider How You Start

My grandmother always told me how you start is how you finish.

Bernie Mac [1]

Sometimes newly appointed leaders share their vision, concentrating on what they will crack down, how they will stamp their mark. But is this the road to success, or will this just stir trouble? “Attack is the best defence” may have its role in battlefields, but is this really the best approach in civilised working environments?

Alex Ferguson, the world famous manager of Manchester United, summarised the tendency of newly started leaders to mark their presence [34]:

> “I know that there is a widely held belief that the leader only has a chance to make his presence felt during his first one hundred days, but it is not something to which I subscribe. It is all well and good, as a new arrival, to feel the impulse to issue your own version of the ten commandments, but actions speak so much louder than words. When a former player, who is new to management, asks me for advice, I usually tell him not to seek confrontation. Whenever you show up in a new role, it will not be long before you have to face trouble in a clash over something. There is nothing to be gained by stirring it up yourself. Trouble will find you quick enough.”

Such an approach may just alienate your followers, and once trust is lost it is difficult to regain. Work towards maintaining trust rather than trying to regain it after you lose it.

8.5.8 Consider How You End

I was the future once.

David Cameron [35]

Being a leader may be a long, sometimes lonely journey. But like most parts of life it may have a start and a finish. Not just the start, the route, but also the end may be challenging.

Sometimes we may have to give up a leadership role or post, even when we are not ready yet, when we dont wish to stop, when we feel we have so much more to offer, when we prefer to keep going on. You may have come to the end of a predetermined term and want to run for another tenure. You may be in the middle of your term but circumstances may have changed, your followers or superiors may feel unsure as to whether you can pull it through, you may be deprived of the resources to achieve what you set out to do.

Leading or managing is not a self-fulfilling purpose, many of us acknowledge that, but it is on occasions hard to just let go. Know when the time has come to move on, and consider letting go.

Moving away from a post which is not working is not a sign of failure but strength, it doesn't mean that you failed as a leader, it could mean that the circumstances or the structures may not be the ones to enable you to show the best of your potential. And even if you realise that you were not cut for that role, well not all of us are, at least you tried and got to find out.

You may feel bitter, disappointed, after all the hard work and achievements you accomplished, and that's understandable and acceptable. But leaders are not only judged by their crowing and reign, but also by their ability to step down at the right time. A leader is not the captain who will not abandon the sinking ship, but someone willing to step aside to let a new pair of hands try and stir the ship away from the rocks, sail to calmer waters.

And if your reputation or pride make it difficult to let go, just recall Hugh Grant's "today's newspapers will be lining tomorrow's waste paper bins" in the Notting Hill's blockbuster [36].

If prime ministers resign [35], presidents step down [37], kings abdicate [38] and even Popes step aside [39] should it not be much easier for a surgeon to let go of a leadership post?

8.6 Undesirable Leadership Behaviours

I have heard your views. They do not harmonize with mine. The decision is taken unanimously.

Charles de Gaulle [1]

We often concentrate on what one should do to be an effective leader, what positive actions to take, but it may be equally important to consider what one should avoid. Several authors have written on leadership and managerial failure in many aspects of life, with their writings showing common themes in the behaviours one ought to avoid.

Bentz [40] examined managerial performance at Sears Roebuck & Company and related the following attributes to failure:

- Lack of business skills
- Inability to deal with complexity
- Being reactive and tactical
- Inability to delegate
- Inability to build a team
- Inability to maintain relationships
- Allowing emotions cloud judgement
- Being slow to learn

Eichinger and Lombardo [41] evaluated more than 1000 managers by relating 360° evaluations to subsequent changes in employment status and associated the following characteristics with failure:

- Problems with interpersonal relationships
- Inadequate administrative skills
- Difficulty in building a team
- Lack of following through
- Poor political skills
- Inability to handle conflict
- Poor integrity
- Low insight

Sydney Finkelstein in his book "Why smart executives fail" [42] described seven habits of unsuccessful leaders:

1. View own-selves as dominating
2. Act as if they own the organisation
3. Have all the answers
4. Eliminate dissent
5. Devote most efforts in managing image
6. Underestimate
7. Relying on outdated practises

Similarly David Dotlich and Peter Cairo in their book "Why CEOs fail" [43] described the following 11 behaviours as leading to failure of top executives:

1. Arrogance
2. Melodrama
3. Volatility
4. Excessive caution
5. Habitual distrust
6. Aloofness
7. Mischievousness
8. Eccentricity
9. Passive resistance
10. Perfectionism
11. Eagerness to please

8.7 Personality and Leadership

My personality is up and down, sassy and cheeky.

Katy Perry [1]

It is thought that personality has an important influence on leadership characteristics being able to shape both effective and ineffective leadership.

Personality has been described as having a bright side and a dark side. The "Big Five" is a description of the bright side traits, attributes of individuals when they are at their best, at the bright side of personality [44]:

1. Openness to experiences (exploratory)
2. Conscientiousness (reliable)
3. Extraversion (outgoing)
4. Agreeableness (cooperative)
5. Neuroticism (emotionally stable)

Judge and Bono [45] examined the relationship between personality traits and transformational leadership, in leaders from over 200 organisations. They reported extraversion and agreeableness as positively predicting transformational leadership.

In contrast, dark side traits are flawed inter-personal behaviours that may prevent leaders from succeeding by impairing their ability to build and run effective teams. Robert Hogan, a personality theorist in the USA, systematically studied failed leadership in various contexts (such as the civil service, business, military) and concluded that leaders fail on one or more of certain inter-personal behaviours. Along with his wife Joyce he went on to relate these behaviours to personality traits to which they referred to as the "the dark side of personality" [46–50], and designed the Hogan development survey to assess them. Hogan describes several attributes that can be disruptive to the surgeon leader, and these were incorporated in Leadership guidance provided by the Royal College of Surgeons of England [51]:

1. Excitable—unpredictable
2. Sceptical—distrustful
3. Cautious—indecisive
4. Reserved—detached, introverted
5. Leisurely—resistant to calls for improvement
6. Bold—arrogant, unrealistic
7. Mischievous—manipulative
8. Colourful—attention seeking
9. Imaginative—eccentric
10. Diligent—over-controlling, micro-managing
11. Dutiful—unable to stand up to others

One may evaluate own-self and own personality to try and avoid harmful behaviours or improve on what may hold them back from reaching their full leadership potential.

8.8 Leadership Posts and Titles

It is not titles that honor men, but men that honor titles.

Niccolo Machiavelli [1]

Much is said about leadership in surgery, with not just chapters but whole books devoted to the subject; surgeons acting as leaders of departments or institutions, leaders of research, leaders of audit, leaders of innovation, the list is endless. Such focus on leadership may put unintentional pressure on one to strive for leadership posts and could even be considered a deficiency if this is not achieved.

But leadership may not be all about taking a leadership post, or even leading others. Having leadership skills does not equate being in a leadership position. Some of us don't have the motivation, interest or resilience to take on such major challenges. It's not a failure if you cannot see yourself wearing the helm.

Many of us may be leaders whatever our position in society or organisation. Most of us lead one way or another, our families our friends our own-selves. One may gain pride in being able to lead own-self into further development, further learning and improvement to allow them to do even better for their patients. One may lead own-self into being a constructive follower who will applaud the achievements of those running the system, and speak up where improvements are necessary.

One may lead themselves into being someone that can be looked up to for their principles and actions, rather than for their titles and posts. You may be aware of politicians having left an inspirational legacy without ever making it to the front bench team. And some may be remembered because despite tasting power and posts, when the times and their principles demanded, they followed their values, returning from the frontline to the backbench.

References

1. Famous Quotes at BrainyQuote. https://www.brainyquote.com/quotes/quotes. Accessed 31 Mar 2017.
2. Popular Quotes. Goodreads. http://www.goodreads.com/quotes. Accessed 1 Apr 2017.
3. Lewin K, Lippit R, White RK. Patterns of aggressive behavior in experimentally created social climates. J Soc Psychol. 1939;10:271–301.
4. Waller DJ, Smith SR, Warnock JT. Situational theory of leadership. Am J Hosp Pharm. 1989;46(11):2335–41.
5. Turner N, Barling J, Epitropaki O, Butcher V, Milner C. Transformational leadership and moral reasoning. J Appl Psychol. 2002;87(2):304–11.
6. Bass BM, Avolio BJ. Improving organizational effectiveness through transformational leadership. Thousand Oaks, CA: SAGE; 1994.
7. Bass BM. From transactional to transformational leadership: Learning to share the vision. Organ Dyn. 1990;18(3):19–31.
8. Kaplan B. In: Kaiser R, editor. The versatile leader: make the most of your strengths without overdoing it: a guide to becoming a versatile leader (J-B US non-franchise leadership) hardcover. 1st ed. New York: John Wiley & Sons; 2006.
9. Kruse K. 100 Best Quotes On Leadership. https://www.forbes.com/sites/kevinkruse/2012/10/16/quotes-on-leadership/#1dcf6c382feb. Accessed 1 Apr 2017.
10. Spinelli RJ. The applicability of Bass's model of transformational, transactional, and laissez-faire leadership in the hospital administrative environment. Hosp Top. 2006 Spring;84(2):11–8.
11. Hogan R, Kaiser RB. What we know about leadership. Rev Gen Psychol. 2005;9(2):169–80.
12. Hogan R. Personality and the fate of organizations. Hillsdale, NJ: Erlbaum; 2007.
13. Rasch R, Shen W, Davies SE, Bono J. The development of a taxonomy of ineffective leadership behaviors. Paper presented at the 23rd annual conference of the Society for Industrial and Organizational Psychology, San Francisco, CA, 2008 April 10–12.
14. Hu YY, Parker SH, Lipsitz SR, Arriaga AF, Peyre SE, Corso KA, Roth EM, Yule SJ, Greenberg CC. Surgeons' leadership styles and team behavior in the operating room. J Am Coll Surg. 2016;222(1):41–51.
15. Leach LS, Myrtle RC, Weaver FA, Dasu S. Assessing the performance of surgical teams. Health Care Manag Rev. 2009;34(1):29–41.
16. Leach LS, Myrtle RC, Weaver FA. Surgical teams: role perspectives and role dynamics in the operating room. Health Serv Manag Res. 2011;24(2):81–90.
17. Kissane-Lee NA, Yule S, Pozner CN, Smink DS. Attending surgeons' leadership style in the operating room: comparing junior residents' experiences and preferences. J Surg Educ. 2016;73(1):40–4.
18. Up in the Air (2009) - Quotes – IMDb. http://www.imdb.com/title/tt1193138/quotes. Accessed 1 Apr 2017.
19. Hogan R, Warrenfeltz R. Educating the modern manager. Acad Manage Learn Educ. 2003;2:74–84.
20. Gordon T. What every leader should know. http://www.gordontraining.com/wp-content/uploads/What%20Every%20Leader%20Should%20Know%202012.pdf. Accessed 1 Apr 2017.
21. Lombardo MM, Ruderman MN, McCauley CD. Explanations of success and derailment in upper-level management positions. J Bus Psychol. 1988;2:199–216.
22. Henrickson Parker S, Flin R, McKinley A, Yule S. The Surgeons' Leadership Inventory (SLI): a taxonomy and rating system for surgeons' intraoperative leadership skills. Am J Surg. 2013;205(6):745–51.
23. Hjortdahl M, Ringen AH, Naess AC, Wisborg T. Leadership is the essential non-technical skill in the trauma team–results of a qualitative study. Scand J Trauma Resusc Emerg Med. 2009;17:48. doi:10.1186/1757-7241-17-48.

24. Patel VM, Warren O, Humphris P, Ahmed K, Ashrafian H, Rao C, Athanasiou T, Darzi A. What does leadership in surgery entail? ANZ J Surg. 2010;80(12):876–83.
25. Edmondson AC. Speaking up in the operating room: how team leaders promote learning in interdisciplinary action teams. J Manag Stud. 2003;40:1419–52.
26. The First Lady of Fleet Street - BBC Newsnight. 2016. Accessed 1 Apr 2017.
27. Inspirational quotes. http://www.inspirational-quotes.info/leadership.html. Accessed 1 Apr 2017.
28. Kübler-Ross E. On death and dying. New York: Macmillan; 1969.
29. Healy JM. How hospital leaders implemented a safe surgery protocol in Australian hospitals. Int J Qual Health Care. 2012;24(1):88–94.
30. Bergs J, Lambrechts F, Simons P, Vlayen A, Marneffe W, Hellings J, Cleemput I, Vandijck D. Barriers and facilitators related to the implementation of surgical safety checklists: a systematic review of the qualitative evidence. BMJ Qual Saf. 2015;24(12):776–86.
31. Lau BD, Arnaoutakis GJ, Streiff MB, Howley IW, Poruk KE, Beaulieu R, Ellison TA, Van Arendonk KJ, Kraus PS, Hobson DB, Holzmueller CG, Black JH 3rd, Pronovost PJ, Haut ER. Individualized performance feedback to surgical residents improves appropriate venous thromboembolism prophylaxis prescription and reduces potentially preventable VTE: a prospective cohort study. Ann Surg. 2016;264(6):1181–7.
32. 10th man theory, the God of devils advocate. https://www.youtube.com/watch?v=AcNK7M2eCI4. Accessed 1 Apr 2017.
33. Bellisle M. Washington is 1st state to sue Trump over immigration order. AP. http://bigstory.ap.org/article/ddb53cbedf694c8c94c2df7c9fa2ecdb/washington-state-sues-trump-overimmigration-order. Accessed 1 Apr 2017.
34. Ferguson A, Moritz M. Leading. 1st ed. London: Hodder & Stoughton; 2015.
35. BBC News. 2016. http://www.bbc.co.uk/news/uk-politics-36781565. Accessed 1 Apr 2017.
36. Notting Hill (1999). Awesome Movie Scripts and Screenplays. Accessed at http://www.awesomefilm.com/script/nothill.txt on 29/6/17
37. President Richard Nixon - Address Announcing Resignation. https://www.youtube.com/watch?v=ZEOGJJ7UKFM. Accessed 1 Apr 2017.
38. BBC Archive. Abdication of King Edward VIII | Farewell Speech from Windsor Castle. 1936. http://www.bbc.co.uk/archive/edward_viii/12937.shtml. Accessed 1 Apr 2017.
39. BBC News. Pope Benedict XVI to resign citing poor health. 2013. http://www.bbc.co.uk/news/world-21411304. Accessed 1 Apr 2017.
40. Bentz VJ. A view from the top: a thirty-year perspective on research devoted to discovery, description, and prediction of executive behavior. Paper presented at the 93rd annual convention of the American Psychological Association, Los Angeles, CA, 1985.
41. Eichinger RW, Lombardo MM. Knowledge summary series: 360-degree assessment. Hum Resour Plan. 2003;26:34–44.
42. Finkelstein SM. Why smart executives fail: and what you can learn from their mistakes. New York: Portfolio; 2003.
43. Dotlich DL, Cairo PC. Why CEOs fail. San Francisco: Jossey-Bass; 2003.
44. John OP. The "Big Five" factor taxonomy: dimensions of personality in the natural language and in questionnaires. In: Pervin LA, editor. Handbook of personality theory and research. New York: Guildford. 1990. p. 66–100.
45. Judge TA, Bono JE, Ilies R, Gerhardt MW. Personality and leadership: a qualitative and quantitative review. J Appl Psychol. 2002;87:765–80.
46. Hogan R, Hogan J. Hogan development survey manual. Tulsa, OK: Hogan Assessment systems; 1997.
47. Hogan R, Curphy GJ, Hogan J. What we know about leadership. Effectiveness and personality. Am Psychol. 1994;49(6):493–504.
48. Hogan R, Raskin R, Fazzini D. The dark side of charisma. In: Clark KE, Clark MB (eds). Measures of leadership. West Orange, NJ: Leadership Library of America; 1990. p. 343–54.

49. Hogan R, Hogan J. Assessing leadership: a view from the dark side. Int J Sel Assess. 2001;9:40–51.
50. Hogan J, Hogan R, Kaiser RB. Management derailment. In: Zedeck S, editor. APA handbook of industrial and organizational psychology, Maintaining, expanding, and contracting the organization, vol. 3. Washington, DC: American Psychological Association; 2011. p. 555–75.
51. The Leadership and Management of Surgical Teams. The Royal College of Surgeons of England. 2007. https://www.google.co.uk/url?url=https://www.rcseng.ac.uk/-/media/files/rcs/library-and-publications/non-journal-publications/leadership--management.pdf. Accessed 1 Apr 2017.

Chapter 9
Surgical Research

Research is defined by the Cambridge dictionary as the "detailed study of a subject, especially in order to discover (new) information or reach a (new) understanding" [1]. Hence, surgical research may be considered as the process which investigates and aims to discover new information on the development, presentation, diagnosis and treatment of surgical conditions.

The Surgeon has a pivotal role in identifying where surgical research is needed, and in the design, delivery and dissemination of surgical research. Through research participation a surgeon appreciates the limitations of research, and the limitations of our knowledge; one begins to understand that the didactic teaching encountered in books is often an over simplification of the current state of knowledge. Participating in research may also enable surgeons to question the validity of methodologies, as well as scrutinise the interpretation of results and accuracy of conclusions presented in research studies.

Surgical research may be basic science, clinical or literature based. Clinical studies may be of multiple designs including retrospective or prospective, controlled, or randomised. Clinical studies evaluating the role of surgical interventions pose particular challenges and these are described in this chapter. The value of feasibility and pilot studies, the role of patient and public involvement, and the importance of surgeon engagement in enhancing the successful completion of surgical clinical studies are also discussed. Ways of facilitating the translation of research findings to clinical practise are also described, and guidance is presented on how to contribute to the assessment of research as part of a Journal Club.

C. Panayiotou Charalambous, *Career Skills for Surgeons*,
DOI 10.1007/978-3-319-57490-5_9

CENTRE
OF
SHAM
SURGERY
We had to change our clinic name following the latest RCT in The Journal

9.1 Accepting the Need for Surgical Research

Medicine is a science of uncertainty and an art of probability.

William Osler [2]

The first step in constructing a clinical research study is to acknowledge that there is a gap in current knowledge, accept that there may be uncertainty as to best practice, that there may be alternatives to what is perceived as acceptable or mainstream.

Clinical equipoise refers to the presence of real uncertainty amongst the medical community as to whether a treatment is efficacious or beneficial [3]. In comparing two treatments, clinical equipoise refers to the presence of real doubt as to whether one treatment is better than the other. The presence of clinical equipoise may justify research to determine if a treatment works, if an intervention is better. But is not the mere presence of equipoise, but also its acceptance by the clinical community, that is vital in setting up and successfully running a research study.

However, such acceptance of uncertainty, may be more difficult amongst the surgical profession. As surgeons we are often trained using the apprentice model, learning through observation and experience, taught by seniors or experts. We may thus develop specific, almost regimented ways of doing things, of approaching and treating surgical conditions.

As surgeons we may be having intuitive rather than analytical evidence based preference for one treatment over another, our ways may have worked well in the hands of our trainers, may be working well in our hands, hence the reluctance for these to be challenged. If we have strong, almost dogmatic views as to the value of one intervention and the ineffectiveness of another, how can we easily accept the need to formally compare the two?

Some interventions may be substantially different—operate versus wait and see, pills versus the scalpel. How can we justify comparing the two if we think we already know the answer?

We may not be comfortable to accept that there is uncertainty in what we do as we like to give our patients definite information and leave little doubt. We may feel that if we introduce uncertainty, our authority or competence may be questioned. Patients may turn to more definite and more confident surgeons. There may be worries of being more prone to medico-legal challenge if our plans went wrong, if adverse events were to happen. And when considering a randomised trial, how can we accept that randomisation or simple chance, can choose the intervention applied? After all how would we feel if when getting on a plane, the pilot proposed throwing a dice to decide between going round or through the upcoming storm?

Hence the first step in surgical research is to question if what we do in our current practise is evidence based, based on good quality, scientifically robust data. If not, should we be re-examining our practise? Should we be trying to come up with the evidence?

And if as researchers we accept that there is clinical equipoise, is that view shared by the wider surgical community? Are there enough surgeons who share these views who can then participate in the proposed research study?

9.2 Defining the Aim

If you aim at nothing, you'll hit it every time.

Zig Ziglar [2]

In setting up a clinical study one must clearly define its aim. This is its main objective, the question the study attempts to answer.

One may consider the Ladder of Interventions described earlier in this book. The least invasive intervention lies at the bottom of the ladder and the most invasive at the top. In ascending the ladder one goes from observation (allowing natural history to follow its course), to non-medical interventions, medical pharmaceutical interventions, minimally invasive non-surgical procedures (such as injections), minimally invasive surgery (arthroscopic, laparoscopic) to extensive open surgery.

One may concentrate at a particular step of this ladder or look at multiple steps, to evaluate one or more interventions. Comparisons of interventions may be between separate steps or between various treatment options within the same step.

Comparisons between management options on different steps may help determine whether there is any benefit of following a step by step treatment approach or whether it is more beneficial to jump up steps. In contrast, comparison between various treatment options within the same step may help determine which one is the preferable option in patients whose lower step treatments have been exhausted.

Hence, the specific aim of the study will determine the groups to be examined, and the "general populations" from which the study groups can be obtained.

Consider a group of patients with osteoarthritis of the knee:

- Within step comparison: Patients may be taken through the non-invasive treatment options and only those who have not responded to these options are considered for knee replacement. Hence, cases who do not respond to non-surgical treatments and who end up needing knee replacement may then be studied to determine the benefits of various types of knee replacement implants.
- Comparison of separate steps: It may be of interest to evaluate whether it is beneficial to exhaust non-invasive options prior to proceeding with knee replacement or whether early knee replacement is better. It may be that if early knee replacement is performed then time wasted trying less invasive options may be avoided, with patients improving faster with regards to pain and function. Potential disadvantages of this however is that patients who could have avoided surgery end up having major surgery.

In comparison studies it is also necessary to specify what you aim to determine [4, 5]:

- Superiority study: is one group superior to the other?
- Non-inferiority study: is one group not inferior to the other?
- Equivalence study: are the two groups equivalent?

To understand the above, you may consider research aiming at evaluating a new surgical implant. In introducing a new implant it may be important to determine whether this new prosthesis is better (superior) with regards to pain and function as compared to other established implants. However, even if not superior, it would be important to establish that the new implant is not any worse (inferior) with regards to these outcomes. If it is not worse but it is less costly it could still be preferable. It may also be of use to know that the new implant is neither any worse nor any better (equivalent), hence allowing a more fair play in implant selection [4, 5].

9.3 Defining the Study Groups

Groups are grammatical fictions; only individuals exist, and each individual is different.

Robert Anton Wilson [2]

The next step is to determine which patients are eligible for participation in the study. To do this certain inclusion/exclusion criteria are applied. Cases that meet these inclusion/exclusion criteria can participate in the study. Such criteria may refer to:

- Characteristics of the patient (demographics).
- Characteristics of the condition in consideration (symptomatology, pathological stage).
- Characteristics of the intervention to be examined (type of surgical procedure).
- Presence of previous failed interventions.
- Other relevant parameters (ability to consent, anaesthetic fitness, centre of treatment, treating surgeon).

A physician evaluating the effect of a new antihypertensive tablet may have to define the study group in terms of type and level of hypertension so that only those with primary severe hypertension can participate. Along similar lines, an orthopaedic surgeon examining the effect of plating on wrist fractures may want to look at two part displaced fractures rather than any wrist fractures (which could vary from un-displaced to very multi-fragmented completely displaced fractures). A vascular surgeon may wish to examine the effect of prophylactic aneurysm surgery on medium sized rather than any size aortic aneurysm.

If the inclusion/exclusion criteria are too strict, it may be difficult to identify the necessary number of patients for participation. In addition, any observations made in such a selected group may not be easily applicable to the wider population having the investigated disorder. In contrast, if too loose inclusion criteria are applied, then you may end up with a very heterogeneous group, like a "bowl of apples and oranges"; this could potentially distort the results that might have been obtained if "bowl" subsets were examined in isolation.

An additional challenge, which may be of particular relevance to surgery, is whether definition of the study groups should be disease or intervention driven. The former approach allows you to specify a disorder, identify a homogeneous group with this disorder, and then examine the effect of an intervention on this group (e.g. the outcomes of knee replacement in a group of patients with primary tri-compartmental knee osteoarthritis).

Surgical studies however often fall in the second category, whereby the study group is defined based on whether they had the intervention (e.g. outcomes of a group of patients who had knee arthroplasty). This may be because the focus is on getting to know how well a procedure is doing, how well a new implant is performing, what are the outcomes of an intervention in the hands of an individual surgeon.

However, if an intervention driven approach is utilised, attention must be paid to avoid ending up with a very distorted heterogeneous treated group; such as evaluating a knee replacement implant in a group of cases in whom the underlying diagnosis includes any cause of arthritis (primary, post-traumatic, inflammatory, metabolic), and any stage of arthritis (uni-/bi-compartmental, tri-compartmental).

9.4 Defining the Study Population

Designers simply can't afford to make dresses in 10 sizes for a sample.

Nigel Barker [2]

Once the intervention group is specified the next challenge is to identify the "general population" from which this group can be drawn. This will help plan as how to approach or screen this population to identify potential study participants.

You may consider the following examples:

- In examining the outcomes of knee replacements that have been performed by an individual surgeon, the general population may be patients who had any type of surgery under that particular surgeon. Hence, screening the listing forms of patients scheduled for any surgery, or screening the theatre logbooks could help identify the group to be examined.
- In comparing the clinical outcomes of different implants of knee replacement (which is yet to be carried out), the general population may be those patients presenting with symptomatic knee arthritis to secondary care and who are scheduled for surgery. Hence, patients may need to be approached for participation in secondary care clinics.
- In comparing physiotherapy versus early knee replacement following an initial presentation with arthritic knee pain, the general population may be those patients presenting with symptomatic knee arthritis to primary care rather than to secondary care. Hence, patients may need to be approached at primary care clinics rather than hospital clinics. If instead of approaching patients in primary care, patients are approached in secondary care clinics, the results of the researched comparison may not be applicable to the wider community of arthritic knees as secondary care patients are a pre-selected group; for some patients treated in primary care might have had resolution of their symptoms, and hence never got referred to secondary care.

9.5 Study Design

I don't start with a design objective, I start with a communication objective. I feel my project is successful if it communicates what it is supposed to communicate.

Mike Davidson [2]

Another challenge in surgical clinical research is which study design to employ. The Research Ladder [6] describes study designs according to how reliable the evidence they present may be considered. At the bottom of the Research Ladder are reports of single or limited number of clinical cases with further steps ascending to retrospective studies, prospective studies, randomised controlled trials and finally systematic reviews and meta-analyses (Fig. 9.1).

Case reports: these describe the events or outcomes of the treatment of one patient or small group of patients. They are limited by their small numbers and lack of comparison groups.

Retrospective studies: these are studies whereby the researcher looks back, in the past, to assess the outcome of an event (in this case a surgical intervention) that has already occurred. They often rely on reviewing medical records or patient recall for data collection, the accuracy and reliability of which may be limited. However, as the intervention in question has already occurred, retrospective studies can be completed more quickly than prospective studies.

Prospective (non-randomised) studies: these are studies whereby the researcher identifies the participants prior to the intervention applied, and follows them into the future to assess as to how they perform following the application of the intervention. They allow more accurate and complete data collection, but could take longer to complete and are resource consuming.

Case control studies: in epidemiological research this design has been mainly used to study risk factors for a disease. These studies compare the rate of one or more risk factors/exposures amongst a sample of individuals with the disease (cases) against the rate of risk factors/exposures amongst a sample of individuals without the disease (controls). Hence, they compare the distribution of risk factors/exposures among diseased and non-diseased groups. Such case control studies could be utilised in the assessment of surgical interventions, such as by considering the intervention as the exposure and a surgical complication as the "disease" (Fig. 9.2).

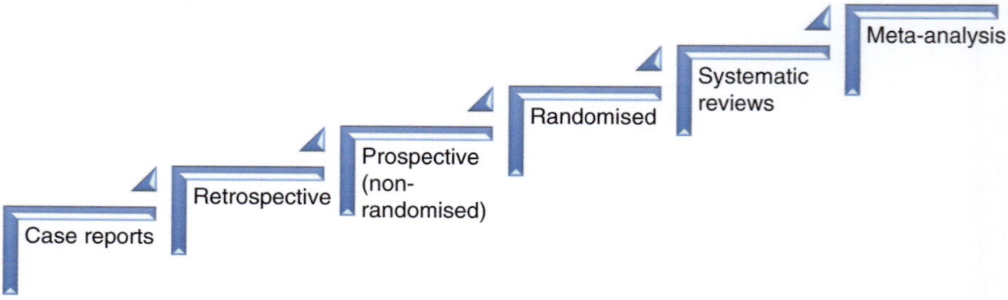

Fig. 9.1 Research ladder. Adapted from Charalambous CP [6]

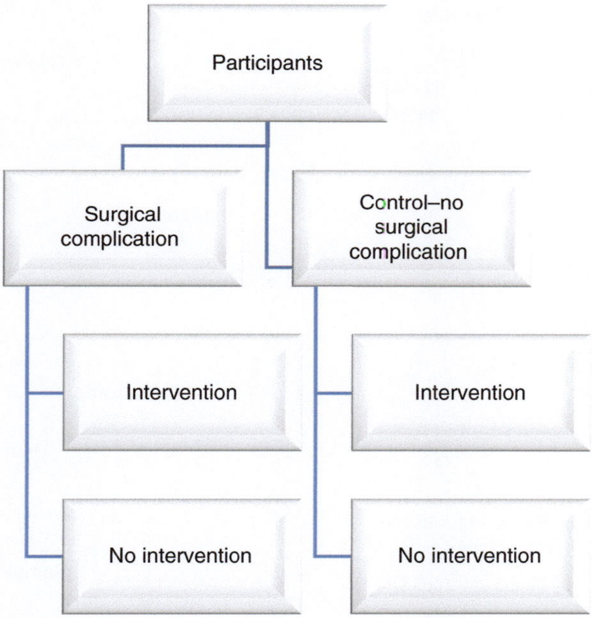

Fig. 9.2 Case control study

In surgery, however, the term "case-control design" has also been used to describe studies whereby patients having an intervention (cases) are compared with those not having the intervention (controls-having no intervention or another intervention), to determine if the intervention in question offers any advantage. Cases and controls are matched with regards to certain parameters that could influence outcomes, to ensure they are as similar as possible, with the main difference between the groups being the application of the examined intervention. However, the decision as to whether an individual is having the intervention is not made randomly, which means that the groups may differ in some other unknown (and hence not matched) influential parameter, which could in turn have accounted for the observed results. Mihailovic et al. [7] examined all self-identified "case–control" studies published between 1995 and 2000 in six surgical journals, and reported that only 35% of these met the epidemiological case-control definition. The rest used the term whilst comparing the clinical outcomes of groups treated with different interventions. This lack of consistency in the terms should be taken into account when reporting or interpreting surgical research.

<u>Cohort studies:</u> these identify a group (cohort) which is then followed up to see how it performs. The cohort may be identified prospectively before it sets off, or by going back in time. It may be that all individuals in the cohort end up having one intervention hence providing no comparison group (similar to a large case series). It is also possible that some individuals have a different intervention or no intervention, allowing comparisons to be made (Fig. 9.3). The decision as to whether an individual is having the intervention is not made randomly, hence the potential limitations this study design confers.

Fig. 9.3 Cohort study

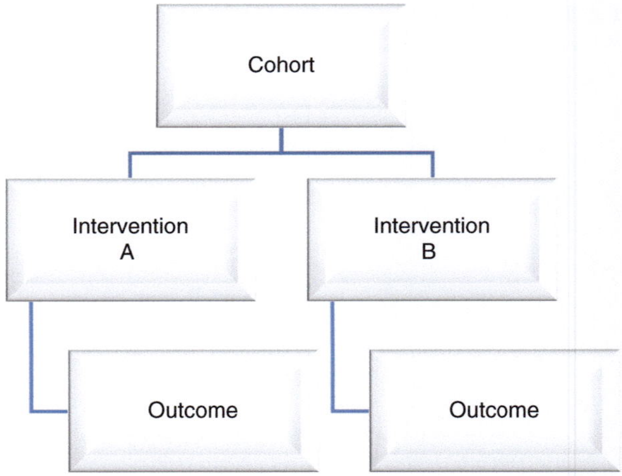

Fig. 9.4 Randomised controlled trial

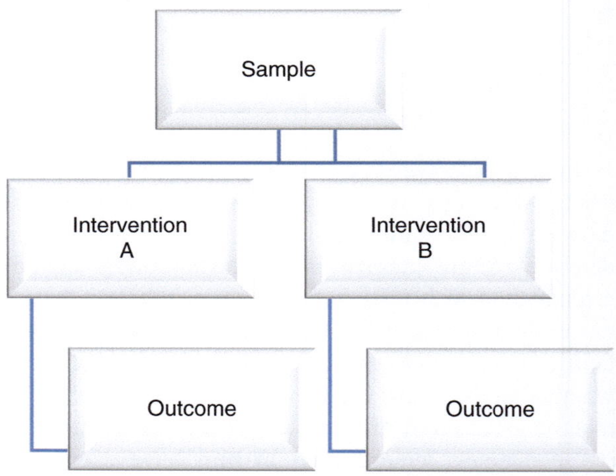

Randomized controlled clinical trials (RCTs): these introduce an intervention to a group of patients and assess its effects upon that group compared to a group having a different or no intervention. Intervention allocation is random which aims to eliminate any bias in assigning treatment, and to match the comparison groups both with regards to known and unknown potentially influential parameters (Fig. 9.4).

Where possible, patient, investigator and assessor are blinded as to the treatment allocation. Blinding in surgical clinical studies is particularly challenging. It is very difficult, if not impossible to blind a surgeon to the administration of a surgical intervention. The same may apply to the patient when a particular surgical intervention is compared with non- surgical treatment. However, in some cases, for instance where different implants may be compared, attempts can be made for the surgeon or patient to be blinded. Nevertheless, the assessors may be blinded such as by having the patient wearing a T-shirt to hide an abdominal scar, to stop the assessor of knowing whether that particular patient had abdominal aortic surgery, whilst examining their peripheral circulation. The utilisation of members of the treatment team in assessing outcomes is to be avoided to minimise assessor bias (a junior trying to please their senior may sub-consciously guide patients into giving positive responses when assessing outcomes of a surgical procedure done by their senior). Hence, independent assessors are preferred.

Systematic reviews: these identify (through a literature search) studies which examined the intervention in question. These studies are reviewed and their results summarized and presented. The results of the included individual studies are not statistically combined. In contrast, meta-analyses use statistical methods to combine the results of the individual studies, and report such results as if a large study had been performed.

RCTs are considered the gold standard in clinical research but it is often not possible to jump straight to an RCT, and RCTs s may be considered as complementing rather than completely replacing all other types of research [6]. Several stages have been described in which a new intervention is assessed, starting from an initial description in the form of a case report or case series followed by good quality cohort studies, to RCTs, to long term surveillance studies. Hence, all types of research have their role in answering vital surgical questions, rather than just RCTs [8–11].

9.6 Defining the Intervention

We don't define the unlimited.

Toba Beta [12]

Unlike pharmaceutical studies where the intervention may be in the form of a tablet or medicine infusion, surgical interventions may be much more complex in their composition and application. The term "surgical intervention" may not just include the surgical procedure that takes place in the operating theatre, but also the pre-operative, intra-operative and post-operative patient care that could potentially influence clinical outcomes.

In evaluating a particular knee replacement implant, clinical outcomes may be influenced not only by how well the implant behaves in terms of durability and resilience, but also by the way the soft tissues were managed, the accuracy with which the bone cuts were made, the pre-operative care of the patients in terms of optimisation for surgery (haemoglobin control, physiotherapy to strengthen the muscles to allow early post-surgical mobilisation, treating urinary tract infections to reduce infection risk), the intra-operative care (use of spinal anaesthesia, local anaesthetic infiltration, use of tourniquet, timing of antibiotic prophylaxis) and post-operative care (timing, type and intensity of physiotherapy, use of cryotherapy to reduce postoperative swelling). Unlike a medicine the administration and dosage of which may be more easily defined, it may be difficult to standardise a surgical procedure, as two cases are not identical, and surgeons may vary as to how exactly they perform a procedure.

In principle two of the potential ways that can be adopted are [13, 14]:

1. Explanatory approach where the various steps are specified and controlled, hence any findings will be directly applicable to those specific conditions.
2. Pragmatic approach whereby the various management components are left to the surgeon as per their normal practice other than what is considered as the major component of the intervention under examination. This means identifying those parts of the intervention that are considered vital in influencing outcomes and ensuring that these are consistently applied by all participants, whilst leaving the remaining parts of care to the discretion of the treating surgeon.

The challenge is that if very restricted criteria are utilised in defining the intervention, the surgeons may find that the intervention is not similar to how themselves carry out the procedure, and hence consider the results of the study irrelevant to their practice. This could mean they are reluctant to adopt any study findings. In contrast, if the definition of the intervention is too loose, there may be such a variation that the quality and relevance of the results may be questioned. The study design should try and find the right balance, to increase acceptability to the general surgical population.

The early stages of a surgical intervention may be assessed with explanatory studies. Once its efficacy is established under strict conditions, a more pragmatic approach may be utilised to determine what the results are when the intervention is applied to the real world.

In line with this, there is a need to consider whether the study should include cases from an individual surgeon, multiple surgeons, single centre, or multiple centres. In reality, this decision may be influenced by the study aim and available resources. Nevertheless, in the early stages where an intervention is introduced and efficacy is assessed under strict conditions, inclusion of a single surgeon or a small group of surgeons (who were involved in developing the intervention) may be preferable. Once the procedure is established, the inclusion of a number of surgeons, in multiple centres will give results that can be applicable to the wider surgical community. On occasions, the results of a procedure amongst the general surgical community are inferior to those reported by the pioneers of the technique, hence the need to carry this further assessment.

9.7 Quality Assurance of Interventions

Quality is not an act, it is a habit.

Aristotle [2]

A further challenge in surgical clinical research is to put in place check mechanisms, to ensure the intervention was applied at an acceptable level. In a pharmaceutical intervention one may ensure that an infusion is administered over a specific timeframe by an infusion pump. The equivalent in surgery, in a knee replacement study would be to determine that a knee replacement procedure was actually performed to satisfactory standards. Ensuring high quality of any examined interventions in clinical studies such as RCTs, is an important factor for the successful uptake and application of RCT findings [15, 16].

Some design approaches that may be employed in ensuring quality of the applied interventions are:

1. Including only surgeons of certain experience, to minimise any learning curve effects.
2. Utilising a study design that allows surgeons to perform only procedures they are comfortable with, without demanding to perform all types of interventions evaluated.
3. Intra-operative video recording which can be then accessed by an independent assessor.
4. Post-intervention by examination of post- operative radiographs by an independent assessor.

Such measures aim to give the reader the confidence that the intervention was applied at a high quality level and avoid comments such as "I don't recognise those results because I would perform the procedure in a much better way".

9.8 Control Groups—Sham Surgery—The Surgical Placebo

What is inefficient is the use of relatively expensive drugs as placebos. It is a pity some enterprising drug company does not produce a wide range of cheap, brightly coloured, non-toxic placebos.

Archibald L Cochrane [17, 18]

With many surgical procedures we are uncertain as to the exact mechanism by which they work and any placebo component that may come into play. Hence, a sham control may help us determine how much of the potential benefit we observe is due to a placebo effect, and how much is truly attributable to the technical aspects of the applied intervention.

In pharmacological studies we often compare an active agent against a placebo agent by administering the latter in a form (such as pill or capsule) which has the same appearance and characteristics as that of the therapeutic agent. However, examining a placebo effect in surgery may be more challenging. Surgical interventions cannot be easily masked, as patients will be aware as to whether they had a cut or not. One way of going round this is to carry out the surgical incision but not the rest of the surgical procedure, but this approach brings the ethical questions of potentially exposing patients to harm without any potential gain to those participants [19, 20].

Such studies have been performed previously. Moseley et al. [21] carried out an RCT to evaluate the role of arthroscopic debridement in knee osteoarthritis, and their results were published in the New England Journal of Medicine in 2002. They compared patients having arthroscopic debridement, with a group having arthroscopic washout, and a group having sham surgery. Several measures were taken to ensure that the sham surgery patients were blinded as to their treatment—they had their knee prepped and draped, and they had three small skin incisions made around the knee, the same that someone having a full debridement would need. The debridement instruments were used in theatre to simulate the noise and motions of surgery and saline was splashed to simulate the sounds of lavage, but the patient's knee was not entered. Sham surgery patients were kept in theatre for as long as those having real debridement, and their post-operative care was the same. In that hallmark study, which changed clinical practise amongst many orthopaedic surgeons, no significant differences in pain or function were observed between the intervention and sham surgery groups.

Along similar lines Freed et al. [22] randomised 40 patients who had severe Parkinson's disease into either receiving a transplant of embryonic dopamine producing nerve tissue into their brain, or sham surgery (that involved sole drilling holes in the skull). They demonstrated some clinical benefit in the transplant group amongst young but not amongst older patients.

Wartolowska et al. [23] carried out a systematic review of RCTs comparing surgical interventions with placebo. In 39 of 53 (74%) trials examined there was improvement in the placebo group, and in 27 trials (51%) the effect of placebo and surgery did not differ. In 26 trials (46%) the effect of surgery was superior to that of placebo but tended to be small.

Another issue to consider when it comes to sham surgery, is the extent to which the results of such comparisons are useful in clinical practice. Even though such comparisons may give an insight as to how surgery exerts its effects, it may not fully guide clinical practise in real life where the sham surgery option is unlikely to be available. In addition, a lack of demonstrable difference between "intervention surgery" and sham surgery may not necessarily imply lack of difference between "intervention surgery" and "application of no intervention".

9.9 Sample Size Calculation

Weather forecast for tonight: dark.

George Carlin [2]

A major challenge in surgical study design is to decide how many patients should be recruited. The greater the number of cases, the more time and resource consuming the study will be. In addition, if more patients were recruited than those required to answer the study question, some patients (through participation in the study) may be exposed to unnecessary interventions, or be denied helpful interventions.

In non-comparative studies we may be limited by the number of cases we have available, such as the number of procedures a surgeon has performed, and accept that even though the numbers are small they are still worthy reporting. In prospective trials we must accurately estimate how many cases we would need to obtain an answer to the research question. This can be achieved by sample size calculation that takes into account [24, 25]:

1. Type I error—concluding that there is a difference between groups when one does not exist. The P value is a measure of the probability of this happening, and is usually set at <0.05.
2. Type II error—concluding that there is no difference when one exists. A study may fail to show a difference between two groups because there is no significant difference or because the study did not have enough patients to allow the demonstration of such difference which exists. The power is the ability of the study to show the difference if one exists, and is often aimed at a level of 80% or greater.
3. Estimated mean and variance of the outcome in the groups.
4. Effect size—level of smallest difference in outcomes that would be considered of clinical importance. Tiny differences between groups may be of statistical significance, but if so tiny may be of no clinical significance.

It is necessary to describe which outcome is used to calculate sample size; usually that would be the primary outcome a study aims to examine. It should be noted that a study may be large enough to evaluate potential differences in the primary outcome but that does not mean it is also large enough to determine differences in other secondary outcomes, unless a sample size calculation was also applied to those. Hence, close scrutiny is necessary of sample size calculation, when studies report that no difference could be demonstrated between study groups in some of the parameters examined.

9.10 Follow Up

Instant gratification takes too long.

Carrie Fisher [2]

Another challenge in clinical surgical studies is the need to follow up participants to assess how well they do after an intervention is applied. The required duration of follow up and how to deal with those lost to follow up need to be considered.

Duration of follow up: In some cases the outcomes evaluated occur early following an intervention, hence patients do not need to be followed up for a long time (outcomes such as early post-operative pain, 30-day infection or mortality rates, surgical wound healing).

Some outcomes, however, need to be assessed several months or years following the applied intervention as the true effects may take time to become apparent. One may take a substantial period of time to recover from the trauma of surgery per se, and any gains from the intervention may not be assessable until after that.

Consider a rugby player that undergoes surgery to stabilise their shoulder to prevent further dislocations. Post-surgery there is a period in which the sutured ligaments heal and during which shoulder movements are limited to protect the repair. This is followed by a period of shoulder mobilisation, strengthening and rehabilitation with gradual return to play. The athlete may be able to return to competitive practise at six to nine months post-surgery. During this time one would expect a further need to fully re-adjust to play, regain confidence and use the arm at the same level as prior to all the shoulder troubles starting. Hence, it may take about two years for the arm to be exposed to activities that may re-dislocate the arm; an assessment of re-dislocation rates may require at least a two year follow up to be meaningful.

Similarly, the effects of surgery may deteriorate with time. Consider the evaluation of an implant used in hip replacement. As time goes by, implants wear, may loosen and hence the rate of revision surgery rises. Early evaluation of the revision rates of an implant may thus be misleading, providing false reassurance.

Lost to follow up: Attrition rate refers to the proportion of patients who do not complete their allocated treatment or assessment and these include lost to follow up. Participants may be lost to follow up and not be assessed for multiple reasons including:

- Relocation—relevant to mobile populations (younger, students) as patients may move away from the area where intervention was applied.
- Change of contact details, not traceable. Obtaining multiple contacts, addresses and phone numbers may help limit this.
- Not available to attend follow ups due to other commitments.
- Failure of intervention—symptoms persistent and sought treatment elsewhere.
- Success of intervention –hence no need to further engage with healthcare and attend any invited appointments.

There are several ways of dealing with those lost to follow up in analysing the results of a study. Lost to follow ups may be excluded, with only those who completed the follow up analysed. Alternatively, those lost to follow up may be considered as "all failures" or as "all successes" in the analysis, which can provide the spectrum of potential outcomes [26].

9.11 Surgical Waiting Lists and Research

I have a new philosophy. I'm only going to dread one day at a time.

Charles M. Schulz [2]

In many healthcare systems there are waiting lists for elective surgery which may vary from days to weeks to months or even longer. This places particular challenges in conducting clinical surgical studies including:

- If a patient is left to wait for long prior to surgical intervention their condition may deteriorate. Hence, an up to date evaluation is needed prior to administering the intervention to record their immediate pre-intervention status.
- Some cases may improve whilst awaiting surgery and no longer needing the intervention, leading to drop outs, which must be accounted for in any subsequent analysis.
- Study participants allocated to one treatment may change their mind and ask for the other treatment in the study. This will have to be accounted for in any subsequent analysis.

9.12 Outcomes

All right everyone, line up alphabetically according to your height.

Casey Stengel [2]

A major challenge of surgical studies is to identify the outcomes by which an intervention is assessed. There are multiple outcomes that may be assessed such as pain, function, mortality, morbidity, complication rates, speed of recovery, psychological impact, patient satisfaction, health economics, the list is endless.

The choice of outcomes will be influenced by the study's aims. Outcomes that are directly relevant to the various stakeholders to which the intervention applies are preferable. Such stakeholders are:

- Patients (patient reported outcomes).
- Surgeons (objective functional scores, infection rates, operative complications).
- Funders (cost-effectiveness).

In choosing outcomes you may consider the need for:

1. Well defined outcomes, which are easy to assess.
2. Outcomes assessed by validated systems.
3. Outcomes that are generic, disease specific, or intervention specific.
4. Outcomes that are commonly used in similar research studies, which can thus allow direct comparison between such studies, and also facilitate the pooling of such studies in subsequent systematic reviews and meta-analyses. Core outcomes may be identified for specific conditions or interventions and be utilised when those are evaluated. The COMET (Core Outcome Measures in Effectiveness Trials) initiative was launched in Liverpool, UK, in 2010, and brings together researchers aiming at achieving standardised sets of outcomes (core outcome sets) for specific conditions; with the aim that these will be the minimum reported in studies evaluating those conditions [27].

9.13 Recruitment

Success is how high you bounce when you hit bottom.

George S. Patton [2]

A major challenge faced by surgical clinical studies, particularly RCTs is the recruitment of patients to the study. Surgical studies, even when well designed, may have difficulties recruiting patients.

Chapman et al., in a 2014 observational study [28], reported early discontinuation of one in five surgical RCTs, with poor recruitment cited as the most common reason, observed in 44% of cases. Similarly, recruitment rates vary widely between orthopaedic trials comparing surgical and non-surgical interventions, ranging from 22 to 86% [29–31].

Poor recruitment may be due to several reasons:

- Patients not meeting inclusion/exclusion criteria.
- Patients not agreeing to participation.
- Surgeons not considering patients for inclusion.

Recruitment rates may be particularly challenging when comparisons are made between substantially different interventions, and when dealing with patients or surgeons who may have strong starting preferences for one line of treatment.

Recruitment may be low if recruiters cannot effectively communicate the clinical uncertainty that exists with regards to differences in effectiveness between treatment options, and hence the need for the research study.

In setting up a research study ways of ensuring recruitment must be considered. Training of surgeons or other recruiters may be needed so they can provide information about the study and interventions assessed in an impartial way. Systems that identify patients who, despite being suitable for recruitment, are not recruited and examination of the reasons behind this, may guide to recruitment improvement strategies.

9.14 Improving Success of Surgical Clinical Research

I'm writing a book. I've got the page numbers done.

Steven Wright [2]

There are several ways of trying to overcome surgical study challenges some of which are:

1. Carrying out a feasibility study prior to the full trial.
2. Carrying out a pilot study prior to the full trial.
3. Patient and public involvement—actively involving patients and the wider public in identifying research priorities, and in the design and delivery of surgical research.
4. Surgeon engagement—engaging surgeons to encourage collaboration and participation in research studies.

Feasibility studies in surgery

Rather than setting off with a definitive trial, with all the resource and time investment that entails, only to find out down the line that the study cannot be completed, one may initially set out to evaluate whether it is possible (feasible) to run such a definitive trial. This may be achieved using a feasibility study [32, 33].

A feasibility study may help [32, 33]:

- Determine recruitment and attrition rates (proportion of eligible patients willing to consent, proportion of patients that surgeons are willing to randomise, withdrawals, lost to follow up) and estimate deviation in outcomes to inform sample size calculation for the definitive trial. This can guide as to the number of centres and length of time necessary for recruiting to a full trial.
- Identify barriers to recruitment and provide information about how the recruitment process could be improved in a subsequent definitive trial. A feasibility study may help identify patient or surgeon concerns about the study that may then be addressed in the future full trial. Along these lines Donovan et al. [34] looked at recruitment barriers such as treatment preferences and perceptions of the recruitment process, in the ProTeCT (Prostate Testing for Cancer and Treatment) study, an RCT comparing surgery versus radiotherapy versus active monitoring for men with localised prostate cancer. The recruitment process was then adjusted based on the feedback received, such as emphasising equivalence of the treatment options, avoiding terms that could be mi-interpreted, and presenting clinical equipoise and randomisation in a more convincing way. This led to a rise in randomisation from 30 to 70% over 12 months. Similarly, Blazeby et al. [35] in a feasibility trial comparing surgical and non-surgical interventions for oesophageal cancer, demonstrated that close investigation of the challenges recruiters faced in articulating proposed treatments, and subsequent relevant training to help manage these challenges, increased recruitment from 0 to 31%.
- Assess ease and rates of completion of outcome measures.
- Assess the feasibility of trial design.

- Assess ability to blind outcome assessors.
- Develop models for health economic analysis.
- Assess acceptability of randomisation amongst surgeons.
- Map variation in patient treatment pathways across participating sites.
- Assess whether measures aiming at safeguarding the high quality of the applied interventions, can be adhered to.

Pilot studies

Pilot studies are a mini-version of the definitive trial which aims to test whether all the components of the study can fall in place, to allow a smooth run [32, 33]. It helps to identify problems that can be resolved prior to setting off with the main trial. Any data collected from the pilot may be used in the definitive study analysis.

Patient and public involvement

It is increasingly recognised that patient and public participation in the conception, design, development and running of research studies has an important role to play [36–38]. Patients are major stakeholders in the outcomes of research. In addition, as patients are the ones experiencing the disease and interventions, they may be able to provide an insight from the participant's perspective, a different angle to what the surgeon or researcher may view.

Patients may participate in the:

- Identification of important outcomes—what is important to the surgeon may not be of equal importance to the patient. A surgeon may consider the degree of angulation of fracture healing as an important outcome, but the patient may value more the ability to return to work early.
- Assessment of the ease of understanding outcomes measures, such as evaluation questionnaires.
- Study design, with an input to various parameters such as the frequency of follow up appointments, length of time needed to consider participation in the research.
- Governance of the study, as members of committees overseeing the study.
- Dissemination of information about the study to facilitate awareness of the study and encourage recruitment.
- Dissemination of the study results, to facilitate translation of any findings into clinical practise.

Surgeon engagement

One of the challenges researchers face is accessibility to potential participants.

In studies involving multiple centres, there is a need for patients to be identified and recruited by each of the participating centres. In each centre a researcher, which in surgical studies is often a surgeon, is allocated the task of overseeing the local running of the study, and is often referred to as the trial's gatekeeper at the site. It is a challenging task to ensure that an adequate number of such gatekeepers are identified to allow participation of the required number of centres, but also to ensure that active participation of such gatekeepers is maintained for the duration of the study [39, 40].

The same may apply in smaller studies, when trying to recruit from multiple surgical teams in a single centre. You may be faced with surgeons who are not interested in participating and make this clear early on, but also with ones who sound very keen in taking part and give lots of promises, but do not actually deliver when recruitment starts.

Several approaches may be employed in surgeon engagement such as:

- Having a strong enthusiastic team, that can effectively communicate and engage with the individual surgeons.
- High quality design, one that promises high quality outputs.
- Getting the support or endorsement of the study by a recognised society or institution, which may again indicate high quality design.
- Simplified design that can be easily incorporated into a surgeon's busy schedule or clinical practise.
- Involving potential collaborators early on in the study design, application of funding, and running of the study. Surgeons may thus feel as stakeholders in the study rather than mere recruiters for someone else's study.
- Recognising their contribution through authorship in any publications that may emerge.

9.15 Putting Evidence into Practise

Dreaming is good but implementation is success.

Paballo Seipei [12]

Another challenge in surgical clinical research is to ensure that any research findings influence or change clinical practice [41–45].

At a surgeon level this involves informing the wider surgical community of the findings, and convincing individual surgeons that study results are reliable and valid, as well as relevant to their own practise. This may be achieved in several ways:

- Study having high external validity (generalizability) –
  - Internal validity refers to the utilisation of safeguards in the design, methodology, and contact of the study, to ensure that the effects tested in the study are truly the result of the intervention applied rather than due to bias or random chance.
  - External validity refers to the extent of which the results of the study can be generalised or applied to a definable group of patients in a particular clinical setting [46, 47].

  Study characteristics such as the healthcare context of the trial, the demographics of participants, differences between the trial protocol and real clinical practice, and the outcome measures examined are some of the parameters which may limit the external validity of a study [47]:

  - The results of a study carried out in a specialised hip centre, comparing hemiarthroplasty versus total hip replacement in young active patients with hip fracture in terms of ability to return to high demand jobs and sports, may have limited generalisability when applied to low demand elderly hip fracture patients operated in district hospitals and in whom pain control and early mobilisation are the main surgical goals.

- Presenting the findings at local, national, or international meetings attended by stakeholders asked to implement change.
- Publicising the results through journal publications in high quality journals, that are accessible and read by the target population.
- Follow up initial publications with review articles or editorials.
- Developing clinical guidelines based on the study findings. Such guidelines may be at local or national level. These may be achieved through local institution policies, or through the development of guidance by surgical societies or national bodies evaluating the effectiveness of healthcare interventions.
- Having healthcare systems and institutional processes that are flexible to allow early incorporation of new research findings into clinical practise.
- Commissioners directing funding and remuneration for healthcare interventions in line with best evidence.
- Publicising the findings to the general public and patient groups, so that they can ask for what is available.

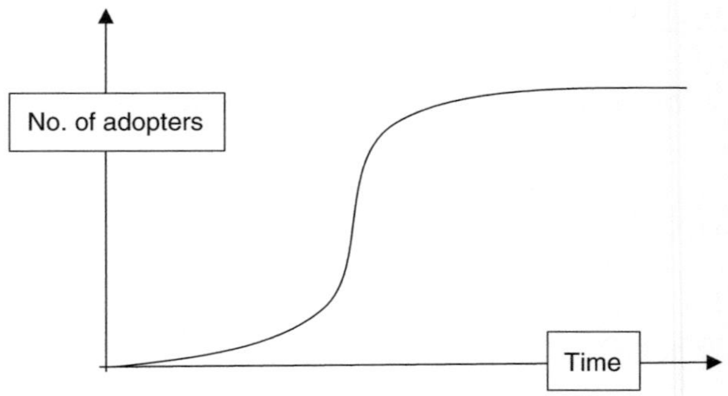

Fig. 9.5 Spread of an innovation over time. Adapted from Rogers E [48]

The challenge of putting research into practise may be explained by the theory of Diffusion of Innovations [48]. Everett Rogers, a professor of communication studies, described the Diffusion of Innovations theory, to describe how new ideas and technologies spread and come into wide use [48]. Spread of an innovation requires sufficient practitioners to take it on, who are described as early adopters, or as early majority, late majority and laggards according to the timing of their acceptance of new ideas. Spread of an innovation tends to follow an S curve when plotted against time (Fig. 9.5) [48].

Rogers [48] proposed that the main factors that influence the spread of a new finding are:

1. The innovation itself—simplicity, advantages, fitting in rather than disrupting established routines.
2. Communication—spread the word, increase awareness.
3. Time.
4. Social system—practitioners, organisations, leadership.

Hence, communication, patience and perseverance are often needed to turn the tide.

Thornley et al. [49] carried out an international survey of orthopaedic surgeons, to assess what factors of research studies are likely to influence a change in orthopaedic practise. They found that RCTs, systematic reviews and meta-analyses are the study designs mostly valued amongst orthopaedic surgeons. Publication of research in reputable journals and research performed by highly skilled subspecialist surgeons, and sample sizes of 101—500 cases were amongst the other factors encouraging adoption of research findings.

Despite the challenges faced with, surgical research can change practise. Costa et al. [50] evaluated the impact of the DRAFFT (Distal Radius Acute Fracture Fixation Trial) study on clinical practise in the UK. The DRAFFT compared volar locking plating (a more invasive and costly procedure) with percutaneous wire fixation (a less invasive and less costly procedure) for distal radial fractures, and showed no difference in functional outcomes between the two [51]. Prior the DRAFFT, 75% of distal radial fractures were plated and only 12% wired. Following publication of the trial, the rate of distal radial fractures wired rose to 42% and the rate of plating fell to 48%.

9.16 Running a Journal Club

Get your facts first, then you can distort them as you please.

Mark Twain [2]

Educating surgeons is an essential activity in breaching the gap between research and clinical practise. Journal Clubs are educational meetings where individuals meet to discuss and critically appraise newly published journal research articles [52, 53].

A Journal Club aims to assess the full article in a systematic way. It is not so much about getting informed of the question the study examined and the answer obtained, but more about evaluating the way in which a particular answer was reached.

As part of your training or teaching commitments you may have to present an article in a Journal Club, to your peers, colleagues, juniors or seniors. The aim is to present a summary of the article, followed by an assessment of each of its components, giving the positives as well as negatives. In presenting a Journal Club article you may consider:

Article distribution

Inform the meeting participants well in advance of the article to be presented, asking them to read it prior to the meeting. This may achieve a discussion rather than a one way presentation.

Article presentation

- Background: State the background of the problem as stated by the authors.
- Aim: State the aim of the article as stated by the authors.
- Methodology: Describe the method that was used, the design, the inclusion, exclusion criteria, the target population, outcomes.
- Results: State the most important results. You may use tables or graphs taken from the actual article.
- Discussion: Describe what the authors discussed and give their final conclusion in their own words.

Article evaluation

Evaluate each step of the article in turn:

- Background: Did the authors give a clear, concise but adequate description of the background that forms the basis of the research article? Did they justify the need of their study?
- Aims: Did the authors give a specific, clearly understood aim?
- Methods: Was the choice of method design appropriate? Were the inclusion/exclusion criteria correct and rigid? Was the target population appropriate? Were the statistical methods relevant to the results and aims? Was there a sample calculation or power analysis?
- Results: Were the results clearly presented through text, graphs, tables or figures? Were the results complete or are some missing? Do the numbers add up?

- Discussion: Was there a balanced discussion given the findings? Did the authors relate their findings to those of other relevant studies?
- Conclusions: Are the authors' conclusions justified, given their aims and results?

At the end of your presentation you need to ask whether the research study is relevant to your practise, or the practise of your audience. If it is relevant to your practise, will the study make you change your practice based on its findings?

- You may decide to incorporate the results of the study in your clinical practise.
- Alternatively, you may decide to stick to what you do; you may not be seeing the patient population examined by the study, you may not agree with the conclusions given the actual study results, you may be doing already what the study authors propose and support.

References

1. Cambridge Dictionary. http://dictionary.cambridge.org/dictionary/english/research. Accessed 2 Apr 2017.
2. Famous Quotes at BrainyQuote. https://www.brainyquote.com/quotes/quotes. Accessed 2 Apr 2017.
3. Freedman B. Equipoise and the ethics of clinical research. N Engl J Med. 1987 Jul 16;317(3):141–5.
4. Lesaffre E. Superiority, equivalence, and non-inferiority trials. Bull NYU Hosp Jt Dis. 2008;66(2):150–4.
5. Vavken P. Rationale for and methods of superiority, noninferiority, or equivalence designs in orthopaedic, controlled trials. Clin Orthop Relat Res. 2011 Sep;469(9):2645–53.
6. Charalambous CP. Research and audit. In: Career skills for doctors. Switzerland: Springer international publishing; 2015. p. 119–37.
7. Mihailovic A, Bell CM, Urbach DR. Users' guide to the surgical literature. Case-control studies in surgical journals. Can J Surg. 2005 Apr;48(2):148–51.
8. McCulloch P, Altman DG, Campbell WB, Flum DR, Glasziou P, Marshall JC, Nicholl J, Aronson JK, Barkun JS, Blazeby JM, Boutron IC, Campbell WB, Clavien PA, Cook JA, Ergina PL, Feldman LS, Flum DR, Maddern GJ, Nicholl J, Reeves BC, Seiler CM, Strasberg SM, Meakins JL, Ashby D, Black N, Bunker J, Burton M, Campbell M, Chalkidou K, Chalmers I, de Leval M, Deeks J, Ergina PL, Grant A, Gray M, Greenhalgh R, Jenicek M, Kehoe S, Lilford R, Littlejohns P, Loke Y, Madhock R, McPherson K, Meakins J, Rothwell P, Summerskill B, Taggart D, Tekkis P, Thompson M, Treasure T, Trohler U, Vandenbroucke J, Balliol Collaboration. No surgical innovation without evaluation: the IDEAL recommendations. Lancet. 2009 Sep 26;374(9695):1105–12.
9. McCulloch P, Cook JA, Altman DG, Heneghan C, Diener MK, IDEAL Group. IDEAL framework for surgical innovation 1: the idea and development stages. BMJ. 2013 Jun 18;f3012:346. doi:10.1136/bmj.f3012.
10. Ergina PL, Barkun JS, McCulloch P, Cook JA, Altman DG, IDEAL Group. IDEAL framework for surgical innovation 2: observational studies in the exploration and assessment stages. BMJ. 2013 Jun 18;f3011:346. doi:10.1136/bmj.f3011.
11. Cook JA, Mcculloch P, Blazeby JM, Beard DJ, Marinac-Dabic D, Sedrakyan A, IDEAL Group. IDEAL framework for surgical innovation 3: randomised controlled trials in the assessment stage and evaluations in the long term study stage. BMJ. 2013 Jun 18;f2820:346. doi:10.1136/bmj.f2820.
12. Popular Quotes. Goodreads. http://www.goodreads.com/quotes. Accessed 2Apr 2017.
13. Sedgwick P. Explanatory trials versus pragmatic trials. BMJ. 2014 Nov 13;349:g6694. doi:10.1136/bmj.g6694.
14. Sackett DL. Explanatory and pragmatic clinical trials: a primer and application to a recent asthma trial. Pol Arch Med Wewn. 2011 Jul-Aug;121(7-8):259–63.
15. Markar SR, Wiggins T, Ni M, Steyerberg EW, Van Lanschot JJ, Sasako M, Hanna GB. Assessment of the quality of surgery within randomised controlled trials for the treatment of gastro-oesophageal cancer: a systematic review. Lancet Oncol. 2015 Jan;16(1):e23–31.
16. Blencowe NS, Boddy AP, Harris A, Hanna T, Whiting P, Cook JA, Blazeby JM. Systematic review of intervention design and delivery in pragmatic and explanatory surgical randomized clinical trials. Br J Surg. 2015 Aug;102(9):1037–47.
17. Archibald L Cochrane 1972. Effectiveness and efficiency. The nuffield trust: Chapter 5, subchapter b, second bullet 1.
18. IDK D. It is a pity some drug company does not produce a wide range of placebos –Archie Cochrane said! BMJ. 2004;329:944.
19. Horng S, Miller FG. Ethical framework for the use of sham procedures in clinical trials. Crit Care Med. 2003 Mar;31(3 Suppl):S126–30.

20. Sutherland ER. Sham procedure versus usual care as the control in clinical trials of devices: which is better? Proc Am Thorac Soc. 2007 Oct 1;4(7):574–6.
21. Moseley JB, O'Malle K, Petersen NJ, Menke TJ, Brody BA, Kuykendall DH, Hollingsworth JC, Ashton CM, Wray NP. A controlled trial of arthroscopic surgery for osteoarthritis of the knee. N Engl J Med. 2002 Jul 11;347(2):81–8.
22. Freed CR, Greene PE, Breeze RE, Tsai WY, DuMouchel W, Kao R, Dillon S, Winfield H, Culver S, Trojanowski JQ, Eidelberg D, Fahn S. Transplantation of embryonic dopamine neurons for severe Parkinson's disease. N Engl J Med. 2001 Mar 8;344(10):710–9.
23. Wartolowska K, Judge A, Hopewell S, Collins GS, Dean BJ, Rombach I, Brindley D, Savulescu J, Beard DJ, Carr AJ. Use of placebo controls in the evaluation of surgery: systematic review. BMJ. 2014 May 21;348:g3253. doi:10.1136/bmj.g3253.
24. Noordzij M, Dekker FW, Zoccali C, Jager KJ. Sample size calculations. Nephron Clin Pract. 2011;118(4):c319–23. doi:10.1159/000322830.
25. Gupta KK, Attri JP, Singh A, Kaur H, Kaur G. Basic concepts for sample size calculation: critical step for any clinical trials. Saudi J Anaesth. 2016 Jul-Sep;10(3):328–31. doi:10.4103/1658-354X.174918.
26. Akl EA, Briel M, You JJ, Sun X, Johnston BC, Busse JW, Mulla S, Lamontagne F, Bassler D, Vera C, Alshurafa M, Katsios CM, Zhou Q, Cukierman-Yaffe T, Gangji A, Mills EJ, Walter SD, Cook DJ, Schünemann HJ, Altman DG, Guyatt GH. Potential impact on estimated treatment effects of information lost to follow-up in randomised controlled trials (LOST-IT): systematic review. BMJ. 2012 May 18;344:e2809. doi:10.1136/bmj.e2809.
27. Gargon E, Williamson PR, Altman DG, Blazeby JM, Tunis S, Clarke M. The COMET Initiative database: progress and activities update (2015). Trials. 2017 Feb 3;18(1):54. doi:10.1186/s13063-017-1788-8.
28. Chapman SJ, Shelton B, Mahmood H, Fitzgerald JE, Harrison EM, Bhangu A. Discontinuation and nonpublication of surgical randomised controlled trials: observational study. BMJ. 2014 Dec 9;349:g6870.
29. Jarvik JG, Comstock BA, Kliot M, Turner JA, Chan L, Heagerty PJ, Hollingworth W, Kerrigan CL, Deyo RA. Surgery versus non-surgical therapy for carpal tunnel syndrome: a randomised parallel-group trial. Lancet. 2009 Sep 26;374(9695):1074–81.
30. Klazen CA, Lohle PN, de Vries J, Jansen FH, Tielbeek AV, Blonk MC, Venmans A, van Rooij WJ, Schoemaker MC, Juttmann JR, Lo TH, Verhaar HJ, van der Graaf Y, van Everdingen KJ, Muller AF, Elgersma OE, Halkema DR, Fransen H, Janssens X, Buskens E, Mali WP. Vertebroplasty versus conservative treatment in acute osteoporotic vertebral compression fractures (Vertos II): an open-label randomised trial. Lancet. 2010 Sep 25;376(9746):1085–92.
31. Kirkley A, Birmingham TB, Litchfield RB, Giffin JR, Willits KR, Wong CJ, Feagan BG, Donner A, Griffin SH, D'Ascanio LM, Pope JE, Fowler PJ. A randomized trial of arthroscopic surgery for osteoarthritis of the knee. N Engl J Med. 2008 Sep 11;359(11):1097–107.
32. NIHR. Feasibility and pilot studies. http://www.nihr.ac.uk/funding-and-support/documents/funding-for-research-studies/research-programmes/RfPB/FAQs/Feasibility_and_pilot_studies.pdf. Accessed 2 Apr 2017.
33. Eldridge SM, Lancaster GA, Campbell MJ, Thabane L, Hopewell S, Coleman CL, et al. Defining feasibility and pilot studies in preparation for randomised controlled trials: development of a conceptual framework. PLoS One. 2016;11(3):e0150205. doi:10.1371/journal.pone.0150205.
34. Donovan JL, Lane JA, Peters TJ, Brindle L, Salter E, Gillatt D, Powell P, Bollina P, Neal DE, Hamdy FC, ProtecT Study Group. Development of a complex intervention improved randomization and informed consent in a randomized controlled trial. J Clin Epidemiol. 2009 Jan;62(1):29–36.
35. Blazeby JM, Strong S, Donovan JL, Wilson C, Hollingworth W, Crosby T, Nicklin J, Falk SJ, Barham CP, Hollowood AD, Streets CG, Titcomb D, Krysztopik R, Griffin SM, Brookes ST. Feasibility RCT of definitive chemoradiotherapy or chemotherapy and surgery for oesophageal squamous cell cancer. Br J Cancer. 2014 Jul 15;111(2):234–40.

36. Brett J, Staniszewska S, Mockford C, Herron-Marx S, Hughes J, Tysall C, Suleman R. Mapping the impact of patient and public involvement on health and social care research: a systematic review. Health Expect. 2014 Oct;17(5):637–50. doi:10.1111/j.1369-7625.2012.00795.x.
37. Brett J, Staniszewska S, Mockford C, Herron-Marx S, Hughes J, Tysall C, Suleman R. A systematic review of the impact of patient and public involvement on service users, researchers and communities. Patient. 2014;7(4):387–95. doi:10.1007/s40271-014-0065-0.
38. Kasivisvanathan V, Williams K. Involving public and patients in research. Ann R Coll Surg Engl (Suppl). 2014;96:148–9.
39. Patterson S, Mairs H, Borschmann R. Successful recruitment to trials: a phased approach to opening gates and building bridges. BMC Med Res Methodol. 2011;11:73. doi:10.1186/1471-2288-11-73.
40. Whicher DM, Miller JE, Dunham KM, Joffe S. Gatekeepers for pragmatic clinical trials. Clin Trials. 2015 Oct;12(5):442–8.
41. Ishii LE. Closing the clinical gap: translating best practice knowledge to performance with guidelines implementation. Otolaryngol Head Neck Surg. 2013 Jun;148(6):898–901.
42. Bero LA, Grilli R, Grimshaw JM, Harvey E, Oxman AD, Thomson MA. Closing the gap between research and practice: an overview of systematic reviews of interventions to promote the implementation of research findings. The cochrane effective practice and organization of care review group. BMJ. 1998 Aug15;317(7156):465–8.
43. Grol R. Successes and failures in the implementation of evidence-based guidelines for clinical practice. Med Care. 2001;39(8 Suppl 2):1146–54.
44. Grimshaw JM, Thomas RE, MacLennan G, Fraser C, Ramsay CR, Vale L, Whitty P, Eccles MP, Matowe L, Shirran L, Wensing M, Dijkstra R, Donaldson C. Effectiveness and efficiency of guideline dissemination and implementation strategies. Health Technol Assess. 2004;8:1–72.
45. Rycroft-Malone J, Seers K, Crichton N, Chandler J, Hawkes CA, Allen C, Bullock I, Strunin L. A pragmatic cluster randomised trial evaluating three implementation interventions. Implement Sci. 2012 Aug 30;7:80.
46. Flather M, Delahunty N, Collinson J. Generalizing results of randomized trials to clinical practice: reliability and cautions. Clin Trials. 2006;3(6):508–12.
47. Rothwell PM. External validity of randomised controlled trials: "to whom do the results of this trial apply?". Lancet. 2005 Jan 1-7;365(9453):82–93.
48. Everett R. Diffusion of Innovations. 3rd ed. New York: Free Press, Macmillan Publishing Co.; 1983.
49. Thornley P, de Sa D, Evaniew N, Farrokhyar F, Bhandari M, Ghert M. An international survey to identify the intrinsic and extrinsic factors of research studies most likely to change orthopaedic practice. Bone Joint Res. 2016 Apr;5(4):130–6.
50. Costa ML, Jameson SS, Reed MR. Do large pragmatic randomised trials change clinical practice?: Assessing the impact of the distal radius acute fracture fixation trial (DRAFFT). Bone Joint J. 2016 Mar;98-B(3):410–3.
51. Costa ML, Achten J, Parsons NR, Rangan A, Griffin D, Tubeuf S, Lamb SE, DRAFFT Study Group. Percutaneous fixation with Kirschner wires versus volar locking plate fixation in adults with dorsally displaced fracture of distal radius: randomised controlled trial. BMJ. 2014 Aug 5;349:g4807. doi:10.1136/bmj.g4807.
52. Dirschl DR, Tornetta P 3rd, Bhandari M. Designing, conducting, and evaluating journal clubs in orthopaedic surgery. Clin Orthop Relat Res. 2003 Aug;413:146–57.
53. Linzer M. The journal club and medical education: over one hundred years of unrecorded history. Postgrad Med J. 1987 Jun;63(740):475–8.

Chapter 10
Surgical Audit

Audit is defined by the Oxford dictionary as "a systematic review or assessment" [1]. In surgery, it may be considered as the process through which we can systematically review our practise, identify those areas where we excel and those where we could do better.

Hence, audit can guide us as to which successful processes to maintain and which deficient areas to improve, to help develop and get better. Understanding the importance of audit in surgical practise, and recognising the strengths and limitations of processes through which surgical audit can be achieved are important skills to develop.

This chapter describes the surgical audit cycle, along with the various types of surgical audit. It also discusses the role of the surgical logbook, mortality and morbidity meetings, and national databases in evaluating surgical outcomes.

C. Panayiotou Charalambous, *Career Skills for Surgeons*,
DOI 10.1007/978-3-319-57490-5_10

Based on my experience I would highly recommend this extremely successful procedure
SUCCESS
100% (n=1)

10.1 Surgical Audit Cycle

Without reflection, we go blindly on our way.

Margaret J. Wheatley [2].

The surgical audit cycle refers to the process whereby we set standards for an "ideal" practise and then compare our current practise against that "ideal". This helps identify areas whereby our current practise falls short of being "ideal". Based on those, we then introduce changes in our practise to allow us to move towards or meet the predefined "ideal". This is the first part of the audit cycle.

After a settling in period (allowing our practise to get used to the changes we introduced and for any initial enthusiasm or despair to reach a steady state) we re-examine our practise to see if we are now closer to the "ideal". This is the second part of the audit or the closing of the audit cycle (Fig. 10.1).

Surgical audit may be described according to what it examines:

- Audit of process—pathways, protocols, procedures
- Audit of resources—human, technological, space, time
- Audit of outcomes—clinical, cost effectiveness

Audit may examine the practice of an individual surgeon, a group of surgeons, a department, an institution, region, the surgical practice across a nation or in multiple countries.

Standards of "ideal" practise may be set (amongst others) by:

- International surgical associations
- National bodies that evaluate any available clinical evidence for the effectiveness of healthcare processes and provide guidance as to their use (such as the National Institute of Clinical Excellence in the UK, or the Agency for Healthcare Research and Quality in the USA)
- Government healthcare bodies or regulators
- Surgical Training Organisations
- Institutions—hospitals, clinics, practises
- Surgical departments, groups or individual surgeons

Audits may be prospective (commence before the practise in question has occurred) or retrospective (looking back at events which have already taken place). Audit should be viewed as an improvement process through which deficiencies can be identified and addressed, as well as through which high quality practices can be appreciated and popularised. Audit is not an exercise utilised simply to improve one's qualifications or aid one's career advancement, or allow an institution to obtain accreditation and government funding.

Audit may be considered as being dynamic rather than static, with the audit cycle viewed like a wind driven turbine rather than a single twist; an audit demonstrating a low infection rate may be reassuring but should be regularly repeated to ensure that the conditions have not changed.

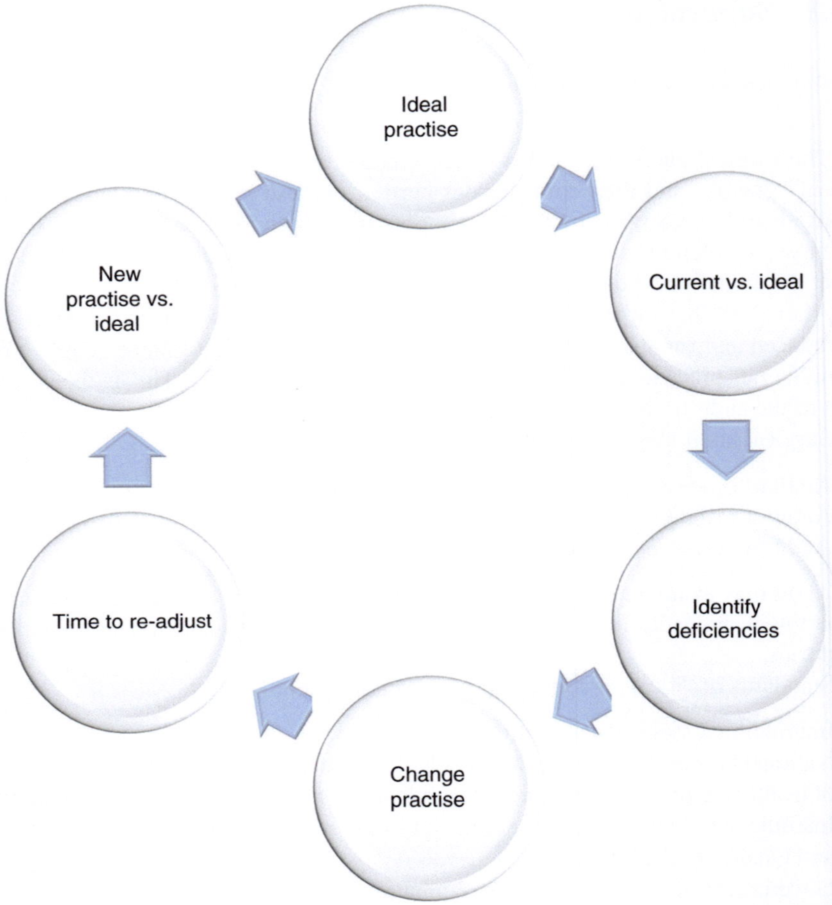

Fig. 10.1 Surgical audit cycle

10.2 Surgical Audit Ladder

Surgical audit may be performed at various levels, which may be described as the ladder of surgical audit (Fig. 10.2). Surgical audit may be performed by keeping a surgical logbook, or through departmental morbidity and mortality (M&M) meetings, systematic local data collection, and regional/national/international registries.

As one ascends the ladder of surgical audit the required resources for audit completion escalate, but any findings may have more widespread influence on healthcare systems.

10.2.1 Surgical Logbook

An individual surgeon may routinely keep (and is advisable to keep) a logbook of surgical cases performed under his/her care. This may be written or electronic. Various formats may be used but information recorded may include:

- Identifiers of the case—without compromising confidentiality and data protection rules
- Date of procedure
- Type of procedure
- Details of procedure
- Level of surgeon's participation in procedure
- Intra-operative and post-operative complications

Complications may be recorded as they are encountered (in theatre, ward, clinic). Such a record may allow surgeons to glance through their records and see how their cases are doing. Surgical logbook is not resource demanding, and may be considered a part of continuous assessment of own results, self-evaluation, self-reflection and professional development. However, surgical logbooks are limited by how much information can be recorded. In addition, recording of complications is often opportunistic rather than systematic as complications are recorded if and when

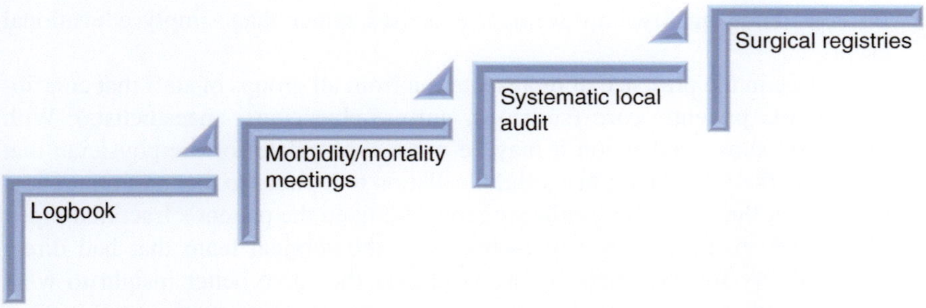

Fig. 10.2 Surgical audit ladder

encountered; this may miss cases who develop complications but are treated by other healthcare providers, hence not captured by a surgeon's logbook. Conclusions reached from surgical logbooks may thus be limited.

One may try to resist looking at own log book results using the 3 Ds approach which could refer to:

- Did it once
- Didn't go well
- Disaster procedure

Alternatively it could refer to

- Did it once
- Did very well
- Discovery

10.2.2 *Mortality and Morbidity Meetings*

These are meetings taking place on a regular basis, which aim to examine complications (morbidity) and deaths (mortality) encountered in the period since the last meeting. In hospitals, these often take place at a departmental level. Cases who developed such unwanted outcomes are examined in detail to help determine:

- What went well in care
- Deficiencies in care
- Factors leading to the unwanted outcome
- Whether the unwanted outcome could have been prevented
- Lessons learnt

Findings of such meetings may provide reassurance that complications and deaths occurred despite all preventable measures taken. Alternatively, such examination may identify care deficiencies, allowing lessons to be learnt and changes in practise to be introduced to minimise the risk of such unwanted events recurring.

M&M meetings should:

- Be considered quality improvement exercises rather than simply educational events.
- Take place in the presence of representation from all groups of staff that contributed to the patients' care (surgeons, nurses, physicians, anaesthetists). With increasing subspecialisation it may be more appropriate for the physician that dealt with the post-infarct fast atrial fibrillation to input as to that part of medical care, rather than the orthopaedic surgeon who fixed the patient's fractured hip.
- Take place in the presence of members of the surgical team that had direct responsibility for the patient's care (and may thus give better insight to what occurred rather than simply having to rely on recorded information).
- Conducted in a "no-blame" spirit where all are supported and helped to develop and learn. Such meetings should not be viewed as opportunities to "blame", to "name and shame" to "get at" longstanding foes. Pinto et al. [3] interviewed

vascular and general surgeons in two National Health Service(NHS) Trusts in London, UK to assess the impact of surgical complications on surgeons. About half of their participants described M&M meetings as being dominated by blame, not being supportive and exacerbating the burden of complications. The words of one registrar in general surgery in that study are striking:

"…morbidity and mortality meetings are supposed to be a forum where you can have an open discussion but if anyone believes that they're only kidding themselves; everybody in that room is very defensive and aggressively pursues an angle that puts them in the best possible light and professional rivalries exist…".

- Deal with a manageable volume of cases for the time devoted, to avoid rushing through.
- Should be performed regularly so that cases are discussed close to their time of occurrence.
- Have mechanisms of adequately identifying cases (theatres, wards, out-patients) to avoid bias in case selection.

A recent survey [4] of 15 general surgical units in Scotland, reported that M&M meetings happened with variable frequency (three-weekly, one-fortnightly, ten-monthly, one-three monthly) and in only one unit the meeting was attended by nursing staff. In only three cases, morbidities identified during follow up appointments in the outpatient clinics were included.

If assigned with the task of presenting such cases in M&M meetings, you may consider the following in your presentation:

- Describe the patient's journey as a sequence of events.
- Be succinct but thorough.
- Obtain up to date information (such as post-mortem report).
- Verify the accuracy of the information you gathered with the team looking after the patient. Simply reviewing the medical records may not give you the whole picture.
- Give an assessment as to:
  - What went well
  - What could have gone better
  - Lessons to be learnt

10.2.3 Systematic Audit at Local Level

Such audits aim to examine practise at local level (one surgeon, group of surgeons, one department or institution). They aim to gather data in a more systematic, thorough and detailed way, particularly with regards to case identification and outcomes examined.

Systematic audits are closely related to research studies with regards to design and data collection. However, unlike research which aims to discover what the "ideal" practise is, in audit one knows what the "ideal" practise is and tries to discover if current practise meets that "ideal".

10.2.4 Audit Databases

Surgical registries may be kept at local/national or international level. They usually aim to record the outcomes of patients or procedures carried out by a large number of surgeons across various institutions.

There are several national audits taking place in the UK [5] evaluating outcomes and processes in various aspects of surgery and these include:

- Hip Fracture Database
- National Joint Replacement Registry(NJR)
- National Vascular Registry
- National Emergency Laparotomy Audit
- Congenital Heart Disease Audit
- Neurosurgical National Audit
- Ophthalmology Cataract Audit

Similar databases exist in the USA and include:

- Society of Thoracic Surgeons Adult Cardiac Surgery Database [6]
- Bariatric Surgery Quality Improvement Program [7]
- American College of Surgeons National Quality Improvement [8]
- Trauma Data Bank [9]
- Vascular Quality Initiative [10]

There are also examples of international databases:

- European registry for abdominal wall hernias [11]

National/international databases offer important advantages such as:

1. Large number of cases (patients, procedures, implants).
2. Large number of surgeons; hence any findings can be more representative of the real world picture and provide a pragmatic presentation of potential outcomes, rather than if evaluation is limited to few specialist centres.
3. Can provide a temporal rather than static assessment as cases are continuously added to the database. This may allow one to examine not only current performance but also assess trends in performance. Trends in practice may also be determined (such as patient selection, preference for particular procedures or implants, outcomes). In orthopaedic surgery national registers have been able to identify early problems with some implants such as metal on metal hip arthroplasties, or patella-femoral knee arthroplasties.

However, when referring to surgical registries certain design challenges may be considered:

Accuracy of data: Data collected and used for analysis must be accurate, as inaccurate data can lead to erroneous results and wrong conclusions. In some situations data may be collected retrospectively from medical records, often by non-medically trained staff, which could limit the accuracy of collected data. Accuracy of data may also be influenced by whether they are collected by someone independent to the provision of treatment; in theory if the person performing the operation is completing the data recording form, the complexity of the procedure and the pre-surgery condition of the patient may be exaggerated consciously or subconsciously. It has been demonstrated that the accuracy of data recorded for the orthopaedic national registries in the UK may be lower than expected [12, 13]. Similarly, Seagrave et al. [14] examined the accuracy of data recorded from two hospitals in the Arthroplasty Clinical Outcomes Registry in Australia. They reported that in one hospital only 28 out of 49 (57%) variables collected met a 95% benchmark for data accuracy, whereas in the other hospital 38 of 51 (75%) variables did so. Inaccuracies involved mainly co-morbid conditions and complications.

Completeness of data: When assessing the results of a surgical registry one must examine whether data collection in the target population is complete, as incomplete data may lead to inaccuracies. Missing data may be due to random factors, in which case such loss may not substantially affect the overall results (if these results are presented as rates rather than absolute numbers). However, missing data may also be due to systematic factors (bias), which could lead to over- or under- representation of certain groups of patients in the population examined, and hence influence the overall results. Such bias may be conscious or subconscious. It may be that if an operation goes less well than expected, the data collection form is deliberately not completed (conscious bias). Alternatively, it may be that the records of cases who developed an infection following surgery are kept in the microbiology department for a long time as part of infection surveillance, or records of cases who died are filed in a different location and hence not made available to the person collecting the data (and who may not be aware of such arrangements). This could lead to the systematic (unconscious bias) under-representation of infected and dead cases in the population under examination and hence under-estimation of such complications.

The NJR in the UK records primary and revision joint replacements. In this way the survival rates of primary hip and knee replacements can be determined and outlier implants and surgeons identified. However, accurate calculation of such survival rates requires recording of all revision procedures and linking them to the initial primary procedure. Sabah et al. [15] compared data recorded by the London Implant Retrieval Centre (LIRC) and the NJR in the UK for hip replacement revision procedures performed between 2003 and 2013. They reported that 213 revision procedures on the LIRC (23% of those examined) could not be found on the NJR. Only 349 (38%) of revision procedures on the LIRC were found to be linked to their primary procedure on the NJR and hence contributed to the survival rate calculations.

Choice of outcomes: Appropriate outcomes must be utilised. Hard outcomes are preferable, ones that cannot be easily disputed—such as mortality rates in assessing emergency laparotomies, revision rates in assessing joint arthroplasty surgery, amputation rates in assessing vascular surgery reconstruction. Outcomes such as infection may be considered "softer" as their occurrence may be subjected to dispute due to the various definitions of "infection" available, and subjectivity that may be encountered in applying such definitions to real clinical practise.

Outcomes should also be able to capture "failure" that is not identified by hard outcome measures. If one looks only at occurrence of revision surgery in assessing failure of hip joint replacements then several cases who are doing badly (such as repeated dislocators, cases with persistent chronic pain) but do not undergo revision (for various factors including being medically unfit, not wishing further surgery, surgeon's view that revision surgery will provide no benefit) will not be considered, giving false reassurance. Inclusion of other outcomes such as patient reported pain or function may be preferable to help capture failures.

Similarly, if one looks at re-dislocation rates following shoulder arthroscopic stabilisation surgery, it would be helpful to describe re-dislocation according to whether it was related to a new injury or happened with minimal or no trauma. If the outcome of surgery is not good enough to enable an athlete to rehabilitate and return to professional rugby, then that athlete may be at less risk of further injury and hence at lower risk of further dislocations. In contrast, if the outcome of surgery is so good to enable an athlete to return back to rugby, then the athlete may sustain further rugby related injuries and hence further dislocations. Hence, if one simply looks at overall re-dislocation rates erroneous conclusions may be reached.

Case adjustments: Failures may be related to multiple factors other than the technical quality of performing the surgical procedure or the type of implant utilised. Such patient related factors may include:

- Patient characteristics—medical fitness, biology of healing
- Patient compliance with post-surgery instructions and rehabilitation [16, 17]
- Secondary gains such as worker's compensation and disability allowances [18, 19]
- Presence of other associated injuries

This raises the need to make adjustments for the case-mix under consideration, for factors that could influence outcomes. Several risk prediction models have been described which may help to predict outcomes such as the:

- EuroSCORE—to predict mortality post cardiac surgery [20]
- POSSUM—to predict morbidity and mortality post general surgical procedures [21]
- Nottingham Hip fracture score—to predict mortality in hip fracture patients [22]

Such prediction models may be used to calculate predicted mortality in specific patients which can aid decision making and counselling of patients about surgery. Prediction models can also be used to calculate the expected mortality in a group of patients and then compare the observed mortality to the expected, allowing evaluation of the quality of care provided by individual surgeons or units. Stratified outcomes are thus preferable to crude outcomes.

Case selection: As patient characteristics and procedure complexity may influence outcomes, it may be that if only straightforward cases are offered surgery (with more complex ones avoided) outcomes are better [23]. Such "gaming" [24] could theoretically happen at any level from an individual surgeon, to departments, institutions, commissioners or wider healthcare systems. There has been a recent trend in the UK and USA for the public reporting of surgeon specific or unit specific results [25, 26]. Such public reporting of outcomes can help identify areas of excellent practice which can then be mimicked or areas of deficient practice which can be improved. In addition, such reporting may encourage transparency and accountability, and can reassure the public for the existence of quality control in delivered care. Nevertheless, unit or surgeon reporting may (in theory) encourage case selection, with risk aversion behaviour, and potentially deny surgery to those cases that need it most.

10.2.4.1 Impact of Surgical Registries

There are several examples of surgical registries having a positive impact in various parameters, some of which are described below.

1. Improving quality of care: By encouraging and monitoring adherence to quality measures, registries may improve the overall quality of care.

The National Hip Fracture Database (NHFD) was introduced in 2007 in England, Wales and Ireland, with the aim of improving the care of patients admitted with a hip fracture. It involved setting standards for care, collection of performance and outcome data and feedback to participating units [27]. Neuburger et al. [28] looked at almost half a million cases of hip fractures in patients older than 60, treated in the NHS in England and compared those treated during 2003–2007 (prior to the introduction of the NHFD) to those treated during 2007–2011. The rate of early surgery increased from 55 to 71% during 2007–2011, whereas it had been stable in the previous years. In relation to this, 30-day mortality decreased from 10.9 to 8.5% during 2007–2011 but only from 11.5 to 10.9% during 2003–2007 (Fig. 10.3). These findings suggest that the NHFD improved hospital care with resultant improvement in survival.

In the USA, the Vascular Quality Initiative collects data on vascular procedures (risk factors, peri-operative management, outcomes) and gives feedback to participating centres allowing them to compare their performance with their peers. Participation in the Vascular Quality initiative was shown to lead to improvements in utilisation of antiplatelet and statin therapy in patients having vascular surgery from 55 to 68%, with centres participating in the initiative for more than three years being 30% more likely to have patients on these medications. Discharge on these medications was associated with an improvement in five year survival, compared with discharge on no such medication (82% vs. 67%) [29].

2. Reducing healthcare costs: Surgical registries may lead to substantial healthcare cost reductions by improving care quality and reducing complication rates. Larsson et al. [30] evaluated the role of the Swedish Hip Arthroplasty Register in reducing hip replacement revision rates and the cost savings of such reduction. By comparing the revision rates seen in Sweden with those encountered in the USA they estimated a cost saving of about $14 million per year.

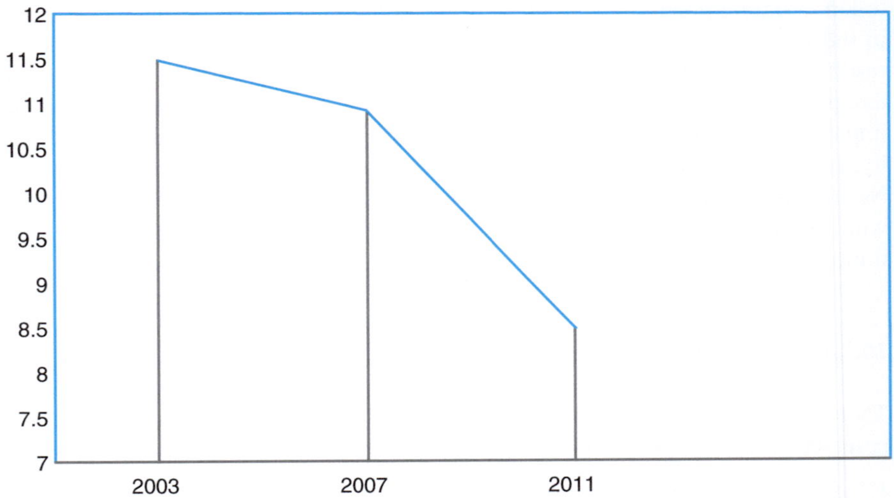

Fig. 10.3 Reduction in 90 day mortality following the introduction of the National Hip Fracture Database (NHFD) in 2007. Adapted from Neuburger et al. [28]

Guillamondegui et al. [31] evaluated the role of participation of Tennessee hospitals in the Tennessee Surgical Quality Collaborative in reducing complication rates in general and vascular procedures. They estimated that through the reduction in complication rates $2.2 million per 10,000 surgical cases were saved each year.

3. Limiting harm: Registries (when thorough and accurate) may aid to identify at an early stage implants, procedures, units or surgeons underperforming and hence by addressing these minimise further harm.
4. Reassuring stakeholders: Through the inclusion of huge numbers of cases, registries can provide meaningful rates on rare complications.

Hunt et al. evaluated about 409,000 total hip replacements [32] and 468,000 total knee replacements [33] carried out for osteoarthritis in the NJR in the UK. They reported that the mortality rates in total hip replacements fell from 0.56 to 0.29% and for total knee replacements from 0.37 to 0.2% between 2003 and 2011. Along similar lines Zaidi et al. [34] reviewed 1522 primary ankle replacements on the NJR in the UK and reported a 0.13% 90-day mortality. The 90-day incidence rate of pulmonary embolism was 0.51% but no deaths occurred as a result of this complication.

Publication of surgeon specific outcomes or unit specific outcomes (even though they may have several limitations) may improve transparency and confidence of the public on the surgical profession.

10.3 Hawthorne Effect

In audit one may consider the possibility of the Hawthorne effect, whereby improvements in performance may be due to the awareness of being observed [35, 36].

This effect was described in the 1950s when analysing the results of studies done at the Hawthorne Works, a factory of the Western Electric Company in Illinois. Those studies assessed ways of improving workers' productivity, but found that productivity increased whatever the intervention applied (changes in levels of light, or even a reversal to the original working conditions). Hence, it was concluded that it was not the intervention per se, but the attention that the participants received (knowing that they were being watched) that improved productivity [35, 36].

The Hawthorne effect may be considered when auditing snap shots of clinical practise. If participants know that during the audit period their performance is monitored (such as their compliance with completing venous thromboembolism assessment forms, hand washing during wards rounds), they may be more likely to behave as they should, leading to a false reassurance of appropriate practise. One may have to go undercover or come up with other novel ways of monitoring performance to counteract this effect [37–40].

In national registries, which provide continuous feedback, a Hawthorne effect may be welcomed.

References

1. English Oxford living dictionaries. https://en.oxforddictionaries.com. Accessed 2 Apr 2017.
2. Famous Quotes at BrainyQuote. www.brainyquote.com. Accessed 2 Apr 2017.
3. Pinto A, Faiz O, Bicknell C, Vincent C. Surgical complications and their implications for surgeons' well-being. Br J Surg. 2013;100(13):1748–55.
4. Khine M, Leung E, McGregor JR. A survey of morbidity and mortality review meetings in the general surgical units of the West of Scotland. Scott Med J. 2015;60(4):244–8.
5. The National Clinical Audit Programme. Healthare Quality Improvement Partnership. http://www.hqip.org.uk/national-programmes/a-z-of-nca/. Accessed 2 Apr 2017.
6. Adult Cardiac Surgery Database. The Society of Thoracic Surgeons. Accessed at http://www.sts.org/sts-national-database/database-managers/adult-cardiac-surgery-database
7. Metabolic and Bariatric Surgery Accreditation and Quality Improvement Program Bariatrics—American College of Surgeons. Accessed at https://www.facs.org/quality%20programs/mbsaqip
8. ACS National Surgical Quality Improvement Program. American College of Surgeons. Accessed at https://www.facs.org/quality-programs/acs-nsqip
9. National Trauma Data Bank—American College of Surgeons. https://www.facs.org/quality-programs/trauma/ntdb
10. The Vascular Quality Initiative. Accessed at http://www.vascularqualityinitiative.org/
11. EuraHS—EUROPEAN REGISTRY OF ABDOMINAL WALL HERNIAS. Accessed at http://www.eurahs.eu/
12. Cundall-Curry DJ, Lawrence JE, Fountain DM, Gooding CR. Data errors in the national hip fracture database: a local validation study. Bone Joint J. 2016;98-B(10):1406–9.
13. Kosy JD, Kassam AA, Hockings M. National Joint Registry data inaccuracy: a threat to proper reporting. Br J Hosp Med (Lond). 2013;74(12):691–3.
14. Seagrave KG, Naylor J, Armstrong E, Leong KM, Descallar J, Harris IA. Data quality audit of the arthroplasty clinical outcomes registry NSW. BMC Health Serv Res. 2014;14:512. doi:10.1186/s12913-014-0512-6.
15. Sabah SA, Henckel J, Koutsouris S, Rajani R, Hothi H, Skinner JA, Hart AJ. Are all metal-on-metal hip revision operations contributing to the national joint registry implant survival curves? : a study comparing the London implant retrieval centre and national joint registry datasets. Bone Joint J. 2016;98-B(1):33–9.
16. Whiting PS, Greenberg SE, Thakore RV, Alamanda VK, Ehrenfeld JM, Obremskey WT, Jahangir A, Sethi MK. What factors influence follow-up in orthopedic trauma surgery? Arch Orthop Trauma Surg. 2015;135(3):321–7.
17. Ahmad S, Haber M, Bokor DJ. The influence of intraoperative factors and postoperative rehabilitation compliance on the integrity of the rotator cuff after arthroscopic repair. J Shoulder Elbow Surg. 2015;24(2):229–35.
18. Djerbi I, César M, Lenoir H, Coulet B, Lazerges C, Chammas M. Revision surgery for recurrent and persistent carpal tunnel syndrome: clinical results and factors affecting outcomes. Chir Main. 2015;34(6):312–7.
19. Henn RF 3rd, Tashjian RZ, Kang L, Green A. Patients with workers' compensation claims have worse outcomes after rotator cuff repair. J Bone Joint Surg Am. 2008;90(10):2105–13.
20. euroSCORE.org. http://www.euroscore.org/. Accessed 2 Feb 2017.
21. Copeland GP, Jones D, Walters M. POSSUM: a scoring system for surgical audit. Br J Surg. 1991;78(3):355–60.
22. Rushton PR, Reed MR, Pratt RK. Independent validation of the Nottingham hip fracture score and identification of regional variation in patient risk within England. Bone Joint J. 2015;97-B(1):100–3.
23. Behrendt K, Groene O. Mechanisms and effects of public reporting of surgeon outcomes: a systematic review of the literature. Health Policy. 2016;120(10):1151–61.
24. Burns EM, Pettengell C, Athanasiou T, Darzi A. Understanding the strengths and weaknesses of public reporting of surgeon-specific outcome data. Health Aff (Millwood). 2016;35(3):415–21.

25. Shahian DM, Grover FL, Prager RL, Edwards FH, Filardo G, O'Brien SM, He X, Furnary AP, Rankin JS, Badhwar V, Cleveland JC Jr, Fazzalari FL, Magee MJ, Han J, Jacobs JP. The society of thoracic surgeons voluntary public reporting initiative: the first 4 years. Ann Surg. 2015;262(3):526–35.
26. Haddad FS, Manktelow AR, Skinner JA. Publication of surgeon level data from registers: who benefits? Bone Joint J. 2016;98-B(1):1–2.
27. The national hip fracture database. Royal college of physicians. Accessed at http://www.nhfd.co.uk/.
28. Neuburger J, Currie C, Wakeman R, Tsang C, Plant F, De Stavola B, Cromwell DA, van der Meulen J. The impact of a national clinician-led audit initiative on care and mortality after hip fracture in England: an external evaluation using time trends in non-audit data. Med Care. 2015;53(8):686–91.
29. De Martino RR, Hoel AW, Beck AW, Eldrup-Jorgensen J, Hallett JW, Upchurch GR, Cronenwett JL, Goodney PP. Vascular quality initiative. Participation in the vascular quality initiative is associated with improved perioperative medication use, which is associated with longer patient survival. J Vasc Surg. 2015;61(4):1010–9.
30. Larsson S, Lawyer P, Garellick G, Lindahl B, Lundström M. Use of 13 disease registries in 5 countries demonstrates the potential to use outcome data to improve health care's value. Health Aff (Millwood). 2012;31(1):220–7.
31. Guillamondegui OD, Gunter OL, Hines L, Martin BJ, Gibson W, Clarke PC, Cecil WT, Cofer JB. Using the national surgical quality improvement program and the tennessee surgical quality collaborative to improve surgical outcomes. J Am Coll Surg. 2012;214(4):709–14.
32. Hunt LP, Ben-Shlomo Y, Clark EM, Dieppe P, Judge A, MacGregor AJ, Tobias JH, Vernon K, Blom AW, National Joint Registry for England, Wales and Northern Ireland. 90-day mortality after 409,096 total hip replacements for osteoarthritis, from the national joint registry for England and Wales: a retrospective analysis. Lancet. 2013;382(9898):1097–104.
33. Hunt LP, Ben-Shlomo Y, Clark EM, Dieppe P, Judge A, MacGregor AJ, Tobias JH, Vernon K, Blom AW, National Joint Registry for England and Wales. 45-day mortality after 467,779 knee replacements for osteoarthritis from the National Joint Registry for England and Wales: an observational study. Lancet. 2014;384(9952):1429–36.
34. Zaidi R, MacGregor A, Cro S, Goldberg A. Pulmonary embolism and mortality following total ankle replacement: a data linkage study using the NJR data set. BMJ Open. 2016;6(6):e011947. doi:10.1136/bmjopen-2016-011947.
35. Roethlisberger FJ, Dickson WJ. Management and the worker. Cambridge, Mass: Harvard University Press; 1939.
36. Mayo E. The human problems of an industrial civilization, vol. Vol. 3. New York: MacMillan; 1993. p. 53–73.
37. Gould DJ, Creedon S, Jeanes A, Drey NS, Chudleigh J, Moralejo D. Impact of observing hand hygiene in practice and research: a methodological reconsideration. J Hosp Infect. 2017;95(2):169–74.
38. Kovacs-Litman A, Wong K, Shojania KG, Callery S, Vearncombe M, Leis JA. Do physicians clean their hands? Insights from a covert observational study. J Hosp Med. 2016;11(12):862–4.
39. Hagel S, Reischke J, Kesselmeier M, Winning J, Gastmeier P, Brunkhorst FM, Scherag A, Pletz MW. Quantifying the hawthorne effect in hand hygiene compliance through comparing direct observation with automated hand hygiene monitoring. Infect Control Hosp Epidemiol. 2015;36(8):957–62.
40. McCarney R, Warner J, Iliffe S, van Haselen R, Griffin M, Fisher P. The Hawthorne effect: a randomised, controlled trial. BMC Med Res Methodol. 2007;7:30.

Chapter 11
Safety in Surgery

The Institute of Medicine defines patient safety as "the prevention of harm to patients", that is "avoiding injuries to patients from the care that is intended to help them" [1].

Patients seek surgical care to improve and get better. Surgeons provide care to help patients recover and pull through. The surgeon's motto remains that above all "do no harm" [2].

Yet, in the process of delivering surgical care patients may be harmed. Understanding potential causes of such harm, the factors that may lead to errors, and recognising processes as well as individual behaviours that may help promote safety, are important skills to develop and are discussed in this chapter.

C. Panayiotou Charalambous, *Career Skills for Surgeons*,
DOI 10.1007/978-3-319-57490-5_11

I am sorry Surgeon, we can't find the right screwdriver
W.H.O.
EQUIPMENT
TOOL BOX

11.1 The Safety Problem

Make a habit of two things: to help; or at least to do no harm.

Hippocrates [3]

It is estimated that about 10% of patients in hospitals in the developed world may suffer an injury related to medical management with 50% of these possibly being preventable [4]. A study in Portugal, reported that 10.8% of hospital patients died and 5.4% suffered permanent disability due to an adverse event [5]. In the USA, lapses in patient safety may cause more than 200,000 deaths each year [6].

Patient harm may occur at any step of care delivery, from diagnosis to treatment. The surgical theatre is prone to harmful events as they are dynamic environments with interaction of multiple players, multiple teams (surgical, nursing, anesthetic), and reliance on technology.

Harmful events may vary in their severity, the resultant damage they confer, or the likelihood of taking place. At one extreme of severity are "Never Events" a term introduced in by the National Quality Forum (NQF) in the USA. The term "Never Event" was used to describe healthcare errors that should never be encountered [7, 8].

The National Health Service (NHS) in the United Kingdom (UK) defines "Never Events" as "serious, largely preventable, patient safety incidents that should not occur if the available preventative measures have been implemented by healthcare providers" [9]. The NQF identifies 29 [10] and the NHS describes 25 [11] such events.

In relation to surgical or invasive procedures, "Never Events" include [10, 11]:

- Wrong site surgery.
- Wrong patient surgery.
- Wrong surgery.
- Implantation of a wrong implant.
- Retention of a foreign object in a patient after surgery.

Such events attract massive attention within healthcare organisations but also amongst the wider public, with press reports referring to wrong kidney removed [12], wrong eye lens implanted [13, 14], swabs or needles left behind [15].

Amongst 9744 paid malpractice settlements and judgements for surgical never events occurring in the USA between 1990 and 2010 Mehtsun et al. [16] reported mortality occurring in 6.6% of patients, permanent injury in 32.9% and temporary injury in 59.2%. Kwaan et al. [17] reviewed wrong site surgery cases reported to a large mal-practise insurer in the USA between 1985 and 2004, and reported an incidence of one in 112,994 operations. Retention of foreign bodies after surgery was in the order of one in 8801 to one in 18,760 operations.

Never events are unacceptable but also rare, and should not distract attention from other more common but equally harmful encounters such as medication errors which in the USA alone may kill up to 7000 patients per year [18, 19].

Hence, healthcare safety is a real issue, a major problem that a surgeon needs to have at the forefront of their mind. Surgery may not help a patient due to treatment resistance or progression of the underlying disorder, but patients should not be further harmed by failures of the treatment process.

11.2 Contributors to Healthcare Safety

Our goal is to make the best devices in the world, not to be the biggest.

Steve Jobs [3]

Safe surgical care is not the responsibility of one individual but of all those involved in healthcare. Various parts of the healthcare system could contribute to safety as below:

- Education—Medical Schools, universities and postgraduate professional development bodies that aim to promote safe care rather than simply transfer knowledge.
- Governments and politicians who develop safe healthcare policies, prioritize safety, take safety into account in describing quality care, incorporating safety measures in tariffs and remuneration of surgeons or institutions.
- Professional bodies who give guidance about safe care and support safe practice.
- National bodies for reporting and investigating unsafe acts.
- Regulatory bodies which inspect institutions to ensure standards are met, which respond effectively and promptly when such standards deteriorate, which investigate and act when things go wrong, which ensure lessons are truly learnt and minimise the possibility of harmful events recurring.
- Leadership that makes safety a priority alongside productivity and cost efficiency.
- Healthcare Industry with manufactures having safety as a main consideration in technology design.
- Processes, pathways, policies aiming at improving safety.
- Culture of safety at individual, departmental, or organizational level. This takes into account the priority given to safety relative to other goals of the organization (such as productivity, financial gains). Such a culture encourages front line workers to speak up, avoids assigning blame as the reflex response to adverse events.
- The individual at the frontline of surgical care—the surgeon, anesthetist, nursing staff, theatre scrub staff.

11.3 Causes of Harm

One of these days, I'm going to stop learning through my errors.

Burgess Owens [3]

Failures in healthcare may lead to no harm, minor harm or major harm. On occasions minor harm if left unchecked may progress to more severe harm (Fig. 11.1).

Surgical care harm may be due to several types of failures:

1. Recognised complications of care: occurring despite all reasonable precautions.
2. Errors: James Reason, Professor of Psychology at the University of Manchester, UK, and author of the book, "Human Error" [20] described error as "the failure of planned actions to achieve their desired goal" [21]. Reason classified errors into slips/lapses and mistakes. Slips and lapses refer to situations where the intended plan is appropriate, but there is failure to execute that plan. Mistakes refer to situations whereby the actions are executed as planned, but the plan is wrong [21].
3. Violations: These were described by Reason [21] as intended "deviations from safe operating practises, procedures, standards, or rules". Violations may refer to cutting corners simply because the opportunity is there, to get the job done, or to get personal gratitude such as to get a kick in a boring situation.
4. Technological breakdowns: A systematic review reported that almost one in four adverse events in the surgical theatre involved medical devices [22]. Doyle et al. [23] reported that, between 2005 and 2012, medical device errors cost the Johns Hopkins Hospital $75.3 million in legal fees and settlement pay-outs.

The outcomes of the above may not differ. A patient may be harmed as a result of any of the above. You may consider the example of a patient suffering an infection of a surgical wound. Infection can have devastating consequences on the individual concerned. Several precautions are taken, pre- intra- and post-operatively to minimize the risk of infection. Despite taking all necessary precautions, infections may still occur, and infection is a recognized complication of surgery, one

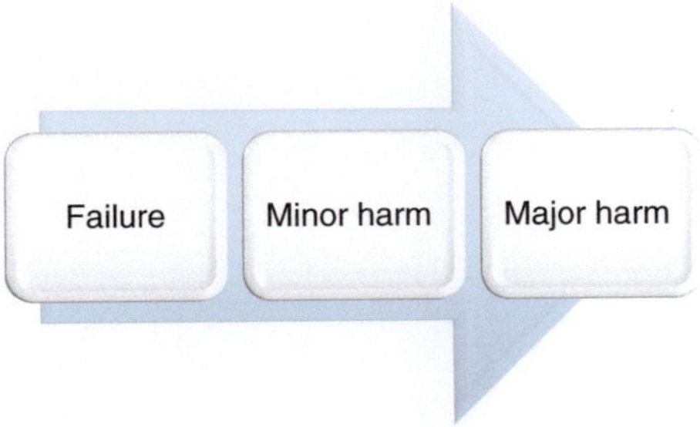

Fig. 11.1 Relationship between healthcare failures and harm

for which patients are appropriately counseled. However, infections may also occur due to care errors: a lapse may lead to omission of administering the correctly prescribed prophylactic antibiotic. Alternatively, the prescribed antibiotic may be administered but there was a mistake in that the wrong antibiotic was chosen in the first place. Infections may also be the result of good practise violation, such as deliberately overlooking the need for frequent ward cleansing, in an attempt to get more patients processed for surgery. Infections may also arise due to technology failures such as breakdown of the sterilization autoclave (that sterilises surgical instruments), which may not become immediately apparent, despite regular checks.

Safety failures may be described as:

1. Active—errors or violations performed by those at the frontline of surgical care. Once these happen they have an immediate effect, and become easily apparent [21, 24].
2. Latent—due to decisions, omissions, actions or inactions by those behind the frontline staff, those higher up in an organisation or general healthcare system, failures in any of the contributors to healthcare safety [21, 24]. The effects of such acts do not become immediately apparent, but may remain dormant and silent for a long time. Such failures may set the context predisposing the frontline staff to err, or may set the context whereby the frontline staff are not prevented from erring.

Active failures are those of the individual, and are the ones that often catch the focus when things go wrong. However, the importance and contribution of latent failures to harm are increasingly recognised [21, 24, 25].

Latent failures are those of the "system" and the "system" can be at any level, such as national, regional, individual hospital, practise, department, ward, operating theatre suite. Latent conditions may exist due to inadequate decisions made by government, healthcare authorities, institution authorities, lack of thorough oversee regulatory bodies, wrong assumptions or dogmatism installed in medical education curricula, design faults of medical devices, deficiencies in policies and procedures.

11.4 Latent Failures in Aviation

Talk sense to a fool and he calls you foolish.

Euripides [3]

The role of latent conditions may be more apparent in the aviation industry where they have been extensively studied with examples of latent failures occurring at various levels:

Education: American Airlines Flight 587, an Airbus A300, crashed in 2001, shortly after taking off from New York's JFK airport, due to break off of one the plane's fins, the vertical stabiliser. One of the factors contributing to the vertical stabiliser's failure was the use by the pilot of excessive rudder inputs, in trying to deal with turbulence caused by another plane flying in front. This pilot's action in using the rudder was considered unusual but it became apparent that education issues could have explained such a response; American Airlines had incorporated into their "Advanced Aircraft Manoeuvring Program" the guidance that rudder could be used in recovering from upsets and unusual attitudes (such as encountered in wake turbulence), but without giving any advice as to the effects of such rapid rudder reversals [26].

Processes: In 1996 a mid-air collision occurred between two planes travelling at similar altitude in opposite directions; a Saudi-Arabian Airlines Boeing 747 leaving from Delhi and a Kazahstan Airlines II-76 plane on the way to Delhi, over the village of Charkhi Dadri. Several factors are thought to have contributed to this world's deadliest mid-air collision, but these included the lack of separate air corridors for departing and arriving planes in the civilian airspace around Delhi, with inbound and outbound flights separated only by differences in flight altitude. Hence, this arrangement, led to two planes colliding when they (by mistake) ended up flying at the same altitude [27].

Regulation: Alaskan Airlines Flight 261, a McDonnell Douglas MD-83 plane, crashed into the Pacific Ocean near California due to loss of pitch control, in 2000. Loss of pitch control was considered secondary to failure of the plane's horizontal stabiliser occurring due to inadequate maintenance (inadequate lubrication of the jackscrew assembly causing screw thread wear). The Federal Aviation Administration, a national authority in the USA that regulates civil aviation, was found to have granted extension in the required intervals of plane maintenance, despite inadequate technical data to support the safety of such extensions. The investigation into the crash concluding that "FAA surveillance of Alaska Airlines had been deficient for at least several years" [28].

Technology designs—British Airways flight 38, a Boeing 777, crashed in 2008 just short of Heathrow runway. During its long journey from Beijing, whilst passing over Siberia and other frozen territories, at extremely low temperatures, ice crystals formed in the fuel lines. As the plane approached to landing, this ice was released obstructing the fuel feed system to the engines. The possibility of ice clogging was not recognised at the time of engine design and initial certification, and eventually led to a redesign of the engine components to allow for the possibility of such an occurrence [29, 30].

Technology interface designs—Helios Airways flight 522, a Boeing 737, crashed in Greece, in 2005. The plane took off with its pressurisation system set to "manual" rather than "auto", which led to failure of the cabin to pressurise, and the crew becoming incapacitated due to oxygen lack. The plane flew for several hours, eventually running out of fuel and crashing. Amongst the potential contributors to the crash was a misidentification of a warning sound. This could have been partly due to a complexity of the pilot–cockpit interface, as the same warning sound was used to inform the crew of two different types of dysfunction—when airborne, the warning sound acted as an altitude warning, prompting the crew to stop ascending, but when on ground the same warning acted as a take-off configuration advising pilots that the plane is not ready for take-off [31]. To improve the pilot–cockpit interface the FAA subsequently released a directive requiring all similar Boeings to install two separate warning lights, one for take-off configuration problems and one for cabin altitude warnings [32].

11.5 Latent Failures in Healthcare

The way to gain a good reputation is to endeavor to be what you desire to appear.

Socrates [3]

But even in healthcare there are examples of latent failures leading to unsafe situations and patient harm and three of these are described below:

1. Thomas Hugh, past Chairman of the Australian Cases Committee of the Medical Defence Union, looked at the factors behind bile duct injury during laparoscopic cholecystectomy [33]. Laparoscopic cholecystectomy is a commonly performed procedure for symptoms of gall bladder stones. In this procedure, the gallbladder is identified, and the cystic duct is dissected, ligated, and divided. However, the hepatic or common bile ducts may be mistaken for the cystic duct and wrongly divided. Bile duct injury is a feared complication due to its associated morbidity and mortality. It is estimated that 1500–2500 patients have bile duct injury in the United States every year [34].

Hugh [33] examined in detail the circumstances of bile duct injury during laparoscopic cholecystectomy in a series of cases. He reported that mis-identification of the biliary anatomy was the major cause of such injury. In 27 of 34 cases examined, mis-identification of the biliary anatomy occurred, usually mistaking the common bile duct for the cystic duct. It was noted that the injury was unrecognised by the operating surgeon in three out of four cases. This suggested spatial disorientation on the part of the surgeon. Hugh [33] questioned the common assertion that bile duct injury is simply the result of inadequate dissection and poor visualisation of anatomic structures and looked for more latent systemic deficits that could contribute to such injuries. Latent errors were identified that contributed to bile duct damage even in the hands of well trained, skilled and experienced surgeons. These included:

- Latent deficiencies in training whereby traditional advice and landmarks for safe dissection were considered as inadequate.
- Latent deficiencies in equipment such as defective laparoscopes, inadequate light illumination, inadequate cameras or defective telescopes that led to poor view, and reluctance of hospital authorities to invest in new equipment.
- Latent deficiencies in the provision of inexperienced assistants who unsettled instead of facilitating the surgeon.

Dekker and Hugh [35] also stressed the importance of under-estimating the risk of potential bile duct injury and visual misperception leading to bile duct misidentification as contributory causes to such harmful events.

Hugh [33] proposed changes in what had been previously taught as how to perform gall bladder dissections. He proposed that principles used in maritime and aviation industries to minimise navigation errors should be employed:

- Starting dissection from a fixed point.
- Knowing where you are at all times.
- Having a clear bearing.

In addition, surgeons were encouraged to refuse inadequate equipment; in one case surgeons suspended laparoscopic operations until the hospital management provided better equipment. It was also emphasised that targeted training in camera holding could improve the quality of assistance provided. Attention was also paid to teaching juniors to be assertive and be prepared to question the actions of their seniors, if they felt there was mis-identification of anatomical structures. In a prospective study of 2000 laparoscopic cholecystectomies, where the above principles were applied, there were no bile duct injuries [33].

2. In the UK, The Bristol Inquiry, chaired by Professor Ian Kennedy, examined the factors that led to unacceptably high death rates in children undergoing heart surgery at the Bristol Royal Infirmary and the Bristol Royal Hospital for Sick Children [36]. It was reported that between 1988 and 1994 the death rate at Bristol hospitals was about double that seen in other centres in 5 out of 7 years, with an estimated 30–35 excess deaths observed in children under one undergoing heart surgery in Bristol between 1991 and 1995. The enquiry found that about one in three children who had open-heart surgery received less than adequate care due to a number of latent failures occurring at multiple levels. These failures included [36]:

   - Processes: "*a service offering paediatric open-heart surgery which was split between two sites*" "*with cardiologists in one hospital and surgeons in another*", "*no dedicated paediatric intensive care beds, no full-time paediatric cardiac surgeon and too few paediatrically trained nurses*" [36].
   - Leadership: "*There was a 'club culture'; an imbalance of power, with too much control in the hands of a few individuals*". "*There was power but no leadership. The environment was one in which problems were neither adequately identified nor addressed*" [36].
   - Management: "*Senior managers were invited to take control, but little or no system existed to monitor what they did in the exercise of that control*" [36].
   - Regulation: "*At a national level there was confusion as to who was responsible for monitoring quality of care*". "*There was no agreed means of assessing the quality of care. There were no standards for evaluating performance*". "*The clinicians in Bristol had no one to satisfy but themselves that the service which they provided was of appropriate quality. There was no systematic mechanism for monitoring the clinical performance of healthcare professionals or of hospitals*". "*Audit was represented as an educational tool, not a mechanism for accountability to the profession, the employer (the NHS) or to the public*". "*When concerns were raised, it took years for them to be taken seriously*" [36].
   - Education: "*There was no requirement on hospital consultants at that time*"…. "*to keep their skills and knowledge up to date*" [36].

3. Wrong side surgery can have devastating consequences for the patient and healthcare staff involved. Pre-operative verification of the side with the involvement of the patient, appropriate marking of the side to be operated, and taking time out to check that we are dealing with the correct side, are recognised requirements for minimising the possibility of side errors.

Underlying latent failures may contribute to wrong side surgery, rather than the surgeon simply putting the knife on the wrong body part. Mulloy and Hughes [37] describe some of the latent system failures that may exist and predispose to such an event occurring:

- Lack of institutional check mechanisms to verify side.
- Reliance only on surgeon for confirming side.
- Lack of continuity of surgical care -multiple surgeons or teams.
- Inadequate communication channels between team members.
- Time pressures to reduce pre-operative preparation time.
- Inadequate staffing levels.

11.6 Isolated Versus Sequence of Events Leading to Harm

Habit is the nursery of errors.

Victor Hugo [3]

Failure of safety leading to patient harm may be due to an isolated event, but in many cases it is due to a sequence or combination of events.

Reason [38] proposed the concept of the Swiss Cheese model of safety barriers. According to this, normally there are several barriers protecting against harm, each of which may have potential holes or weaknesses. Failure in one barrier may be prevented from causing harm by the next barrier. However, if the weaknesses of consecutive barriers coincide, then harm may occur. Consider the following two hypothetical scenarios:

1. A patient may be having surgery on the index finger for tendon triggering:
    - The patient may be correctly consented, marked, pass all the verification checks, prepped and draped.
    - But the surgeon instead of making the incision on the index finger, opens up the ring finger instead.

In this situation harm occurs due to a single error occurring at the terminal step of care delivery, when no more safety barriers are left.

2. In contrast, wrong side surgery may be due to failures at multiple barriers. A patient with established dementia may be admitted with a left hip fracture, needing a hip hemi-arthroplasty:
    - The junior doctor admitting the patient in the Emergency Department, has just clerked three other hip fracture cases and by mistake recalls the wrong radiograph when it comes to marking the leg, marking the "right" instead of the "left". The patient cannot self-consent, screaming with pain on moving either leg.
    - The attending surgeon the next morning simply checks the side mark on the leg, and countersigns the consent form; the attending surgeon cannot recall the actual radiograph as in the admission meeting several cases of hip fractures were reviewed, and it is hard to recall "who is who".
    - The trauma coordinator (in line with the junior doctor's handover sheet) lists the patient for "right hip" surgery.
    - The checks in the ward confirm that the leg mark, consent and listing agree.
    - The checks on arrival at the surgical theatres' suite confirm that the leg mark, consent and listing agree. The patient is not questioned as they cannot give a reliable answer and there is no family or carer around.
    - The team pre-op briefing, confirms the listing and the type of procedure to be performed.
    - The final verification, prior to "knife to skin", confirms the presence of the correct patient and that the marking, listing and consent agree. The surgical theatre monitors are down and a final examination of the radiograph cannot occur.
    - The surgeon opens up the wrong leg.

In this situation harm occurs due to an initial marking error, early on in care delivery, but also failures of multiple subsequent barriers where the error could have been identified and prevented from causing harm.

11.7 Strategies for Improving Safety

Don't spend time beating on a wall, hoping to transform it into a door.

Coco Chanel [39]

In improving safety, we aim to prevent failures from occurring and also aim to recover from such failures if they were to occur (Fig. 11.2).

Recovering from failures may mean having mechanisms in place for identifying such failures early, and stopping them from propagating and causing harm or causing worse harm than what may have already occurred. Such mechanisms may include multiple verifications, or checklists; if a medicine has been prescribed wrongly on the drug chart, a routine check by the pharmacist may identify the error, hence stopping the administration of the wrong medicine and potentially preventing harm from occurring. Alternatively, even if a dose has been given, identification of the error may stop further administrations hence minimising harm.

Recovery, may also mean having plans to counteract the effect of failures to stop or minimise the harm they may cause. Early identification of a hip replacement wound infection may lead to appropriate antibiotic treatment, stopping it from becoming deep, involving the prosthetic implant and leading to revision surgery.

In considering how to improve safety one may look at addressing the "system" and the behaviours or actions of the individual.

Howell et al. [40] carried out a systematic review of studies evaluating interventions used to reduce adverse events in surgery. They reported that of the interventions examined the following were shown to have a beneficial effect in decreasing morbidity and mortality:

1. Use of safety check lists.
2. Improving nurse to patient ratios.
3. Intensive care unit physician involvement in post- operative care.
4. Sub-specialisation in surgery.
5. Submission of outcome data to national audit.
6. Adherence to care pathways.
7. Certain safe technologies and team training.

As an individual surgeon, you may not be able to radically alter the system you find yourself practising in. Nevertheless, individuals and front line staff have a central role to play in minimising the possibility of harm. Mehtsun et al. [16] in their review of the federal repository of medical malpractice claims, estimated that about 4000 surgical never event claims occur every year in the USA. Of interest, 12.4% of physicians named in a surgical never event claim were also named in at least one further surgical never event claim.

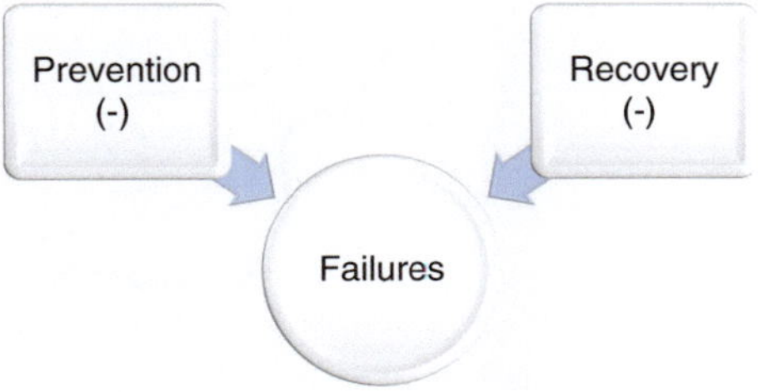

Fig. 11.2 Minimising the effect of healthcare failures

11.8 The Iceberg of Error

What we think, we become.

Buddha [3]

One may think of "system" errors as the iceberg which lurks round the corner, and the skipper the individual which may have a last chance to spot it, to redirect and save the ship (Fig. 11.3). Sometimes, no matter how alert, able and skillful the skipper is, there may be fog, a rough sea, the iceberg tip maybe hardly visible and can't avoid. But on situations the skipper may not be alert, may be tired or sleepy, may be distracted, may be entertaining rather than looking. There may also be times when the skipper is aware of the iceberg and still decides to risk it, take chances, and sail close to the wind. And then, there may be times when the skipper can't see but one of the stewards calls the iceberg out and saves the ship.

And even if the unlikely were to happen and the ship hits an iceberg, the skipper may help avert disaster by shutting the compartment barriers to isolate the flooding, or by coordinating a safe transfer of all to the life vessels.

Appreciate your vital position as the last layer of protection, the last line of defence.

Clarke et al. [41] looked at near misses of wrong side surgery in Pennsylvania, USA. They reported that surgeon involvement in the verification process in the pre-operative holding bay helped prevent wrong side surgery in 19%. Nurses carrying out verification in the pre-operative holding area were also effective in catching errors as did some of the scrub nurses in the surgical theatre.

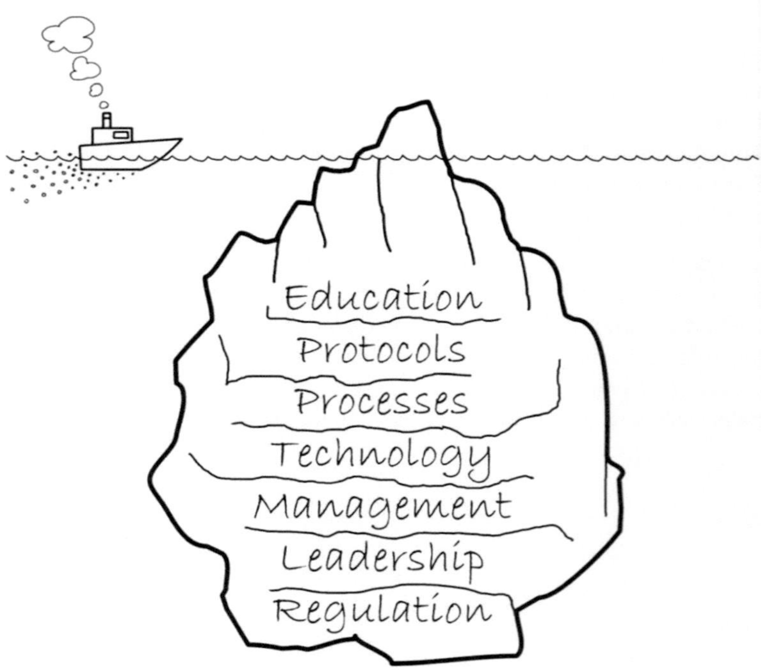

Fig. 11.3 The iceberg of error

11.9 Individual's Actions to Improve Safety

Just as courage is the danger of life, so is fear its safeguard.

Leonardo da Vinci [3]

Hugh [33] described certain attitudes thought to characterise a surgeon at low risk for error as someone who:

- Anticipates the order of events.
- Accepts feedback and input from other staff.
- Considers and modifies the working hypothesis as the need for information arises.
- Recognises the limitations of own-self such as self-fatigue, time pressures.

In addition to these attributes there may be other behaviours and actions that an individual surgeon could exhibit to help promote safety, and these are described next.

11.9.1 Be Risk Aware

One is not exposed to danger who, even when in safety is always on their guard.

Publilius Syrus [3]

One of the initial steps in encouraging a safe practise may be to consider risks and the harmful effects they could have:

- "What could go wrong? What if it doesn't work out?"
- "What if the "unlikely" happens?"
- "What would be the consequences?"

Considering the possibility of things going wrong, may be the first step in looking for what could go wrong, and for planning preventive measures or ways out.

- One may need to consider the risk of being run over if walking down the street in the middle of the night dressed in pitch black, in order to consider putting a shiny vest on.
- If you stay in a hotel for the first time, considering the possibility of fire breaking out, no matter how rare that may be, may let you look for a fire exit in case you have to run out.
- One may need to consider the risk of being struck by lightning if seeking shelter under a tree in the mid of a thunderstorm, to think about searching for another cover from rain.

Dekker and Hugh [35] reported that a chronic pre-operative underestimation of risk amongst experienced surgeons may contribute to bile duct injury during laparoscopic surgery. They postulate that such underestimation may be based on experienced surgeons failing to accept that it could happen to them. They also suggest that risk underestimation may be based on previous success in avoiding the error and stress that such past success cannot guarantee future safety. They propose that surgeons should deliberately heighten their awareness for potential risks, and keep reminding themselves "this could be the one".

Adverse events will happen no matter how skilled, careful, dedicated or thoughtful we are, but sometimes what determines their outcome is how we respond to those, how we recover from error, how we intervene to rectify and prevent further harm.

Being risk aware may enable you to prepare back-up plans and ensure the resources are available if those plans had to materialise:

- If your initial plan of fixing the fracture does not work out, do you have the equipment for replacing the broken hip, or would it mean having to stitch up the wound and return to theatre another day?
- If you hit the popliteal artery whilst drilling in the back of the knee do you have a vascular surgeon on standby? How long away will they be if you were to need them?

11.9.2 Avoid Taking Unnecessary Risks

I can resist everything except temptation.

Oscar Wilde [3]

Avoid taking unnecessary risks, cutting corners, ignoring well established processes. Stick to safe practises all the time. Once safe practises, rules and procedures are established, keep to those in a consistent manner. Avoid cutting corners even when you could easily do.

The equipment you asked for may not be ready, the company representative that you needed to guide you with using the new instruments may not be able to stay, there may not be backup equipment in case you drop that pelvic retractor on the floor. Should you be taking risks and proceeding, hoping it will all be good, hoping that you can get the job done just with the equipment on the shelf, that the scrub staff will manage to work out the new equipment, that you will be careful and won't drop the retractor?

The pressures may be high, not wanting to cancel a patient who has made plans and has been fasted from the night before; pressures from management who are worried about the patient breaching and theatre time under-utilised, or from yourself who may be reimbursed by case and a cancellation may mean loss of valuable income for that week or month. You may feel as being "on a mission" of getting the job done. You may be thinking it's a "bravery act", with you "the hero" saving the day.

It is a balance of risks and every case needs to be considered on its merits, but keep safety as one of the main factors in your thoughtful consideration.

11.9.3 Identify Error Prone Situations

Once we know our weaknesses they cease to do us any harm.

George C Lichtenberg [3]

Develop the ability to recognise situations with a high probability of things going wrong, of errors occurring. Reason [38] proposed the three bucket model for recognising situations that have a high potential for error and harm. Bad stuff in each bucket should make alarm bells ring, concentrate attention and focus minds:

- Bucket 1—the current state of yourself.
- Bucket 2—the context in which the task occurs.
- Bucket 3—the error potential of the task per se.

One may consider a task like driving and the potential for accidents and errors. What if you just finished a long shift and you hit the road with your eyes half closed? What if you are short sighted but you forgot your glasses at home? What if the car is due for service and the brakes or tyres are worn out? What if you are used to a manual gearbox and you have been given an automatic? How reliable is your co-driver, can you trust the directions they are giving? Has the co-driver been round London many times before as they claim to have done? What if it is raining, foggy or snowing, do you have to take that into account? And what if you are used to driving on an empty stretch of road in the Atacama Desert, and then you find yourself in the middle of the M25 motorway going round London?

Similar considerations may be applied in surgery. Recognise your own weaknesses that may predispose to error. You may be tired at the end of a full day theatre list, you may be recovering from bad flu, you may be upset that your last case will be cancelled due to lack of ward beds, dismayed for the lack of essential equipment. Appreciate that such factors may alter your behaviour, impair your performance and predispose to error.

Recognise the context in which you are expected to perform. The theatre environment is often dynamic with theatre staff changing, new faces you never came across before, junior staff not familiar with the tasks you are about to carry out. It may be that your assistant is very junior, just started in surgery and needs guidance to ensure that sterility is maintained. It may be that your assistant has surgical experience but has not in this type of operation and needs guidance about retracting. It may be that the scrub nurse is not familiar with the equipment being used and needs guidance as what instrument to give you next. It may be the radiographer has never screened for this type of fracture and needs guidance as how to position the image intensifier, where to position the viewing screen, how to move the intensifier in and out without getting in your way. It may be the anaesthetist has little experience managing patients in a sitting position, maintaining low blood pressure to reduce bleeding, and needs guidance as to what you are looking for. You may have a heavy list and may be pressed for time. It may be that two cases on your list share the same name, or there may be a mixture of left and right knees to be operated upon.

Recognise the complexity of the task. It may be that you are carrying out a challenging lengthy operation, in a challenging case, a medically unfit patient, a patient with difficult access or exposure.

Recognise error prone situations and adjust your behaviour and actions appropriately. Anticipate the storm and plan appropriately.

11.9.4 Check and Check Again

There is no harm in repeating a good thing.

Plato [3]

Carry out multiple checks to ensure the information you rely on is accurate. If needed check your self rather than simply relying on the word of others. Don't rely on a single port of information as it may be wrong, use multiple sources, cross check the information for accuracy. Even patients themselves on occasions may not be a reliable source of information. Clarke et al. [41] looked at near misses of wrong site surgery in Pennsylvania, USA and reported that the patient or their family provided incorrect information in 11 wrong site procedures. You may consider:

- In admitting patients for surgery, ensure that you have available and see all relevant documentation from the patient's journey, the clinic letters, investigation results, the listing form for surgery.
- You may practise in an environment where there is no continuity in administrative support, where your waiting list administrator changes every couple of weeks, where separate administrators put together your Monday and Saturday lists. Lists may be put together just before surgery, patients may be rushed onto the list to replace short notice cancellations, cases may even be rung in the morning to see if they could come in. The typing of clinic letters may lack behind, not all cases may have been listed by you in person.

Avoid cutting corners and proceeding without the necessary information. Simply relying on the information you gather from the patient on the day may not be adequate. You may have previously identified in clinic consultations important issues, such as the previous history of deep venous thrombosis, the strong metal allergy, the previous reaction to local anaesthetic, and you may not be able to elicit this information again on the surgery day, in the rushed environment of the theatre suite.

Clarke et al. [41] in their study looking at near misses or initiation of wrong site surgery, reported that failures in documentation, review or verification of patient information (such as in diagnostic tests, clinic records, consent forms, patient records, theatre listing forms) contributed to a substantial proportion of the cases examined. Non-availability of clinic records to allow verification on the day of surgery contributed to four wrong site procedures in their report. Errors in scheduling were frequently mentioned but implicated in seven never events involving patient contact.

11.9.5 Use Checklists Appropriately

I watch a lot of astronaut movies....Mostly Star Wars. And even Han and Chewie use a checklist.

Jon Stewart [42]

Checklists refer to a list of items to be checked or tasks to be done. They help to avoid reliance on memory, and to provide consistency and completeness in performing vital tasks. Checklists are well known in aviation industry, but their potential value is increasingly recognised in healthcare [43]. The ability of intensive care units in Michigan to reduce the incidence of catheter related blood stream infections from 2.7 per 1000 catheter days to zero by using an evidence base five point check list is well recognised [44].

There are several aims of checklists such as:

- Facilitate communication.
- Ensure all steps are completed.
- Ensure all items are available.

Facilitate communication: to ensure that all information is communicated to all team members, at briefs, debriefs, handovers. If you leave it to memory you may forget to let the anaesthetist know that your patient has a history of deep venous thrombosis and that thrombo-prophylaxis is warranted.

To ensure all steps are done: to ensure that antibiotics have been administered, venous thrombo-prophylaxis has been initiated, correct preparation of a patient for theatre has been performed, all outstanding jobs are completed in the surgical ward round.

Crisis check lists for rare events: to ensure all steps are executed and in the correct order, when emergencies or rare events are encountered. Flight 1549 took off from the La Guardia airport in New York, on a nice bright day in January of 2009, but two minutes into the flight both engines were burnt out by a geese strike. The heroic pilots who successfully landed the plane on Hudson River took out the emergency checklist, trying to follow step by step what they were to do in such an eventuality [45]. In theatre you may face a rare event. A patient starts bleeding, you fractured the patella whilst doing a knee replacement, you dropped the graft on the floor during a tendon reconstruction, you perforated the aorta during an oesophagectomy. These will be stressful times, when everyone is on the edge, when all may be looking to the surgeon for a solution. Rather than relying on memory for a bail-out plan, a checklist may ensure a consistent approach. It may also give you the confidence that you are following the correct steps.

To ensure all items are available—such as checking that all equipment, instruments, and implants are available, sterilised and ready, rather than realising in the middle of an operation that some are missing, and having to send to the nearby hospital for them.

Weiser et al. [46] evaluated the impact of the World Health Organisation (WHO) Surgical Safety Checklist in reducing complications and mortality in urgent surgical cases in eight hospitals round the globe. They reported a reduction in complication rate from 18.4% to 11.7% and a reduction in death rate from 3.7% to 1.4% following introduction of the checklist. Borchard et al. [47] carried out a systematic review assessing the effectiveness and compliance of the use of safety checklists in surgery. They reported that surgical checklists reduced both mortality (relative risk 0.57) and complications (relative risk 0.63). However the mean compliance rate of surgical checklists was low at 75% (range 12–100%).

Checklists should be appropriate and relevant to the individual practise in which they are employed, and can be adjusted or modified to achieve that. Checklists should also be considered as complimentary rather than a complete substitute of all other practises, a substitute of common sense; pre-surgery briefing of the team using checklists does not totally substitute informal communication between team members to discuss patients or other issues.

Accept the importance of checklists and use checklists appropriately. They ought to be viewed not as a ritual or a box ticking exercise, but an opportunity to improve safety, to allow the team to stop, think, slow down, to concentrate minds. Checklists are as good as their operators. Do not simply go through the list for the sake of the process. If you identify a discrepancy during your checks, stop and evaluate it fully. Keep thinking of the consequences to patient and others if any adverse event were to occur, and the potential of checklists to help avoid those.

Appreciate the value of checklists. If one does not rely on memory for the weekly shopping but takes a shopping list along to the grocer's, should we be relying on memory for vital safety surgical tasks?

11.9.6 Recognise Critical Steps

The aim of the wise is not to secure pleasure, but to avoid pain.

Aristotle [3]

There may be critical situations, steps at which minds must be focussed, when all resources must be available, steps which if do not go as planned can have a huge impact, steps whose failure may not be easily salvageable. The ability to recognise such critical steps is an important skill to develop.

During surgery you may be supervising a junior surgeon, they may have done the procedure before, may be confident in what they are doing. Yet, there may be steps when you want to be physically present, when you need to double check that they are doing the right thing, difficult steps when your input will be of real value.

- A junior may be doing a Dynamic Hip Screw fixation for a hip fracture. Is the fracture adequately reduced? Is the initial guide wire central?
- You may be supervising a junior surgeon in doing a bowel resection in a case with multiple previous abdominal surgeries, with lots of intra-abdominal adhesions. They may have opened and closed the abdomen multiple times before, you

may have watched them opening the abdomen several times previously, but you may well need to scrub up from the beginning in this one.

- You are about to open an implant. Are we opening the correct one? Correct type, size, side? Is it still within date? Routine checklists at the beginning of surgery are unlikely to stop the use of the wrong implant. Double checking means neither simply assuming that the person who has checked it is correct, nor simply accepting what they are saying, but checking your-self and confirming that your readings match.

11.9.7 Aim for Effective Error Reduction Measures

The best car safety device is a rear-view mirror with a cop in it.

Dudley Moore [3]

Measures taken to improve safety may differ in their effectiveness in achieving their aim of preventing errors and hence harm. One may consider the Ladder of Effectiveness of Safety Interventions, with protocols, processes, education and warnings at the lower part of the ladder with technology imposing safe behaviour higher up (Fig. 11.4). The specific ladder order may vary according to what you are trying to achieve but this emphasises a concept that should be considered in safety planning.

If we look at the use of car seat belts, one may set processes whereby drivers and passengers are informed that seat belts have to be worn. Drivers may be educated, be made aware of the benefits and risks, to make them buy in the process, change their behaviour. The process could be enforced with heavy penalties and fines, but one cannot be looking all the time. Technology may be employed to warn the driver, such as the constant beeping when we set off without a seat belt on, but the driver can still ignore it. Alternatively, technology can be introduced which prevents the car from setting off without the driver's seat belt on [48].

Redesigning equipment to force safe behaviour can have a profound effect on safety. In contrast, re-educating staff, changing a policy or adding a warning label or sign may have a more limited effectiveness in promoting safety.

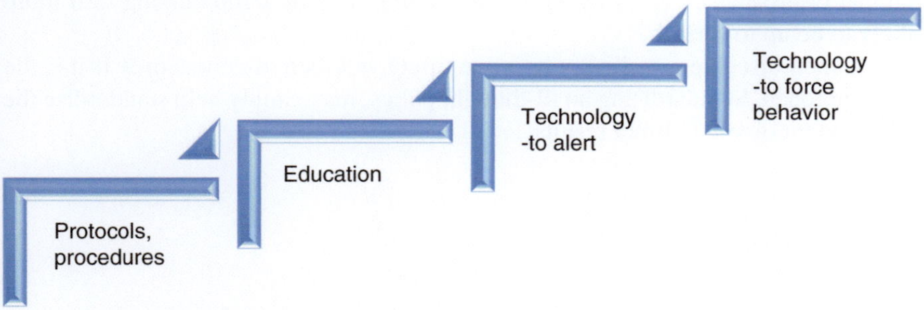

Fig. 11.4 Ladder of effectiveness of safety interventions

Consider the possibility of a wrong implantation of a right sided knee implant on to a left knee.

- A policy may be put together of having to check the side writing on the implant box before opening and handing over the implant to the surgeon.
- Staff may be informed of this policy and educated of the possibility of inserting the wrong side implant and of the need to pause and double check prior to opening the component and prior to implantation.
- A warning sign may be put on the implant box reminding the operator to check the implant side.
- The behaviour of staff may be monitored to ensure that they carry out the checks, with disciplinary action taken for those deviating from good practise.
- In contrast, the implant could be designed such as to make it physically impossible to insert a right sided implant into a left knee and vice versa.

Which of the above is likely to be more reliable?

11.9.8 Aim for Standardisation

There's no harm being a copycat. If someone else is doing a good job, copy. It's free.

Uday Kotak [3]

Standardisation of procedures, policies or protocols may help improve safety. Standardisation of certain actions may help avoid ambiguity and confusion, may ensure that actions are performed rather than avoided, and may minimise the possibility of error.

Consider a department where each attending surgeon has a different protocol for VTE, or a different protocol for peri-operative antibiotic prophylaxis, versus a department where there is one protocol for all. If there is a high turn-over of junior staff who are asked to assess and prescribe for VTE prophylaxis or antibiotic prophylaxis which setting is likely to produce more reproducible results?

Consider a department where each surgeon prefers to use a different knee arthroplasty prosthesis, versus one where there is agreement for the use of one or a small number of prostheses. In which of the two settings are the scrub nursing staff more likely to be up to scratch?

A standardised approach may improve safety, but then of course only if it is the right approach. Standardising an ill-thought policy, may simply help standardise the achievement of sub-optimal results.

11.9.9 Aim for Consistency

If we can play like that every week we'll get some level of consistency.

Alex Ferguson [3]

Safety measures should be consistently applied rather than a la carte. Avoid cutting corners just because you can. Safety measures should be fully applied at all times, to help encourage a culture where safety is the norm. Develop the mental strength to resist the temptation of cutting corners even if you have got away with it before. If you start cutting corners where will you stop? How effective is team briefing if utilised by all surgeons apart from the odd one? Or if you religiously apply team briefing first thing in the morning, but not after the lunchtime break when you are in a rush?

When are the red traffic lights be more likely to be effective? When they are respected by all drivers who stop at all times, or when they are simply considered as guidance, and ignored when one cannot see another car coming?

11.9.10 Recognise the Impact of Minor Failures

We are all of us failures - at least, the best of us are.

James M. Barrie [3]

We often pay attention to big failures, huge errors. Yet minor failures may accumulate and impair outcomes.

de Leval et al. [49] evaluated the role of human factors on surgical outcomes in 243 cases of major cardiac surgery. Major and minor human failures during surgical procedures were recorded. Major failures were considered as potentially life threatening events whereas minor failures were events that in isolation were not expected to have serious consequences on patient care. Major events were related to death. Similarly the total number of minor events per case was closely related to death and/or near misses. Some of the minor events recorded were positioning errors by surgical assistants, instrument handing errors by scrub nurses, distracters, inappropriate task delegation to inexperienced assistants, problems with theatre equipment, problems with communicating information between team members, absence of a senior team member in theatre at critical safety times, surgical fixation to one particular aspect of the procedure at the expense of monitoring, human resource problems such as no assistant available.

Catchpole et al. [50] observed 24 paediatric cardiac and 18 orthopaedic operations, to identify negative intraoperative events. They found that for every three minor problems (defined as "negative events that were seemingly innocuous") encountered above what was expected for the operation, intra-operative performance decreased by 1% and the duration of surgery increased by 10 min. Hence, reducing such minor events could lead to safer and faster operations.

Recognising the impact of minor failures may help concentrate minds, pay attention to detail, aim for a smooth ride.

11.9.11 Learn from Near Misses

The crashes people remember, but drivers remember the near misses.

Mario Andretti [44]

Learn from near misses, give them the attention they deserve, investigate what exactly happened, learn and improve. Don't simply feel relieved for real harm was avoided, but also feel relieved that you have been made aware of what could go wrong and given the opportunity to put barriers in place to help prevent real harm in the future. Investigating near misses and learning from them may help to identify latent systemic flaws which if addressed could prevent the occurrence of an actual adverse event occurring.

You may find yourself driving on the outer lane in a busy motorway, and whilst suddenly switching to the middle lane inadvertently cutting the way of the car behind you. You may have forgotten putting the indicator on, properly looking at your mirror, relying too much on your mirror and not turning your head to ensure a car is not too close for the rear or side mirrors to detect. The driver behind, even though upset, could simply slow down, shake their head in despair and just keep driving, without you ever becoming aware of how close to an accident you may have just come. Alternatively, the driver behind you may sound the horn, not to punish you, but to alert you of what has just happened. You may then drive away simply relieved that it was a near miss, or use the event as a wake-up call, a reminder of how careful one ought to be when changing lanes.

11.9.12 Use Incident Reporting Systems Appropriately

In giving advice seek to help, not to please, your friend.

Solon [3]

Incident reporting systems are commonly utilised in institutions, to record unwanted events where harm was done, or near misses where there was a lapse of performance but without actual harm occurring. Any filed reports are then looked into, examined, and where necessary actions are taken to learn from the event, and bring about changes that minimise the risk of similar events recurring. For any incident reporting system to effectively work it must be used as it is intended by those reporting and those investigating such incidents. You may consider:

- Don't use incident reporting systems for threatening others, that "if you don't send for the patient I will file an incident form" or use them just for getting at foes. And if you are on such a receiving end, with others using the incident filing system to put pressure on you, that "If you don't do the discharge summary I will do an incident form", then try to educate them that that's not what incident reports are about.
- If you find yourself unfairly named or accused in such a report, engage with the system, provide the correct facts and help to resolve the situation. It may be dismayed at

the ease with which others can file a report, but is it not better to have a culture where individuals feel free to report (even if on occasions get it wrong) rather than a culture where they don't dare to speak up and unsafe circumstances are not brought to light? It may just be that someone with good intentions misunderstood the situation, at least you know that a system of feedback and improvement is in place and working.

- If you find yourself in a situation where you examine any such incident reports ensure they are not simply put aside, closed without thorough investigation and action. What message would that send? Would it inspire confidence to its users, would it encourage them to report, would it help to alter behaviours and practise?

11.9.13 Lead in Safety by Example

Put safety high up on your agenda. Raise and act upon concerns. Prepare to stand up and be counted when it comes to safety matters. Have the courage to take big decisions and, where needed, to halt, rethink and reconsider. Be the first to do what you expect from others. How can you expect all to wear their face mask in theatre, if you are not wearing yours?

11.9.14 Encourage Flat Hierarchy

There's no sort of hierarchy at 'Corrie.' The crew get on.

Katherine Kelly [3]

Aim for an easy going environment which may empower all to speak up, to raise concerns, to warn of risks and unsafe acts, where they will be praised for noticing your mishaps rather than made to feel they should stay quiet. Do not underestimate the "power of hierarchy", the power of "authority gradients" [51] in inhibiting free communication and predisposing to error.

There are multiple examples in the aviation industry of airplane crashes where the first officers are thought to have correctly read the situation and could have averted disaster by challenging their captains; yet their perception of authority and hierarchy stopped them from doing so [52, 53].

Would you not like your junior to be able to inform you that you just put your subcutaneous suture through the skin and you need to take it out and start again, remind you that you are planning to fix the medial part of the clavicle hence the incision you are about to make is too lateral and must be reconsidered, advice you that your bone cut for knee arthroplasty is not fully smooth and you need to run the saw again, warn you that the tourniquet time is getting too lengthy?

Kobayashi et al. [51] surveyed USA and Japanese residents to identify barriers to questioning or challenging their superiors. They reported that the decision not to challenge a superior (in a given general surgery scenario in the surgical theatre setting) was mostly influenced by the relationship with their superior and their superior's personality. Friedman et al. [54] randomised 34 trainee anaesthetists into facing a consultant anaesthetist in a strict/exclusive or open/inclusive communication dynamic. They showed a better performance in their ability to challenge clearly wrong decisions of their superior in dealing with a difficult intubation, in the flattened hierarchy as compared to the steep hierarchy group (although statistical significance was not reached).

If all feel that they have a say, then frontline staff may look out, identify risks, and come up with solutions and actions to address them.

"STOP me if about to err" may encourage your juniors to speak up and avert harm.

11.9.15 Aim for a No-Blame Culture

To err is human, to forgive, divine.

Alexander Pope [3]

Aim for a no-blame culture, to empower others to admit to, report and address minor lapses, that if ignored could accumulate and lead to major harm. Don't blame the junior for touching your back and desterilising, but praise them for letting you know and for taking time out to re-scrub.

A no-blame culture doesn't mean lack of accountability; repeated active violations, or repeated disregard of processes and rules need to be looked at and addressed. Nevertheless, it is a fine balance and a challenging task between rushing for the stick when things go wrong and trying to explore the real cause.

A driver may accept a parking ticket when parked on a double yellow line, but may not be so understanding or forgiving of the traffic wardens hiding round the corner to catch them out.

11.9.16 Communicate for Safety

Words empty as the wind are best left unsaid.

Homer [3]

Appreciate the role of communication in maintaining safety.

The Joint Commission for Hospital Accreditation in the USA, reported that failures in communication contributed to more than 60% of 7194 never events [55]. Hu et al. [56] evaluated 22 hours of operating in six complex surgical procedures, and observed communication failures in all, occurring at a rate of one in eight minutes. The most common failures were due to communication exchanges occurring in the absence of key individuals, and due to exchanges not leading to resolution of the issues raised.

Several aspects of communication may help improve safety, and these are described below.

Use similar language

This refers to using structured systems in communication, which all participants can comprehend and follow. These include the ALS (Advanced Life Support) [57] and ATLS (Advanced Trauma Life Support) [58] systems used in the resuscitation of a collapsed or trauma patient.

Use understandable language

Avoid technical terms or abbreviations that may not be understood or may be misunderstood by others.

Use closed loop communication

Ask the receiver to repeat the message you have just given them, to confirm they have understood what you said and that they are on board. Warn that you will ask them to repeat your message prior to giving them the information.

Expect a response to a question or a request. Respond yourself to a question or a request of others. If you are asking for a surgical instrument in theatre, expect an answer whether that is available or not or someone has gone to get it. If you asked how much more time you need to finish the procedure, give a response than leaving others to guess as to whether you have heard them or whether it is too long to say.

Brief/debrief

Brief and debrief prior to commencing and after completing a task to help ensure that all participants share information, and have a similar picture of what will happen or of what has happened. Briefing allows all to contribute to what will take place, and debriefing allows all to learn from what has taken place.

Such briefs/debriefs may occur at various levels:

- Between all surgical theatre team members.

The WHO Surgical Safety Checklist provides a structured briefing between the various members of the surgical theatre team, prior to commencing surgery and debriefing at the end of surgery [59, 60]. In such briefings:

- Aim to develop a shared plan of what will happen, give specifics as to what will be done.
- Share all considerations you have.
- Share your concerns, don't paint a rosy picture, if anticipating difficulties let all the team know. An action in response to an event may be considered "a plan" if it was expected, but may be considered a "reaction" if it was unexpected; expectation may distinguish between a "forced water landing" on the Hudson and a plane's "river crash" [61].
- Share the problems you have encountered, to learn for the next time, and warn others of similar problems.

- Between surgeons–supervisor and trainee.
  - Discuss the procedure.
  - Potential problems, anticipated difficulties.
  - Distribution of roles –who will do what.
  - In the end what went wrong, what well, what has been learnt.

Communicate over and over again

Avoid under-estimating the need for communication or over-estimating what the communication exchanges have achieved. Communicate the message multiple times in various ways or means. Enquire whether it is all clear for your co-workers or whether there are still any doubts that ought to be resolved. Wauben et al. [62] evaluated the perceptions of communication amongst surgical theatre team members in five hospitals in Netherlands. Surgeons rated the current state of communication (exchange of information, establishing shared understanding, co-ordination) significantly higher than scrub nurses and anaesthetists did. What may look crystal clear to you may still look blurred to others.

Have a system for alerting about unsafe acts

You may need to communicate your immediate concerns when things are not right. The receiver of your concerns may be a senior making such communication more difficult and challenging. Several warning systems have been described, to help the expression of concerns. The CUS (Concerned, Uncertain, Safety) technique puts a structure on how to escalate concerns and attract attention to those [63]. It was developed by the Agency for Healthcare Research and Quality in the USA as illustrated below:

C-oncern – "I am concerned we should not be doing this".
U-nsure – "I am uncertain that we should be doing this".
S-afety – "We shouldn't be doing this, it is unsafe. We should stop".

Use eligible written information

Use clear, eligible writing in constructing a theatre list, a consent form. Use capital letters for key words, avoid abbreviations. Do not let others having to guess as to what you are trying to say.

Ask for information, not confirmation

Avoid asking others to confirm what you are saying. They may reply "yes" or "no" without fully appreciating your question. They may want to please you with an answer, they may be embarrassed to tell you they cannot understand what you say or that they cannot recognise your accent.

- When confirming if you have the right person in front of you, instead of asking "Are you Mr. John Smith?" you may ask "Can you please tell me your name?"
- When checking if the correct procedure is documented for the correct patient instead of stating "I am doing your right knee arthroscopy today" you may initially ask "What am I doing for you today?" "Which knee am I doing?"

Alert others of potential trouble

- At handover of ward cases, of admissions.
- In theatre, of problems you encountered with particular surgical equipment.

Use team inclusion words

Sexton and Helmreich [64] evaluated the relation of language content with performance and error rates, in a flight simulator. They showed that use of the first person plural (WE) and words indicating strive towards a successful outcome (ACHIEVE) were positively related to cockpit performance as well as communication and negatively related to error rates. Use of achievement words by the captain improved subsequent performance and reduced errors by the flight engineer.

Hear the warnings

If someone is questioning your decision, then stop, listen, take on board what they are trying to say and reconsider.

- "Why are they saying that?"
- "What are they worried about?"
- "What are they trying to tell me?"
- "Am I missing something?"

Listen, no matter how junior the challenger is, whether a doctor or not.

11.9.17 Avoid Disruptive Behaviours

I was a disruptive child.

Allyson Felix [3]

Avoid behaviours that put others out of ease, which distract rather than concentrating minds. You may alert others without shouting, swearing, fast walking, jumping up and own, kicking anaesthetic room doors, throwing instruments around. All that such behaviours may achieve is to stress even more those around you, making them less able to help you out.

Cochran and Elder [65, 66] explored the perspectives of 19 individuals, who had diverse occupations in the peri-operative setting, on the effects of disruptive behaviours by surgeons. They described that such behaviours compromised safety by shifting the attention from the patient to the surgeon, and increasing the risk of mistakes during surgical procedures.

After all what would you think if the scrub nurse started shouting every time you asked for a surgical instrument, did not reply to your requests, or kept refusing to look for an instrument you needed in a challenging case? Similarly how would you feel if the anaesthetist started shouting that you should rush to scrub up, or simply stared through you each time you tried to discuss a medical matter. Would such behaviours help you focus or would they make the procedure an unnerving experience?

11.9.18 Plan to Recover from Failures

Any man is liable to err, only a fool persists in error.

Marcus Tullius Cicero [3]

Plan for the possibility of failures, and prepare as to how to deal with them to help avoid or minimise harm. Develop mental strategies, institutional protocols or pathways.

Failures may lead to stressful situations. Startle has been used to describe a "fight or flight" response that occurs in encountering unexpected, highly stressful significant events. It is thought that startle can impair information processing and decision making, leading to poor handling of such occurrences [67].

The startle reaction has been used to explain inappropriate reactions amongst pilots leading to catastrophic failures. Air France Flight 447, an Airbus A330, flying from Rio de Janeiro to Paris in 2009, crashed in the Atlantic Ocean after going into an aerodynamic stall. A subsequent investigation suggested that the first officer continued to pull back rather than push forwards the control stick (equivalent to hitting the brakes rather than accelerating) almost all the way down to the ocean [68].

When faced with an unexpected intra-operative event one may:

- Freeze, unable to decide what to do or take action.
  - You may be manipulating a pelvic fracture to apply your fixation plate when suddenly you see a gush of blood.
  - You may be inserting the laparoscopy camera when you suddenly realise you entered the aorta. Can you respond appropriately or do you freeze?
- Persevere with the original plan of action, rushing to finish the task:
  - You just drilled the femur for inserting a screw and a gush of blood is coming out of the screw hole. Do you simply continue as planned and insert the screw or do you stop and re-valuate the situation, looking for a major vessel injury behind the femur?
- Take wrong action

If faced with an unexpected emergency try to:

- Maintain composure.
- Evaluate what you are facing.
- Take appropriate action.
- Communicate with the team.
- Emphasise urgency.

Expect the unexpected, to minimise the effect of startle when the unexpected occurs. A complication may be "unlikely", but if you are not prepared for the "unlikely" how can you deal with it if and when it arose.

References

1. Kohn L, Corrigan J, Donaldson ME. To err is human. Washington DC: National Academy Press; 1999.
2. Leung A, Luu S, Regehr G, Murnaghan ML, Gallinger S, Moulton CA. “First, do no harm”: balancing competing priorities in surgical practice. Acad Med. 2012;87(10):1368–74.
3. Famous Quotes at BrainyQuote. www.brainyquote.com. Accessed 2 Apr 2017.
4. Vincent C, Neale G, Woloshynowych M. Adverse events in British hospitals: preliminary retrospective record review. BMJ. 2001;322(7285):517–9.
5. Sousa P, Uva AS, Serranheira F, Nunes C, Leite ES. Estimating the incidence of adverse events in Portuguese hospitals: a contribution to improving quality and patient safety. BMC Health Serv Res. 2014;14:311. doi:10.1186/1472-6963-14-311.
6. James JT. A new, evidence-based estimate of patient harms associated with hospital care. J Patient Saf. 2013;9(3):122–8.
7. National Quality Forum. Serious reportable events in healthcare: a consensus report. Washington, DC: National Quality Forum; 2002.
8. Kizer W and Stegun MB. Serious reportable adverse events in health care. In: Henriksen K, Battles JB, Lewin DI (eds.). Advances in patient safety: From research to implementation. Rockville (MD), Agency for Healthcare Research and Quality (US); 2003
9. Never events. Annual report 2009/2010. National Patient Safety Agency. 2010. http://www.nrls.npsa.nhs.uk/resources/collections/never-events/?entryid45=83319. Accessed 2 Apr 2017.
10. Serious Reportable Events. NQF. http://www.qualityforum.org/Topics/SREs/List_of_SREs.aspx. Accessed 2 Apr 2017.
11. The “never events” list 2011/2012. Department of health. 2011. https://www.gov.uk/government/uploads/system/uploads/attachment_data/file/216724/dh_124580.pdf. Accessed 2 Apr 2017.
12. ‘Shock’ as wrong kidney removed. BBC News. 2002. http://news.bbc.co.uk/1/hi/wales/2049839.stm. Accessed 2 Apr 2017.
13. Kelly SP, Jalil A. Wrong intraocular lens implant; learning from reported patient safety incidents. Eye (Lond). 2011;25(6):730–4.
14. AVVO. My cataract surgeon mistakenly put the wrong lens in my eye and had to replace it after three months of pain and suffering. https://www.avvo.com/legal-answers/my-cataract-surgeon-mistakenly-put-the-wrong-lens--1732312.html. Accessed 2 Apr 2017.
15. Hagan P. The (frighteningly common) medical horror stories: patients left with needles, swabs and even razor-sharp SCISSORS inside them after surgery. Mailonline. 2016. http://www.dailymail.co.uk/health/article-3605482/The-frighteningly-common-medical-horror-stories-Patients-left-needles-swabs-razor-sharp-SCISSORS-inside-surgery.html. Accessed 2 Apr 2017.
16. Mehtsun WT, Ibrahim AM, Diener-West M, Pronovost PJ, Makary MA. Surgical never events in the United States. Surgery. 2013;153(4):465–72.
17. Kwaan MR, Studdert DM, Zinner MJ, Gawande AA. Incidence, patterns, and prevention of wrong-site surgery. Arch Surg. 2006;141(4):353–7.
18. Phillips DP, Christenfeld N, Glynn LM. Increase in US medication-error deaths between 1983 and 1993. Lancet. 1998;351:643–4.
19. Starfield B. Is US health really the best in the world? JAMA. 2000;284(4):483–5.
20. Reason J. Human error. New York: Cambridge University Press; 1990.
21. Reason J. Safety in the operating theatre—Part 2: human error and organisational failure. Qual Saf Health Care. 2005;14(1):56–60.
22. Weerakkody RA, Cheshire NJ, Riga C, Lear R, Hamady MS, Moorthy K, Darzi AW, Vincent C, Bicknell CD. Surgical technology and operating-room safety failures: a systematic review of quantitative studies. BMJ Qual Saf. 2013;22(9):710–8.
23. Doyle PA, Gurses AP, Pronovost PJ. Mastering medical devices for safe use. Am J Med Qual. 2017;32(1):100–2.

24. Reason J. Human error: models and management. BMJ. 2000;320(7237):768–70.
25. Reason JT, Carthey J, de Leval MR. Diagnosing “vulnerable system syndrome”: an essential prerequisite to effective risk management. Qual Health Care. 2001;10(Suppl 2):ii21–5.
26. AirCraft Accident report. Inflight separation of vertical stabilizer. American Airlines flight 587. National Transportation Safety Board. Washington. 2004. https://www.ntsb.gov/investigations/AccidentReports/Reports/AAR0404.pdf. Accessed 3 Apr 2017.
27. Aviation Knowledge. Saudi Arabian Airlines Flight 763 & Air Kazakhstan Flight 1907: head-on collision. http://aviationknowledge.wikidot.com/asi:air-kazakhstan-1907. Accessed 3 Apr 2017.
28. Aircraft Accident report. Loss of control and impact with Pacific ocean, Alaska Airlines Flight 261. National Transportation Safety Board. Washington. 2002. https://www.ntsb.gov/investigations/AccidentReports/Reports/AAR0201.pdf. Accessed 3 Apr 2017.
29. Aircraft Accident Report AAR 1/2010 - Boeing 777-236ER, G-YMMM, 17 January 2008. Air Accidents Investigation Branch. 2010. https://assets.publishing.service.gov.uk/media/5422f3dbe5274a1314000495/1-2010_G-YMMM.pdf. Accessed 3 Apr 2017.
30. National Transportation Safety Board Office of Public Affairs. NTSB issues urgent safety recommendation to address engine thrust rollback events on B-777 Aircraft. 2009. https://www.ntsb.gov/news/press-releases/Pages/NTSB_Issues_Urgent_Safety_Recommendation_to_Address_Engine_Thrust_Rollback_Events_on_B-777_Aircraft.aspx. Accessed 3 Apr 2017.
31. Aircraft Accident report. Helios Airways Flight HCY522. Hellenic Republic Ministry of Transport and Communications. 2006. http://www.aaiu.ie/sites/default/files/Hellenic%20Republic%20Accident%20Helios%20Airways%20B737-31S%20HCY522%20Grammatiko%20Hellas%202005-085-14.pdf. Accessed 3 Apr 2017.
32. Hradecky S. FAA requires separate configuration and cabin altitude warning lights on Boeing 737s. The Aviation Herald. 2011. http://avherald.com/h?article=43778c6b. Accessed 3 Apr 2017.
33. Hugh TB. New strategies to prevent laparoscopic bile duct injury—surgeons can learn from pilots. Surgery. 2002;132(5):826–35.
34. Strasberg SM, Hertl M, Soper NJ. An analysis of the problem of biliary injury during laparoscopic cholecystectomy. JAm Coll Surg. 1995;180:101–25.
35. Dekker SW, Hugh TB. Laparoscopic bile duct injury: understanding the psychology and heuristics of the error. ANZ J Surg. 2008;78(12):1109–14.
36. Learning from Bristol. The report of the public inquiry into children's heart surgery at the Bristol Royal Infirmary1984–1995. Kennedy I, 2001. http://webarchive.nationalarchives.gov.uk/20090811143745/http:/www.bristol-inquiry.org.uk. Accessed 1 Apr 2017.
37. Mulloy DF, Hughes RG. Wrong-site surgery: a preventable medical error. In: Hughes RG, editor. Patient safety and quality: an evidence-based handbook for nurses. Rockville, MD: Agency for Healthcare Research and Quality, US; 2008.
38. Reason J. Beyond the organisational accident: the need for "error wisdom" on the frontline. Qual Saf Health Care. 2004;(Suppl 2):ii28–33.
39. Popular Quotes. Goodreads. http://www.goodreads.com/quotes. Accessed 1 Apr 2017.
40. Howell AM, Panesar SS, Burns EM, Donaldson LJ, Darzi A. Reducing the burden of surgical harm: a systematic review of the interventions used to reduce adverse events in surgery. Ann Surg. 2014;259(4):630–41.
41. Clarke JR, Johnston J, Finley ED. Getting surgery right. Ann Surg. 2007;246(3):395–403.
42. AZ Quotes. http://www.azquotes.com/quotes/topics/self-reflection.html. Accessed 2 Apr 2017.
43. Clay-Williams R, Colligan L. Back to basics: checklists in aviation and healthcare. BMJ Qual Saf. 2015;24(7):428–31.
44. Pronovost P, Needham D, Berenholtz S, Sinopoli D, Chu H, Cosgrove S, Sexton B, Hyzy R, Welsh R, Roth G, Bander J, Kepros J, Goeschel C. An intervention to decrease catheter-related bloodstream infections in the ICU. N Engl J Med. 2006;355(26):2725–32.

45. Aviation Accident Report 10_03 – NTSB. National Transportation Safety Board. Washington. 2010. https://www.ntsb.gov/investigations/AccidentReports/Reports/AAR1003.pdf. Accessed 4 Apr 2017.
46. Weiser TG, Haynes AB, Dziekan G, Berry WR, Lipsitz SR, Gawande AA, Safe Surgery Saves Lives Investigators and Study Group. Effect of a 19-item surgical safety checklist during urgent operations in a global patient population. Ann Surg. 2010;251(5):976–80.
47. Borchard A, Schwappach DL, Barbir A, Bezzola P. A systematic review of the effectiveness, compliance, and critical factors for implementation of safety checklists in surgery. Ann Surg. 2012;256(6):925–33.
48. George P. New GM System won't let you drive unless you're buckled. 2014. http://jalopnik.com/new-gm-system-wont-let-you-drive-unless-youre-buckled-1580033614. Accessed 4 Apr 2017.
49. de Leval MR, Carthey J, Wright DJ, Farewell VT, Reason JT. Human factors and cardiac surgery: a multicenter study. J Thorac Cardiovasc Surg. 2000;119(4 Pt 1):661–72.
50. Catchpole KR, Giddings AE, Wilkinson M, Hirst G, Dale T, de Leval MR. Improving patient safety by identifying latent failures in successful operations. Surgery. 2007; 142(1):102–10.
51. Kobayashi H, Pian-Smith M, Sato M, Sawa R, Takeshita T, Raemer DA. cross-cultural survey of residents' perceived barriers in questioning/challenging authority. Qual Saf Health Care. 2006;15(4):277–83.
52. Controlled Flight Into Terrain Korean Air Flight 801 Boeing 747 – NTSB. National Transportation Safety Board. Washington. 1997. https://www.ntsb.gov/investigations/AccidentReports/Reports/AAR0001.pdf. Accessed 4 Apr 2017.
53. Weick KE. The vulnerable system. An analysis of the Tenerife air disaster. J Manag. 1990;16:571–93.
54. Friedman Z, Hayter MA, Everett TC, Matava CT, Noble LM, Bould MD. Power and conflict: the effect of a superior's interpersonal behaviour on trainees' ability to challenge authority during a simulated airway emergency. Anaesthesia. 2015;70(10):1119–29.
55. Joint Commission on Accreditation of Healthcare Organizations. Sentinel Event Statistics Data - Root Causes by Event Type (2004–2013). http://www.jointcommission.org. Accessed 18 Aug 2014.
56. YY H, Arriaga AF, Peyre SE, Corso KA, Roth EM, Greenberg CC. Deconstructing intraoperative communication failures. J Surg Res. 2012;177(1):37–42.
57. Advanced Life Support. Resuscitation Council (UK), 6th edition, 2011
58. ATLS: Advanced Trauma Life Support for Doctors (Student Course Manual). American College of Surgeons, 8th ed, 2008; ISBN-10: 1880696312.
59. World Alliance for Patient Safety. WHO surgical safety checklist and implementation manual. 2008. http://www.who.int/patientsafety/safesurgery/ss_checklist/en/. Accessed 6 Apr 2017.
60. Haynes AB, Weiser TG, Berry WR, Lipsitz SR, Breizat AH, Dellinger EP, Herbosa T, Joseph S, Kibatala PL, Lapitan MC, Merry AF, Moorthy K, Reznick RK, Taylor B, Gawande AA, Safe Surgery Saves Lives Study Group. A surgical safety checklist to reduce morbidity and mortality in a global population. N Engl J Med. 2009;360(5):491–9.
61. Sully: miracle on the Hudson (2016). Warner Bros. Pictures.
62. Wauben LS, Dekker-van Doorn CM, van Wijngaarden JD, Goossens RH, Huijsman R, Klein J, Lange JF. Discrepant perceptions of communication, teamwork and situation awareness among surgical team members. Int J Qual Health Care. 2011;23(2):159–66.
63. Agency for Healthcare Research and Quality. http://teamstepps.ahrq.gov/. Accessed 4 Apr 2017.
64. Sexton JB, Helmreich RL. Analyzing cockpit communications: the links between language, performance, error, and workload. Hum Perf Extrem Environ. 2000;5(1):63–8.
65. Cochran A, Elder WB. Effects of disruptive surgeon behavior in the operating room. Am J Surg. 2015;209(1):65–70.

66. Cochran A, Elder WB. A model of disruptive surgeon behavior in the perioperative environment. J Am Coll Surg. 2014;219(3):390–8.
67. Martin WL, Murray PS, Bates PR. The effects of startle on pilots during critical events: a case study analysis. Accessed at http://www98.griffith.edu.au/dspace/bitstream/handle/10072/54072/82496_1.pdf?sequence=1. Accessed 4 Apr 2017.
68. Final Report – BEA. 2012. https://www.bea.aero/docspa/2009/f-cp090601.en/pdf/f-cp090601.en.pdf. Accessed 4 Apr 2017

Chapter 12
When Things Go Wrong

Despite all good intentions, hard work and perseverance, implementation of safety nets and precautions, it is unlikely to achieve absolute surgical safety. On occasions things are likely to go wrong.

Adverse events are likely to happen, that may or may not cause harm, recognised complications or potentially avoidable errors, adverse events that our patients may become aware of or require disclosure, isolated occurrences or clusters of adverse events.

The ability to deal with an adverse event, disclose its occurrence, explore and investigate its underlying causes, alter practise to prevent similar events occurring in the future, and the ability to communicate any lessons learnt to the wider surgical community are important skills to develop, and are discussed in this chapter. Ways of evaluating clusters of uncommon, yet recognised, surgical complications are also described.

C. Panayiotou Charalambous, *Career Skills for Surgeons*,
DOI 10.1007/978-3-319-57490-5_12

This umbrella was supposed to keep me dry

12.1 Managing Adverse Events

It is better to destroy one's own errors than those of others.

Democritus [1]

As a surgeon you may encounter or be directly involved in an adverse event, where things have gone wrong. The World Health Organisation defines an adverse event as "an injury related to medical management, in contrast to complications of disease", with medical management referring to all aspects of care, including surgical care [2].

When faced with an adverse event in its acute presentation, one has to ensure that harm is minimised, and any harm that has already occurred is dealt with. The events leading to the harmful event are then examined to determine what went wrong, and how its recurrence could be avoided.

An approach used by the author and described using the acronym SAFEST (Stop, Antidote, Find, Explain, Sorry, Transform) [3], may provide guidance in dealing with a harmful event. This involves:

S-top any further harm from occurring
A-ntidote any harm that already took place
F-ind why this happened. Root cause analysis
E-xplain to the patient or relatives- disclose
S-orry Sympathise for what went wrong and apologise for any errors
T-ranform your practise to minimise possibility of recurrence

12.2 Root Cause Analysis of an Adverse Event

When things go right it's hard to figure out why, but when things go wrong it's really easy.
Steven Soderbergh [1]

An adverse event may be the result of a recognised complication of treatment (such as complications of surgical procedures), or occur due to medical error, due to violation of rules and processes, or other factors (such as failure of equipment or technology). On occasions the distinction is obvious, but in some it may be difficult to define.

Once harm is halted, and if possible reversed, it is important to get to the bottom as to what has happened, get to the root of the cause, a process known as root cause analysis.

The examination of the causal factors of an adverse event may be considered as having four components (Fig. 12.1):

1. Defining the adverse event, defining the problem.
2. Defining the patient pathway which one would expect to encounter under normal conditions, and locating the occurrence of the adverse event in that pathway.
3. Cause effect analysis.
4. Introducing changes that will minimise the possibility of the adverse event recurring.

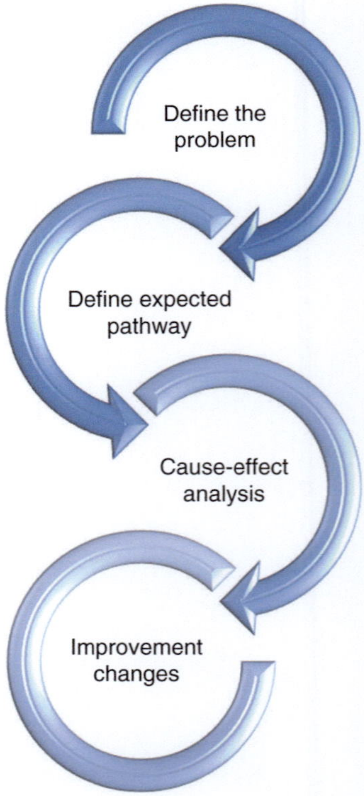

Fig. 12.1 Learning from an adverse event

A cause-effect analysis, is a method that attempts to determine what factors led to an adverse event occurring. This may be a single factor or multiple factors. Cause-effect analysis was initially described by Professor Kaoru Ishikawa at the University of Tokyo [4]. It considers potential factors contributing to the adverse event such as the personnel involved, patient pathways and processes, equipment, management, supervision, surrounding environment, available resources. It then examines each in detail, to determine if there were any deficiencies that could have led to the unwanted event.

A fish bone diagram is a way of demonstrating the cause-effect analysis in graphic form [4]. Horizontal lines point to the adverse event with branches and sub branches arising from this, pointing to potential causes (Fig. 12.2). Such an analysis may be part of a formal investigation into what went wrong, but also employed in a surgeon's individual practise to examine complications, to self-reflect and learn.

A cause effect analysis can only examine what it is asked to look into. Consider a complication that could arise due to pre-operative deficiencies in preparing patients for admission, deficiencies in the ward, or in the recovery suite; a cause effects analysis that examines only the surgical theatre environment may fail to give the true answer. If an adverse event could be due to understaffing, but resources and management pressures are not considered in the cause effect analysis, then one may not get to the bottom of the cause. Hence, in circumstances where the various

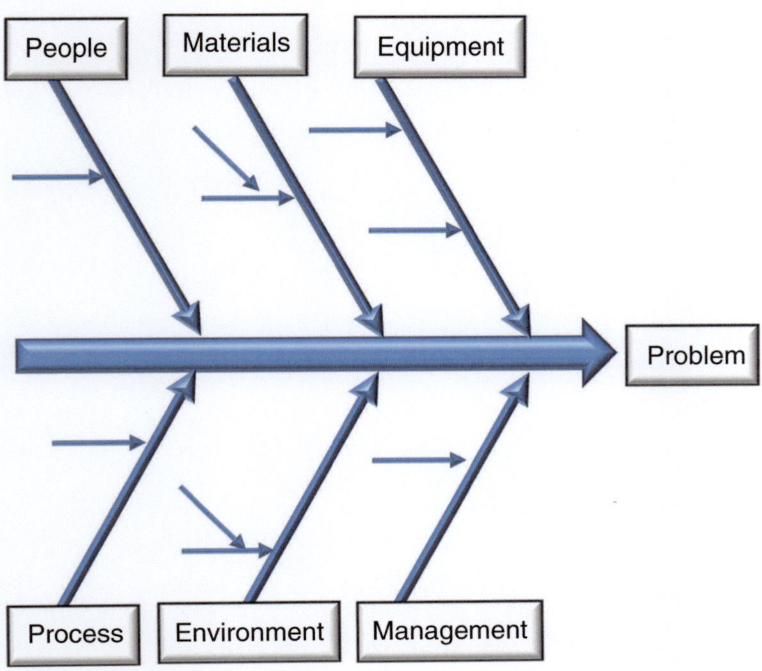

Fig. 12.2 Fish-bone analysis

potential causes may arise from multiple environments (pre-theatre ward, theatre, post-theatre, ward) each of these may have to be analysed in order to identify where the problem lies.

The same problem may arise due to one or more failures of possible many. Clarke et al. [5] looked at reports of wrong side surgery in Pennsylvania, USA. Amongst the causes of such failures were:

- Failures of the time out process.
- Failures of documenting, reviewing, or verifying patients' information.
- Problems with the diagnostic tests leading to surgery.
- Incorrect consents.
- Incorrect information on the medical clinic records.
- Clinic records not available.
- Errors in positioning for surgery, preparing the operative side, applying a tourniquet.
- Failure to mark the site, obliteration of marks, incorrect markings.
- Errors in scheduling the procedure.
- Incorrect information provided by patient or relatives.

Root cause analyses may identify a single factor or multiple factors, or may suggest as to the possible causes rather than fully confirming the actual cause. But then, even painstaking investigations into aviation accidents may not fully confirm the cause, and there may be disputes years later as to whether it was the plane or pilot to blame.

12.3 Dealing with a Cluster of Recognised Complications

Sometimes when you lose your way, you find YOURSELF.

Mandy Hale [6]

In some cases the adverse incident may be well defined, such as wrong side surgery, but in some cases it may be uncertain whether the observed events constitute a system breakdown. There may be a cluster of recognised complications which are seen at a higher than expected rate. This could occur simply by statistical chance, or could be due to dealing with high risk cases. However, such an occurrence may also be related to a breakdown of the normal precautions or barriers in the patient pathway that aim to prevent such complications, or due to sub-optimal performance of the surgeon or healthcare team.

A systematic approach in analysing such a cluster of adverse events may ensure that the root cause is found and appropriate action taken.

<u>Cluster of surgical wound infections</u>

Infection is the invisible enemy of the surgeon, one that becomes apparent only once it is established and settled. Much of what we do aims at minimising the risk of infection occurring. From the moment a patient is listed for surgery to the time they fully recover from their operation, several actions are taken to try and protect against infection. These may be pre-operative, intra-operative or post-operative.

At some point in your career you may face a cluster of surgical infections; multiple patients may develop a surgical wound infection in a short space of time, many more than you would expect to encounter. This may arise simply by chance, a statistical fluke, or may signify something more sinister; hence the need to determine if indeed a problem truly exists:

- What is the rate of infections?
- How does it compare with historical controls?
- How does it compare with other institutions?

If such analysis confirms an unexpected high infection rate, its causes must be examined.

1. <u>Define the problem</u>
    - Identify the cases which developed the infection.
    - Were they patients with an inherent high infection risk?
    - Were the infections superficial or deep?
    - What organisms were involved (strain, antibiotic sensitivities)?
    - What was the timing of presentation in relation to surgery?
    - Did they affect one surgeon's team or multiple teams?
    - One particular theatre or multiple theatres?
    - One or multiple procedures, one or many parts of the body?
    - Are the infections encountered in one or multiple specialties?

2. Define the normal patient's pathway

   - Explore the patients' pathway from presentation, to clinic, to surgery, to the infection presenting.
   - Has anything recently changed in the patient's pathway, especially in the steps taken to reduce the possibility of infection? For each step explore whether there has been any recent change in practise.

3. Carry out a cause effect analysis

Infection may arise due to failure at any of the various steps in the patient's pathway from the pre-operative assessment, to intra- operative events such as equipment and theatre sterilisation processes, post-surgery infection prophylaxis and nursing, to post hospital discharge and management in the community. A cause effect analysis may be drawn for each of these steps. This can be achieved either by grouping all components together or by using a different fishbone diagram for every step in the patient's journey (the pre-op, the intra-op the post op, the community, the ward, the theatre).

Haessler et al. [7] described their experience of an increased level of surgical site infections over nine consecutive months in 2007, and the subsequent "cluster investigation" they performed. This investigation involved review of the medical records, microbiological results, and inspection of the practises in the instrument sterilisation areas, surgical theatres and storage areas. As part of their investigation they determined that various types of bacteria with varying antibiotic sensitivities were responsible for the encountered infections, but the antibiotic prophylaxis choice was appropriate and the antibiotic administration timing was adhered to. There was no clustering of the infections around any particular surgical theatre or any particular personnel (surgeons or other staff). There were no changes in the sterilisation process of equipment or any obvious breaches of sterility. Investigators carried out interviews with the surgeons and other personnel, who expressed concerns about the amount of traffic through the surgical theatre and also the practise of wearing scrubs outside theatres. They also reported confusion of the appropriate use of the agents used in hand antisepsis, which led the investigators to directly observe the hand antisepsis practises. They noted improper use of one of two hand scrub agents used (an alcohol-based antiseptic); they noticed that there was limited soap and water wash before the alcohol was applied, the nail picks were not utilised and the alcohol was insufficiently applied. Removal of that antiseptic agent and reminding all staff of the importance of infection control, led to a reduction in infection rates to historical levels.

Septic arthritis following arthroscopic surgery is rare, in the order of 1/500 to 1/1000 [8–10]. Blevins et al. [11] described their encounter of three cases of septic arthritis following an arthroscopic meniscal repair of the knee, within a four day period at the same institution. These arthroscopies were performed by two surgeons, in different surgical theatres, with different assistants and other staff involved. All cases were positive for coagulase negative staphylococci. As part of their investigation they carried out molecular typing of the bacteria's DNA which showed that in two of the cases the infecting organisms shared common DNA patterns, which would be consistent with a common inoculation source. They identified that the same set of meniscal repair cannulas was used in all cases. Inspection of these cannulas after they had been cleansed, sterilised, and packaged ready for surgery, identified dried organic material residues in their lumen, and samples taken from these cannulas tested positive for coagulase negative staphylococci. Following a change in the sterilisation process of such cannulas, the authors encountered only one case of septic arthritis in a further 89 meniscal repairs.

Cluster of intra-operative cardio-vascular arrests

One may encounter a cluster of intra operative adverse events such as patients with hip fractures having cardio-vascular collapse or arrests during hemi-arthroplasty surgery. This could be a co-incidence, occurring by chance, but should be investigated thoroughly to ensure there is no other underlying cause.

1. Define the problem
    - What exactly happened?
    - Population affected?
    - Timing of arrest?
    - One surgeon versus multiple teams?
    - One versus multiple anaesthetic teams?
    - One or more theatres involved?
    - What was the patient's condition during those events, how were they managed, what was the eventual outcome?
2. Define the normal patient pathway
    - What is the journey of patients presenting with a hip fracture from presentation, to hospital admission, to ward preparation for surgery, to surgery and the intra operative events? Each of these steps is investigated to look whether there has been a change in practice, as well as plausible causes which could have led to the observed adverse events.

12.4 Stop Whilst You Investigate

Your philosophy determines whether you will go for the disciplines or continue the errors.
Jim Rohn [6]

When faced with serious adverse events, one may have to pause whilst investigating the cause, rather than continuing with their usual practise and potentially expose more patients to harm. At which point one pauses is a challenging decision, and needs to be carefully considered on a situation by situation basis.

Marc de Leval [12] a cardiac surgeon described his experience of encountering a high number of post-operative deaths following complex cardiac surgery (arterial switch operations for arterial transposition) between 1987 and 1993. In his first 52 patients he encountered only one death, but, suddenly, amongst his subsequent 16 cases he encountered seven deaths. When faced with this cluster of post-operative deaths, the surgeon visited a centre of low mortality to see what could be done differently, but following the seventh death he recognised that there could be a performance issue and decided to retrain (and follow to the finest detail the practise of a surgeon with low mortality rates). Following this, only one death was encountered in the subsequent 36 cases. Subsequent formal analysis suggested that the risk factors of the individual cases could not account for the mortality rates observed, and that there was an element of suboptimal performance which was addressed by re-training.

Maguire et al. [13] described their experience of encountering a cluster of six unexplained deaths in a paediatric intensive care unit (PICU) between 1989 and 1990. These deaths were seen in children who had undergone heart surgery for congenital heart conditions. During the investigation into the cause of these deaths the unit was temporarily closed. Despite extensive effort, no cause for the encountered high mortality rate could be established. The PICU was re-opened only when it was established that there was no scientific evidence to justify its continued closure.

Sometimes in such events an absolute cause may not be found, and multiple factors may need to be improved, several controls may need to be tightened to address the problem.

12.5 Near Misses and No Harm Events

Failure is instructive. The person who really thinks learns quite as much from his failures as from his successes.

John Dewey [6]

A near miss may be considered as an unplanned unwanted event that is stopped from completing its course at the last moment, either because its recognised or by chance. A no harm event is an unplanned unwanted event that is not stopped, but does not result in harm, although it could have done so. It may be that real harm was avoided just due to pure luck, or because the medical student watching the surgery tried to speak up.

Herbert William Heinrich, an engineer and pioneer of industrial safety, reported in his 1931 book entitled "Industrial accident prevention, a scientific approach" [14] that for every accident that leads to major injury, there are 29 accidents that cause only minor injuries and 300 that cause no injuries. This 1-29-300 ratio is now known as Heinrcih's Law or Triangle (Fig. 12.3).

Even though the exact figures, and the validity of Heinrcih's reasoning have been questioned [15, 16], several messages may be drawn from this concept:

- Near misses or events that lead to minor harm are much more common than those leading to major harm. Hence much more information may be gathered by investigating these more common events, with greater opportunities for learning.
- If the mechanism of occurrence of near misses or less harmful events is similar to those leading to major failures, then by minimising the former one may eliminate the latter.
- Staff may be more willing to report, and be more open and transparent in dealing with events where no harm was done, facilitating learning from such occurrences.
- Near misses may help identify those barriers that stopped full harm from occurring, and hence identify potential areas that could be strengthened further to avoid real harm occurring in the future.

Reporting and investigating each near miss may lead to unbearable workload. To minimise this, such reports may be assessed in terms of their frequency, and potential risk for real harm, in order to prioritise which ones to fully investigate and act upon.

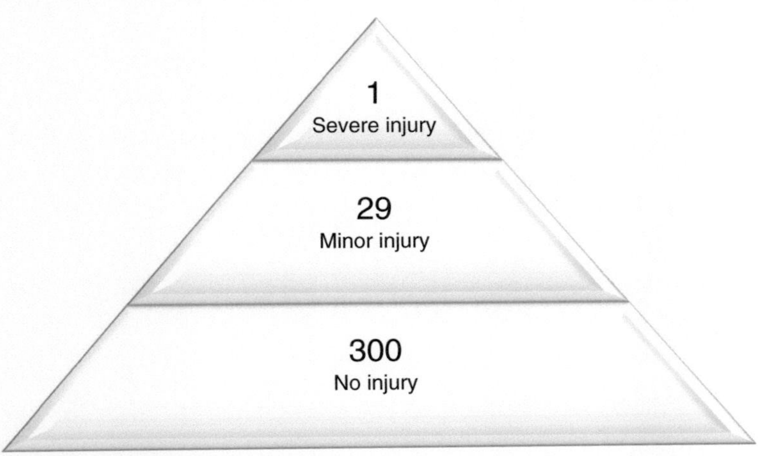

Fig. 12.3 Heinrcih's triangle. Adapted from Heinrich HW [14]

12.6 Learning from Adverse Events

I never made a mistake in my life; at least, never one that I couldn't explain away afterwards.
Rudyard Kipling [6]

Once the root cause analysis is done, and the causal factors are identified, practical changes may have to be made to minimise the risk of adverse events recurring. Aim for changes that can have a substantial impact, high up on the ladder of safety interventions, rather than simply for changes in policies or protocols which may have minimal effect.

Despite increasing numbers of root cause analyses performed in healthcare, it is estimated that only a small proportion of these lead to implementation of changes that will minimise similar errors recurring. Instead in many cases such changes are weak, with no probability of reducing further risks.

It is increasingly recognised that root cause analyses carried out at individual institutions do not have the impact that they could possibly have [17, 18], hence centralising the analysis of adverse events may have a more substantial effect. Such an approach might ensure that various safety stakeholders are brought together, including manufacturers of medical devices or technology industries that would help produce and provide more robust safety mechanisms.

12.7 Informing the Wider Surgical Community

Nothing is more intolerable than to have admit to yourself your own errors.

Ludwig van Beethoven [6]

As part of the learning process, the outcomes of any investigation into adverse events ought to be shared not just with the participants of the event, but also the wider surgical community. At a local institutional level this may be achieved through:

- Correspondence—newsletters.
- Ground rounds.
- Departmental morbidity and mortality meetings.

In addition, there may be central agencies which need to be informed, which monitor trends in adverse events and can thus identify wider system problems.

In the UK there are several reporting mechanisms:

1. The National Reporting and Learning System (NRLS) for adverse and patient safety incidents—NHS trusts have to report all incidents resulting in patient's death or severe harm. This provides a central agency, which can analyse the various reports to identify trends, risks, and potential areas for improvement. It is estimated that since the NRLS was established in 2003, more than four million incidents have been submitted [19].
2. The Yellow Card scheme for reporting adverse drug reactions [20].
3. Medical devices regulation and safety notification system, for events involving medical devices [21].

In the USA adverse incidents involving medical devices are reported to the Food and Drug Administration Agency (FDA). This is mandatory for manufacturers, device user facilities and importers, or voluntary for healthcare professionals, patients, and consumers [22].

Yet, there still seems to be lack of sharing of information, filtering of findings from one department to another, from one institution to the other. Healthcare seems to lag far behind the aviation and other high risk industries in both reporting and learning.

12.8 Disclosure

I make mistakes; I'll be the second to admit it.

Jean Kerr [6]

A vital component of the management of adverse events is informing the affected patients or their relatives of their occurrence, disclosing complications or medical errors. If something has gone wrong better to hear it from you with an explanation, rather than the patient finding out later from others. Anything said afterwards may sound more like an excuse rather an explanation.

In the USA, the American College of Physicians' ethics manual states that "physicians should disclose to patients information about procedural and judgement errors made in the course of care, if such information significantly affects the care of the patient" [23]. In the UK, the General Medical Council (GMC) guides that "every healthcare professional must be open and honest with patients when something that goes wrong with their treatment or care causes, or has the potential to cause, harm or distress" [24] This guidance advises that healthcare professionals should inform the patient when something wrong has happened, apologise, offer an appropriate solution to improve the situation (if applicable) and explain the short and long term consequences of the harmful event [24].

In the UK, in addition to the professional duty of candour set by the GMC for individual healthcare workers [24], the Care Quality Commission has introduced an organisational duty of candour whereby healthcare providers are required to inform patients and apologise to them about adverse events [25].

12.8.1 Challenges of Disclosure

Nothing is so firmly believed as what we least know.

Michel de Montaigne [6]

Even though doctors recognise the importance of disclosing and informing patients of things gone wrong, patients are often not told.

Wu et al. [26] reported that junior doctors (house officers) told patients about serious medical mistakes in less than a quarter of cases and informed their senior physicians about such mistakes in only half of the cases encountered.

The observed low disclosure rates may not come as a surprise when one considers the challenges that disclosure of medical errors entails, especially in cases where the patients or their families may not be aware that such errors have occurred. There may be worry of the reactions of patients and their relatives to such news who may become upset and angry, worry about the harm it may do to the patient-surgeon relationship, concern of legal action and law suits, anxiety about the impact on career progression [27–30].

In disclosing complications, no matter how well recognised the occurrence of these is, you may be afraid of getting criticised, concerned that you are having troubles when others are thriving, worried that the complications are due to personal under-performance, anxious that your practise may get restricted, you may dread that foes will capitalise on these to get at you. You may fear that your reputation or authority are at stake. Such emotions may even slow your ability to identify and hence deal with complications; sometimes one may rush to recognise complications of others, but deny or fail to accept that these could be happening to them.

Disclosure is a hard and difficult task, but may become easier if one develops the ability to recognise that:

- Informing the patient when things go wrong is the right thing to do, because it helps maintain honesty and trust that is central in the patient-surgeon relationship. Patients who are not informed of medical errors may feel let down and betrayed when they eventually find out. Trust and honesty are central in a surgeon-patient relationship [27–30] and it is necessary to put patients' interests above our own interests, our career progression or professional well-being.
- Disclosure may allow patients to better understand what has happened, rather than second guess of what went wrong, or keep wondering why their condition is not improving. Disclosure may also allow patients to seek further essential treatment in a timely fashion. In addition, compensation can be sought to cover further care and lost earnings incurred due to the injury sustained [27–30].
- Humans are fallible, and is it important to recognise and accept that mistakes are part of clinical practice. This may help create a culture whereby emotional and practical support is provided to those who have made a mistake, and help those surgeons involved navigate through a turbulent period.
- One may see a mistake as an opportunity to self- reflect, learn and get better and the ability to disclose adverse events as the just and noble action to take. Facing the real facts may help one improve. And if that means you have to get more training, one may accept that life is a process of continuous learning.

Most surgeons are technically sound and when complications occur it may be factors other than the surgeon to blame.

12.8.2 Delivering the Disclosure

In the UK, the GMC states that the most appropriate team member for disclosing medical errors will usually be the accountable clinician for the patient's care [24]. Nevertheless, when errors involve not just an individual but a whole system or institution then members of the wider team of the institution hierarchy may participate in the disclosure.

If you are a junior, involve your seniors at an early stage of the adverse event, and it may be advisable to disclose the error to the patient together. If you are a senior clinician it may be advisable to involve the appropriate hospital authorities who may accompany you at the time of disclosure [31]. Disclosure should take place as soon as possible once established that an adverse event has taken place, to avoid giving the impression that there is an attempt of cover up.

When faced with such adverse events seek help and advice from your seniors, peers and colleagues. They may help you analyse the situation better and help you prepare for what you will say and how you will say it. Albert Wu, Professor in the John Hopkins School of Public Health, and his team give practical guidance as to how to disclose medical errors to patients as below [31]:

- Begin by stating that you have made a mistake.
- Describe the mistake.
- Explain in detail the events surrounding the error.
- Describe the consequences of the mistake.
- Describe actions taken or will be taken to remedy the error.
- Express personal regret and apologise.
  Enquire about any questions or concerns.

12.8.3 Disclosure When Another Surgeon Is Involved

A difficult situation is when one is aware that another physician has made a mistake. Wu et al. [31] describe the options in this situation:

1. Wait for the physician involved to disclose the mistake.
2. Advise the physician to disclose the mistake.
3. Arrange a joint meeting with the patient and the physician who made the mistake to facilitate disclosure.
4. Tell the patient yourself about the mistake.

It is indeed a challenging situation when one faces this dilemma. One has to balance the need for disclosure, whilst at the same time respecting the relationship the patient has with the physician concerned. It is an awkward situation due to the fear of damaging relationships with colleagues, worry of betraying their trust. If is not certain whether a mistake occurred avoid jumping to conclusions. Speak to the physician involved, in a non-judgemental way, and gain more information to see whether a mistake did occur. As part of this process you may facilitate and encourage the physician involved to disclose the mistake (if a mistake took place). This will ensure that the relationship between patient and physician is not disturbed but at the same time disclosure does occur. One may view the disclosure of someone else's error as the last resort, which must be done in an impartial unbiased way. If necessary, one may, at the same time, inform the relevant authorities of the Institution (such as the Quality Assurance Personnel) to facilitate disclosure [32].

References

1. Famous Quotes at BrainyQuote. www.brainyquote.com. Accessed 5 Apr 2017.
2. World Alliance For Patient Safety. WHO draft guidelines for adverse event reporting and learning systems. From information to action. Geneva: World Health Organization; 2005.
3. Charalambous CP. When things go wrong. In: Career skills for doctors. Switzerland: Springer International Publishing; 2015. P. 203–217.
4. Ishikawa K. Guide to quality control. Tokyo: Asian Productivity Organisation; 1976.
5. Clarke JR, Johnston J, Finley ED. Getting surgery right. Ann Surg. 2007;246(3):395–403.
6. Popular Quotes. Goodreads. http://www.goodreads.com/quotes. Accessed 5 Apr 2017.
7. Haessler S, Connelly NR, Kanter G, Fitzgerald J, Scales ME, Golubchik A, Albert M, Gibson C. A surgical site infection cluster: the process and outcome of an investigation--the impact of an alcohol-based surgical antisepsis product and human behavior. Anesth Analg. 2010;110(4):1044–8.
8. Salzler MJ, Lin A, Miller CD, Herold S, Irrgang JJ, Harner CD. Complications after arthroscopic knee surgery. Am J Sports Med. 2014;42:292–6.
9. Yeranosian MG, Petrigliano FA, Terrell RD, Wang JC, McAllister DR. Incidence of post-operative infections requiring reoperation after arthroscopic knee surgery. Arthroscopy. 2013;29:1355–61.
10. Coudane H, Buisson P. Les complications de l'arthroscopie. In: editors. Symposium de la Société française d'arthroscopie.
11. Blevins FT, Salgado J, Wascher DC, Koster F. Septic arthritis following arthroscopic meniscus repair: a cluster of three cases. Arthroscopy. 1999;15(1):35–40.
12. de Leval MR, François K, Bull C, Brawn W, Spiegelhalter D. Analysis of a cluster of surgical failures. Application to a series of neonatal arterial switch operations. J Thorac Cardiovasc Surg. 1994;107(3):914–23.
13. Maguire H, Watson J, George R, Taylor L, Murray V, Penny D, Rigby M, Taylor P, Gaya H, Durman L. A multidisciplinary investigation of a cluster of deaths on a paediatric intensive care unit. J Hosp Infect. 1993;24(2):95–108.
14. Heinrich HW. Industrial accident prevention: a scientific approach. New York: McGraw-Hill; 1931.
15. Manuele FA. Heinrich revisited: truisms or myths. 2002. National Safety Council, USA.
16. Manuele FA. Reviewing Heinrich. Dislodging two myths. From the practice of safety. Prof Saf. 2011;52–61. http://www.coshnetwork.org/sites/default/files/Reviewing%20Heinrich%20(Manuele%20PS%2010-2011).pdf. Accessed 5 Apr 17.
17. Wu AW, Lipshutz AK, Pronovost PJ. Effectiveness and efficiency of root cause analysis in medicine. JAMA. 2008;299(6):685.
18. Mills PD, Neily J, Luan D, Stalhandske E, Weeks WB. Using aggregate root cause analysis to reduce falls. Jt Comm J Qual Patient Saf. 2005;31(1):21–31.
19. NRLS reporting. https://report.nrls.nhs.uk/nrlsreporting/. Accessed 5 Apr 2017.
20. Yellow Card Scheme-MHRA. https://yellowcard.mhra.gov.uk/. Accessed 5 Apr 2017.
21. Notify MHRA about a clinical investigation for a medical device. https://www.gov.uk/guidance/notify-mhra-about-a-clinical-investigation-for-a-medical-device. Accessed 5 Apr 2017.
22. Medical device reporting. https://www.fda.gov/medicaldevices/safety/reportaproblem/. Accessed 5 Apr 2017.
23. American College of Physicians. American College of Physicians Ethics Manual, 3rd edition. Ann Intern Med. 1992;117:947–60.
24. General Medical Council. Openness and honesty when things go wrong: the professional duty of candour. http://www.gmc-uk.org/DoC_guidance_englsih.pdf_61618688.pdf. Accessed 5 May 2017.
25. Care Quality Commission. Duty of candour. http://www.cqc.org.uk/content/regulation-20-duty-candour#full-regulation. Accessed 5 Apr 2017.

26. Wu AW, Folkman S, McPhee SJ, Lo B. Do house officers learn from their mistakes? JAMA. 1991;265(16):2089–94.
27. Chamberlain CJ, Koniaris LG, Wu AW, Pawlik TM. Disclosure of "nonharmful" medical errors and other events: duty to disclose. Arch Surg. 2012 Mar;147(3):282–6.
28. Wu AW, Huang IC, Stokes S, Pronovost PJ. Disclosing medical errors to patients: it's not what you say, it's what they hear. J Gen Intern Med. 2009;24(9):1012–7.
29. Wu AW, Boyle DJ, Wallace G, Mazor KM. Disclosure of adverse events in the United States and Canada: an update, and a proposed framework for improvement. J Public Health Res. 2013;2(3):e32. doi:10.4081/jphr.2013.e32.
30. Cole AP, Block L, Wu AW. On higher ground: ethical reasoning and its relationship with error disclosure. BMJ Qual Saf. 2013;22(7):580–5.
31. Wu AW, Cavanaugh TA, McPhee SJ, Lo B, Micco GP. To tell the truth: ethical and practical issues in disclosing medical mistakes to patients. J Gen Intern Med. 1997;12(12):770–5.
32. AHC Media. Ethical approaches to disclose errors made by other clinicians. https://www.ahcmedia.com/articles/136978-ethical-approaches-to-disclose-errors-made-by-other-clinicians. Accessed 5 Apr 2017.

Chapter 13
Surgical Interviews

Surgical career progression often involves multiple applications for training or non-training posts, clinical, research, or academic placements. Such interviews may be highly competitive with one having to shine to impress.

Hence, the ability to successfully perform at surgical interviews is an important skill to develop.

This chapter discusses considerations in choosing a surgical specialty or sub-specialty, which may help justify in an interview an application for a post. It also provides advice as how to construct a surgical CV and put together a surgical portfolio, in preparation for surgical interviews.

Finally, potential topics and questions that may be encountered in surgical interviews are presented.

C. Panayiotou Charalambous, *Career Skills for Surgeons*,
DOI 10.1007/978-3-319-57490-5_13

Dear interviewers what would you like to see first, my surgical logbook or self-reflections?
LOG BOOK
SELF Reflections

13.1 Choosing a Surgical Specialty or Sub-Specialty

As a child my family's menu consisted of two choices: take it or leave it.

Buddy Hackett [1]

Choosing surgery as a long term career is only the first step in a long journey. Soon you may have to decide which surgical specialty to follow. The days of one surgeon undertaking a wide variety of procedures, sometimes ranging from abdominal to vascular, orthopaedic to gynaecological, are in many countries well gone and one will be asked to choose one path to follow amongst specialties; Cardiothoracic Surgery, Abdominal Surgery, Neurosurgery, Orthopaedic Surgery, Head and Neck, Paediatric Surgery, Plastic, Urology, Breast, Endocrinological, Vascular Surgery.

And even if you know the broad area where you wish to practise, what sub-specialty do you next choose? Sub-specialty division may be based on an anatomical area, type of surgical procedures (open or minimally invasive—arthroscopic, laparoscopic), the underlying disorders treated (neoplasia, endocrinological disorders), patients' age (adults or kids). You may decide to do Orthopaedics but you may then have to choose a joint, a limb, or a procedure to specialise in. You may decide to become an abdominal surgeon but then you have to choose between the upper and lower gut or the hepatobiliary tract. Specialties and sub-specialties may vary in different areas or countries.

Apart from the above, in deciding which career path to follow you may also consider:

- Will I enjoy that surgical specialty? Not just the operating, not just the cutting, but also clinics, ward work, perioperative care.
- What does the specialty involve? One stop treatment delivery or life-long care?
- What are the on calls like? Resident or from home, up all night or very rarely called in, covering admissions ats one hospital or jetting round the country for organ harvesting?
- How physically demanding is surgery? Does it involve quick operations or lengthy prolonged procedures with in between lunch breaks, sitting down or standing up, using fine instruments, needle and forceps, or heavy power tools, mallets and hammers?
- What skills are required and are you good in those? Fine suturing skills, ability to operate under the microscope, arthroscopic or laparoscopic triangulation skills.
- How competitive is it to get into the specialty? Is there a need for special qualifications, postgraduate degrees?
- How competitive is it to be employed at the end of training? But be aware that what is in demand now may not be so by the time you are looking for a job.
- What support do I need for operating? Surgeon and scrub nurse or whole team?
- Where could I be employed? Public Hospital (secondary care or super-specialist centre), private clinic, own clinic.
- What equipment would I need if I decided to practise independently?
- Work intensity and pattern.
- Remuneration.
- Lifestyle.

You may already know what specialty you are after, you may be clear as to your destiny, but it is also possible that you want to experience different areas before you commit. Take your time, is there so much rush?

To help you decide you may:

- Speak to those already in the specialty and those out of the specialty to get a balanced view. We did those in the specialty choose it? Why did those out of specialty avoid it?
- Spend time in specialty, get to see what the juniors but also seniors do.
- See all aspects of the specialty—the theatre, ward, clinics, multi-disciplinary meetings, on call shifts.

And if still undecided and not sure what to do, take time, consider carefully, spend time in a post where you can gain basic surgical skills that can become of use whichever pathway you follow.

13.2 Surgical CV

Resume: a written exaggeration of only the good things a person has done in the past, as well as a wish list of the qualities a person would like to have.

Bo Bennett [1]

CV format

The exact format and layoutof your CV is a personal preference but the following may be considered:

- Keep it simple and short.
- Use succinct sentences, avoiding unnecessary text.
- Choose easy to read font and size and use this consistently.
- Use plain text avoiding excessive bolds, italics, underlining.
- Consistent use of subheadings.
- Present posts in reverse chronological order starting with your current post.
- Give dates of any described events (research publications, audits, qualifications, courses etc.).
- Consider starting important CV sections on a new page, where they can be easily identified.

CV content

The content of your CV will depend on the stage of your career but consider including the following:

- Personal details: name, address, phone, email.
- Qualifications: professional exams, postgraduate and undergraduate degrees.
- Academic achievements: awards, honours.
- Hospital posts:
  - Current and previous posts.
    You may list your duties for each post (on-call commitments, attendance of theatres or clinics, ward cover, peri-operative care).
    You may combine posts that were part of a training program (internship, residency, basic surgical training, higher surgical training).
- Postgraduate education: courses, conferences, visitations.
- Membership of professional bodies:
  - Postgraduate surgical college or society, medical association, medical indemnity/insurance body.
- Research:
  - Publications—give full reference with its serial number in any indexed database such as PubMed.
  - Articles submitted for publication, or in preparation, ongoing research (state stage of research development).

For each study state the names of all authors (with your name included) in the order they would appear in any future publication.

- Audit:
  - Initial audit, re-audit.
    Consider providing a summary of each audit with standards assessed, achievements, change in practise.
- Teaching and other educational activities—students, format, target population (institution, regional, national, international).
- Administrative/managing experience.
  - Duties—rota master, doctors mess treasurer, management courses.
- Interests: anything out of the ordinary?
- Referees:
  - Type of referee—academic, clinical.
  - Names.
  - Position.
  - Contact details.

Be accurate and honest in what you include in your CV, do not overplay or exaggerate, have the evidence to back it up.

13.3 Surgical Portfolio

Portfolio is a collection of documents providing proof and evidence of one's personal development, training and progression [2, 3]. It helps to support the contents of your CV but may also contain information not provided in the CV.

Portfolio format

- The portfolio should be thorough yet short and succinct. It should be thorough in providing the necessary supportive information, but succinct to allow its quick, easy assessment by an interviewer. Huge pieces of documentation, resulting in multiple portfolio volumes may not make such an assessment easy, hence selection of the most relevant documentation may be more appropriate.
- The portfolio should look professional in terms of presentation; invest in a robust folder, high quality printouts.
- Use easy to read text, consistent use of font and size, format of headings and subheadings.
- Include indexing with initial contents page.
- Use a filing system that allows the reader to access information, without having to take pages out of the folder and then replace them.
- Information that an interviewer may be specifically looking for, may be highlighted with a marker to facilitate its identification.
- In line with data privacy rules and confidentiality, ensure no information is contained that may identify a patient (such as on case presentations, radiographs on presentation slides, logbook of procedures)
- Use original certificates or certified copies, clearly demonstrating the authority issuing them. If including a letter from a hospital or senior, a hospital stamp may be necessary to confirm authenticity. Avoid letting the reader wonder whether a document is genuine or not.

You may consider the following sections, along with examples of supportive documents for each:

- Personal details (passport or identity card copy).
- Qualifications and other awards: university and other certificates.
- Personal statement about career interests and aspirations, commitment to surgery.
- Hospital posts—including voluntary, honorary, observation attachments.
  - Confirmation of post completion from institution (human resources, supervising senior doctor, training program).
- Postgraduate courses: attendance certificates.
- Assessments of competence—completed work based assessments.
- Surgical logbook: summary of surgical procedures you participated:

- Record date of surgery, type of surgical procedure, level of participation (performed unsupervised, performed with senior scrubbed or present, 1st assistant, 2nd assistant, observed).
- Complications (intra-operative, post-operative).

- Membership of professional bodies/societies.
  - Certificates of membership.
- Research:
  - Publications: copies of articles published (full copy or 1st page).
  - Presentations—a copy of conference program confirming presentation, and a copy of the presentation slides.
- Audit:
  - Copy of presentation slides if presented.
  - Confirmation letter from the audit department or lead.
  - Summary—standards, methodology, findings, change in practise, re-audit, findings.
- Teaching:
  - Letter from senior overseeing activities.
  - Copy of teaching slides.
  - Summary of teaching—aim, target student population, format, date, number of students.
  - Feedback with regards to effectiveness of teaching session from students.
- Leadership/managing experience:
  - Confirmation letters from seniors or institution.
  - Short description of role.
  - Description of projects undertaken.
- Self –reflections
  - Summary of event, how it was addressed, learning points, change in practice.
- Feedback:
  - Colleagues, co-workers—structured feedback or letters.
  - Patients—formal survey results, complaints (summary of complaints received, action taken and outcomes), compliments—thank you cards, letters.

- Appraisals: documents summarising formal appraisals.
- Revalidation/relicensing: certificate from regulatory body.
- Other evidence showing commitment to surgery.
  - Surgical journal subscriptions.
  - Surgical Elective at Medical School.
- Personal Interests—supporting certificates such as golf club membership!

In some interviews there may be a pre-defined marking scheme for portfolio that the candidates are allowed to access prior to the interview. Such scheme may give marks to domains that rely on activities from multiple areas (such as commitment to surgery domain, team skills domain, leadership domain). Consider adding to your portfolio separate sections reflecting such domains and gathering relevant information from all areas in your portfolio (surgical research, audit etc.) to those domains. This could make it easier for an interviewer to extract evidence hence allocate marks, rather than having to go back and forth through the folder to find information for each domain.

13.4 Surgical Interview Questions

It is better to ask a question that is deep and strange and unexpected than having an answer to everything.

Werner Herzog [4]

Surgical interviews aim to examine multiple aspects of your ability to be a surgeon or your qualities to successfully perform in the post you are applying for.

Surgical interviews may be in the form of multiple stations, each examining an area, and run by different interviewers. Alternatively, they can take place as a single interview, in front of a single panel, where multiple themes and areas are covered.

Whatever the interview format, certain themes are commonly encountered, and some of these are presented next.

13.4.1 CV—Portfolio Questions

The interviewers may ask you questions related to your CV or portfolio. Ensure you can talk about them and justify all you claim in these.

<u>What have you done up to now?</u>

You may start by reminding the interviewers as to which Medical school you graduated from and what year, and then summarise the posts you did since then. You may group certain posts e.g., under "core training," "higher specialist training."

<u>What is your surgical experience</u>?

You may split this into:

- Surgical posts—duration, level, sub-specialty.
- Surgical operating: may summarise it in terms of overall number of procedures, with numbers assisted or performed, or may split it into emergency procedures versus elective procedures.

Depending on the post you are applying for you may list the numbers of procedures you have done that are directly relevant to that post.

<u>What surgical procedures can you do?</u>

Split your response into what you can do with minimal supervision, and what you could do unsupervised. If you are at a junior level, and do not have much surgical operating experience, you may refer to procedures done in the ward (cannulation, catheterisation, central line insertion). Be pragmatic rather than exaggerating your abilities, as they will be soon tested if you are successful in getting the post.

<u>What is the most complex procedure you have performed?</u>

This may vary, if applying for a junior post it could simply be an incision and drainage of an abscess, excision of "lumps and bumps."

Do you consider yourself to be a good surgeon? How do you know you are a good surgeon? You may refer to feedback you have received from trainers and seniors, the good outcomes with few complications you encountered amongst those you have operated upon.

Why did you go into surgery?

- The ability to use technical skills, to help to cure.
- Combines technical and non-technical interventions.
- Good with my hands!
- Well defined problems that can then be tackled.
- Enjoyed surgical attachments.

Why this specialty?

- Requires team work and multi-disciplinary approach.
- Well defined surgical problems in otherwise medically well patients.
- Enjoy technical aspect—power tools, microscope, laparoscope.
- Mind stimulating with multiple potential solutions to one medical problem.
- Restore sight, save limb or life, improve pain or function.

Why this post? What are you aiming to get out of this post?

- Acquisition of generic skills such as suturing.
- Acquisition of specific skills—related to the specific surgical sub-specialty.
- Learning new skills or the strengthening of established skills.
  The post may be research orientated or academic.

13.4.2 Surgical Clinical Scenario Questions

You may be given a clinical scenario and asked how you would deal with it. This is often an acute or emergency surgical presentation, but a scenario of a chronic presentation (that you may expected to deal with in that post) may also be given.

In Orthopaedics this may be related to the management of a trauma patient, supracondylar fracture in a child with or without pulses, management of an open fracture, management of compartment syndrome, open book pelvic injury, displaced intra-capsular fracture in the young adult, septic arthritis in a child, cauda equina, fracture with neurovascular compromise.

In General Surgery it may be related to an acute abdominal presentation (appendicitis, cholecystitis, viscus perforation, bowel obstruction), gastrointestinal bleeding, or related to the management of a trauma patient (splenic, kidney, liver or other viscus injury).

In Vascular Surgery it may be related to a ruptured aneurysm, acute limb ischemic limb, or related to the management of a trauma patient.

The scenario examined may not require surgical management but may be related to medical conditions that can be encountered in surgical patients—such as post-surgery chest infections, peri-operative myocardial infarction, anticoagulation for atrial fibrillation, alcohol withdrawal symptoms following hospitalisation.

The aim of a clinical scenario is to determine that you:

- Adopt a systematic approach in assessing a patient, to ensure that you do not concentrate on one problem and end up missing another.
- Look at the patient as a whole rather than concentrating on the presenting complain.
- Involve appropriately the various members of the team required in patient care.

Try to demonstrate that you are safe, that you are aware of the steps in clinical assessment and management, speak as if you are living it in real life, give the minor details that differentiates one who has previously experienced the scenario as compared to one that has simply read about it, demonstrate an ability of lateral thinking.

In describing how you would approach an acute scenario consider:

- Start with the basics.
- Follow protocols: Advanced Trauma Life Support (ATLS) protocol for trauma patient [5], Advanced Life Support (ALS) [6] for a non-trauma patient.
- Resuscitate and stabilise an unstable patient alongside your clinical evaluation: large bore cannulas, IV fluids.
- Take history and examine.
- Examination—start from basic observations blood pressure, pulse, temperature, neurovascular status, compartment syndrome, skin integrity. State what positives you are looking for such as "tenderness and guarding in the right iliac fossa."
- Investigations: start with basic investigations (FBC, UEs, plain radiograph) and work up to more complex ones (Amylase, CT). Explain what you are looking for in these investigations: FBC looking for an elevated WCC, urea and electrolytes looking for low calcium, erect plain radiograph looking for free air, CT scan looking for an intra-abdominal abscess.
- State your preferred treatment option, but be aware of other options, and be prepared to justify your choice.
- If stating that you need to take a patient to theatre consider mentioning:
  - Consent.
  - Inform theatres.
  - Keep nil by mouth.
  - Inform anaesthetics and other specialties, your senior, the radiographer.
  - Positioning, incision, approach, procedure, postoperative care.
- If you are not able to operate you may state "I would ask my senior to operate but the principles of surgery would be to……"

In the emergency scenario, when you need to take a patient to theatre consider situations where not all is going smoothly:

- The theatre is blocked by another case and will not be free soon. Would you consider opening another theatre, transferring patient out, can patient wait?

- What if it is a shared theatre list and another specialty is refusing to give your case priority, even though their patient is not as unwell?
- What if patient is a child?
- What if the patient is not fasted?
- What if refusing treatment?
- Refusing blood transfusion?

13.4.3 Leadership/Management Questions

Only the guy who isn't rowing has time to rock the boat.

Jean-Paul Sartre [1]

Generic questions:

- Types or qualities of leader/manager.
- Differences between manager and leader.
- Does a surgeon need such skills?

Questions specific to you:

- Do you consider yourself to be a leader? And if yes why?
- Examples where you demonstrated your leadership/managerial skills?

Scenarios:

- You are tasked with introducing a new rota—how would you bring it about?
- Two of your juniors are not getting in and keep fighting. What would you do?
- A junior keeps turning up late to theatre. How would you address that?
- How do you achieve compliance with a new thromboprophylaxis protocol?

13.4.4 Awkward Scenarios

If you're in an awkward position, feel comfortable enough to walk away.

Beau Mirchoff [1].

As part of the interview you may be given a scenario of an awkward situation, and asked how you would deal with it.

You are facing difficulties at the workplace—how will you deal with it?

- You are not getting enough training.
- Your rota keeps changed at short notice.
- Being pushed around by your senior, a scrub nurse in theatre, a ward nurse.
- Your junior keeps turning up late.

Be able to address the issue with the person concerned but if not getting the appropriate response escalate it to the right authorities.

There is a need to demonstrate that you are aware of the different hierarchy structures that various co-workers may be answerable to and hence the pathways for escalation:

- For a training issue this may be: clinical supervisor educational supervisor, post-graduate director of your hospital, regional head of training, postgraduate dean.
- For a non-training issue this may be: clinical supervisor, head of department, department manager, human resources department.

<u>You become aware of improper behaviour—what will you do?</u>

You may be given a scenario whereby you are faced with improper behaviour and you are asked as what you would do. You may consider:

- You turn up to theatre and your senior smells of alcohol and behaves oddly.
- You become aware that your senior is mis-documenting the surgical procedures, to increase the remuneration of the institution or for personal financial gains.
- You see your senior, helping themselves to an implant from the hospital's theatres store room, to take away.
- You see your senior helping themselves from the opioids store cupboard.
- You walk into your seniors office to find them looking at illegal, inappropriate websites.

The aim is to determine if:

1. You recognise what the issue is—patient safety issue vs. fraud/criminality.
2. You take appropriate action—you don't leave the matter rest but you report/escalate to the appropriate authorities.

If patient safety is at stake, then the first priority is to ensure patient safety: by removing the person endangering patients.

If it is an issue of fraud/criminality you may confront the individual first or directly report it to the Human Resources or fraud office of your institution and to appropriate law enforcement authorities (according to the institution or healthcare system you are practising in). Offer that you may seek advice from your medical protection society or a senior colleague as such events can be challenging and testing.

13.4.5 Communication Questions

If you can't convince them, confuse them.

Harry S Truman [1]

You may be given a scenario where you have to show your communication skills. This could involve speaking to patients, relatives, another colleague or co-worker. You may be asked to pass on news, break bad news, disclose an error or an adverse event, explore why a patient is refusing treatment. You may face an anxious patient, an upset patient, an angry relative, a non-cooperative colleague.

The exact approach of how to communicate will depend on the scenario and exact circumstances but you may consider the following:

- Be courteous and polite.
- Introduce your-self, your role.
- Determine what needs to be communicated.
- Sympathise and apologise if necessary.
- Empathise, be understanding, show you appreciate the worries, do not be dismissive.
- Give clear plans for action.
- Check if anything else to be said, Any questions? Any remaining queries?
- If dealing with complaint offer to escalate, or provide information about the available escalation routes.
- Ensure you pass the message, it may be easy to keep going round the topic and not transferring the information you have been asked to communicate.

In an interview the process may feel very artificial, especially if it involves role playing with an actor. Treat it as if you are in the ward, in clinic, as if it is just another day at work.

13.4.6 Political Issues Questions

In a democracy, dissent is an act of faith.

J. William Fulbright [1]

You may be asked about your opinions with regards to an issue of the day affecting surgery.

The aim is to determine:

- If you are aware of challenges your specialty or the profession faces.
- What your views are.
- Your approach in dealing with controversial topics (dogmatic or more conciliatory way).

Political issues examined may include:

- Seven day working [7].
- Role of physician assistants [8–10].
- Role of surgical assistants [8–10].
- Non-medically qualified surgeons doing surgical procedures—such as podiatrists [11, 12].
- Shortening of surgical training [13, 14].
- Role of general surgical training prior to specialist training.
- Targets from referral to surgery [15–18].
- Commissioning public contracts to private hospitals [19, 20].
- Bundled payment initiatives [21].

If you are aware of the issue you may consider:

- Avoiding strong, passionate views, as the interviewer may be on the other side of the argument. Aim for a more balanced approach.
- State a summary of the issue.
- Give one side of the argument- positive site e.g., the benefits of 7 day working—increased productivity, better access of patients to healthcare.
- Give the other side of the argument—negative side—difficulties and challenges of introducing 7 day working in an under-staffed healthcare system.
- Summarise the arguments, and offer a half way solution e.g., 7 day working would be welcomed once adequate resources become available.

If you haven't come across the issue the interviewer refers to, ask the interviewer to explain "I'm not aware of the issue you describe, can you please elaborate?"

References

1. Famous Quotes at BrainyQuote. www.brainyquote.com. Accessed 2 Apr 2017.
2. Haldane T. "Portfolios" as a method of assessment in medical education. Gastroenterol Hepatol Bed Bench. 2014;7(2):89–93.
3. Webb TP, Merkley TR. An evaluation of the success of a surgical resident learning portfolio. J Surg Educ. 2012;69(1):1–7.
4. BBC Two – Newsnight, 12/10/2016.
5. Advanced life support. Resuscitation council (UK), 6th ed.; 2011.
6. ATLS: Advanced Trauma Life Support for Doctors (Student Course Manual). American College of Surgeons, 8th ed.; 2008. ISBN-10: 1880696312.
7. Scott HR, Isles CJ, Fisher BM, Long J, Dunn FG. Adopting seven-day working in practice: a report by the Royal College of Physicians and surgeons of Glasgow. Scott Med J. 2014;59(4):193–7.
8. Fisher MG. Who are physicians' assistants (anaesthesia) and where do they work? J Perioper Pract. 2015;25(4):67.
9. Phillips MJ, Harris M. What's in a name? An evaluation of public perceptions around job role titles within the modern anaesthesia team. J Perioper Pract. 2015;25(4):68–71.
10. Abraham J, Whiteman B, Coad J, Kneafsey R. Development and implementation of non-medical practitioners in acute care. Br J Nurs. 2016;25(20):1129–34.
11. Maher A. Podiatry: an illustration of surgery provided by allied health professionals. J Perioper Pract. 2013;23(10):218–21.
12. Isaac A, Gwilym SE, Reilly IN, Kilmartin TE, Ribbans WJ. Interprofessional relationships between orthopaedic and podiatric surgeons in the UK. Ann R Coll Surg Engl. 2008;90(8):663–70.
13. Sayana MK, Ashraf M, O'Byrne J. Modernising the higher surgical training in trauma and orthopaedic surgery in Ireland: taking the middle path approach. Ir J Med Sci. 2009;178(4):389–92. doi:10.1007/s11845-009-0322-8.
14. Inaparthy PK, Sayana MK, Maffulli N. Evolving trauma and orthopedics training in the UK. J Surg Educ. 2013;70(1):104–8.
15. NHS makes good progress on waits. BBC News. 2008. http://news.bbc.co.uk. Accessed on 26 Sep 2014.
16. Gubb J. Have targets done more harm than good in the English NHS? Yes. BMJ. 2009;338:a3130.
17. Bevan G. Have targets done more harm than good in the English NHS? No. BMJ. 2009; 338:a3129.
18. White C. Hunt announces drive to clear 12 month waiting list. BMJ. 2014;349:g5011.
19. Greaves F, Laverty AA, Pape U, Ratneswaren A, Majeed A, Millett C. Performance of new alternative providers of primary care services in England: an observational study. J R Soc Med. 2015;108(5):171–83.
20. Kula N, Fryatt RJ. Public-private interactions on health in South Africa: opportunities for scaling up. Health Policy Plan. 2014;29(5):560–9.
21. Doran JP, Beyer AH, Bosco J, Naas PL, Parsley BS, Slover J, Zabinski SJ, Zuckerman JD, Iorio R. Implementation of bundled payment initiatives for Total joint Arthroplasty: decreasing cost and increasing quality. Instr Course Lect. 2016;65:555–66.

Chapter 14
Emotions in Surgery

Surgeons are sometimes perceived as acting in an emotionless way, cold executors of tasks and procedures, but the reality may be different. In your surgical training or career you may be faced with a roller coaster of emotions, both positive and negative.

Considering your emotions, and appreciating their potential effects upon your wellbeing and performance are important skills to develop. They may assist you in dealing or coping with such emotions and help you recognise that the experience of those emotions may make you part of the norm rather than the odd one out. Considering emotions may also assist you in appreciating what colleagues, and co-workers faced with similar situations are going through, which may guide your actions and behaviours towards them.

This chapter discusses the concept of emotional intelligence, addresses the emotions that a surgeon may face in different scenarios, and suggests ways in which one may deal with those. It discusses emotions one may experience when under excessive pressure and stress, when encountering complications or errors, dealing with work environment challenges, and when faced with an established stereotype and trying to fit in.

C. Panayiotou Charalambous, *Career Skills for Surgeons*,
DOI 10.1007/978-3-319-57490-5_14

BEST ROBOTIC SURGEON
BEST SURGEON·COM
SCALPEL OF THE YEAR
I am simply the BEST

14.1 Types of Emotions

Flowers are restful to look at. They have neither emotions nor conflicts.

Sigmund Freud [1]

Several names or labels have been used to describe emotions, with many theories available on the subject. In describing the emotional world, one may consider basic emotions and more complex emotions. Robert Plutchick, a well-known Psychologist and Professor Emeritus at the Albert Einstein College of Medicine in New York, described eight basic emotions as pairs of opposites: joy versus sadness, anger versus fear, trust versus disgust, and surprise versus anticipation. These emotions may vary in intensity, or may interact to form more complex emotions such as optimism, contempt or disapproval, a concept Plutchick successfully presented in his Wheel of Emotions (Fig. 14.1) [2]. Even though Plutchick's theory may have limitations, such as by not including "pride" as an emotion, it can nevertheless help one to appreciate the complexities of our emotional world.

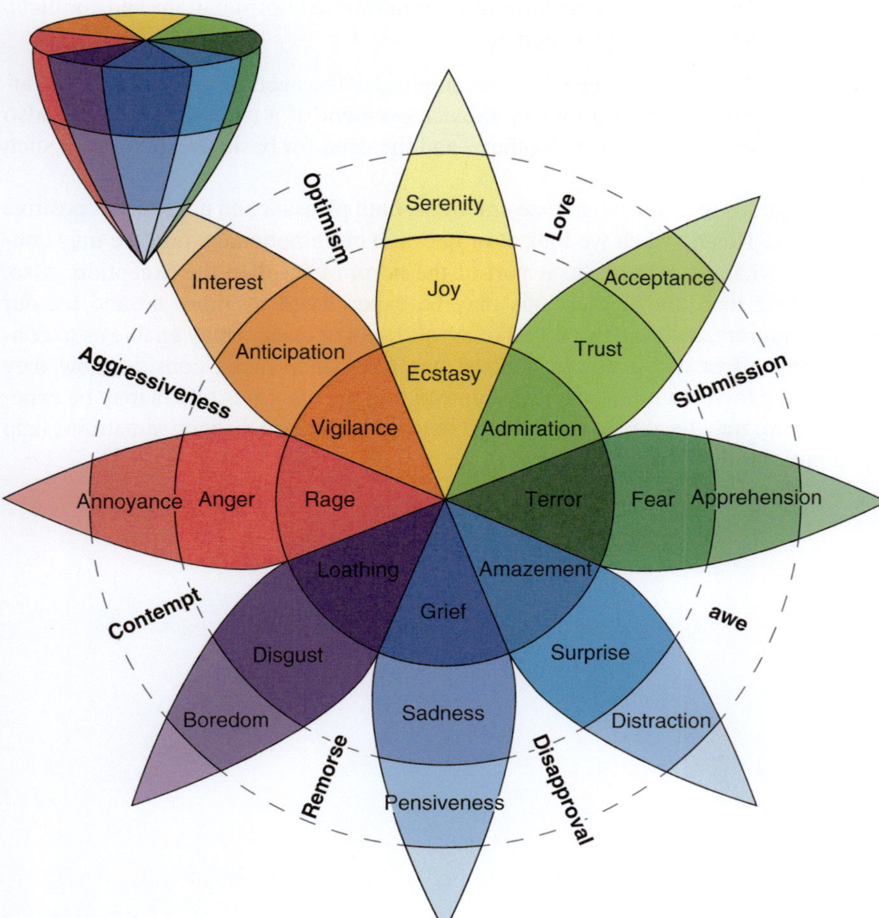

Fig. 14.1 Plutchick's Wheel of Emotions. From Plutchik R [2]. Reprinted with permission

14.2 Emotional Intelligence

The audience likes their emotions to be touched. They want to laugh and cry and feel good.

Magic Johnson [1]

Emotional Intelligence has been described as the ability to perceive, understand and manage one's own and other people's emotions [3]. It's increasingly recognised as an important skill to develop, as it may facilitate inter-personal interactions and relationships. Emotional intelligence has been described as consisting of five components [4]:

- Empathy—the ability to consider the feelings of others to guide behaviour and actions.
- Motivation—drive to achieve.
- Self-awareness—ability to recognise own emotions and their impact upon others.
- Self-control—ability to control own potentially disruptive emotions and impulses.
- Social skills—to manage relationships, to minimise confrontations and conflicts, to constructively engage, negotiate.

Emotional intelligence may help to emphasise the need of not just concentrating on the completion of a task or the achievement of a technical goal, but also consider how one interacts with others, and the need for basic social skills in such interactions.

As surgeons we may experience emotions, both pleasant and unpleasant, positive or negative. Hence, when we look at or question our emotional world we may consider that what we experience is part of the norm rather than the exception. Also, appreciating that similar emotions may be experienced by those around us, our trainees, juniors, colleagues, co-workers, followers or leaders may enable us to consider what effect our words or actions may have upon them, consider how they might make them feel. Similarly, by appreciating the emotions others may be experiencing, we may be able to support and help them through difficult situations, help them pull through challenging times.

14.3 Emotions and Stress

James Donovan: Aren't you worried?
Rudolf Abel: Would it help?

Bridge of Spies [5]

Stress is defined by the Oxford dictionary as "a state of mental or emotional strain or tension resulting from adverse or very demanding circumstances" [6]. It is essentially a reaction to outside demands or pressures.

Stress may be due to heavy workload, lack of senior support, institutional pressures, organisational dysfunction, a disruptive colleague, a challenging situation, a change from one post to another, from one training level to the next, may be due to workplace or personal life pressures.

A small amount of stress may make us more alert, but if stress is excessive our performance may be impaired. Stress can lead to negative emotions, anxiety, agitation, unhappiness, moodiness, thought preoccupation, loneliness, loss of enjoyment, lack of concentration, feeling overwhelmed (Fig. 14.2). Excessive stress may have physical effects leading to inability to fall asleep or early awaking, sometimes almost acting like a "paralysing" force impairing physical skills or actions [7–11].

Prolonged high levels of stress may lead to burnout that is characterised by [12, 13]:

- Emotional exclusion
- Depersonalisation
- Feelings of insufficiency

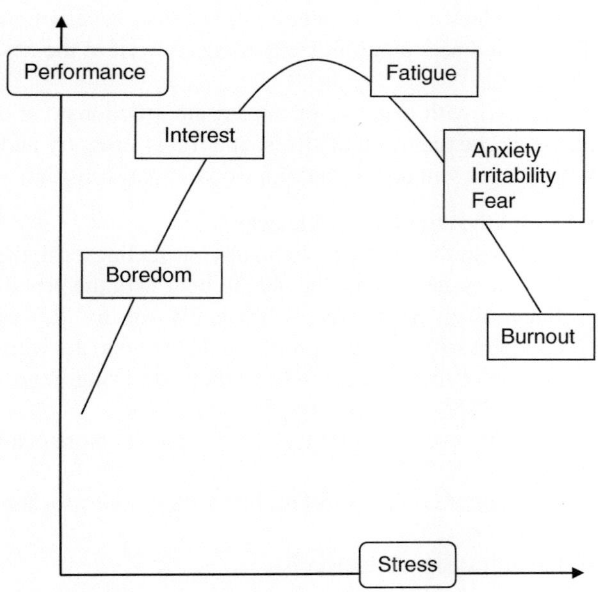

Fig. 14.2 Stress-Performance curve and associated emotions. Adapted from Bradberry [11]

But can we really be impaired by stress? Are surgeons truly susceptible to such pressures? After all, the surgeon is often pictured as strong, tough, as the one who "takes risks, defies death, comes close to the edge and carries it off" [14].

Nevertheless, several studies report high levels of stress and burnout amongst surgeons. Yoo et al. [15] surveyed General Surgery residents in New England. Of 166 respondents, 92% reported moderate to extreme work related stress which in 72% affected their overall wellbeing. Arora et al. [16], in a review of published studies, evaluated burnout levels amongst orthopaedic surgeons and reported this as affecting 50–60%, with the highest levels seen in residents, followed by departmental chairs and faculty members.

Klimo et al. [17] reported 27% burnout rate amongst neurosurgeons in the USA whereas Businger et al. [18] reported a 39% moderate or high burnout rate amongst surgical residents and surgeons in Switzerland. O'Kelly et al. [19] surveyed 575 urologists, members of the Irish Society of Urology and the British Associations of Urological Surgeons. They showed moderate levels of emotional exhaustion and de-personalisation on burnout assessing questionnaires. Fifteen percent of the respondents reported self-medication with non-prescription drugs and alcohol to counteract symptoms and signs of burnout, whilst 8% sought professional help for such symptoms. Shanafelt et al. [20] surveyed 7905 members of the American College of Surgeons, of which 6.3% reported suicidal ideation during the previous 12 months, which was related to symptoms of burnout and depression.

Orri et al. [21] evaluated through interviews the emotions of 27 pancreatic and liver French surgeons before, during and after surgery. Their emotional world was influenced by interactions with patients and their relatives, their co-workers, and the framework of their institutions. Time pressures, fatigue and the need to conform to the expected image of a surgeon influenced their emotions. They reported emotions of pleasure when operating, but also negative emotions anxiety, fear, guilt, distress and accountability, experienced even in the absence of any adverse events occurring.

If faced with negative or unpleasant emotions one may consider as to whether that could be the result of stress, and if it is stress try and deal with it rather than just letting it get you down. In doing so you may consider:

- Avoid the stressor—walk away.
- Deal with the stressor—explain your feelings, ask them to change attitude, escalate the issue to someone able to help with the problem.
- Put it all in perspective—is it really worthy the fuss? How do my difficulties compare with all other problems that trouble the world? Is worrying an appropriate use of my intellect, when others are using theirs brains to send a mission to Mars [22]?
- Find interests and activities that encourage more positive emotions, that help you relax.
- Concentrate on the present, block the rankles for the past and the worries about the future [23].
- Eat and sleep well, exercise.
- Avoid unhealthy coping mechanisms (caffeine or alcohol).

14.4 Facing Complications

People who do no work, make no mistakes.
People who make no mistakes get promoted [24]

A complication may be described as an unfavourable, an unwanted consequence of an intervention. Complications may occur even when all possible care and precautions are taken; the infected hip replacement despite all antibiotic prophylaxis, the wound breaking down despite proper layer stitching, the grafted artery thrombosis despite good intraoperative flow, the bowel anastomosis leaking despite meticulous suturing, the nerve palsy despite careful dissection and nerve retraction.

There may be times when you are riding high and all are satisfied and pleased. But then there may be times when all you touch seems to go wrong, with complications coming in one after the other. You may be perplexed as to why in situations where you thought you did very well problems arose, but where you were worried recovery happened fast.

The effects of complications can be devastating both on the wellbeing of patients and on the emotional world of surgeons. Patients with complications may stay in the mind of a surgeon for a really long time, may recall their details years or decades later. As in many aspects of surgery we deliver care not simply by decision making but with our own two hands, one may feel a greater ownership, personal responsibility for the occurrence of complications.

Complications may bring the constant worry about your patients not doing well, the constant questioning whether you could have done better. They may also bring the worry of legislation, having to face the coroner or even the press. The pressures may be mounting of the personal repercussions, the loss of face, reputation, good name, the damage to one's career prospects, loss of a job. One mistake, a judgement error may transform one from "pioneer" to "maverick"; one's working environment may be threatening, with minimal support, you may be in constant fear that they are all out to get you. You may be anxious of the breakdown of the relationship you had with a patient and their close ones, of them feeling that you let them down. You may feel regret—could you have done more, could you have accepted less, could you have taken a different course of action?

Pinto et al. [25] examined the impact of major surgical complications on 27 general and vascular surgeons practising in the UK. The emotions they experienced are shown in Fig. 14.3. Many felt that strong blame culture and inadequate support by their institutions hastened their response in those difficult times. In a subsequent study [26] the authors reported that one third of 47 surgeons assessed, exhibited acute traumatic stress of clinical concern a month after their patients sustained a major complication whilst under their care.

	No. of participants (n=27)
Emotional impact	26
Guilt	15
Crisis of confidence	8
Worry about reputation	8
Worry for the patient	6
Anger (at self or at seniors for inadequate supervision)	6
Anxiety	6
Disappointment	3
Sadness	3
Behavioural impact	21
Surgical practice affected (become more cautious/risk-averse)	18
Increasing efforts to improve	4
Becoming aggressive in interactions with colleagues	4
Cognitive impact	10
Rumination	6
Reflection on what went wrong	6
Loss of concentration	3
Social impact	8
Interference with personal life	6
Relationships with colleagues enhanced	3
Other impact	12
Learning	11
Physical reactions (feeling sick)	2

Fig. 14.3 Emotions experienced due to major surgical complications amongst general and vascular surgeons practising in the UK. From Pinto et al. [25]. Reprinted with permission

Cristancho et al. [27] evaluated how surgeons perceive the context in which they find themselves practising, through interviews, drawing sessions and observations. One surgeon's words from that study demonstrate the emotional pressures faced with in decision making,

> *"people may be standing around saying, why did you even bother operating on this person? What were you thinking when you did this, this is ridiculous? You've created a problem where there was no problem"* [27].

Complications are part of surgery and will occur, no matter how careful one is; we aim to minimise complications as it is unlikely we can fully abolish them. It may be that the only way to completely avoid surgical complications is for one not to operate.

When faced with complications you may consider:

- Discuss with supportive colleagues of peers. Someone reliable who will take your worries seriously, will be fair and just in their comments, and will neither dismiss them, just telling you what they think you want to hear, or make you feel even more guilty.
- Break down the complication. Could something else have been done, and even if something else were done would the outcome have been any different?

But it is also important to recognise that bad outcomes do not always relate to how good technical job one did. Poor outcomes may be:

- Due to the natural history of the underlying disorder with progression despite intervention.
- Poor biology to allow a favourable response to our intervention—a well sutured bowel anastomoses may break down if the blood supply is not adequate, a well fixed bone fracture may fall apart if the biology of healing is not there.
- Poor compliance with treatment.
- Factors such as ongoing compensation claims, psychological status [28–30].

Give complications the consideration they deserve, examine them fully, learn from complications, progress and develop, but try not let them get you down.

14.5 Dealing with Errors

I've safely delivered one million passengers over 40 years and tomorrow I'm going to be judged over 208 seconds.

Chesley Sully Sullenberger [31].

Medical errors can have a devastating effect on the doctors involved who may struggle to recover, if they can recover at all. The term "second victim" has been used to describe such individuals, emphasising that in addition to the patient who is the first victim of any error and harm, the physician can also be a victim [32, 33].

The culture of perfection whereby human fallibility and mistakes have no place may be encountered in Medicine but more particularly amongst surgeons. Hence when faced with a medical mistake, feelings of failure and despair may occur along with questioning one's own clinical skills and even career choice.

Scott et al. [34] interviewed 31 health care professionals who were second victims following adverse patient events and described six stages in the second victim phenomenon:

1. Initial shock.
2. Intrusive reflections characterised by re-enactments repeatedly asking what if a different pathway had been taken, if different actions had happened.
3. Support from a trusted colleague, friend or relative was sought to help restore personal integrity along with prevalence of self- doubt and worry about the future.
4. Enduring the inquisition—a stage marked by concerns about litigation, career and professional repercussions.
5. Emotional support sought.
6. Moving on—characterised by an attempt to move on and leave the past behind. These could be achieved in various ways such as:

   - Dropping out—leaving the profession, changing professional role, or moving to a different location to practice.
   - Surviving—continues to work and function keeps haunted by the event.
   - Thriving—tries to see that something good has come out from the adverse event, to improve the service or the individual as a professional or person.

Surgeons, exhibit similar responses to adverse events. Luu et al. [35] evaluated the emotional reactions of surgeons to adverse events and reported four phases in their response:

1. The kick—characterised by feelings of failure, self-doubt, and a substantial physiological response.
2. The fall—characterised by a feelings of chaos and an attempt to assess own role in the adverse event.
3. The recovery—characterised by attempts to reflect and learn from the adverse even, to move on.
4. Long term impact.

Like complications, adverse events can haunt the surgeon for years to come. As Rene Leriche, a French Surgeon that practised in the early nineteenth century once wrote:

> "*Every surgeon carries about him a little cemetery, in which from time to time he goes to pray, a cemetery of bitterness and regret, of which he seeks the reason for certain of his failures*" [36].

In helping to deal with adverse events you may consider:

- Life is a long learning process.
- Things can go wrong even when our intentions are right, even when we set out to do good.
- What we do has limitations, surgery has its limits too.
- The fallibility of humans, and that surgeons are humans too.
- Having insight as to what happened, recognising what went wrong, constructively self-reflecting and learning can be a positive light amongst the perceived darkness.
- Even though we may find ourselves at the receiving end of legislation, one may accept that those harmed are entitled to seek mend and compensation for what went wrong.

14.6 Coping with the Stereotype

In the John Wayne movies, the Indians were savages that were trying to scalp you.

Ricky Schroder [1].

Stereotypes is described as "a widely held but fixed and oversimplified image or idea of a particular type of person or thing" [6]. As for other professions, and medical specialties, the surgeon stereotype has been repeatedly described, with such descriptions sharing common themes:

- *"He is big bold, beer swilling rugby fan, and when it comes to clinical practise he prefers to cut and run, rather than communicate with patients"* [37].
- *"Decisive, well organised, practical, hardworking, but also cantankerous, dominant, arrogant, hostile, impersonal, egocentric, and poor communicator"* [38].
- *"Dominant and arrogant doctor, aggressive, energetic, and very concerned with his own prestige. He is seen as decisive, efficient, and a realist, but as an impersonal and autocratic person. Egocentric, he is more interested in rapid actions and immediate results than in interpersonal relationships"* [39].
- *"…admires cars, sports, speed, competence"* [40].

Surgeons may have a stereotype of how they are perceived, one framed by many years of surgical practise, thought of by the public, perpetuated by colleagues of other medical specialties [41], television dramas [42], soap operas, the media and the press [43]. Given the stereotype you may be anxious, worried, concerned as whether you could be it. "How could I fit in? How could I be accepted?"

Try not to be put off by the stereotype, try not to rush to decide that surgery it's not for you, that you could not fit in. Professions can benefit from individuals of all talents, sexes, ethnicities, personalities, backgrounds or beliefs. You could be the fresh air, the breath of welcomed change. Surgery has struggled to recruit female trainees, and the perceived macho stereotype may be at least partly to blame [44]. But there is evidence to suggest that even male trainees that have spent time in the specialty may feel not measured up to the masculine stereotype, leading to inability to fit in and a struggle to engage with the profession [45].

Avoid trying to act or behave in certain ways simply to match the stereotype, the perceived norm, in an attempt to be accepted as part of the team. Be genuine, be your-self, be the real you, avoid trying to pretend being what you may think is expected from you.

Many of us may have been in situations where we felt that others did not look at us as if we were up to scratch, as if we could make it, that our behaviours, personalities and actions did not conform to the departmental norm, and sometime later, we may gladly say, we proved them wrong. Some of us may have even been on the other side, looking at early starters and thinking that they most likely won't make it, that they may just scrape through, that they just don't have what it takes to be a surgeon, simply to be later glad that we were proved wrong.

One does not have to act or behave in a specific way in order to fit in. Professional behaviour doesn't mean changing one's own identity. A workplace, an institution or profession may be enriched by actors of various talents, different personalities, communication skills, backgrounds, aspirations.

Stereotypes are often perceptions, and to perceptions there can be exceptions. The most noisy, the most pushy, the most outrageous behaviours may be more easily remembered, more easily recalled, and hence attributed to the so called “norm.” But you may also recall peers, seniors, trainers or teachers who were thought of as of low tone, quiet and content, but excelled through hard work, perseverance and respect. Luis Toledo-Pereyra, Editor in Chief of the Journal of Investigative Surgery writes of the “Gentleman Surgeon” a someone who “demonstrates that he/she cares, respects and helps all of those involved in the profession”, co-workers, students and patients alike [46].

Don’t fall for the stereotype, it may be out of date or just an illusion. Stay yourself, do it your way.

14.7 Struggles in the Workplace

If you expect the world to be fair with you because you are fair, you're fooling yourself. That's like expecting the lion not to eat you because you did not eat him.

Pravinee Hurbungs [47].

As a surgeon you may be lucky enough to practise in a workplace where you can have real influence and control in what you do, in a well organised system, where all share the same vision, where mutual respect prevails. But on occasions, whether self-employed or an employee, you may find yourself in a less calm, less peaceful, less well functioning workplace or more general working environment, where behaviours and actions of those around you may adversely affect your emotional world.

You may feel anger that your rota is constantly altered and you cannot plan your life, or that you get asked to hold the on call pager at short notice losing out again on that long awaited theatre session.

You may feel let down for even though you followed the instructions of your immediate senior to accept and admit the referred patient, they are staying silent when you get grilled the next morning by the most senior as to the reasoning behind your admission.

You may feel upset that you are being placed more in ward cover and others are placed in theatre most of the time, or that you are asked for a business case each time you request a new suture whilst others get with real ease any surgical kit they wish.

You may feel bewildered that whilst you are courteous and considerate, others try to step on your toes, that even though you are careful not to infringe on anyone's space others try to break into yours, that although you stick to your theatre list other trainees jump into yours.

You may feel disheartened for whilst you try to lift others up, others try to put you down, for even though you don't bother with others, others try to wind you up, even though you don't gossip, others keep talking behind your back. You may be worried that you are targeted and picked on due to association with others, rather than for whom you are, for what you stand for.

You may feel disbelief of the actions of others, that their priorities, or motives don't match yours, saddened that bad politics, targets, financial incentives, struggles for power, competition to impress seniors, personal gains are put before high quality care, dispirited of the constant struggle you have to face in trying to make good, of the multiple heads deciding but the responsibility still residing with you.

You may feel betrayed for those you once bailed out are now turning against you, those you supported are ignoring you, for others are rushing to say you did wrong, without even bothering to check the facts or rules.

You may be disappointed that your work is recognised round the globe but you are dismissed at home, your work admired by strangers but belittled by those close, that what you may have built others try to flatten, what you may introduced others try to abolish.

One may try to accept the human nature, accept that we cannot be all the same, have the same personalities, uphold the same principles or values, think and act in similar ways. Bad politics, plots and intrigues, unpleasant behaviours, hidden agendas, uncompromising egos, struggles for power, may be encountered in any workplace and the surgical field cannot stay immune from those.

Instead of drowning in your emotions you may consider:

- Try to better the working environment, start by reflecting upon your own behaviours and actions.
- Appreciate that disagreement with your proposals, or reversal of what you previously set up, is not necessarily criticism of you or your efforts, but just a different approach to a challenging problem. Even "Obamacare" was revoked on the first day of the new chief in town [48].
- Stand up for yourself, your rights, fight your own corner. Let others know what effects their actions have, make it clear that you are upset. Make it clear that what matters to you most is YOU [49].
- Avoid, despite the pressures, taking sides, joining factions, talking down some to please others.
- Concentrate at what you do best, to train, to learn, to care.
- Argue and fight for causes, not for or against personalities.
- If possible avoid those that upset you, they may quietly go away.
- And if you can't fit in, if you have had enough, move on, search for new shores, leave them to it, life is too short.

14.8 Be Proud

I am very proud, I am very proud. It's a hard honest work, you get an accomplishment, You make something and you see the final product.

Pat Eslich, Steel man, Canton, Ohio [50].

Becoming a surgeon is a long, challenging journey. You may face obstacles, barriers, hurdles, may have to work hard and persevere, make personal sacrifices, devote time and effort.

Be proud for what you have achieved, the knowledge and experiences you have gained, for the acquisition of fine motor skills, the critical clinical thinking you may have attained, for reaching the peak of your learning curve, building up the cases in your logbook, getting faster in operating, cracking the nerve racking interviews, passing the really hard exams, stopping being the most junior for the first time, getting to operate unsupervised for the first time, reaching a senior position, getting your first research article and every subsequent one published. Be proud for the status in the community you may have gained, for the respect, the reputation, the remuneration.

But also be proud for those in need that you helped, the fallen that you picked up, the hurt you comforted, the worried you reassured. Be proud for keeping patients in the centre of what you do, in the centre of your actions and motives. Be proud for recognising that slowing down your ward round to prop up someone sliding down their hospital bed, to place another pillow under their head, to push their tea cup closer so they can reach it, to have a social chat to make their day, to help them arrange care for their dog who has been left home alone, to get them a phone to ring their son who must be really worried, may be as valuable and important as the clever robotic surgery, the minimally invasive techniques, the latest durable implants that you may employ.

Be proud for providing humane care alongside skilled technical surgical care.

References

1. Famous Quotes at BrainyQuote. www.brainyquote.com. Accessed 05 Apr 2017.
2. Plutchik R. Human emotions have deep evolutionary roots, a fact that may explain their complexity and provide tools for clinical practice. The Nature of Emotions. 2001;89(4):344–50.
3. Mayer JD, Salovey P. What is emotional intelligence? In: Salovey P, Sluyter D, editors. Emotional development and emotional intelligence: implications for educators. New York: Basic Books; 1997. p. 3–31.
4. Goleman D. Working with emotional intelligence. Bloomsbury: St Ives; 1998.
5. Bridge of Spies. 2015. http://www.quotes.net/mquote/981862. Accessed 05 Apr 2017.
6. English Oxford Living dictionaries. https://en.oxforddictionaries.com/definition/. Accessed 05 Apr 2017.
7. Anderson CR. Coping behaviours as intervening mechanisms in the inverted-U stress-performance relationship. J Appl Psychol. 1976 Feb;61(7):30–4.
8. Kalia M. Assessing the economic impact of stress—the modern day hidden epidemic. Metabolism. 2002 Jun;51(6 Suppl 1):49–53.
9. Gemignani A, Piarulli A, Menicucci D, Laurino M, Rota G, Mastorci F, Gushin V, Shevchenko O, Garbella E, Pingitore A, Sebastiani L, Bergamasco M, L'Abbate A, Allegrini P, Bedini R. How stressful are 105 days of isolation? Sleep EEG patterns and tonic cortisol in healthy volunteers simulating manned flight to Mars. Int J Psychophysiol. 2014 Aug;93(2):211–9.
10. Dorrian J, Paterson J, Dawson D, Pincombe J, Grech C, Rogers AE. Sleep, stress and compensatory behaviors in Australian nurses and midwives. Rev Saude Publica. 2011 Oct;45(5):922–30.
11. Bradberry T. How successful people stay calm. Accessed at https://www.forbes.com/sites/travisbradberry/2014/02/06/how-successful-people-stay-calm/#1767bf882f79.
12. Maslach C, Schaufeli WB, Leiter MP. Job burnout. Annu Rev Psychol. 2001;52:397–422.
13. Maslach C, Jackson SE. The measurement of experienced burnout. J Occup Behav. 1981;2:99–113.
14. Cassell J. The temperament of surgeons. In: Expected miracles: surgeons at work. Philadelphia: Temple University Press; 1991. p. 33–59.
15. Yoo PS, Tackett JJ, Maxfield MW, Fisher R, Huot SJ, Longo WE Personal and professional well-being of surgical residents in New England. J Am Coll Surg. 2017. pii: S1072-7515(16)31728-8.
16. Arora M, Diwan AD, Harris IA. Burnout in orthopaedic surgeons: a review. ANZ J Surg. 2013 Jul;83(7–8):512–5.
17. Klimo P Jr, DeCuypere M, Ragel BT, McCartney S, Couldwell WT, Boop FA. Career satisfaction and burnout among U.S. neurosurgeons: a feasibility and pilot study. World Neurosurg. 2013 Nov;80(5):e59–68.
18. Businger A, Stefenelli U, Guller U. Prevalence of burnout among surgical residents and surgeons in Switzerland. Arch Surg. 2010 Oct;145(10):1013–6.
19. O'Kelly F, Manecksha RP, Quinlan DM, Reid A, Joyce A, O'Flynn K, Speakman M, Thornhill JA. Rates of self-reported 'burnout' and causative factors amongst urologists in Ireland and the UK: a comparative cross-sectional study. BJU Int. 2016 Feb;117(2):363–72.
20. Shanafelt TD, Balch CM, Dyrbye L, Bechamps G, Russell T, Satele D, Rummans T, Swartz K, Novotny PJ, Sloan J, Oreskovich MR. Special report: suicidal ideation among American surgeons. Arch Surg. 2011 Jan;146(1):54–62.
21. Orri M, Revah-Lévy A, Farges O. Surgeons' emotional experience of their everyday practice – a qualitative study. PLoS One. 2015;10(11):e0143763. doi:10.1371/journal.pone.0143763. eCollection 2015
22. Danes G. NASA's Journey to Mars | NASA. 2015. https://www.nasa.gov/content/nasas-journey-to-mars. Accessed 05 Apr 2017.
23. Esch JC. Stress management and balance for the orthopaedic surgeon: mindfulness. Instr Course Lect. 2011;60:627–31.
24. Raj V. Intellectual jokes and riddles. Better yourself books. 2008.
25. Pinto A, Faiz O, Bicknell C, Vincent C. Surgical complications and their implications for surgeons' well-being. Br J Surg. 2013 Dec;100(13):1748–55.

26. Pinto A, Faiz O, Bicknell C, Vincent C. Acute traumatic stress among surgeons after major surgical complications. Am J Surg. 2014 Oct;208(4):642–7.
27. Cristancho SM, Bidinosti SJ, Lingard LA, Novick RJ, Ott MC, Forbes TL. What's behind the scenes? Exploring the unspoken dimensions of complex and challenging surgical situations. Acad Med. 2014 Nov;89(11):1540–7.
28. Djerbi I, César M, Lenoir H, Coulet B, Lazerges C, Chammas M. Revision surgery for recurrent and persistent carpal tunnel syndrome: clinical results and factors affecting outcomes. Chir Main. 2015 Dec;34(6):312–7.
29. Henn RF 3rd, Tashjian RZ, Kang L, Green A. Patients with workers' compensation claims have worse outcomes after rotator cuff repair. J Bone Joint Surg Am. 2008 Oct;90(10):2105–13.
30. Britteon P, Cullum N, Sutton M. Association between psychological health and wound complications after surgery. Br J Surg. 2017 Feb 14; doi:10.1002/bjs.10474.
31. Sully: Miracle on the Hudson. 2016. Warner Bros. Pictures.
32. Wu AW. Medical error: the second victim. The doctor who makes the mistake needs help too. BMJ. 2000 Mar 18;320(7237):726–7.
33. Wu AW, Steckelberg RC. Medical error, incident investigation and the second victim: doing better but feeling worse? BMJ Qual Saf. 2012 Apr;21(4):267–70.
34. Scott SD, Hirschinger LE, Cox KR, McCoig M, Brandt J, Hall LW. The natural history of recovery for the healthcare provider "second victim" after adverse patient events. Qual Saf Health Care. 2009 Oct;18(5):325–30.
35. Luu S, Patel P, St-Martin L, Leung AS, Regehr G, Murnaghan ML, Gallinger S, Moulton CA. Waking up the next morning: surgeons' emotional reactions to adverse events. Med Educ. 2012 Dec;46(12):1179–88.
36. La philosophie de la chirurgie, René Leriche (1951).
37. Oxtoby K. Do the classic specialty stereotypes still hold true for today's doctors? BMJ Careers. http://careers.bmj.com/careers/advice/view-article.html?id=20015762. Accessed 05 Apr 2017.
38. Cherrington L. I'm a surgeon: respect me! Student. BMJ. 2008;1:62–3.
39. Bellodi PL. The general practitioner and the surgeon: stereotypes and medical specialities. Rev Hosp Clin Fac Med Sao Paulo. 2004;59:15–24.
40. Cassell J. Dismembering the image of god: surgeons, heroes, wimps and miracles. Anthropology Today. 1986;2(2):13–5.
41. Orthopaedics vs. Anaesthesia. https://www.youtube.com/watch?v=3rTsvb2ef5k. Accessed 05 Apr 2017.
42. Nip-Tuck. 2003–2010. http://www.imdb.com/title/tt0361217/. Accessed 05 Apr 2017.
43. Wen S. Is the quest to build a kinder, gentler surgeon misguided? Pacific standard. 2014. https://psmag.com/is-the-quest-to-build-a-kinder-gentler-surgeon-misguided-3950dee829fa. Accessed 05 Apr 2017.
44. Peters K, Ryan M. Machismo in surgery is harming the specialty. BMJ Careers. http://careers.bmj.com/careers/advice/Machismo_in_surgery_is_harming_the_specialty. Accessed 05 Apr 2017.
45. Peters K, Ryan MK, Haslam SA. Marines, medics, and machismo: lack of fit with masculine occupational stereotypes discourages men's participation. Br J Psychol. 2015 Nov;106(4):635–55.
46. Toledo-Pereyra LH. Gentleman surgeon. J Investig Surg. 2009;22(1):1–3.
47. Popular Quotes. Goodreads. http://www.goodreads.com/quotes. Accessed 05 Apr 2017.
48. Lee MJ, Luhby T. Trump issues executive order to start rolling back Obamacare. CNN. 2017 http://edition.cnn.com/2017/01/20/politics/trump-signs-executive-order-on-obamacare/. Accessed 05 Apr 2017.
49. Ldsponderings. Quotes by Shannon L. Alder. 2013. https://ldsponderings.wordpress.com/2013/09/23/wise-words-from-shannon-l-alder-a-mazing/. Accessed 05 Apr 2017.
50. America's left behind voters. Channel 4 news, 19/10/16 http://www.channel4.com/news/catch-up. Accessed 20 Oct 2016.

Index

A
Advanced Life Support (ALS), 239
Advanced Trauma Life Support (ATLS), 239
Adverse events
  disclosure, 260
    another surgeon involvement, 262
    challenges of, 260–261
    delivering, 261–262
  learning from, 258
  management, 249
  root cause analysis of, 250–252
  whilst investigation, 256
  wider surgical community, information sharing, 259
Aggregate bias, 127
ALS. *See* Advanced Life Support (ALS)
Analytical decision making, 124
Anchoring bias, 128
ATLS. *See* Advanced Trauma Life Support (ATLS)
Availability bias, 128

B
Body language, 50
British Medical Journal, 25

C
Cluster, recognised complications, 253–255
Commission bias, 127
Communication
  basic concepts in, 49
  body language, 50
  communicating urgency, 55
  consent, 57, 60–61
  in consenting, 57–60
  cues, 52–53
  decision making, 56
  definition, 47
  empathy, 53–54
  information, 52
  informed consent, 61–63
  introduce self and get to know others, 49
  listen, 50–51
  repeat in different ways, 51
  summarise, 52
  timing of obtaining consent, 63
  unanswered questions, 51
  words and tone, choosing, 49–50
Confirmation bias, 128
Creative decision making, 124

D
Decision making
  biases in, 127–129
  choice of, 125
  communication, 57
  and emotions, 134–135
  factors influencing surgical decision making, 130
  improving decision making, 136
  in offering surgery, 131–133
  spectrum of, 123
  surgeons, mind changing, 137
  switching, 126
  types of, 124
Dynamic Hip Screw (DHS) fixation, 126

E
Effective communication, 1
Ego bias, 127
Emotions
  coping with the stereotype, 294–295
  dealing with errors, 292–293
  emotional intelligence, 286

C. Panayiotou Charalambous, *Career Skills for Surgeons*,
DOI 10.1007/978-3-319-57490-5

Emotions (*cont.*)
  facing complications, 289–291
  and stress, 287–288
  types of, 285
  workplace, struggles in, 296–297

F
Framing bias, 129

H
Hawthorne effect, 209
Healthcare safety, 216. *See also* Safety

I
IDEAL pathway to innovation, 39
Interventions, 7
Intuitive decision making, 125

K
Kolb's cycle of learning, 27

L
Leadership
  alternative voices, 154
  behaviours, 148–149, 157–158
  challenges of surgeon leader, 147
  core values, 150
  by example, 150–151
  follow the rules, 154–155
  leader *vs.* manager, 143
  leading change, 151–154
  matters, 145–146
  and personality, 159
  posts and titles, 160
  start and finish, 155–156
  types of, 144
Long term memory, 112

N
National Health Service (NHS), 20
National Hip Fracture Database (NHFD), 207

O
Omission bias, 127
Operating surgeon, 60
Outcome bias, 129

P
Post-operative course, 60
Professionalism, 1
  acceptable conflict, 23–24
  accurate communication, 35
  acknowledge uncertainty, 34
  act upon concerns, 17
  anaesthetic techniques, 29
  choose, 40–41
  consider emotions, 16
  continuity of care, 28
  core principles, 18
  craft, 43
  experience, 27
  feedback, 32
  innovation, 38–39
  limitations, 30
    of technology, 42
  procedures, 36
  profession, stand up for, 22
  respect all staff, 13
  respect colleagues, 14
  self-reflect, 33
  situations, 31
  skill, development, 19
  support colleagues, 15
  surgery for right reasons, 7–8
  technical part of surgery, 21
  times, 20
  trust of other staff, 11–12
  trust of patients, 9–10
  unacceptable conflict, 25–26

R
Root cause analysis, adverse
    events, 250–252
Rule based decision making, 124

S
SA. *See* Situation awareness (SA)
Safety
  aim
    for consistency, 235
    no-blame culture, 238
    for standardisation, 234
  checklists appropriately, 231–232
  communicate for, 238–241
  critical steps, 232–233
  cross check, 230
  disruptive behaviours, 241
  encourage flat hierarchy, 237–238

error prone situations, 229–230
error reduction measures, 233–234
by example, 237
harm
causes of, 217–218
isolated *vs.* sequence of events, 224
healthcare safety, 216
iceberg of error, 226
improving strategies, 225
incident reporting systems, 236–237
individual's actions to, 227
latent failures
in aviation, 219–220
in healthcare, 221–223
learn from near misses, 236
minor failures, impact of, 235
problem, 215
recover from failures, 242
risk awareness, 227–228
unnecessary risks, 228
Search satisfying, 128
Situation awareness (SA)
barriers to, 107
components of, 105
vs. decision making, 106
failure
aviation, 108
surgery, 109–110
vehicle driving, 109
improvement, 111
information gathering, 111
levels of, 104
manageable workload, 111–112
memory utilisation, 112–113
recognising loss of, 116–117
stress on, 113
team situation awareness, 114–115
Status quo bias, 127
Stress, 287–288
Surgeon-patient relationship, 9
Surgery
avoiding early finishes, 77
avoiding interruptions, 77–78
listing for, 72
patient's journey to, 69–71
planning for, 83–84
equipment choice, 83
equipment positioning, 84
general considerations, 83
incision, 84
patient positioning, 83
personnel choice, 83
post-operative care, 84
prepping and draping, 84
procedure, 84
staff positioning, 84
theatre choice, 83
wound closure, 84
prompt start, 76
surgical care, 73–74
surgical list
location, 74–75
order of, 75
theatre list, 73
utilisation, 76
Surgical audit
audit databases, 204–208
cycle, 199–200
Hawthorne effect, 209
ladder, 201–202
local level, systematic audit, 203
mortality and morbidity meetings, 202–203
surgical logbook, 201–202
surgical registries, impact of, 207–208
Surgical conditions, management of, 2
Surgical interviews
awkward scenarios, 277–278
communication questions, 278–279
political issues questions, 279–280
surgical CV, 269–273
portfolio questions, 274–275
surgical clinical scenario questions, 275–277
surgical interview questions, 274
surgical specialty or sub-specialty, 267–268
Surgical research
defining
the aim, 168–169
the intervention, 176–177
the study groups, 170
the study population, 171
evidence into practise, 189–190
follow up, 182
interventions, quality assurance of, 178
journal club, 191–192
need for, 167
outcomes, 184
recruitment, 185
sample size calculation, 181
study design, 172–175
success of, 186–188
the surgical placebo, 179–180
surgical waiting lists and research, 183
Surgical skills, limitations of, 97

T

Technical skills
- active assistant, 94
- deliberate practise, 89–90
- distributed practise, 90–91
- feedback, 92–93
- learning curve, 88–89
- levels of participation, 93–94
- mental practise, 91
- operating theatre environment, 86–87
  - development, 88
  - motor skills acquisition, 88
- piecemeal learning, 90
- planning for surgery, 83–84
- simulation, 91–92
- surgical skills, limitations of, 97
- surgical technique, 95
- technically fluent, 96

W

Working memory, 112

World Health Organisation (WHO), 231, 249

MIX
Papier aus verantwortungsvollen Quellen
Paper from responsible sources
FSC® C105338

If you have any concerns about our products,
you can contact us on
ProductSafety@springernature.com

In case Publisher is established outside the EU,
the EU authorized representative is:
Springer Nature Customer Service Center GmbH
Europaplatz 3, 69115 Heidelberg, Germany

Printed by Libri Plureos GmbH
in Hamburg, Germany